VOLUME FIVE HUNDRED AND ELEVEN

Methods in
ENZYMOLOGY

RNA Helicases

METHODS IN ENZYMOLOGY

Editors-in-Chief

JOHN N. ABELSON AND MELVIN I. SIMON

*Division of Biology
California Institute of Technology
Pasadena, California*

Founding Editors

SIDNEY P. COLOWICK AND NATHAN O. KAPLAN

VOLUME FIVE HUNDRED AND ELEVEN

METHODS IN ENZYMOLOGY

RNA Helicases

EDITED BY

ECKHARD JANKOWSKY
Center for RNA Molecular Biology
School of Medicine
Case Western Reserve University
Cleveland, Ohio
USA

AMSTERDAM • BOSTON • HEIDELBERG • LONDON
NEW YORK • OXFORD • PARIS • SAN DIEGO
SAN FRANCISCO • SINGAPORE • SYDNEY • TOKYO
Academic Press is an imprint of Elsevier

Academic Press is an imprint of Elsevier
525 B Street, Suite 1900, San Diego, CA 92101-4495, USA
225 Wyman Street, Waltham, MA 02451, USA
The Boulevard, Langford Lane, Kidlington, Oxford, OX51GB, UK

First edition 2012

Copyright © 2012, Elsevier Inc. All Rights Reserved.

No part of this publication may be reproduced, stored in a retrieval system or transmitted in any form or by any means electronic, mechanical, photocopying, recording or otherwise without the prior written permission of the publisher

Permissions may be sought directly from Elsevier's Science & Technology Rights Department in Oxford, UK: phone (+44) (0) 1865 843830; fax (+44) (0) 1865 853333; email: permissions@elsevier.com. Alternatively you can submit your request online by visiting the Elsevier web site at http://elsevier.com/locate/permissions, and selecting *Obtaining permission to use Elsevier material*

Notice
No responsibility is assumed by the publisher for any injury and/or damage to persons or property as a matter of products liability, negligence or otherwise, or from any use or operation of any methods, products, instructions or ideas contained in the material herein. Because of rapid advances in the medical sciences, in particular, independent verification of diagnoses and drug dosages should be made

For information on all Academic Press publications
visit our website at elsevierdirect.com

ISBN: 978-0-12-396546-2
ISSN: 0076-6879

Printed and bound in United States of America
12 13 14   10 9 8 7 6 5 4 3 2 1

Working together to grow
libraries in developing countries

www.elsevier.com | www.bookaid.org | www.sabre.org

ELSEVIER   BOOK AID International   Sabre Foundation

Contents

Contributors	xiii
Preface	xix
Volumes in Series	xxi

1. Analysis of Duplex Unwinding by RNA Helicases Using Stopped-Flow Fluorescence Spectroscopy — 1
Andrea Putnam and Eckhard Jankowsky

1. Introduction	2
2. Experimental Considerations for Stopped-Flow Fluorescence Experiments	3
3. Substrate Design and Preparation	7
4. Designing and Performing Stopped-Flow Fluorescence Unwinding Experiments	13
5. Data Fitting and Analysis	20
Acknowledgments	24
References	24

2. Analyzing ATP Utilization by DEAD-Box RNA Helicases Using Kinetic and Equilibrium Methods — 29
Michael J. Bradley and Enrique M. De La Cruz

1. Introduction	30
2. Reagents and Equipment	31
3. Important Considerations for Initial DBP Characterization	34
4. Steady-State ATPase Measurements	39
5. Transient Kinetic Analysis of the DBP ATPase Cycle	46
6. ATP Hydrolysis Reversibility Measured by Isotope Exchange	54
7. RNA Unwinding Assays and ATPase Coupling	55
8. Kinetic Simulations	56
9. Putting It All Together: Quantitative Analysis of the DBP ATPase Cycle	56
Acknowledgments	58
References	58

3. **Oxygen Isotopic Exchange Probes of ATP Hydrolysis by RNA Helicases**   65

David D. Hackney

1. Introduction   66
2. Determination of the k_{-2}/k_3 Ratio by Intermediate Exchange During Net ATP Hydrolysis with Pyruvate Kinase and PEP to Regenerate ATP   67
3. Derivatization of the Pi to Volatile Triethyl Phosphate   68
4. Isotopic Analysis by GC–MS   70
5. Fitting Intermediate Exchange Data to Obtain $R = k_{-2}/k_3$ 70
6. Medium Pi=HOH Medium Exchange   72
7. Complication of Concurrent Medium Pi=HOH Exchange During Net Hydrolysis   72

Acknowledgment   73
References   73

4. **Conformational Changes of DEAD-Box Helicases Monitored by Single Molecule Fluorescence Resonance Energy Transfer**   75

Alexandra Z. Andreou and Dagmar Klostermeier

1. Introduction   76
2. Theoretical Background of smFRET   80
3. Preparing DEAD-box Proteins for smFRET Studies   83
4. smFRET Experiments and the Calculation of Corrected smFRET Histograms   87
5. Extracting Distance Information from smFRET Measurements: Distance Histograms   91
6. The Conformational Cycle of DEAD-Box Proteins   94
7. Perspective   103

Acknowledgments   104
References   104

5. **RNA Catalysis as a Probe for Chaperone Activity of DEAD-Box Helicases**   111

Jeffrey P. Potratz and Rick Russell

1. Introduction   112
2. Catalytic Activity as a Probe of RNA Folding   113
3. Self-splicing as a Readout for Native State Formation   115
4. Substrate Cleavage as a Readout for Native State Formation   118
5. Other Applications of the Discontinuous Assay   124

Acknowledgments   126
References   127

6. Molecular Mechanics of RNA Translocases — 131
Steve C. Ding and Anna Marie Pyle

1. Introduction — 132
2. Major Examples of Monomeric RNA Translocases — 134
3. Concluding Remarks — 144

Acknowledgments — 145
References — 145

7. Analysis of Helicase–RNA Interactions Using Nucleotide Analog Interference Mapping — 149
Annie Schwartz, Makhlouf Rabhi, Emmanuel Margeat, and Marc Boudvillain

1. Introduction — 150
2. Materials and Reagents — 152
3. Methods — 154
4. Conclusions — 166

Acknowledgments — 167
References — 167

8. Crystallization and X-ray Structure Determination of an RNA-Dependent Hexameric Helicase — 171
Nathan D. Thomsen and James M. Berger

1. Introduction — 172
2. Biochemical Foundation — 173
3. Protein Expression, Purification, and Storage — 175
4. Ligand Preparation and Storage — 176
5. Trapping Rho–RNA–Nucleotide Complexes — 177
6. Design of a Substrate-Centric Crystal-Screening Strategy — 177
7. Crystallization of Rho Bound to RNA and Adenosine Nucleotides — 178
8. Conclusion — 185
9. Additional Methods — 186

Acknowledgment — 187
References — 187

9. Structural Analysis of RNA Helicases with Small-Angle X-ray Scattering — 191
Manja A. Behrens, Yangzi He, Cristiano L. P. Oliveira, Gregers R. Andersen, Jan Skov Pedersen, and Klaus H. Nielsen

1. Introduction — 192
2. Basic Principles of Solution Scattering and Initial Data Treatment — 193
3. Model-Independent Analysis — 194

4.	Shape Determination Without Use of *A Priori* Information	198
5.	Atomic Resolution Models and Solution Scattering	201
6.	Using Atomic Resolution Models to Investigate Solution Structures	203
7.	Analysis of Protein Complexes	206
8.	Concluding Remarks	210
	Acknowledgments	210
	References	210

10. Analysis of Cofactor Effects on RNA Helicases — 213

Crystal Young and Katrin Karbstein

1.	Introduction	214
2.	How to Study Cofactors	219
3.	Reagents	221
4.	Protocols	224
	Acknowledgments	233
	References	233

11. Analysis of DEAD-Box Proteins in mRNA Export — 239

Ben Montpetit, Markus A. Seeliger, and Karsten Weis

1.	Introduction	240
2.	Purification of Dbp5, Gle1, and Nup159	242
3.	Steady-State ATPase Assay	244
4.	Use of Fluorescence Polarization to Monitor RNA or Nucleotide Binding and Release	246
5.	Concluding Remarks	252
	Acknowledgments	252
	References	252

12. Biochemical Characterization of the RNA Helicase UPF1 Involved in Nonsense-Mediated mRNA Decay — 255

Francesca Fiorini, Fabien Bonneau, and Hervé Le Hir

1.	Introduction	256
2.	Preparation of Active UPF1 and UPF1–UPF2 Complex	257
3.	Complex Assembly	259
4.	RNAse Protection Assay	264
5.	ATPase Assay	266
6.	Unwinding Assay	268
7.	Conclusions	272
	Acknowledgments	273
	References	273

13. Identification of RNA Helicase Target Sites by UV Cross-Linking and Analysis of cDNA — 275

Markus T. Bohnsack, David Tollervey, and Sander Granneman

1. Introduction — 276
2. Material — 277
3. Methods — 279
4. Protocol Adaptation and Trouble-Shooting — 286
Acknowledgments — 287
References — 287

14. *In Vivo* Approaches to Dissecting the Function of RNA Helicases in Eukaryotic Ribosome Assembly — 289

David C. Rawling and Susan J. Baserga

1. Introduction — 290
2. Experimental Strategies Used to Evaluate RB Helicases — 291
3. Determining Where an RNA Helicase Acts in the RB Pathway — 296
4. Elucidating the Supermolecular Context of an RB Helicase: Protein–Protein Interaction Studies — 308
References — 316

15. Analysis of RNA Helicases in P-Bodies and Stress Granules — 323

Angela Hilliker

1. Introduction — 324
2. RNA Helicases in Cytoplasmic mRNP Granules — 325
3. RNA Helicases That Affect P-Bodies or SGs — 329
4. Determining Whether an RNA Helicase Can Localize to Cytoplasmic mRNP Granules — 331
5. Determining Whether an RNA Helicase Affects Cytoplasmic mRNP Granules — 337
6. Discussion and Perspective — 340
Acknowledgment — 341
References — 341

16. DEAD-Box RNA Helicases as Transcription Cofactors — 347

Frances V. Fuller-Pace and Samantha M. Nicol

1. Introduction — 348
2. Analysis of Interactions Between RNA Helicases and Transcription Factors *In Vitro* and in Cell Lines — 349
3. Analysis of p68/p72 Sumoylation in Cell Lines — 358
4. Analysis of Transcriptional Coactivator/Corepressor Activity of p68 and p72 — 361

5. Summary	363
Acknowledgments	365
References	365

17. DEAD-Box RNA Helicases in Gram-Positive RNA Decay   369
Peter Redder and Patrick Linder

1. Introduction	369
2. Measuring mRNA Decay	371
3. Phenotypic Readouts	377
4. Concluding Remarks	381
Acknowledgments	381
References	381

18. RNA Helicases in Cyanobacteria: Biochemical and Molecular Approaches   385
George W. Owttrim

1. Introduction	386
2. Preparation of Cyanobacterial Extracts	387
3. Alteration of RNA Helicase Expression in Cyanobacteria	390
4. Northern Analysis of Transcript Levels	394
5. Western Analysis of Protein Levels	396
6. Cellular Ultrastructure and *In Situ* RNA Helicase Localization	397
Acknowledgments	400
References	400

19. Determination of Host RNA Helicases Activity in Viral Replication   405
Amit Sharma and Kathleen Boris-Lawrie

1. Introduction	406
2. Methods Used To Study Cell-Associated RNA Helicase in Cultured Mammalian Cells	409
3. Biochemical and Biophysical Methods to Study RNA Helicase	417
4. Methods Used to Study Virion-Associated RNA Helicase in Cultured Mammalian Cells	423
5. Concluding Remarks	430
Acknowledgments	431
References	431

20. Inhibitors of Translation Targeting Eukaryotic Translation Initiation Factor 4A — 437

Regina Cencic, Gabriela Galicia-Vázquez, and Jerry Pelletier

1. Introduction — 438
2. Small Molecule Inhibitors of RNA Helicases — 441
3. Small Molecule Inhibitors of eIF4A ATPase Activity — 444
4. Conclusions — 455
Acknowledgments — 456
References — 456

21. Identification and Analysis of Inhibitors Targeting the Hepatitis C Virus NS3 Helicase — 463

Alicia M. Hanson, John J. Hernandez, William R. Shadrick, and David N. Frick

1. Introduction — 464
2. The Need for Additional HCV Drug Targets — 465
3. Targeting the NS3 Helicase — 466
4. HTS for HCV Helicase Inhibitors — 467
5. Expression and Purification of NS3h — 469
6. The Molecular Beacon-Based Helicase Assay (MBHA) — 470
7. An RNA-Based Split Beacon Helicase Assay (SBHA) — 477
8. Discussion — 479
Acknowledgments — 480
References — 480

Author Index — *485*
Subject Index — *513*

Contributors

Gregers R. Andersen
Centre for mRNP Biogenesis and Metabolism, and Department of Molecular Biology and Genetics, Aarhus University, Aarhus, Denmark

Alexandra Z. Andreou
Institute for Physical Chemistry, University of Muenster, Muenster, Germany

Susan J. Baserga
Department of Molecular Biophysics and Biochemistry; Department of Genetics, and Department of Therapeutic Radiology, Yale University School of Medicine, New Haven, Connecticut, USA

Manja A. Behrens
Department of Chemistry and iNANO Interdisciplinary Nanoscience Centre, and Centre for mRNP Biogenesis and Metabolism, Aarhus University, Aarhus, Denmark

James M. Berger
Department of Molecular and Cell Biology and The Institute for Quantitative Biosciences, University of California, Berkeley, California, USA

Markus T. Bohnsack
Cluster of Excellence Macromolecular Complexes, Institute for Molecular Biosciences, Goethe University Frankfurt, Frankfurt, and Centre for Biochemistry and Molecular Cell Biology, Göttingen University, Göttingen, Germany

Fabien Bonneau
Max-Planck-Institute of Biochemistry, Department of Structural Cell Biology, Martinsried, Germany

Kathleen Boris-Lawrie
Department of Veterinary Biosciences; Center for Retrovirus Research; Center for RNA Biology; Comprehensive Cancer Center, Ohio State University, Columbus, Ohio, USA

Marc Boudvillain
CNRS UPR4301, Centre de Biophysique Moléculaire, Orléans cedex 2, France

Michael J. Bradley
Department of Molecular Biophysics and Biochemistry, Yale University, New Haven, Connecticut, USA

Regina Cencic
Department of Biochemistry, McGill University, Montreal, Quebec, Canada

Enrique M. De La Cruz
Department of Molecular Biophysics and Biochemistry, Yale University, New Haven, Connecticut, USA

Steve C. Ding
Infectious and Inflammatory Disease Center, Sanford-Burnham Medical Research Institute, La Jolla, California, USA

Francesca Fiorini
Institut de Biologie de l'Ecole Normale Supérieure, CNRS UMR8197, INSERM U1024, Paris Cedex 05, France

David N. Frick
Department of Chemistry and Biochemistry, University of Wisconsin-Milwaukee, Milwaukee, Wisconsin, USA

Frances V. Fuller-Pace
Division of Cancer Research, Medical Research Institute, University of Dundee, Ninewells Hospital and Medical School, Dundee, United Kingdom

Gabriela Galicia-Vázquez
Department of Biochemistry, McGill University, Montreal, Quebec, Canada

Sander Granneman
Centre for Systems Biology at Edinburgh, University of Edinburgh, Edinburgh, United Kingdom

David D. Hackney
Department of Biological Sciences, The Center for Nucleic Acid Science and Technology, Carnegie Mellon University, Pittsburgh, Pennsylvania, USA

Alicia M. Hanson
Department of Chemistry and Biochemistry, University of Wisconsin-Milwaukee, Milwaukee, Wisconsin, USA

Yangzi He
Centre for mRNP Biogenesis and Metabolism, and Department of Molecular Biology and Genetics, Aarhus University, Aarhus, Denmark

John J. Hernandez
Department of Chemistry and Biochemistry, University of Wisconsin-Milwaukee, Milwaukee, Wisconsin, USA

Angela Hilliker
Department of Biology, The University of Richmond, Richmond, Virginia, USA

Eckhard Jankowsky
Center for RNA Molecular Biology, School of Medicine, Case Western Reserve University, Cleveland, Ohio, USA

Katrin Karbstein
Department of Cancer Biology, The Scripps Research Institute, Jupiter, Florida, USA

Dagmar Klostermeier
Institute for Physical Chemistry, University of Muenster, Muenster, Germany

Hervé Le Hir
Institut de Biologie de l'Ecole Normale Supérieure, CNRS UMR 8197, INSERM U1024, Paris Cedex 05, France

Patrick Linder
Department of Microbiology and Molecular Medicine, University of Geneva, Genève, Switzerland

Emmanuel Margeat
CNRS UMR 5048, Centre de Biochimie Structurale, INSERM, and Universités Montpellier, Montpellier, France

Ben Montpetit
Department of Molecular and Cell Biology, University of California, Berkeley, California, USA

Samantha M. Nicol
Division of Cancer Research, Medical Research Institute, University of Dundee, Ninewells Hospital and Medical School, Dundee, United Kingdom

Klaus H. Nielsen[1]
Centre for mRNP Biogenesis and Metabolism, and Department of Molecular Biology and Genetics, Aarhus University, Aarhus, Denmark

Cristiano L.P. Oliveira
Instituto de Física, Universidade de São Paulo, São Paulo, Brazil

George W. Owttrim
Department of Biological Sciences, University of Alberta, Edmonton, Alberta, Canada

Jan Skov Pedersen
Department of Chemistry and iNANO Interdisciplinary Nanoscience Centre, and Centre for mRNP Biogenesis and Metabolism, Aarhus University, Aarhus, Denmark

---

[1] Current address: Department of Molecular Genetics and Cell Biology, University of Chicago, IL, Chicago, USA

Jerry Pelletier
Department of Biochemistry, and The Rosalind and Morris Goodman Cancer Research Center, McGill University, Montreal, Quebec, Canada

Jeffrey P. Potratz
Department of Chemistry and Biochemistry, Institute for Cellular and Molecular Biology, University of Texas at Austin, Austin, Texas, USA

Andrea Putnam
Department of Biochemistry, Center for RNA Molecular Biology, School of Medicine, Case Western Reserve University, Cleveland, Ohio, USA

Anna Marie Pyle
Department of Molecular, Cellular, and Developmental Biology, and Howard Hughes Medical Institute, Yale University, New Haven, Connecticut, USA

Makhlouf Rabhi
CNRS UPR4301, Centre de Biophysique Moléculaire, Orléans cedex 2, and Ecole doctorale Sciences et Technologies, Université d'Orléans, Orléans, France

David C. Rawling
Department of Molecular Biophysics and Biochemistry, Yale University School of Medicine, New Haven, Connecticut, USA

Peter Redder
Department of Microbiology and Molecular Medicine, University of Geneva, Genève, Switzerland

Rick Russell
Department of Chemistry and Biochemistry, Institute for Cellular and Molecular Biology, University of Texas at Austin, Austin, Texas, USA

Annie Schwartz
CNRS UPR4301, Centre de Biophysique Moléculaire, Orléans cedex 2, France

Markus A. Seeliger
Department of Pharmacological Sciences, State University of New York at Stony Brook, Stony Brook, New York, USA

William R. Shadrick
Department of Chemistry and Biochemistry, University of Wisconsin-Milwaukee, Milwaukee, Wisconsin, USA

Amit Sharma
Department of Veterinary Biosciences; Center for Retrovirus Research; Center for RNA Biology; Comprehensive Cancer Center, Ohio State University, Columbus, Ohio, USA

Nathan D. Thomsen[1]
Department of Molecular and Cell Biology and The Institute for Quantitative Biosciences, University of California, Berkeley, California, USA

David Tollervey
Wellcome Trust Centre for Cell Biology, University of Edinburgh, Edinburgh, United Kingdom

Karsten Weis
Department of Molecular and Cell Biology, University of California, Berkeley, California, USA

Crystal Young
Department of Chemistry, University of Michigan, Ann Arbor, Michigan, and Department of Cancer Biology, The Scripps Research Institute, Jupiter, Florida, USA

---

[1] Current address: Department of Pharmaceutical Chemistry, University of California, San Francisco, California, USA

Preface

RNA helicases have fascinated and puzzled investigators for the three decades that have passed since the first report on a "protein involved in ATP-dependent binding of mRNA." Since then, it has become clear that RNA helicases are ubiquitous, highly conserved enzymes that participate in virtually all aspects of RNA metabolism from bacteria to human. In eukaryotic RNA metabolism, RNA helicases are the largest class of enzymes, and many of these proteins have emerged as central players in the regulation of gene expression at the RNA level.

Given the critical and diverse roles of RNA helicases in biology, detailed knowledge of molecular and cellular functions of these enzymes is important for delineating the molecular basis of RNA metabolism. The past decade has seen remarkable progress in our understanding of these essential enzymes on all fronts: several dozen RNA helicase structures have been solved, enzymatic mechanisms have been illuminated, and specific cellular roles of many of these enzymes have come into better focus. We have come to realize that RNA helicases use ATP not only to unwind RNA duplexes but also to remodel and assemble RNA structures and RNA–protein complexes. It is not yet clear how molecular mechanisms exactly translate into cellular functions for these enzymes. However, answers to this question may be within reach for many RNA helicases, given the available repertoire of biochemical, biophysical, molecular biological, and cellular approaches. This is an exciting time for research on RNA helicases.

Interest in these enzymes comes from all sides of the methodological spectrum of modern biology, from biophysicists to cell biologists. This volume thus aims to provide a reference for the diverse, powerful tools used to analyze RNA helicases. The wonderful work contributed to this volume covers the broad scope of methods in the research on these enzymes. Several chapters describe quantitative biophysical and biochemical approaches to study molecular mechanisms and conformational changes of RNA helicases. Further chapters cover structural analysis, examination of cofactor effects on several representative examples, and the analysis of cellular functions of select enzymes. Two chapters outline the analysis of inhibitors that target RNA helicases.

Investigators who are new to the field now have a focused collection of detailed protocols for an entire range of methods tailored toward RNA helicases. Investigators with previous experience in the field will find the volume useful for the perspective on complementary approaches and for the plethora of conceptual and practical tips that are often invaluable for a successful examination of nucleotide-dependent enzymes that work on RNA. I am deeply grateful to the dedicated group of authors who have written a series of outstanding protocols and guides. It is my hope that this unique compilation encourages investigators and students to join the vibrant research on RNA helicases and that the volume inspires new angles of inquiry for investigators already in the field.

ECKHARD JANKOWSKY

METHODS IN ENZYMOLOGY

VOLUME I. Preparation and Assay of Enzymes
Edited by SIDNEY P. COLOWICK AND NATHAN O. KAPLAN

VOLUME II. Preparation and Assay of Enzymes
Edited by SIDNEY P. COLOWICK AND NATHAN O. KAPLAN

VOLUME III. Preparation and Assay of Substrates
Edited by SIDNEY P. COLOWICK AND NATHAN O. KAPLAN

VOLUME IV. Special Techniques for the Enzymologist
Edited by SIDNEY P. COLOWICK AND NATHAN O. KAPLAN

VOLUME V. Preparation and Assay of Enzymes
Edited by SIDNEY P. COLOWICK AND NATHAN O. KAPLAN

VOLUME VI. Preparation and Assay of Enzymes *(Continued)*
Preparation and Assay of Substrates
Special Techniques
Edited by SIDNEY P. COLOWICK AND NATHAN O. KAPLAN

VOLUME VII. Cumulative Subject Index
Edited by SIDNEY P. COLOWICK AND NATHAN O. KAPLAN

VOLUME VIII. Complex Carbohydrates
Edited by ELIZABETH F. NEUFELD AND VICTOR GINSBURG

VOLUME IX. Carbohydrate Metabolism
Edited by WILLIS A. WOOD

VOLUME X. Oxidation and Phosphorylation
Edited by RONALD W. ESTABROOK AND MAYNARD E. PULLMAN

VOLUME XI. Enzyme Structure
Edited by C. H. W. HIRS

VOLUME XII. Nucleic Acids (Parts A and B)
Edited by LAWRENCE GROSSMAN AND KIVIE MOLDAVE

VOLUME XIII. Citric Acid Cycle
Edited by J. M. LOWENSTEIN

VOLUME XIV. Lipids
Edited by J. M. LOWENSTEIN

VOLUME XV. Steroids and Terpenoids
Edited by RAYMOND B. CLAYTON

VOLUME XVI. Fast Reactions
Edited by KENNETH KUSTIN

VOLUME XVII. Metabolism of Amino Acids and Amines (Parts A and B)
Edited by HERBERT TABOR AND CELIA WHITE TABOR

VOLUME XVIII. Vitamins and Coenzymes (Parts A, B, and C)
Edited by DONALD B. MCCORMICK AND LEMUEL D. WRIGHT

VOLUME XIX. Proteolytic Enzymes
Edited by GERTRUDE E. PERLMANN AND LASZLO LORAND

VOLUME XX. Nucleic Acids and Protein Synthesis (Part C)
Edited by KIVIE MOLDAVE AND LAWRENCE GROSSMAN

VOLUME XXI. Nucleic Acids (Part D)
Edited by LAWRENCE GROSSMAN AND KIVIE MOLDAVE

VOLUME XXII. Enzyme Purification and Related Techniques
Edited by WILLIAM B. JAKOBY

VOLUME XXIII. Photosynthesis (Part A)
Edited by ANTHONY SAN PIETRO

VOLUME XXIV. Photosynthesis and Nitrogen Fixation (Part B)
Edited by ANTHONY SAN PIETRO

VOLUME XXV. Enzyme Structure (Part B)
Edited by C. H. W. HIRS AND SERGE N. TIMASHEFF

VOLUME XXVI. Enzyme Structure (Part C)
Edited by C. H. W. HIRS AND SERGE N. TIMASHEFF

VOLUME XXVII. Enzyme Structure (Part D)
Edited by C. H. W. HIRS AND SERGE N. TIMASHEFF

VOLUME XXVIII. Complex Carbohydrates (Part B)
Edited by VICTOR GINSBURG

VOLUME XXIX. Nucleic Acids and Protein Synthesis (Part E)
Edited by LAWRENCE GROSSMAN AND KIVIE MOLDAVE

VOLUME XXX. Nucleic Acids and Protein Synthesis (Part F)
Edited by KIVIE MOLDAVE AND LAWRENCE GROSSMAN

VOLUME XXXI. Biomembranes (Part A)
Edited by SIDNEY FLEISCHER AND LESTER PACKER

VOLUME XXXII. Biomembranes (Part B)
Edited by SIDNEY FLEISCHER AND LESTER PACKER

VOLUME XXXIII. Cumulative Subject Index Volumes I–XXX
Edited by MARTHA G. DENNIS AND EDWARD A. DENNIS

VOLUME XXXIV. Affinity Techniques (Enzyme Purification: Part B)
Edited by WILLIAM B. JAKOBY AND MEIR WILCHEK

VOLUME XXXV. Lipids (Part B)
Edited by JOHN M. LOWENSTEIN

VOLUME XXXVI. Hormone Action (Part A: Steroid Hormones)
Edited by BERT W. O'MALLEY AND JOEL G. HARDMAN

VOLUME XXXVII. Hormone Action (Part B: Peptide Hormones)
Edited by BERT W. O'MALLEY AND JOEL G. HARDMAN

VOLUME XXXVIII. Hormone Action (Part C: Cyclic Nucleotides)
Edited by JOEL G. HARDMAN AND BERT W. O'MALLEY

VOLUME XXXIX. Hormone Action (Part D: Isolated Cells, Tissues, and Organ Systems)
Edited by JOEL G. HARDMAN AND BERT W. O'MALLEY

VOLUME XL. Hormone Action (Part E: Nuclear Structure and Function)
Edited by BERT W. O'MALLEY AND JOEL G. HARDMAN

VOLUME XLI. Carbohydrate Metabolism (Part B)
Edited by W. A. WOOD

VOLUME XLII. Carbohydrate Metabolism (Part C)
Edited by W. A. WOOD

VOLUME XLIII. Antibiotics
Edited by JOHN H. HASH

VOLUME XLIV. Immobilized Enzymes
Edited by KLAUS MOSBACH

VOLUME XLV. Proteolytic Enzymes (Part B)
Edited by LASZLO LORAND

VOLUME XLVI. Affinity Labeling
Edited by WILLIAM B. JAKOBY AND MEIR WILCHEK

VOLUME XLVII. Enzyme Structure (Part E)
Edited by C. H. W. HIRS AND SERGE N. TIMASHEFF

VOLUME XLVIII. Enzyme Structure (Part F)
Edited by C. H. W. HIRS AND SERGE N. TIMASHEFF

VOLUME XLIX. Enzyme Structure (Part G)
Edited by C. H. W. HIRS AND SERGE N. TIMASHEFF

VOLUME L. Complex Carbohydrates (Part C)
Edited by VICTOR GINSBURG

VOLUME LI. Purine and Pyrimidine Nucleotide Metabolism
Edited by PATRICIA A. HOFFEE AND MARY ELLEN JONES

VOLUME LII. Biomembranes (Part C: Biological Oxidations)
Edited by SIDNEY FLEISCHER AND LESTER PACKER

VOLUME LIII. Biomembranes (Part D: Biological Oxidations)
Edited by SIDNEY FLEISCHER AND LESTER PACKER

VOLUME LIV. Biomembranes (Part E: Biological Oxidations)
Edited by SIDNEY FLEISCHER AND LESTER PACKER

VOLUME LV. Biomembranes (Part F: Bioenergetics)
Edited by SIDNEY FLEISCHER AND LESTER PACKER

VOLUME LVI. Biomembranes (Part G: Bioenergetics)
Edited by SIDNEY FLEISCHER AND LESTER PACKER

VOLUME LVII. Bioluminescence and Chemiluminescence
Edited by MARLENE A. DELUCA

VOLUME LVIII. Cell Culture
Edited by WILLIAM B. JAKOBY AND IRA PASTAN

VOLUME LIX. Nucleic Acids and Protein Synthesis (Part G)
Edited by KIVIE MOLDAVE AND LAWRENCE GROSSMAN

VOLUME LX. Nucleic Acids and Protein Synthesis (Part H)
Edited by KIVIE MOLDAVE AND LAWRENCE GROSSMAN

VOLUME 61. Enzyme Structure (Part H)
Edited by C. H. W. HIRS AND SERGE N. TIMASHEFF

VOLUME 62. Vitamins and Coenzymes (Part D)
Edited by DONALD B. MCCORMICK AND LEMUEL D. WRIGHT

VOLUME 63. Enzyme Kinetics and Mechanism (Part A: Initial Rate and Inhibitor Methods)
Edited by DANIEL L. PURICH

VOLUME 64. Enzyme Kinetics and Mechanism
(Part B: Isotopic Probes and Complex Enzyme Systems)
Edited by DANIEL L. PURICH

VOLUME 65. Nucleic Acids (Part I)
Edited by LAWRENCE GROSSMAN AND KIVIE MOLDAVE

VOLUME 66. Vitamins and Coenzymes (Part E)
Edited by DONALD B. MCCORMICK AND LEMUEL D. WRIGHT

VOLUME 67. Vitamins and Coenzymes (Part F)
Edited by DONALD B. MCCORMICK AND LEMUEL D. WRIGHT

VOLUME 68. Recombinant DNA
Edited by RAY WU

VOLUME 69. Photosynthesis and Nitrogen Fixation (Part C)
Edited by ANTHONY SAN PIETRO

VOLUME 70. Immunochemical Techniques (Part A)
Edited by HELEN VAN VUNAKIS AND JOHN J. LANGONE

VOLUME 71. Lipids (Part C)
Edited by JOHN M. LOWENSTEIN

VOLUME 72. Lipids (Part D)
Edited by JOHN M. LOWENSTEIN

VOLUME 73. Immunochemical Techniques (Part B)
Edited by JOHN J. LANGONE AND HELEN VAN VUNAKIS

VOLUME 74. Immunochemical Techniques (Part C)
Edited by JOHN J. LANGONE AND HELEN VAN VUNAKIS

VOLUME 75. Cumulative Subject Index Volumes XXXI, XXXII, XXXIV–LX
Edited by EDWARD A. DENNIS AND MARTHA G. DENNIS

VOLUME 76. Hemoglobins
Edited by ERALDO ANTONINI, LUIGI ROSSI-BERNARDI, AND EMILIA CHIANCONE

VOLUME 77. Detoxication and Drug Metabolism
Edited by WILLIAM B. JAKOBY

VOLUME 78. Interferons (Part A)
Edited by SIDNEY PESTKA

VOLUME 79. Interferons (Part B)
Edited by SIDNEY PESTKA

VOLUME 80. Proteolytic Enzymes (Part C)
Edited by LASZLO LORAND

VOLUME 81. Biomembranes (Part H: Visual Pigments and Purple Membranes, I)
Edited by LESTER PACKER

VOLUME 82. Structural and Contractile Proteins (Part A: Extracellular Matrix)
Edited by LEON W. CUNNINGHAM AND DIXIE W. FREDERIKSEN

VOLUME 83. Complex Carbohydrates (Part D)
Edited by VICTOR GINSBURG

VOLUME 84. Immunochemical Techniques (Part D: Selected Immunoassays)
Edited by JOHN J. LANGONE AND HELEN VAN VUNAKIS

VOLUME 85. Structural and Contractile Proteins (Part B: The Contractile Apparatus and the Cytoskeleton)
Edited by DIXIE W. FREDERIKSEN AND LEON W. CUNNINGHAM

VOLUME 86. Prostaglandins and Arachidonate Metabolites
Edited by WILLIAM E. M. LANDS AND WILLIAM L. SMITH

VOLUME 87. Enzyme Kinetics and Mechanism (Part C: Intermediates, Stereo-chemistry, and Rate Studies)
Edited by DANIEL L. PURICH

VOLUME 88. Biomembranes (Part I: Visual Pigments and Purple Membranes, II)
Edited by LESTER PACKER

VOLUME 89. Carbohydrate Metabolism (Part D)
Edited by WILLIS A. WOOD

VOLUME 90. Carbohydrate Metabolism (Part E)
Edited by WILLIS A. WOOD

VOLUME 91. Enzyme Structure (Part I)
Edited by C. H. W. HIRS AND SERGE N. TIMASHEFF

VOLUME 92. Immunochemical Techniques (Part E: Monoclonal Antibodies and General Immunoassay Methods)
Edited by JOHN J. LANGONE AND HELEN VAN VUNAKIS

VOLUME 93. Immunochemical Techniques (Part F: Conventional Antibodies, Fc Receptors, and Cytotoxicity)
Edited by JOHN J. LANGONE AND HELEN VAN VUNAKIS

VOLUME 94. Polyamines
Edited by HERBERT TABOR AND CELIA WHITE TABOR

VOLUME 95. Cumulative Subject Index Volumes 61–74, 76–80
Edited by EDWARD A. DENNIS AND MARTHA G. DENNIS

VOLUME 96. Biomembranes [Part J: Membrane Biogenesis: Assembly and Targeting (General Methods; Eukaryotes)]
Edited by SIDNEY FLEISCHER AND BECCA FLEISCHER

VOLUME 97. Biomembranes [Part K: Membrane Biogenesis: Assembly and Targeting (Prokaryotes, Mitochondria, and Chloroplasts)]
Edited by SIDNEY FLEISCHER AND BECCA FLEISCHER

VOLUME 98. Biomembranes (Part L: Membrane Biogenesis: Processing and Recycling)
Edited by SIDNEY FLEISCHER AND BECCA FLEISCHER

VOLUME 99. Hormone Action (Part F: Protein Kinases)
Edited by JACKIE D. CORBIN AND JOEL G. HARDMAN

VOLUME 100. Recombinant DNA (Part B)
Edited by RAY WU, LAWRENCE GROSSMAN, AND KIVIE MOLDAVE

VOLUME 101. Recombinant DNA (Part C)
Edited by RAY WU, LAWRENCE GROSSMAN, AND KIVIE MOLDAVE

VOLUME 102. Hormone Action (Part G: Calmodulin and Calcium-Binding Proteins)
Edited by ANTHONY R. MEANS AND BERT W. O'MALLEY

VOLUME 103. Hormone Action (Part H: Neuroendocrine Peptides)
Edited by P. MICHAEL CONN

VOLUME 104. Enzyme Purification and Related Techniques (Part C)
Edited by WILLIAM B. JAKOBY

VOLUME 105. Oxygen Radicals in Biological Systems
Edited by LESTER PACKER

VOLUME 106. Posttranslational Modifications (Part A)
Edited by FINN WOLD AND KIVIE MOLDAVE

VOLUME 107. Posttranslational Modifications (Part B)
Edited by FINN WOLD AND KIVIE MOLDAVE

VOLUME 108. Immunochemical Techniques (Part G: Separation and Characterization of Lymphoid Cells)
Edited by GIOVANNI DI SABATO, JOHN J. LANGONE, AND HELEN VAN VUNAKIS

VOLUME 109. Hormone Action (Part I: Peptide Hormones)
Edited by LUTZ BIRNBAUMER AND BERT W. O'MALLEY

VOLUME 110. Steroids and Isoprenoids (Part A)
Edited by JOHN H. LAW AND HANS C. RILLING

VOLUME 111. Steroids and Isoprenoids (Part B)
Edited by JOHN H. LAW AND HANS C. RILLING

VOLUME 112. Drug and Enzyme Targeting (Part A)
Edited by KENNETH J. WIDDER AND RALPH GREEN

VOLUME 113. Glutamate, Glutamine, Glutathione, and Related Compounds
Edited by ALTON MEISTER

VOLUME 114. Diffraction Methods for Biological Macromolecules (Part A)
Edited by HAROLD W. WYCKOFF, C. H. W. HIRS, AND SERGE N. TIMASHEFF

VOLUME 115. Diffraction Methods for Biological Macromolecules (Part B)
Edited by HAROLD W. WYCKOFF, C. H. W. HIRS, AND SERGE N. TIMASHEFF

VOLUME 116. Immunochemical Techniques
(Part H: Effectors and Mediators of Lymphoid Cell Functions)
Edited by GIOVANNI DI SABATO, JOHN J. LANGONE, AND HELEN VAN VUNAKIS

VOLUME 117. Enzyme Structure (Part J)
Edited by C. H. W. HIRS AND SERGE N. TIMASHEFF

VOLUME 118. Plant Molecular Biology
Edited by ARTHUR WEISSBACH AND HERBERT WEISSBACH

VOLUME 119. Interferons (Part C)
Edited by SIDNEY PESTKA

VOLUME 120. Cumulative Subject Index Volumes 81–94, 96–101

VOLUME 121. Immunochemical Techniques (Part I: Hybridoma Technology and Monoclonal Antibodies)
Edited by JOHN J. LANGONE AND HELEN VAN VUNAKIS

VOLUME 122. Vitamins and Coenzymes (Part G)
Edited by FRANK CHYTIL AND DONALD B. MCCORMICK

VOLUME 123. Vitamins and Coenzymes (Part H)
Edited by FRANK CHYTIL AND DONALD B. MCCORMICK

VOLUME 124. Hormone Action (Part J: Neuroendocrine Peptides)
Edited by P. MICHAEL CONN

VOLUME 125. Biomembranes (Part M: Transport in Bacteria, Mitochondria, and Chloroplasts: General Approaches and Transport Systems)
Edited by SIDNEY FLEISCHER AND BECCA FLEISCHER

VOLUME 126. Biomembranes (Part N: Transport in Bacteria, Mitochondria, and Chloroplasts: Protonmotive Force)
Edited by SIDNEY FLEISCHER AND BECCA FLEISCHER

VOLUME 127. Biomembranes (Part O: Protons and Water: Structure and Translocation)
Edited by LESTER PACKER

VOLUME 128. Plasma Lipoproteins (Part A: Preparation, Structure, and Molecular Biology)
Edited by JERE P. SEGREST AND JOHN J. ALBERS

VOLUME 129. Plasma Lipoproteins (Part B: Characterization, Cell Biology, and Metabolism)
Edited by JOHN J. ALBERS AND JERE P. SEGREST

VOLUME 130. Enzyme Structure (Part K)
Edited by C. H. W. HIRS AND SERGE N. TIMASHEFF

VOLUME 131. Enzyme Structure (Part L)
Edited by C. H. W. HIRS AND SERGE N. TIMASHEFF

VOLUME 132. Immunochemical Techniques (Part J: Phagocytosis and Cell-Mediated Cytotoxicity)
Edited by GIOVANNI DI SABATO AND JOHANNES EVERSE

VOLUME 133. Bioluminescence and Chemiluminescence (Part B)
Edited by MARLENE DELUCA AND WILLIAM D. MCELROY

VOLUME 134. Structural and Contractile Proteins (Part C: The Contractile Apparatus and the Cytoskeleton)
Edited by RICHARD B. VALLEE

VOLUME 135. Immobilized Enzymes and Cells (Part B)
Edited by KLAUS MOSBACH

VOLUME 136. Immobilized Enzymes and Cells (Part C)
Edited by KLAUS MOSBACH

VOLUME 137. Immobilized Enzymes and Cells (Part D)
Edited by KLAUS MOSBACH

VOLUME 138. Complex Carbohydrates (Part E)
Edited by VICTOR GINSBURG

VOLUME 139. Cellular Regulators (Part A: Calcium- and Calmodulin-Binding Proteins)
Edited by ANTHONY R. MEANS AND P. MICHAEL CONN

VOLUME 140. Cumulative Subject Index Volumes 102–119, 121–134

VOLUME 141. Cellular Regulators (Part B: Calcium and Lipids)
Edited by P. MICHAEL CONN AND ANTHONY R. MEANS

VOLUME 142. Metabolism of Aromatic Amino Acids and Amines
Edited by SEYMOUR KAUFMAN

VOLUME 143. Sulfur and Sulfur Amino Acids
Edited by WILLIAM B. JAKOBY AND OWEN GRIFFITH

VOLUME 144. Structural and Contractile Proteins (Part D: Extracellular Matrix)
Edited by LEON W. CUNNINGHAM

VOLUME 145. Structural and Contractile Proteins (Part E: Extracellular Matrix)
Edited by LEON W. CUNNINGHAM

VOLUME 146. Peptide Growth Factors (Part A)
Edited by DAVID BARNES AND DAVID A. SIRBASKU

VOLUME 147. Peptide Growth Factors (Part B)
Edited by DAVID BARNES AND DAVID A. SIRBASKU

VOLUME 148. Plant Cell Membranes
Edited by LESTER PACKER AND ROLAND DOUCE

VOLUME 149. Drug and Enzyme Targeting (Part B)
Edited by RALPH GREEN AND KENNETH J. WIDDER

VOLUME 150. Immunochemical Techniques (Part K: *In Vitro* Models of B and T Cell Functions and Lymphoid Cell Receptors)
Edited by GIOVANNI DI SABATO

VOLUME 151. Molecular Genetics of Mammalian Cells
Edited by MICHAEL M. GOTTESMAN

VOLUME 152. Guide to Molecular Cloning Techniques
Edited by SHELBY L. BERGER AND ALAN R. KIMMEL

VOLUME 153. Recombinant DNA (Part D)
Edited by RAY WU AND LAWRENCE GROSSMAN

VOLUME 154. Recombinant DNA (Part E)
Edited by RAY WU AND LAWRENCE GROSSMAN

VOLUME 155. Recombinant DNA (Part F)
Edited by RAY WU

VOLUME 156. Biomembranes (Part P: ATP-Driven Pumps and Related Transport: The Na, K-Pump)
Edited by SIDNEY FLEISCHER AND BECCA FLEISCHER

VOLUME 157. Biomembranes (Part Q: ATP-Driven Pumps and Related Transport: Calcium, Proton, and Potassium Pumps)
Edited by SIDNEY FLEISCHER AND BECCA FLEISCHER

VOLUME 158. Metalloproteins (Part A)
Edited by JAMES F. RIORDAN AND BERT L. VALLEE

VOLUME 159. Initiation and Termination of Cyclic Nucleotide Action
Edited by JACKIE D. CORBIN AND ROGER A. JOHNSON

VOLUME 160. Biomass (Part A: Cellulose and Hemicellulose)
Edited by WILLIS A. WOOD AND SCOTT T. KELLOGG

VOLUME 161. Biomass (Part B: Lignin, Pectin, and Chitin)
Edited by WILLIS A. WOOD AND SCOTT T. KELLOGG

VOLUME 162. Immunochemical Techniques (Part L: Chemotaxis and Inflammation)
Edited by GIOVANNI DI SABATO

VOLUME 163. Immunochemical Techniques (Part M: Chemotaxis and Inflammation)
Edited by GIOVANNI DI SABATO

VOLUME 164. Ribosomes
Edited by HARRY F. NOLLER, JR., AND KIVIE MOLDAVE

VOLUME 165. Microbial Toxins: Tools for Enzymology
Edited by SIDNEY HARSHMAN

VOLUME 166. Branched-Chain Amino Acids
Edited by ROBERT HARRIS AND JOHN R. SOKATCH

VOLUME 167. Cyanobacteria
Edited by LESTER PACKER AND ALEXANDER N. GLAZER

VOLUME 168. Hormone Action (Part K: Neuroendocrine Peptides)
Edited by P. MICHAEL CONN

VOLUME 169. Platelets: Receptors, Adhesion, Secretion (Part A)
Edited by JACEK HAWIGER

VOLUME 170. Nucleosomes
Edited by PAUL M. WASSARMAN AND ROGER D. KORNBERG

VOLUME 171. Biomembranes (Part R: Transport Theory: Cells and Model Membranes)
Edited by SIDNEY FLEISCHER AND BECCA FLEISCHER

VOLUME 172. Biomembranes (Part S: Transport: Membrane Isolation and Characterization)
Edited by SIDNEY FLEISCHER AND BECCA FLEISCHER

VOLUME 173. Biomembranes [Part T: Cellular and Subcellular Transport: Eukaryotic (Nonepithelial) Cells]
Edited by SIDNEY FLEISCHER AND BECCA FLEISCHER

VOLUME 174. Biomembranes [Part U: Cellular and Subcellular Transport: Eukaryotic (Nonepithelial) Cells]
Edited by SIDNEY FLEISCHER AND BECCA FLEISCHER

VOLUME 175. Cumulative Subject Index Volumes 135–139, 141–167

VOLUME 176. Nuclear Magnetic Resonance (Part A: Spectral Techniques and Dynamics)
Edited by NORMAN J. OPPENHEIMER AND THOMAS L. JAMES

VOLUME 177. Nuclear Magnetic Resonance (Part B: Structure and Mechanism)
Edited by NORMAN J. OPPENHEIMER AND THOMAS L. JAMES

VOLUME 178. Antibodies, Antigens, and Molecular Mimicry
Edited by JOHN J. LANGONE

VOLUME 179. Complex Carbohydrates (Part F)
Edited by VICTOR GINSBURG

VOLUME 180. RNA Processing (Part A: General Methods)
Edited by JAMES E. DAHLBERG AND JOHN N. ABELSON

VOLUME 181. RNA Processing (Part B: Specific Methods)
Edited by JAMES E. DAHLBERG AND JOHN N. ABELSON

VOLUME 182. Guide to Protein Purification
Edited by MURRAY P. DEUTSCHER

VOLUME 183. Molecular Evolution: Computer Analysis of Protein and Nucleic Acid Sequences
Edited by RUSSELL F. DOOLITTLE

VOLUME 184. Avidin-Biotin Technology
Edited by MEIR WILCHEK AND EDWARD A. BAYER

VOLUME 185. Gene Expression Technology
Edited by DAVID V. GOEDDEL

VOLUME 186. Oxygen Radicals in Biological Systems (Part B: Oxygen Radicals and Antioxidants)
Edited by LESTER PACKER AND ALEXANDER N. GLAZER

VOLUME 187. Arachidonate Related Lipid Mediators
Edited by ROBERT C. MURPHY AND FRANK A. FITZPATRICK

VOLUME 188. Hydrocarbons and Methylotrophy
Edited by MARY E. LIDSTROM

VOLUME 189. Retinoids (Part A: Molecular and Metabolic Aspects)
Edited by LESTER PACKER

VOLUME 190. Retinoids (Part B: Cell Differentiation and Clinical Applications)
Edited by LESTER PACKER

VOLUME 191. Biomembranes (Part V: Cellular and Subcellular Transport: Epithelial Cells)
Edited by SIDNEY FLEISCHER AND BECCA FLEISCHER

VOLUME 192. Biomembranes (Part W: Cellular and Subcellular Transport: Epithelial Cells)
Edited by SIDNEY FLEISCHER AND BECCA FLEISCHER

VOLUME 193. Mass Spectrometry
Edited by JAMES A. MCCLOSKEY

VOLUME 194. Guide to Yeast Genetics and Molecular Biology
Edited by CHRISTINE GUTHRIE AND GERALD R. FINK

VOLUME 195. Adenylyl Cyclase, G Proteins, and Guanylyl Cyclase
Edited by ROGER A. JOHNSON AND JACKIE D. CORBIN

VOLUME 196. Molecular Motors and the Cytoskeleton
Edited by RICHARD B. VALLEE

VOLUME 197. Phospholipases
Edited by EDWARD A. DENNIS

VOLUME 198. Peptide Growth Factors (Part C)
Edited by DAVID BARNES, J. P. MATHER, AND GORDON H. SATO

VOLUME 199. Cumulative Subject Index Volumes 168–174, 176–194

VOLUME 200. Protein Phosphorylation (Part A: Protein Kinases: Assays, Purification, Antibodies, Functional Analysis, Cloning, and Expression)
Edited by TONY HUNTER AND BARTHOLOMEW M. SEFTON

VOLUME 201. Protein Phosphorylation (Part B: Analysis of Protein Phosphorylation, Protein Kinase Inhibitors, and Protein Phosphatases)
Edited by TONY HUNTER AND BARTHOLOMEW M. SEFTON

VOLUME 202. Molecular Design and Modeling: Concepts and Applications (Part A: Proteins, Peptides, and Enzymes)
Edited by JOHN J. LANGONE

VOLUME 203. Molecular Design and Modeling: Concepts and Applications (Part B: Antibodies and Antigens, Nucleic Acids, Polysaccharides, and Drugs)
Edited by JOHN J. LANGONE

VOLUME 204. Bacterial Genetic Systems
Edited by JEFFREY H. MILLER

VOLUME 205. Metallobiochemistry (Part B: Metallothionein and Related Molecules)
Edited by JAMES F. RIORDAN AND BERT L. VALLEE

VOLUME 206. Cytochrome P450
Edited by MICHAEL R. WATERMAN AND ERIC F. JOHNSON

VOLUME 207. Ion Channels
Edited by BERNARDO RUDY AND LINDA E. IVERSON

VOLUME 208. Protein–DNA Interactions
Edited by ROBERT T. SAUER

VOLUME 209. Phospholipid Biosynthesis
Edited by EDWARD A. DENNIS AND DENNIS E. VANCE

VOLUME 210. Numerical Computer Methods
Edited by LUDWIG BRAND AND MICHAEL L. JOHNSON

VOLUME 211. DNA Structures (Part A: Synthesis and Physical Analysis of DNA)
Edited by DAVID M. J. LILLEY AND JAMES E. DAHLBERG

VOLUME 212. DNA Structures (Part B: Chemical and Electrophoretic Analysis of DNA)
Edited by DAVID M. J. LILLEY AND JAMES E. DAHLBERG

VOLUME 213. Carotenoids (Part A: Chemistry, Separation, Quantitation, and Antioxidation)
Edited by LESTER PACKER

VOLUME 214. Carotenoids (Part B: Metabolism, Genetics, and Biosynthesis)
Edited by LESTER PACKER

VOLUME 215. Platelets: Receptors, Adhesion, Secretion (Part B)
Edited by JACEK J. HAWIGER

VOLUME 216. Recombinant DNA (Part G)
Edited by RAY WU

VOLUME 217. Recombinant DNA (Part H)
Edited by RAY WU

VOLUME 218. Recombinant DNA (Part I)
Edited by RAY WU

VOLUME 219. Reconstitution of Intracellular Transport
Edited by JAMES E. ROTHMAN

VOLUME 220. Membrane Fusion Techniques (Part A)
Edited by NEJAT DÜZGÜNEŞ

VOLUME 221. Membrane Fusion Techniques (Part B)
Edited by NEJAT DÜZGÜNEŞ

VOLUME 222. Proteolytic Enzymes in Coagulation, Fibrinolysis, and Complement Activation (Part A: Mammalian Blood Coagulation Factors and Inhibitors)
Edited by LASZLO LORAND AND KENNETH G. MANN

VOLUME 223. Proteolytic Enzymes in Coagulation, Fibrinolysis, and Complement Activation (Part B: Complement Activation, Fibrinolysis, and Nonmammalian Blood Coagulation Factors)
Edited by LASZLO LORAND AND KENNETH G. MANN

VOLUME 224. Molecular Evolution: Producing the Biochemical Data
Edited by ELIZABETH ANNE ZIMMER, THOMAS J. WHITE, REBECCA L. CANN, AND ALLAN C. WILSON

VOLUME 225. Guide to Techniques in Mouse Development
Edited by PAUL M. WASSARMAN AND MELVIN L. DEPAMPHILIS

VOLUME 226. Metallobiochemistry (Part C: Spectroscopic and Physical Methods for Probing Metal Ion Environments in Metalloenzymes and Metalloproteins)
Edited by JAMES F. RIORDAN AND BERT L. VALLEE

VOLUME 227. Metallobiochemistry (Part D: Physical and Spectroscopic Methods for Probing Metal Ion Environments in Metalloproteins)
Edited by JAMES F. RIORDAN AND BERT L. VALLEE

VOLUME 228. Aqueous Two-Phase Systems
Edited by HARRY WALTER AND GÖTE JOHANSSON

VOLUME 229. Cumulative Subject Index Volumes 195–198, 200–227

VOLUME 230. Guide to Techniques in Glycobiology
Edited by WILLIAM J. LENNARZ AND GERALD W. HART

VOLUME 231. Hemoglobins (Part B: Biochemical and Analytical Methods)
Edited by JOHANNES EVERSE, KIM D. VANDEGRIFF, AND ROBERT M. WINSLOW

VOLUME 232. Hemoglobins (Part C: Biophysical Methods)
Edited by JOHANNES EVERSE, KIM D. VANDEGRIFF, AND ROBERT M. WINSLOW

VOLUME 233. Oxygen Radicals in Biological Systems (Part C)
Edited by LESTER PACKER

VOLUME 234. Oxygen Radicals in Biological Systems (Part D)
Edited by LESTER PACKER

VOLUME 235. Bacterial Pathogenesis (Part A: Identification and Regulation of Virulence Factors)
Edited by VIRGINIA L. CLARK AND PATRIK M. BAVOIL

VOLUME 236. Bacterial Pathogenesis (Part B: Integration of Pathogenic Bacteria with Host Cells)
Edited by VIRGINIA L. CLARK AND PATRIK M. BAVOIL

VOLUME 237. Heterotrimeric G Proteins
Edited by RAVI IYENGAR

VOLUME 238. Heterotrimeric G-Protein Effectors
Edited by RAVI IYENGAR

VOLUME 239. Nuclear Magnetic Resonance (Part C)
Edited by THOMAS L. JAMES AND NORMAN J. OPPENHEIMER

VOLUME 240. Numerical Computer Methods (Part B)
Edited by MICHAEL L. JOHNSON AND LUDWIG BRAND

VOLUME 241. Retroviral Proteases
Edited by LAWRENCE C. KUO AND JULES A. SHAFER

VOLUME 242. Neoglycoconjugates (Part A)
Edited by Y. C. LEE AND REIKO T. LEE

VOLUME 243. Inorganic Microbial Sulfur Metabolism
Edited by HARRY D. PECK, JR., AND JEAN LEGALL

VOLUME 244. Proteolytic Enzymes: Serine and Cysteine Peptidases
Edited by ALAN J. BARRETT

VOLUME 245. Extracellular Matrix Components
Edited by E. RUOSLAHTI AND E. ENGVALL

VOLUME 246. Biochemical Spectroscopy
Edited by KENNETH SAUER

VOLUME 247. Neoglycoconjugates (Part B: Biomedical Applications)
Edited by Y. C. LEE AND REIKO T. LEE

VOLUME 248. Proteolytic Enzymes: Aspartic and Metallo Peptidases
Edited by ALAN J. BARRETT

VOLUME 249. Enzyme Kinetics and Mechanism (Part D: Developments in Enzyme Dynamics)
Edited by DANIEL L. PURICH

VOLUME 250. Lipid Modifications of Proteins
Edited by PATRICK J. CASEY AND JANICE E. BUSS

VOLUME 251. Biothiols (Part A: Monothiols and Dithiols, Protein Thiols, and Thiyl Radicals)
Edited by LESTER PACKER

VOLUME 252. Biothiols (Part B: Glutathione and Thioredoxin; Thiols in Signal Transduction and Gene Regulation)
Edited by LESTER PACKER

VOLUME 253. Adhesion of Microbial Pathogens
Edited by RON J. DOYLE AND ITZHAK OFEK

VOLUME 254. Oncogene Techniques
Edited by PETER K. VOGT AND INDER M. VERMA

VOLUME 255. Small GTPases and Their Regulators (Part A: Ras Family)
Edited by W. E. BALCH, CHANNING J. DER, AND ALAN HALL

VOLUME 256. Small GTPases and Their Regulators (Part B: Rho Family)
Edited by W. E. BALCH, CHANNING J. DER, AND ALAN HALL

VOLUME 257. Small GTPases and Their Regulators (Part C: Proteins Involved in Transport)
Edited by W. E. BALCH, CHANNING J. DER, AND ALAN HALL

VOLUME 258. Redox-Active Amino Acids in Biology
Edited by JUDITH P. KLINMAN

VOLUME 259. Energetics of Biological Macromolecules
Edited by MICHAEL L. JOHNSON AND GARY K. ACKERS

VOLUME 260. Mitochondrial Biogenesis and Genetics (Part A)
Edited by GIUSEPPE M. ATTARDI AND ANNE CHOMYN

VOLUME 261. Nuclear Magnetic Resonance and Nucleic Acids
Edited by THOMAS L. JAMES

VOLUME 262. DNA Replication
Edited by JUDITH L. CAMPBELL

VOLUME 263. Plasma Lipoproteins (Part C: Quantitation)
Edited by WILLIAM A. BRADLEY, SANDRA H. GIANTURCO, AND JERE P. SEGREST

VOLUME 264. Mitochondrial Biogenesis and Genetics (Part B)
Edited by GIUSEPPE M. ATTARDI AND ANNE CHOMYN

VOLUME 265. Cumulative Subject Index Volumes 228, 230–262

VOLUME 266. Computer Methods for Macromolecular Sequence Analysis
Edited by RUSSELL F. DOOLITTLE

VOLUME 267. Combinatorial Chemistry
Edited by JOHN N. ABELSON

VOLUME 268. Nitric Oxide (Part A: Sources and Detection of NO; NO Synthase)
Edited by LESTER PACKER

VOLUME 269. Nitric Oxide (Part B: Physiological and Pathological Processes)
Edited by LESTER PACKER

VOLUME 270. High Resolution Separation and Analysis of Biological Macromolecules (Part A: Fundamentals)
Edited by BARRY L. KARGER AND WILLIAM S. HANCOCK

VOLUME 271. High Resolution Separation and Analysis of Biological Macromolecules (Part B: Applications)
Edited by BARRY L. KARGER AND WILLIAM S. HANCOCK

VOLUME 272. Cytochrome P450 (Part B)
Edited by ERIC F. JOHNSON AND MICHAEL R. WATERMAN

VOLUME 273. RNA Polymerase and Associated Factors (Part A)
Edited by SANKAR ADHYA

VOLUME 274. RNA Polymerase and Associated Factors (Part B)
Edited by SANKAR ADHYA

VOLUME 275. Viral Polymerases and Related Proteins
Edited by LAWRENCE C. KUO, DAVID B. OLSEN, AND STEVEN S. CARROLL

VOLUME 276. Macromolecular Crystallography (Part A)
Edited by CHARLES W. CARTER, JR., AND ROBERT M. SWEET

VOLUME 277. Macromolecular Crystallography (Part B)
Edited by CHARLES W. CARTER, JR., AND ROBERT M. SWEET

VOLUME 278. Fluorescence Spectroscopy
Edited by LUDWIG BRAND AND MICHAEL L. JOHNSON

VOLUME 279. Vitamins and Coenzymes (Part I)
Edited by DONALD B. MCCORMICK, JOHN W. SUTTIE, AND CONRAD WAGNER

VOLUME 280. Vitamins and Coenzymes (Part J)
Edited by DONALD B. MCCORMICK, JOHN W. SUTTIE, AND CONRAD WAGNER

VOLUME 281. Vitamins and Coenzymes (Part K)
Edited by DONALD B. MCCORMICK, JOHN W. SUTTIE, AND CONRAD WAGNER

VOLUME 282. Vitamins and Coenzymes (Part L)
Edited by DONALD B. MCCORMICK, JOHN W. SUTTIE, AND CONRAD WAGNER

VOLUME 283. Cell Cycle Control
Edited by WILLIAM G. DUNPHY

VOLUME 284. Lipases (Part A: Biotechnology)
Edited by BYRON RUBIN AND EDWARD A. DENNIS

VOLUME 285. Cumulative Subject Index Volumes 263, 264, 266–284, 286–289

VOLUME 286. Lipases (Part B: Enzyme Characterization and Utilization)
Edited by BYRON RUBIN AND EDWARD A. DENNIS

VOLUME 287. Chemokines
Edited by RICHARD HORUK

VOLUME 288. Chemokine Receptors
Edited by RICHARD HORUK

VOLUME 289. Solid Phase Peptide Synthesis
Edited by GREGG B. FIELDS

VOLUME 290. Molecular Chaperones
Edited by GEORGE H. LORIMER AND THOMAS BALDWIN

VOLUME 291. Caged Compounds
Edited by GERARD MARRIOTT

VOLUME 292. ABC Transporters: Biochemical, Cellular, and Molecular Aspects
Edited by SURESH V. AMBUDKAR AND MICHAEL M. GOTTESMAN

VOLUME 293. Ion Channels (Part B)
Edited by P. MICHAEL CONN

VOLUME 294. Ion Channels (Part C)
Edited by P. MICHAEL CONN

VOLUME 295. Energetics of Biological Macromolecules (Part B)
Edited by GARY K. ACKERS AND MICHAEL L. JOHNSON

VOLUME 296. Neurotransmitter Transporters
Edited by SUSAN G. AMARA

VOLUME 297. Photosynthesis: Molecular Biology of Energy Capture
Edited by LEE MCINTOSH

VOLUME 298. Molecular Motors and the Cytoskeleton (Part B)
Edited by RICHARD B. VALLEE

VOLUME 299. Oxidants and Antioxidants (Part A)
Edited by LESTER PACKER

VOLUME 300. Oxidants and Antioxidants (Part B)
Edited by LESTER PACKER

VOLUME 301. Nitric Oxide: Biological and Antioxidant Activities (Part C)
Edited by LESTER PACKER

VOLUME 302. Green Fluorescent Protein
Edited by P. MICHAEL CONN

VOLUME 303. cDNA Preparation and Display
Edited by SHERMAN M. WEISSMAN

VOLUME 304. Chromatin
Edited by PAUL M. WASSARMAN AND ALAN P. WOLFFE

VOLUME 305. Bioluminescence and Chemiluminescence (Part C)
Edited by THOMAS O. BALDWIN AND MIRIAM M. ZIEGLER

VOLUME 306. Expression of Recombinant Genes in Eukaryotic Systems
Edited by JOSEPH C. GLORIOSO AND MARTIN C. SCHMIDT

VOLUME 307. Confocal Microscopy
Edited by P. MICHAEL CONN

VOLUME 308. Enzyme Kinetics and Mechanism (Part E: Energetics of Enzyme Catalysis)
Edited by DANIEL L. PURICH AND VERN L. SCHRAMM

VOLUME 309. Amyloid, Prions, and Other Protein Aggregates
Edited by RONALD WETZEL

VOLUME 310. Biofilms
Edited by RON J. DOYLE

VOLUME 311. Sphingolipid Metabolism and Cell Signaling (Part A)
Edited by ALFRED H. MERRILL, JR., AND YUSUF A. HANNUN

VOLUME 312. Sphingolipid Metabolism and Cell Signaling (Part B)
Edited by ALFRED H. MERRILL, JR., AND YUSUF A. HANNUN

VOLUME 313. Antisense Technology
(Part A: General Methods, Methods of Delivery, and RNA Studies)
Edited by M. IAN PHILLIPS

VOLUME 314. Antisense Technology (Part B: Applications)
Edited by M. IAN PHILLIPS

VOLUME 315. Vertebrate Phototransduction and the Visual Cycle (Part A)
Edited by KRZYSZTOF PALCZEWSKI

VOLUME 316. Vertebrate Phototransduction and the Visual Cycle (Part B)
Edited by KRZYSZTOF PALCZEWSKI

VOLUME 317. RNA–Ligand Interactions (Part A: Structural Biology Methods)
Edited by DANIEL W. CELANDER AND JOHN N. ABELSON

VOLUME 318. RNA–Ligand Interactions (Part B: Molecular Biology Methods)
Edited by DANIEL W. CELANDER AND JOHN N. ABELSON

VOLUME 319. Singlet Oxygen, UV-A, and Ozone
Edited by LESTER PACKER AND HELMUT SIES

VOLUME 320. Cumulative Subject Index Volumes 290–319

VOLUME 321. Numerical Computer Methods (Part C)
Edited by MICHAEL L. JOHNSON AND LUDWIG BRAND

VOLUME 322. Apoptosis
Edited by JOHN C. REED

VOLUME 323. Energetics of Biological Macromolecules (Part C)
Edited by MICHAEL L. JOHNSON AND GARY K. ACKERS

VOLUME 324. Branched-Chain Amino Acids (Part B)
Edited by ROBERT A. HARRIS AND JOHN R. SOKATCH

VOLUME 325. Regulators and Effectors of Small GTPases
(Part D: Rho Family)
Edited by W. E. BALCH, CHANNING J. DER, AND ALAN HALL

VOLUME 326. Applications of Chimeric Genes and Hybrid Proteins
(Part A: Gene Expression and Protein Purification)
Edited by JEREMY THORNER, SCOTT D. EMR, AND JOHN N. ABELSON

VOLUME 327. Applications of Chimeric Genes and Hybrid Proteins
(Part B: Cell Biology and Physiology)
Edited by JEREMY THORNER, SCOTT D. EMR, AND JOHN N. ABELSON

VOLUME 328. Applications of Chimeric Genes and Hybrid Proteins (Part C: Protein–Protein Interactions and Genomics)
Edited by JEREMY THORNER, SCOTT D. EMR, AND JOHN N. ABELSON

VOLUME 329. Regulators and Effectors of Small GTPases (Part E: GTPases Involved in Vesicular Traffic)
Edited by W. E. BALCH, CHANNING J. DER, AND ALAN HALL

VOLUME 330. Hyperthermophilic Enzymes (Part A)
Edited by MICHAEL W. W. ADAMS AND ROBERT M. KELLY

VOLUME 331. Hyperthermophilic Enzymes (Part B)
Edited by MICHAEL W. W. ADAMS AND ROBERT M. KELLY

VOLUME 332. Regulators and Effectors of Small GTPases (Part F: Ras Family I)
Edited by W. E. BALCH, CHANNING J. DER, AND ALAN HALL

VOLUME 333. Regulators and Effectors of Small GTPases (Part G: Ras Family II)
Edited by W. E. BALCH, CHANNING J. DER, AND ALAN HALL

VOLUME 334. Hyperthermophilic Enzymes (Part C)
Edited by MICHAEL W. W. ADAMS AND ROBERT M. KELLY

VOLUME 335. Flavonoids and Other Polyphenols
Edited by LESTER PACKER

VOLUME 336. Microbial Growth in Biofilms (Part A: Developmental and Molecular Biological Aspects)
Edited by RON J. DOYLE

VOLUME 337. Microbial Growth in Biofilms (Part B: Special Environments and Physicochemical Aspects)
Edited by RON J. DOYLE

VOLUME 338. Nuclear Magnetic Resonance of Biological Macromolecules (Part A)
Edited by THOMAS L. JAMES, VOLKER DÖTSCH, AND ULI SCHMITZ

VOLUME 339. Nuclear Magnetic Resonance of Biological Macromolecules (Part B)
Edited by THOMAS L. JAMES, VOLKER DÖTSCH, AND ULI SCHMITZ

VOLUME 340. Drug–Nucleic Acid Interactions
Edited by JONATHAN B. CHAIRES AND MICHAEL J. WARING

VOLUME 341. Ribonucleases (Part A)
Edited by ALLEN W. NICHOLSON

VOLUME 342. Ribonucleases (Part B)
Edited by ALLEN W. NICHOLSON

VOLUME 343. G Protein Pathways (Part A: Receptors)
Edited by RAVI IYENGAR AND JOHN D. HILDEBRANDT

VOLUME 344. G Protein Pathways (Part B: G Proteins and Their Regulators)
Edited by RAVI IYENGAR AND JOHN D. HILDEBRANDT

VOLUME 345. G Protein Pathways (Part C: Effector Mechanisms)
Edited by RAVI IYENGAR AND JOHN D. HILDEBRANDT

VOLUME 346. Gene Therapy Methods
Edited by M. IAN PHILLIPS

VOLUME 347. Protein Sensors and Reactive Oxygen Species (Part A: Selenoproteins and Thioredoxin)
Edited by HELMUT SIES AND LESTER PACKER

VOLUME 348. Protein Sensors and Reactive Oxygen Species (Part B: Thiol Enzymes and Proteins)
Edited by HELMUT SIES AND LESTER PACKER

VOLUME 349. Superoxide Dismutase
Edited by LESTER PACKER

VOLUME 350. Guide to Yeast Genetics and Molecular and Cell Biology (Part B)
Edited by CHRISTINE GUTHRIE AND GERALD R. FINK

VOLUME 351. Guide to Yeast Genetics and Molecular and Cell Biology (Part C)
Edited by CHRISTINE GUTHRIE AND GERALD R. FINK

VOLUME 352. Redox Cell Biology and Genetics (Part A)
Edited by CHANDAN K. SEN AND LESTER PACKER

VOLUME 353. Redox Cell Biology and Genetics (Part B)
Edited by CHANDAN K. SEN AND LESTER PACKER

VOLUME 354. Enzyme Kinetics and Mechanisms (Part F: Detection and Characterization of Enzyme Reaction Intermediates)
Edited by DANIEL L. PURICH

VOLUME 355. Cumulative Subject Index Volumes 321–354

VOLUME 356. Laser Capture Microscopy and Microdissection
Edited by P. MICHAEL CONN

VOLUME 357. Cytochrome P450, Part C
Edited by ERIC F. JOHNSON AND MICHAEL R. WATERMAN

VOLUME 358. Bacterial Pathogenesis (Part C: Identification, Regulation, and Function of Virulence Factors)
Edited by VIRGINIA L. CLARK AND PATRIK M. BAVOIL

VOLUME 359. Nitric Oxide (Part D)
Edited by ENRIQUE CADENAS AND LESTER PACKER

VOLUME 360. Biophotonics (Part A)
Edited by GERARD MARRIOTT AND IAN PARKER

VOLUME 361. Biophotonics (Part B)
Edited by GERARD MARRIOTT AND IAN PARKER

VOLUME 362. Recognition of Carbohydrates in Biological Systems (Part A)
Edited by YUAN C. LEE AND REIKO T. LEE

VOLUME 363. Recognition of Carbohydrates in Biological Systems (Part B)
Edited by YUAN C. LEE AND REIKO T. LEE

VOLUME 364. Nuclear Receptors
Edited by DAVID W. RUSSELL AND DAVID J. MANGELSDORF

VOLUME 365. Differentiation of Embryonic Stem Cells
Edited by PAUL M. WASSAUMAN AND GORDON M. KELLER

VOLUME 366. Protein Phosphatases
Edited by SUSANNE KLUMPP AND JOSEF KRIEGLSTEIN

VOLUME 367. Liposomes (Part A)
Edited by NEJAT DÜZGÜNEŞ

VOLUME 368. Macromolecular Crystallography (Part C)
Edited by CHARLES W. CARTER, JR., AND ROBERT M. SWEET

VOLUME 369. Combinational Chemistry (Part B)
Edited by GUILLERMO A. MORALES AND BARRY A. BUNIN

VOLUME 370. RNA Polymerases and Associated Factors (Part C)
Edited by SANKAR L. ADHYA AND SUSAN GARGES

VOLUME 371. RNA Polymerases and Associated Factors (Part D)
Edited by SANKAR L. ADHYA AND SUSAN GARGES

VOLUME 372. Liposomes (Part B)
Edited by NEJAT DÜZGÜNEŞ

VOLUME 373. Liposomes (Part C)
Edited by NEJAT DÜZGÜNEŞ

VOLUME 374. Macromolecular Crystallography (Part D)
Edited by CHARLES W. CARTER, JR., AND ROBERT W. SWEET

VOLUME 375. Chromatin and Chromatin Remodeling Enzymes (Part A)
Edited by C. DAVID ALLIS AND CARL WU

VOLUME 376. Chromatin and Chromatin Remodeling Enzymes (Part B)
Edited by C. DAVID ALLIS AND CARL WU

VOLUME 377. Chromatin and Chromatin Remodeling Enzymes (Part C)
Edited by C. DAVID ALLIS AND CARL WU

VOLUME 378. Quinones and Quinone Enzymes (Part A)
Edited by HELMUT SIES AND LESTER PACKER

VOLUME 379. Energetics of Biological Macromolecules (Part D)
Edited by JO M. HOLT, MICHAEL L. JOHNSON, AND GARY K. ACKERS

VOLUME 380. Energetics of Biological Macromolecules (Part E)
Edited by JO M. HOLT, MICHAEL L. JOHNSON, AND GARY K. ACKERS

VOLUME 381. Oxygen Sensing
Edited by CHANDAN K. SEN AND GREGG L. SEMENZA

VOLUME 382. Quinones and Quinone Enzymes (Part B)
Edited by HELMUT SIES AND LESTER PACKER

VOLUME 383. Numerical Computer Methods (Part D)
Edited by LUDWIG BRAND AND MICHAEL L. JOHNSON

VOLUME 384. Numerical Computer Methods (Part E)
Edited by LUDWIG BRAND AND MICHAEL L. JOHNSON

VOLUME 385. Imaging in Biological Research (Part A)
Edited by P. MICHAEL CONN

VOLUME 386. Imaging in Biological Research (Part B)
Edited by P. MICHAEL CONN

VOLUME 387. Liposomes (Part D)
Edited by NEJAT DÜZGÜNEŞ

VOLUME 388. Protein Engineering
Edited by DAN E. ROBERTSON AND JOSEPH P. NOEL

VOLUME 389. Regulators of G-Protein Signaling (Part A)
Edited by DAVID P. SIDEROVSKI

VOLUME 390. Regulators of G-Protein Signaling (Part B)
Edited by DAVID P. SIDEROVSKI

VOLUME 391. Liposomes (Part E)
Edited by NEJAT DÜZGÜNEŞ

VOLUME 392. RNA Interference
Edited by ENGELKE ROSSI

VOLUME 393. Circadian Rhythms
Edited by MICHAEL W. YOUNG

VOLUME 394. Nuclear Magnetic Resonance of Biological Macromolecules (Part C)
Edited by THOMAS L. JAMES

VOLUME 395. Producing the Biochemical Data (Part B)
Edited by ELIZABETH A. ZIMMER AND ERIC H. ROALSON

VOLUME 396. Nitric Oxide (Part E)
Edited by LESTER PACKER AND ENRIQUE CADENAS

VOLUME 397. Environmental Microbiology
Edited by JARED R. LEADBETTER

VOLUME 398. Ubiquitin and Protein Degradation (Part A)
Edited by RAYMOND J. DESHAIES

VOLUME 399. Ubiquitin and Protein Degradation (Part B)
Edited by RAYMOND J. DESHAIES

VOLUME 400. Phase II Conjugation Enzymes and Transport Systems
Edited by HELMUT SIES AND LESTER PACKER

VOLUME 401. Glutathione Transferases and Gamma Glutamyl Transpeptidases
Edited by HELMUT SIES AND LESTER PACKER

VOLUME 402. Biological Mass Spectrometry
Edited by A. L. BURLINGAME

VOLUME 403. GTPases Regulating Membrane Targeting and Fusion
Edited by WILLIAM E. BALCH, CHANNING J. DER, AND ALAN HALL

VOLUME 404. GTPases Regulating Membrane Dynamics
Edited by WILLIAM E. BALCH, CHANNING J. DER, AND ALAN HALL

VOLUME 405. Mass Spectrometry: Modified Proteins and Glycoconjugates
Edited by A. L. BURLINGAME

VOLUME 406. Regulators and Effectors of Small GTPases: Rho Family
Edited by WILLIAM E. BALCH, CHANNING J. DER, AND ALAN HALL

VOLUME 407. Regulators and Effectors of Small GTPases: Ras Family
Edited by WILLIAM E. BALCH, CHANNING J. DER, AND ALAN HALL

VOLUME 408. DNA Repair (Part A)
Edited by JUDITH L. CAMPBELL AND PAUL MODRICH

VOLUME 409. DNA Repair (Part B)
Edited by JUDITH L. CAMPBELL AND PAUL MODRICH

VOLUME 410. DNA Microarrays (Part A: Array Platforms and Web-Bench Protocols)
Edited by ALAN KIMMEL AND BRIAN OLIVER

VOLUME 411. DNA Microarrays (Part B: Databases and Statistics)
Edited by ALAN KIMMEL AND BRIAN OLIVER

VOLUME 412. Amyloid, Prions, and Other Protein Aggregates (Part B)
Edited by INDU KHETERPAL AND RONALD WETZEL

VOLUME 413. Amyloid, Prions, and Other Protein Aggregates (Part C)
Edited by INDU KHETERPAL AND RONALD WETZEL

VOLUME 414. Measuring Biological Responses with Automated Microscopy
Edited by JAMES INGLESE

VOLUME 415. Glycobiology
Edited by MINORU FUKUDA

VOLUME 416. Glycomics
Edited by MINORU FUKUDA

VOLUME 417. Functional Glycomics
Edited by MINORU FUKUDA

VOLUME 418. Embryonic Stem Cells
Edited by IRINA KLIMANSKAYA AND ROBERT LANZA

VOLUME 419. Adult Stem Cells
Edited by IRINA KLIMANSKAYA AND ROBERT LANZA

VOLUME 420. Stem Cell Tools and Other Experimental Protocols
Edited by IRINA KLIMANSKAYA AND ROBERT LANZA

VOLUME 421. Advanced Bacterial Genetics: Use of Transposons and Phage for Genomic Engineering
Edited by KELLY T. HUGHES

VOLUME 422. Two-Component Signaling Systems, Part A
Edited by MELVIN I. SIMON, BRIAN R. CRANE, AND ALEXANDRINE CRANE

VOLUME 423. Two-Component Signaling Systems, Part B
Edited by MELVIN I. SIMON, BRIAN R. CRANE, AND ALEXANDRINE CRANE

VOLUME 424. RNA Editing
Edited by JONATHA M. GOTT

VOLUME 425. RNA Modification
Edited by JONATHA M. GOTT

VOLUME 426. Integrins
Edited by DAVID CHERESH

VOLUME 427. MicroRNA Methods
Edited by JOHN J. ROSSI

VOLUME 428. Osmosensing and Osmosignaling
Edited by HELMUT SIES AND DIETER HAUSSINGER

VOLUME 429. Translation Initiation: Extract Systems and Molecular Genetics
Edited by JON LORSCH

VOLUME 430. Translation Initiation: Reconstituted Systems and Biophysical Methods
Edited by JON LORSCH

VOLUME 431. Translation Initiation: Cell Biology, High-Throughput and Chemical-Based Approaches
Edited by JON LORSCH

VOLUME 432. Lipidomics and Bioactive Lipids: Mass-Spectrometry–Based Lipid Analysis
Edited by H. ALEX BROWN

VOLUME 433. Lipidomics and Bioactive Lipids: Specialized Analytical Methods and Lipids in Disease
Edited by H. ALEX BROWN

VOLUME 434. Lipidomics and Bioactive Lipids: Lipids and Cell Signaling
Edited by H. ALEX BROWN

VOLUME 435. Oxygen Biology and Hypoxia
Edited by HELMUT SIES AND BERNHARD BRÜNE

VOLUME 436. Globins and Other Nitric Oxide-Reactive Protiens (Part A)
Edited by ROBERT K. POOLE

VOLUME 437. Globins and Other Nitric Oxide-Reactive Protiens (Part B)
Edited by ROBERT K. POOLE

VOLUME 438. Small GTPases in Disease (Part A)
Edited by WILLIAM E. BALCH, CHANNING J. DER, AND ALAN HALL

VOLUME 439. Small GTPases in Disease (Part B)
Edited by WILLIAM E. BALCH, CHANNING J. DER, AND ALAN HALL

VOLUME 440. Nitric Oxide, Part F Oxidative and Nitrosative Stress in Redox Regulation of Cell Signaling
Edited by ENRIQUE CADENAS AND LESTER PACKER

VOLUME 441. Nitric Oxide, Part G Oxidative and Nitrosative Stress in Redox Regulation of Cell Signaling
Edited by ENRIQUE CADENAS AND LESTER PACKER

VOLUME 442. Programmed Cell Death, General Principles for Studying Cell Death (Part A)
Edited by ROYA KHOSRAVI-FAR, ZAHRA ZAKERI, RICHARD A. LOCKSHIN, AND MAURO PIACENTINI

VOLUME 443. Angiogenesis: *In Vitro* Systems
Edited by DAVID A. CHERESH

VOLUME 444. Angiogenesis: *In Vivo* Systems (Part A)
Edited by DAVID A. CHERESH

VOLUME 445. Angiogenesis: *In Vivo* Systems (Part B)
Edited by DAVID A. CHERESH

VOLUME 446. Programmed Cell Death, The Biology and Therapeutic Implications of Cell Death (Part B)
Edited by ROYA KHOSRAVI-FAR, ZAHRA ZAKERI, RICHARD A. LOCKSHIN, AND MAURO PIACENTINI

VOLUME 447. RNA Turnover in Bacteria, Archaea and Organelles
Edited by LYNNE E. MAQUAT AND CECILIA M. ARRAIANO

VOLUME 448. RNA Turnover in Eukaryotes: Nucleases, Pathways and Analysis of mRNA Decay
Edited by LYNNE E. MAQUAT AND MEGERDITCH KILEDJIAN

VOLUME 449. RNA Turnover in Eukaryotes: Analysis of Specialized and Quality Control RNA Decay Pathways
Edited by LYNNE E. MAQUAT AND MEGERDITCH KILEDJIAN

VOLUME 450. Fluorescence Spectroscopy
Edited by LUDWIG BRAND AND MICHAEL L. JOHNSON

VOLUME 451. Autophagy: Lower Eukaryotes and Non-Mammalian Systems (Part A)
Edited by DANIEL J. KLIONSKY

VOLUME 452. Autophagy in Mammalian Systems (Part B)
Edited by DANIEL J. KLIONSKY

VOLUME 453. Autophagy in Disease and Clinical Applications (Part C)
Edited by DANIEL J. KLIONSKY

VOLUME 454. Computer Methods (Part A)
Edited by MICHAEL L. JOHNSON AND LUDWIG BRAND

VOLUME 455. Biothermodynamics (Part A)
Edited by MICHAEL L. JOHNSON, JO M. HOLT, AND GARY K. ACKERS (RETIRED)

VOLUME 456. Mitochondrial Function, Part A: Mitochondrial Electron Transport Complexes and Reactive Oxygen Species
Edited by WILLIAM S. ALLISON AND IMMO E. SCHEFFLER

VOLUME 457. Mitochondrial Function, Part B: Mitochondrial Protein Kinases, Protein Phosphatases and Mitochondrial Diseases
Edited by WILLIAM S. ALLISON AND ANNE N. MURPHY

VOLUME 458. Complex Enzymes in Microbial Natural Product Biosynthesis, Part A: Overview Articles and Peptides
Edited by DAVID A. HOPWOOD

VOLUME 459. Complex Enzymes in Microbial Natural Product Biosynthesis, Part B: Polyketides, Aminocoumarins and Carbohydrates
Edited by DAVID A. HOPWOOD

VOLUME 460. Chemokines, Part A
Edited by TRACY M. HANDEL AND DAMON J. HAMEL

VOLUME 461. Chemokines, Part B
Edited by TRACY M. HANDEL AND DAMON J. HAMEL

VOLUME 462. Non-Natural Amino Acids
Edited by TOM W. MUIR AND JOHN N. ABELSON

VOLUME 463. Guide to Protein Purification, 2nd Edition
Edited by RICHARD R. BURGESS AND MURRAY P. DEUTSCHER

VOLUME 464. Liposomes, Part F
Edited by NEJAT DÜZGÜNEŞ

VOLUME 465. Liposomes, Part G
Edited by NEJAT DÜZGÜNEŞ

VOLUME 466. Biothermodynamics, Part B
Edited by MICHAEL L. JOHNSON, GARY K. ACKERS, AND JO M. HOLT

VOLUME 467. Computer Methods Part B
Edited by MICHAEL L. JOHNSON AND LUDWIG BRAND

VOLUME 468. Biophysical, Chemical, and Functional Probes of RNA Structure, Interactions and Folding: Part A
Edited by DANIEL HERSCHLAG

VOLUME 469. Biophysical, Chemical, and Functional Probes of RNA Structure, Interactions and Folding: Part B
Edited by DANIEL HERSCHLAG

VOLUME 470. Guide to Yeast Genetics: Functional Genomics, Proteomics, and Other Systems Analysis, 2nd Edition
Edited by GERALD FINK, JONATHAN WEISSMAN, AND CHRISTINE GUTHRIE

VOLUME 471. Two-Component Signaling Systems, Part C
Edited by MELVIN I. SIMON, BRIAN R. CRANE, AND ALEXANDRINE CRANE

VOLUME 472. Single Molecule Tools, Part A: Fluorescence Based Approaches
Edited by NILS G. WALTER

VOLUME 473. Thiol Redox Transitions in Cell Signaling, Part A Chemistry and Biochemistry of Low Molecular Weight and Protein Thiols
Edited by ENRIQUE CADENAS AND LESTER PACKER

VOLUME 474. Thiol Redox Transitions in Cell Signaling, Part B Cellular Localization and Signaling
Edited by ENRIQUE CADENAS AND LESTER PACKER

VOLUME 475. Single Molecule Tools, Part B: Super-Resolution, Particle Tracking, Multiparameter, and Force Based Methods
Edited by NILS G. WALTER

VOLUME 476. Guide to Techniques in Mouse Development, Part A Mice, Embryos, and Cells, 2nd Edition
Edited by PAUL M. WASSARMAN AND PHILIPPE M. SORIANO

VOLUME 477. Guide to Techniques in Mouse Development, Part B Mouse Molecular Genetics, 2nd Edition
Edited by PAUL M. WASSARMAN AND PHILIPPE M. SORIANO

VOLUME 478. Glycomics
Edited by MINORU FUKUDA

VOLUME 479. Functional Glycomics
Edited by MINORU FUKUDA

VOLUME 480. Glycobiology
Edited by MINORU FUKUDA

VOLUME 481. Cryo-EM, Part A: Sample Preparation and Data Collection
Edited by GRANT J. JENSEN

VOLUME 482. Cryo-EM, Part B: 3-D Reconstruction
Edited by GRANT J. JENSEN

VOLUME 483. Cryo-EM, Part C: Analyses, Interpretation, and Case Studies
Edited by GRANT J. JENSEN

VOLUME 484. Constitutive Activity in Receptors and Other Proteins, Part A
Edited by P. MICHAEL CONN

VOLUME 485. Constitutive Activity in Receptors and Other Proteins, Part B
Edited by P. MICHAEL CONN

VOLUME 486. Research on Nitrification and Related Processes, Part A
Edited by MARTIN G. KLOTZ

VOLUME 487. Computer Methods, Part C
Edited by MICHAEL L. JOHNSON AND LUDWIG BRAND

VOLUME 488. Biothermodynamics, Part C
Edited by MICHAEL L. JOHNSON, JO M. HOLT, AND GARY K. ACKERS

VOLUME 489. The Unfolded Protein Response and Cellular Stress, Part A
Edited by P. MICHAEL CONN

VOLUME 490. The Unfolded Protein Response and Cellular Stress, Part B
Edited by P. MICHAEL CONN

VOLUME 491. The Unfolded Protein Response and Cellular Stress, Part C
Edited by P. MICHAEL CONN

VOLUME 492. Biothermodynamics, Part D
Edited by MICHAEL L. JOHNSON, JO M. HOLT, AND GARY K. ACKERS

VOLUME 493. Fragment-Based Drug Design
Tools, Practical Approaches, and Examples
Edited by LAWRENCE C. KUO

VOLUME 494. Methods in Methane Metabolism, Part A
Methanogenesis
Edited by AMY C. ROSENZWEIG AND STEPHEN W. RAGSDALE

VOLUME 495. Methods in Methane Metabolism, Part B
Methanotrophy
Edited by AMY C. ROSENZWEIG AND STEPHEN W. RAGSDALE

VOLUME 496. Research on Nitrification and Related Processes, Part B
Edited by MARTIN G. KLOTZ AND LISA Y. STEIN

VOLUME 497. Synthetic Biology, Part A
Methods for Part/Device Characterization and Chassis Engineering
Edited by CHRISTOPHER VOIGT

VOLUME 498. Synthetic Biology, Part B
Computer Aided Design and DNA Assembly
Edited by CHRISTOPHER VOIGT

VOLUME 499. Biology of Serpins
Edited by JAMES C. WHISSTOCK AND PHILLIP I. BIRD

VOLUME 500. Methods in Systems Biology
Edited by DANIEL JAMESON, MALKHEY VERMA, AND HANS V. WESTERHOFF

VOLUME 501. Serpin Structure and Evolution
Edited by JAMES C. WHISSTOCK AND PHILLIP I. BIRD

VOLUME 502. Protein Engineering for Therapeutics, Part A
Edited by K. DANE WITTRUP AND GREGORY L. VERDINE

VOLUME 503. Protein Engineering for Therapeutics, Part B
Edited by K. DANE WITTRUP AND GREGORY L. VERDINE

VOLUME 504. Imaging and Spectroscopic Analysis of Living Cells
Optical and Spectroscopic Techniques
Edited by P. MICHAEL CONN

VOLUME 505. Imaging and Spectroscopic Analysis of Living Cells
Live Cell Imaging of Cellular Elements and Functions
Edited by P. MICHAEL CONN

VOLUME 506. Imaging and Spectroscopic Analysis of Living Cells
Imaging Live Cells in Health and Disease
Edited by P. MICHAEL CONN

VOLUME 507. Gene Transfer Vectors for Clinical Application
Edited by THEODORE FRIEDMANN

VOLUME 508. Nanomedicine
Cancer, Diabetes, and Cardiovascular, Central Nervous System, Pulmonary and Inflammatory Diseases
Edited by NEJAT DÜZGÜNEŞ

VOLUME 509. Nanomedicine
Infectious Diseases, Immunotherapy, Diagnostics, Antifibrotics, Toxicology and Gene Medicine
Edited by NEJAT DÜZGÜNEŞ

VOLUME 510. Cellulases
Edited by HARRY J. GILBERT

VOLUME 511. RNA Helicases
Edited by ECKHARD JANKOWSKY

CHAPTER ONE

ANALYSIS OF DUPLEX UNWINDING BY RNA HELICASES USING STOPPED-FLOW FLUORESCENCE SPECTROSCOPY

Andrea Putnam *and* Eckhard Jankowsky

Contents

1. Introduction	2
2. Experimental Considerations for Stopped-Flow Fluorescence Experiments	3
2.1. Duplex concentration	4
2.2. Verification of the correlation of fluorescence changes to duplex unwinding	6
2.3. Background fluorescence and inner filter effects	7
3. Substrate Design and Preparation	7
3.1. RNA labeling approaches	8
3.2. RNA labeled with Cy dyes	9
3.3. RNA labeled with 2-AP	10
3.4. Assessing the impact of the fluorophore on the unwinding reaction	12
4. Designing and Performing Stopped-Flow Fluorescence Unwinding Experiments	13
4.1. Identification of optimal wavelengths for excitation and emission	13
4.2. Pre-steady-state approaches to mechanistically characterize unwinding: Single and multiple cycle regimes	14
5. Data Fitting and Analysis	20
5.1. Data fitting	21
5.2. Determining unwinding rate constants	22
5.3. Determining functional equilibrium constants	23
Acknowledgments	24
References	24

Department of Biochemistry, Center for RNA Molecular Biology, School of Medicine, Case Western Reserve University, Cleveland, Ohio, USA

Methods in Enzymology, Volume 511
ISSN 0076-6879, DOI: 10.1016/B978-0-12-396546-2.00001-2
© 2012 Elsevier Inc.
All rights reserved.

Abstract

The characterization of unwinding reactions by RNA helicases often requires the determination of rate constants that are too fast to be measured by traditional, manual gel-based methods. Stopped-flow fluorescence measurements allow access to fast unwinding rate constants. In this chapter, we outline strategies and experimental considerations for the design of stopped-flow fluorescence experiments to monitor duplex unwinding by RNA helicases, with focus on DEAD-box helicases. We discuss advantages, disadvantages, and technical considerations for stopped-flow approaches, as well as substrate design. In addition, we list protocols and explain functional information obtained with these experiments.

1. INTRODUCTION

The ability to separate RNA duplexes in an ATP-dependent fashion is the namesake activity associated with RNA helicases (Tanner and Linder, 2001). ATP-driven RNA unwinding appears to be critical for the cellular roles of at least a subset of RNA helicases, although it is clear that the functional repertoire of these enzymes often exceeds simple strand separation and may involve remodeling of RNAs, RNPs, or the modulation of associated proteins (Jankowsky, 2011; Jankowsky and Fairman, 2007; Jia et al., 2011). Which exact RNA or RNP rearrangements most RNA helicases catalyze in the cell is not yet known.

It is, however, well established that physiological functions of RNA helicases correlate with their ability to unwind RNA duplexes in an ATP-dependent fashion *in vitro* (Jankowsky and Fairman, 2007; Tanner and Linder, 2001). Mutations that impair unwinding activity of a given RNA helicase *in vitro* often diminish the physiological function of the protein (e. g., Hilliker et al., 2011; Montpetit et al., 2011; Tanaka and Schwer, 2006). This correlation exists not necessarily because a given RNA helicase unwinds a duplex in the cell, but because both RNA/RNP remodeling and duplex unwinding rely on the ability of the helicase to modulate RNA affinity and the conformation of bound RNA in response to the various stages of the ATP hydrolysis cycle (Jankowsky, 2011). ATP-dependent duplex unwinding *in vitro* is thus perhaps most appropriately viewed as a reflection of the capacity of a given RNA helicase to remodel RNA in an ATP-driven reaction (Jankowsky, 2011). Currently, the most straightforward way to study how RNA helicases use ATP to remodel RNA is therefore the kinetic dissection of the multistep duplex unwinding reaction and the determination of rate constants for basic reaction steps (see also Chapter 2).

Duplex unwinding by RNA helicases has been almost exclusively measured by electrophoretic mobility shift assays (EMSA) using radiolabeled RNA. EMSA approaches utilize equipment, materials, and techniques already present in most laboratories equipped for molecular biology work. It is also straightforward to adapt EMSA-based approaches for quantitative studies of unwinding reactions (Jankowsky and Putnam, 2010). However, gel-based methods are usually limited by the frequency with which individual datapoints can be taken manually. As a consequence, only rate constants smaller than roughly $k_{obs} < 6–10$ min^{-1} are experimentally accessible (Jankowsky and Putnam, 2010). Yet, mechanistic investigations often require access to faster rate constants. For this reason, we and others have adapted stopped-flow fluorescence measurements to measure duplex unwinding (Henn et al., 2010). This technique allows the continuous monitoring of unwinding reactions that are much faster than those accessible by manual EMSA approaches. Utilization of stopped-flow fluorescence is common in the analysis of a wide range of enzymes, including DNA helicases (Lucius et al., 2004; Raney et al., 1994). For RNA helicases, this fluorescence technique has only been used sparingly to study duplex unwinding (Henn et al., 2010). Nonetheless, the use of stopped-flow fluorescence to measure duplex unwinding critically expands the scope of methods available for the functional characterization of RNA helicases.

In this chapter, we outline general strategies and experimental considerations for the design of stopped-flow fluorescence experiments to monitor duplex unwinding by RNA helicases. We discuss advantages, disadvantages, and technical considerations for stopped-flow approaches to the analysis of unwinding reactions, as well as substrate design. Finally, we describe protocols and explain functional information obtained with these experiments.

2. EXPERIMENTAL CONSIDERATIONS FOR STOPPED-FLOW FLUORESCENCE EXPERIMENTS

The stopped-flow fluorescence approach involves rapid mixing of small volumes of reactants, followed by the flow of the mixture into a cell that allows fluorescence measurements, and a subsequent, sudden stop of the flow (Fig. 1.1). Changes in fluorescence following this "stopped-flow" are recorded, usually continuously and in real time (Fig. 1.1). The measurements take place in a dedicated stopped-flow fluorimeter. The dead time of these devices is limited by the mixing time, which is usually around 0.5 ms (Nayak et al., 2012). Thus, rate constants of $k_{obs} < 10^2$ s^{-1} can be reliably measured, an increase of several orders of magnitude over rate constants accessible through manual EMSA approaches.

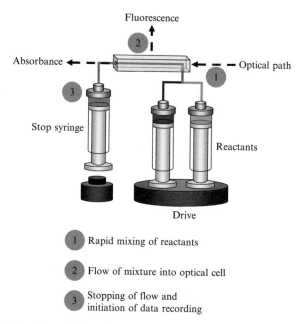

Figure 1.1 Stopped-flow fluorescence setup. Schematic representation.

Typical stopped-flow measurements require volumes of hundreds of microliters of reactants in the reservoir syringes. This volume usually allows five to six individual measurements. The amount of protein and RNA required for a set of experiments with multiple concentrations of ATP and enzyme is thus considerably larger than for EMSA-based measurements (Anderson et al., 2008). In addition, the RNA needs to be fluorescently labeled and then tested for functional integrity (Section 3). Because of the material requirements and the necessary tests of RNA integrity, the use of stopped-flow techniques is probably only advised when unwinding parameters need to determined that cannot be measured by slowing down fast reactions to bring observed rate constants into the range accessible by EMSA.

2.1. Duplex concentration

As noted, monitoring duplex unwinding by fluorescence changes of labeled RNA requires duplex concentrations that are significantly higher than those used in EMSA-based assays (Anderson et al., 2008). The lowest possible RNA concentration is determined by the number of emitted photons that can be collected over the reaction time. This limit thus depends on the sensitivity of the fluorimeter, on excitation and emission settings, on the reaction time, and on the type of fluorophore used. In our hands, stopped-flow experiments with a 2-aminopurine (2-AP)-labeled RNA require duplex concentrations upward of 50 nM for measured rate constants in the range of $k_{obs} = 10$ s^{-1} (Section 4).

2.1.1. Strand annealing

At duplex concentrations needed for fluorescence measurements, the rate of strand annealing of the unwound duplex may equal or even exceed the rate of the unwinding reaction. If this occurs, the amplitude of the unwinding reaction decreases, which diminishes or even eliminates the unwinding signal (Yang and Jankowsky, 2005). In addition, the rate constant for strand annealing impacts the overall reaction rate seen for the unwinding process. If the actual unwinding rate constant is to be extracted from the measurements, the rate constant for strand annealing under the reaction conditions has to be determined (Yang and Jankowsky, 2005).

Under typical unwinding reaction conditions, spontaneous reannealing of complementary strands occurs with rate constants of approximately $k_{on} \sim 10^{-7}$ min^{-1} (Jankowsky and Putnam, 2010). Observed pseudo-first-order annealing rate constants scale with the square of the duplex concentration, and half-lives for duplex formation quickly decrease from $t_{1/2} \sim 20$ min^{-1} for duplex concentrations of [Duplex] = 5 nM to only $t_{1/2} \sim 0.2$ min^{-1} for duplex concentrations of [Duplex] = 50 nM. In addition, many RNA helicases accelerate strand annealing and often raise annealing rate constants by two to three orders of magnitude, thus dramatically decreasing half-lives for duplex formation (Chamot et al., 2005; Halls et al., 2007; Rossler et al., 2001; Uhlmann-Schiffler et al., 2006; Valdez, 2000; Yang and Jankowsky, 2005).

2.1.2. Strand exchange

Given the complications caused by simultaneously occurring unwinding and annealing, it is usually desirable to eliminate strand annealing. This is possible by inclusion of a scavenger RNA or DNA that pairs with one of the duplex strands, preferably with the strand that does not bear a fluorescent label. However, inclusion of a scavenger alters the overall concentration of RNA in the reaction and potentially sequesters a fraction of the helicase. In addition, the scavenger changes the reaction regime from a pure unwinding reaction to a strand exchange (Yang et al., 2007b). This change makes it necessary to determine whether the observed rate constants for the strand exchange exclusively reflect ATP-dependent unwinding by the helicase, or whether helicase- and ATP- independent strand exchange contributes as well.

Interaction of the helicase with the scavenger alters the effective concentration of the helicase. This concentration change has to be considered in the calculation of reaction parameters and in the interpretation of obtained data (Jankowsky and Putnam, 2010). This complication can be avoided if it is possible to select a scavenger strand that does not significantly impact the effective helicase concentration. In our hands, RNA oligonucleotides with 10 nt were found to not significantly alter the effective concentration of the RNA helicase Ded1p at concentrations used in unwinding experiments (Putnam and Jankowsky, unpublished data). The impact of a scavenger on the helicase concentration is assessed by competition experiments, examining the effect of increasing scavenger concentrations on the unwinding reaction of a duplex that does not interact with the scavenger.

Strand exchange that is not connected to helicase-mediated unwinding can impact reaction kinetics if the duplex strands dissociate with rate constants comparable or higher to those of the helicase-mediated unwinding. This effect is more pronounced for short and thus unstable duplexes. Moreover, strand exchange can be facilitated by helicases in an ATP-independent fashion (Rossler et al., 2001). These ATP-independent strand exchange reactions are probed by omitting either helicase or ATP from the reaction. If the strand exchange reaction is not exclusively limited by ATP-dependent duplex unwinding by the helicase, rate constants of competing strand exchange reactions need to be determined and considered when calculating actual unwinding rate constants.

2.2. Verification of the correlation of fluorescence changes to duplex unwinding

Fluorescence measurements of unwinding reactions rely on changes in the fluorescence properties of the labeled RNA. However, such changes can also be caused by processes other than strand separation. For example, mere binding of the helicase to the RNA can alter the fluorescence. Although such fluorescence changes might ultimately prove beneficial for elucidating molecular mechanisms, for unwinding measurements it is necessary to verify that measured changes in fluorescence exclusively reflect the strand separation process. It is often not possible to deconvolute the contributions of unwinding and other processes to fluorescence changes by only fluorescence measurements. Clarification can be obtained by comparing unwinding rate constants measured by fluorescence to rate constants determined with a different method, for example, EMSA. For obvious reasons, reaction conditions have to be chosen where unwinding can be monitored by both approaches.

Fluorescence changes unrelated to the unwinding signal can also be caused by photobleaching of the fluorophore over the reaction time. Photobleaching, the collective term for light-induced decrease in fluorescence emission, is identified by measuring fluorescence emission of only labeled RNA over the expected reaction time course. It is possible to suppress photobleaching by decreasing the excitation intensity, but the unwinding signal then also decreases. This reduction in signal might be offset by an increase in the concentration of labeled RNA. Photobleaching can also be reduced by depleting oxygen from the solution (Hubner et al., 2001). Nevertheless, it is frequently not practical or possible to completely eliminate photobleaching. In these cases, photobleaching has to be considered in the calculation of rate constants (Section 5.1).

2.3. Background fluorescence and inner filter effects

Sensitive fluorescence measurements require the minimization of background fluorescence, which can significantly decrease the signal-to-noise ratio. Commonly used buffer components such as oxidized DTT or detergents such as Triton-X100 have high absorbance or background fluorescence in the excitation and emission range of fluorophores including 2-AP (Iyer and Klee, 1973; Tiller et al., 1984). Oxidized DTT has a strong absorbance band near 280 nm. In its reduced form, there is little absorbance above 275 nm. When DTT cannot be avoided, it is advisable to add it immediately before the reaction to minimize the amount of oxidization. Triton-X100 has a strong fluorescence emission peak at 302 nm, and it is advisable to substitute this component with nonfluorescent detergents, such as IGEPAL (Tiller et al., 1984).

It is also important to minimize inner filter effects. These are caused by reaction components absorbing excitation or emission light, thereby decreasing the fluorescence intensity. Such absorption can be associated with the fluorophore itself, in which case the effect scales with the concentration of the fluorophore. Perhaps more importantly, the RNA itself, and especially ATP at low millimolar concentrations, cause significant inner filter effects for fluorophores with excitation or emission maxima at or below 350 nm. Corrections for inner filter effects, if they cannot be avoided, have been described in detail (Gu and Kenny, 2009; Roy, 2004).

3. Substrate Design and Preparation

Principles for the design of duplex substrates to study RNA helicases that unwind with defined polarity as well as those that unwind without defined polarity have been discussed in detail in a previous publication (Jankowsky and Putnam, 2010). We encourage readers to consult this chapter before designing RNA substrates for a particular RNA helicase, since duplex length and stability and features of unpaired regions all impact observed unwinding rate constants (Yang and Jankowsky, 2006; Yang et al., 2007a). Considerations for substrate design are also outlined in Chapter 2.

Measuring duplex unwinding by fluorescence requires the introduction of one or more fluorescent labels into the RNA. The most commonly used labeling techniques are incorporation of one or more modified RNA bases into the sequence, or the covalent attachment of a fluorophore to a reactive functional group in the RNA. Modified bases, reactive functional groups, or both can be introduced into RNAs that have been transcribed *in vitro* (e.g., Wilson and Szostak, 1999). It is, however, generally more practical to utilize chemically synthesized RNA, which can be purchased with already attached fluorophores or reactive functional groups (Section 3.2). Other methods to attach fluorophores to RNA have also been described (e.g., Roy *et al.*, 2008), but will not been discussed here. As long as the fluorophore is stably associated with the RNA, any labeling method can be used. However, as outlined below (Section 3.3), great care has to be taken to assess the impact of the label on the unwinding reaction.

3.1. RNA labeling approaches

Duplex unwinding can be monitored by fluorescence according to two strategies: (i) fluorescence resonance energy transfer (FRET) or (ii) fluorescence quenching (Fig. 1.2). FRET requires two fluorophores (donor and acceptor) on the RNA, preferably one on each strand (Fig. 1.2A). FRET and suitable donor and acceptor pairs have been described extensively in the literature (e.g., Clegg, 1992; Lilley and Wilson, 2000; Wu and Brand, 1994). Commonly used pairs include Fluorescein/Rhodamine and Cy3/Cy5 (e.g., Cardullo *et al.*, 1988; Lucius *et al.*, 2004). FRET principles are discussed in more detail in Chapter 4.

In the duplex, both fluorophores are in close proximity, which results in high FRET (i.e., low donor emission, high acceptor emission). As the strands are separated, the dyes move apart from each other and FRET decreases (i.e., high donor emission and low or zero acceptor emission, Fig. 1.2A). This decrease in emission is monitored over time. The FRET approach allows the simultaneous monitoring of both donor and acceptor

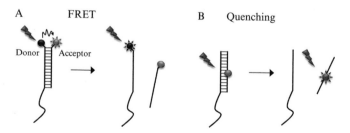

Figure 1.2 Measuring duplex unwinding by fluorescence. (A) Monitoring duplex unwinding by FRET. Circles show acceptor and donor fluorophores, as marked. (B) Monitoring duplex unwinding by fluorescence quenching of an internal fluorophore, (e.g., 2-AP), as marked.

fluorescence, provided the fluorimeter is equipped for dual wavelength detection. Monitoring the two anti-correlated signals enhances the signal-to-noise ratio, compared to the use of a single fluorophore. If emission of both fluorophores can be measured simultaneously, lower RNA concentrations can be used with FRET than with approaches that involve only a single fluorophore.

Duplex unwinding can also be followed through fluorescence changes associated with the quenching of a single fluorophore (Fig. 1.2B). We and others have used 2-AP, which functions as nucleobase within the RNA sequence and only minimally affects helix geometry and stability (discussed in more detail in Section 3.3). In the duplex, 2-AP fluorescence is quenched, due to stacking with neighboring nucleobases (Jean and Hall, 2001). Upon unwinding, 2-AP fluorescence increases (Fig. 1.2B). Quenching approaches can also be used with extraneous fluorophores, such as a single Cy3, provided that strand separation results in a change in fluorescence. These approaches have been employed to measure duplex unwinding by DNA helicases (Fischer et al., 2004).

Quenching techniques are typically associated with smaller fluorescence changes than those seen with FRET. As a consequence, quenching approaches require higher RNA concentrations than FRET. However, it is often more laborious to attach the two extraneous fluorophores required for FRET measurements, compared with the introduction of only a single fluorophore for quenching studies. In addition, many fluorophores used for FRET have the tendency stack to nucleobases, which impacts duplex stability (Iqbal et al., 2008b; Norman et al., 2000). Many fluorophores used for FRET also have the potential to interact with the helicase (Lucius et al., 2004). The decision whether to use FRET or quenching to monitor unwinding thus depends on the specific insight expected from the system under study, and on the available fluorimeter.

3.2. RNA labeled with Cy dyes

Cy dyes and their newer cousins, Alexa fluorphors, have large extinction coefficients and high quantum yields and are comparably resistant to photobleaching (Roy et al., 2008). These fluorophores can thus be used at low concentrations (~ 10 nM in a typical fluorimeter). Moreover, excitation and emission wavelengths of these dyes are well above the absorbance maxima of protein and nucleotide, eliminating inner filter effects. Disadvantages of these dyes include impacts of the large fluorophores on duplex stability (Moreira et al., 2005). Cy3 and Cy5 have been shown to stack on the terminal nucleotide (Iqbal et al., 2008a,b; Norman et al., 2000). This stacking can quench the fluorescence and stabilize the duplex (Moreira et al., 2005). However, the effects on duplex stability can be measured and might not be critical for the planned experiments (Section 3.4). Cy and Alexa dyes may directly interact with protein, which, as discussed above, can cause fluorescence changes unrelated to the actual unwinding process, and may impact the binding of the helicase (Anderson et al., 2008).

3.3. RNA labeled with 2-AP

2-AP can be directly incorporated into the oligonucleotide, which avoids several of the issues associated with covalently attached dyes. 2-AP forms Watson–Crick basepairs with uracils or thymidines, without significantly altering helix geometry (Nordlund et al., 1989). The absorption maximum of 2-AP is 303 nm and the emission maximum is typically at ~ 370 nm. Thus, the fluorescence may be affected by inner filter effects caused by nucleotide and protein. 2-AP has a lower quantum yield ($Q \sim 0.02$, in single-stranded DNA) and smaller extinction coefficient ($\varepsilon^{303nm} = 3600$ cm^{-1} M^{-1}) than Cy or Alexa dyes (for Cy3, $Q \sim 0.3$ in single-stranded DNA, $\varepsilon^{547nm} = 136{,}000$ cm^{-1} M^{-1}). 2-AP thus cannot be used at low concentrations (Iqbal et al., 2008a; Jean and Hall, 2001). It is possible to incorporate multiple 2-AP in a single strand of RNA, which increases the fluorescence signal. 2-AP affects duplex stability, compared to adenosine, since the basepairs formed are not as stable as authentic A:U or A:T pairs (Law et al., 1996). It is important to place 2-AP internally in a duplex, because spontaneous fraying at the ends of a duplex can also cause changes in fluorescence (Jose et al., 2009).

3.3.1. Preparation and purification of fluorescently labeled duplexes

Duplexes with labeled RNA (both 2-AP and extraneous fluorophores) are prepared essentially like radiolabeled duplexes (Jankowsky and Putnam, 2010). In principle, it should be possible to simply combine both strands in roughly equimolar concentrations and then perform the unwinding reaction. However, we have found that this approach never yields 100% duplex. In addition, excess single strand often complicates data analysis and precludes straightforward interpretations of reaction amplitudes. Moreover, exact concentrations of duplex and single-stranded RNA in the reaction need to be known to accurately account for strand annealing, if this is allowed to occur simultaneously. In our hands, the outlined PAGE-based duplex purification has provided highly reproducible results.

Purification of individual RNA strands

1. Dilute a minimum of 0.5 μg of RNA (required for visualization by UV shadowing, assuming a 1-cm well) into 10 μL in nuclease free water.
2. Add 2 μL of 5× denaturing loading buffer without loading dyes (80% formamide) and heat to 95 °C for 2 min to remove any secondary

structure. Loading dyes such as bromophenol blue (BPB) and xylene cyanol (XC) can interfere with UV shadowing.

3. Pre-run a 20% denaturing gel (20% Acrylymide:Bis 19:1, 7 M Urea, 1× TBE (8.9 mM Tris Base, 8.9 mM Boric Acid, 0.2 mM EDTA), 0.8 mm thick) for roughly 30 min to reach separation temperature ($\sim$50 °C).
4. Load sample on gel. A lane containing denaturing loading buffer containing 0.1% BPB and XC adjacent to the RNA is recommended as a marker. Run at $\sim$30 V/cm for at least 2 h, depending on the length of the RNA.
5. Remove the gel from the glass plates and cover with plastic wrap. Place the gel onto a thin layer chromatography (TLC) plate with a fluorescent indicator.
6. Using a UV-handheld lamp, briefly expose the gel to short-wave (254 nm) light. The bands containing RNA will appear as a black shadow on the green fluorescent background of the TLC plate. Mark the appropriate bands to cut out later. Avoid long exposure times to avoid damaging the nucleic acid.
7. Cut out the appropriate bands, cover with 600 µL of gel elution buffer (300 mM NaOAc, 1 mM EDTA, 0.5% SDS), and shake gently overnight at 4 °C. The buffer must cover the gel slice.
8. Divide elution supernatants into two 1.5-mL tubes and add 3× volume of 100% EtOH and 1 µL of 1 mg/mL glycogen.
9. Place on dry ice for 1 h and centrifuge (16,000 × g) for 30 min at 4 °C. For short RNAs ($\leq$10 nt), this step may be repeated to improve precipitation efficiency.
10. Remove supernatant and dry pellet in SpeedVac for at least 15 min.
11. Recombine pellets by mixing one pellet in 10 µL water, resuspend, and transfer to another tube.

Duplex preparation and purification

12. In a volume of 20–30 µL, combine single-stranded RNAs required for duplex formation in roughly equimolar fashion. Add 2 µL of 10× duplex annealing buffer (100 mM MOPS, pH 6.5, 10 mM EDTA, 0.5 M KCl).
13. Heat to 95°C and gradually cool to room temperature over at least 30 min. For short RNA duplexes ($\leq$10 bp), place on ice for several minutes.
14. After room temperature is reached, add 4 µL of 5× nondenaturing gel loading buffer without dyes (50% glycerol) and load on 15% nondenaturing gel (15% Acrylymide:Bis 19:1, 0.5× TBE, 0.8 mm thick). In an adjacent lane, load nondenaturing gel loading buffer containing 0.1% BPB and XC as a marker.

15. Run gel at room temperature (~20 V/cm) for at least 1 h (longer for duplexes exceeding 30 basepairs).
16. Follow steps 5–11.
17. Dissolve pellet in buffer (50 mM MOPS, pH 6.0, 50 mM KCl).
18. Measure concentration by UV spectroscopy. The concentration of the duplex may also be determined by using the absorbance of the incorporated fluorophore, which is frequently more sensitive than measurements of nucleotides at 260 nm.

3.4. Assessing the impact of the fluorophore on the unwinding reaction

As noted, fluorophores in or on the RNA often alter duplex stability, impact helicase binding, or both (Anderson et al., 2008; Moreira et al., 2005). For these reasons, it is critical to assess to which extent the fluorophore alters the observed unwinding rate constants, compared to the unlabeled RNA. Changes in duplex stability are especially important for studies of DEAD-box helicases, where rate constants of strand separation strongly scale with duplex stabilities (Rogers et al., 1999, 2001a,b; Yang and Jankowsky, 2005). Unwinding rate constants measured with RNA helicases of other families also often scale with the duplex stability (e.g., Mtr4p of the Ski2-like family, Bernstein et al., 2008), although the exact reasons for this correlation are not yet understood.

As mentioned, many external fluorophores have the tendency to stack on the ends of helices (Iqbal et al., 2008a). This stacking typically increases duplex stability by roughly the contribution of an additional basepair, resembling the duplex stabilization of "dangling ends" (Moreira et al., 2005). Incorporation of 2-AP into an RNA duplex somewhat decreases duplex stability (on average $\Delta\Delta G \sim 0.5$ kcal mol^{-1}, Law et al., 1996). As a general rule, the effect of the fluorophore on duplex stability is greatest for duplexes with low inherent stability.

Many extraneous fluorophores, including Cy-or Alexa dyes, contain or are based on aromatic compounds that somewhat resemble nucleobases (Iqbal et al., 2008b; Norman et al., 2000). As a result, these fluorophores have the tendency to directly interact with nucleic acid binding proteins, including helicases (Lucius et al., 2004; Myong et al., 2009). This interaction can lead to changes in fluorescence unrelated to the unwinding signal, as noted above (Lucius et al., 2004). In fact, the interaction of helicases with fluorophores is exploited in a single molecule fluorescence approach termed protein-induced fluorescence enhancement, which measures the time a helicase resides in close proximity to the fluorophore (Myong et al., 2009). 2-AP circumvents the direct impact of an additional fluorophore on the helicase, since this fluorophore substitutes for adenosine within the RNA (Law *et al.*, 1996).

To assess changes in unwinding caused by fluorophore-induced changes in duplex stability or by effects of the fluorophore on helicase binding to the substrate, it is advisable to compare unwinding reactions of labeled and unlabeled duplex by EMSA-based methods as a function of the helicase concentration. Reactions have to be conducted under conditions where unwinding rate constants can be measured by manual EMSA-based methods. This is typically accomplished at low ATP or helicase concentrations.

It is also possible to measure the impact of the label on helicase affinity through competition experiments that examine effects of increasing amounts of duplex with and without the fluorophore on the unwinding of an unrelated substrate. This approach avoids potentially lengthy optimizations to identify appropriate conditions for each new duplex, since well-established conditions and duplexes can be used. However, these competition experiments cannot assess the impact of the fluorophore on the unwinding rate constant.

4. Designing and Performing Stopped-Flow Fluorescence Unwinding Experiments

Having prepared and characterized the labeled RNA substrate, the stopped-flow unwinding experiments are designed and performed. As the first step, it is advisable to identify optimal wavelengths for excitation and emission signals, in order to maximize the fluorescence signal resulting from unwinding. In essence, this task involves the recording of emission spectra before and after unwinding and is thus best performed with a steady-state fluorimeter. The protocol below describes the measurements for a duplex labeled with 2-AP (Fig. 1.3).

4.1. Identification of optimal wavelengths for excitation and emission

1. In a 1×1 cm quartz cuvette, mix 60 μL of $10\times$ helicase reaction buffer ($10\times$ HRB: 400 mM Tris, pH 8.0, 5 mM MgCl$_2$, 0.1% IGEPAL, 20 mM DTT), 6 μL of 5 μM 2-AP duplex, 6 μL of 100 μM RNA scavenger, 60 μL of $10\times$ helicase diluted with helicase storage buffer to desired concentration, 60 μL of 20 mM ATP/Mg, and 408 μL of water for a total volume of 600 μL.
2. Monitor the fluorescence of the sample by exciting the sample at 310 nm and scanning emission wavelengths from 340 to 480 nm. The emission

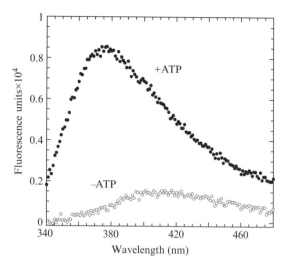

Figure 1.3 RNA unwinding by the DEAD-box protein Ded1p measured by steady-state fluorescence. Emission spectra of a 10-bp RNA duplex containing an internal 2-aminopurine and a 25-nt single-stranded region 3' to the duplex. Reactions containing 50 nM RNA duplex; 1 μM 10 nt unlabeled RNA scavenger and 0.6 μM Ded1p were incubated for 5 min with (●) and without (○) 2 mM ATP/Mg, as indicated. Fluorescence was measured with an excitation wavelength of 310 nm, and emission was monitored from 340 to 480 nm at 5 nm band width in a Jobin-Yvon FluoroMax-2 fluorimeter.

maxima should be centered near 370 nm. Slit widths should be small (1–5 nm) and increased only if necessary to improve the signal intensity.
3. The following additional reactions are recommended:
    a. Omit helicase and 2-AP duplex to determine the background fluorescence of the sample.
    b. Omit ATP from the reaction to measure any ATP-independent changes in fluorescence.
    c. Heat the reaction to 95 °C and allow to cool to reaction temperature to measure the 100% unwound fluorescence and compare to the helicase-mediated signal.

4.2. Pre-steady-state approaches to mechanistically characterize unwinding: Single and multiple cycle regimes

Having identified optimal fluorimeter settings, the actual kinetic measurements are performed. In the design of the kinetic experiments, it is important to carefully consider the specific characteristics of the two different

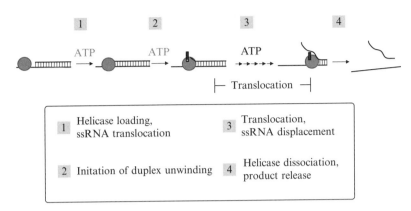

Figure 1.4 Translocation-based unwinding. The closed circle represents the enzyme unwinding unit, which can be an oligomer. The black rectangle represents ATP.

unwinding types of RNA helicase. As extensively reviewed (Fairman-Williams *et al.*, 2010; Jankowsky, 2011; Myong and Ha, 2010; Pyle, 2008; Singleton *et al.*, 2007), one type of helicase unwinds duplexes by ATP-driven translocation along one of the helix strands (Fig. 1.4). This type is represented by viral NPH-II/HCV helicases, likely by the DEAH/RHA helicases, and possibly by further helicase families (Jankowsky, 2011). Unwinding by these helicases is largely identical to the unwinding by canonical DNA helicases and involves multiple consecutive translocation steps, typically using one ATP per translocated nucleotide (Gu and Rice, 2010; Pyle, 2008; Singleton *et al.*, 2007). For a detailed overview of RNA "translocases," see also Chapter 6.

The other type of RNA helicase is represented by DEAD-box helicases, the largest helicase family in eukaryotes (Anantharaman *et al.*, 2002). Unwinding by DEAD-box proteins does not involve translocation of the protein on the RNA (Fig. 1.5). Instead, DEAD-box proteins directly load to the duplex region and then pry the duplex strands apart in an ATP-dependent manner (Bizebard *et al.*, 2004; Tijerina *et al.*, 2006; Yang and Jankowsky, 2006; Yang *et al.*, 2007a). This unwinding mode has been termed local strand separation. A strand separation event occurs in a single step and requires only a single ATP, regardless of the number of basepairs in the duplex (Chen *et al.*, 2008, Liu *et al.*, 2008). However, with increasing duplex length, the rate constant for strand separation strongly decreases (Rogers *et al.*, 2001a,b) and the number of futile ATP turnovers increases dramatically (Bizebard *et al.*, 2004, Chen *et al.*, 2008).

Despite the fundamental differences of the two unwinding modes, both involve a series of basic reaction steps, including helicase binding, overall strand separation, and helicase dissociation (Figs. 1.4 and 1.5). Experimental strategies to measure individual reaction steps in multistep unwinding reactions typically

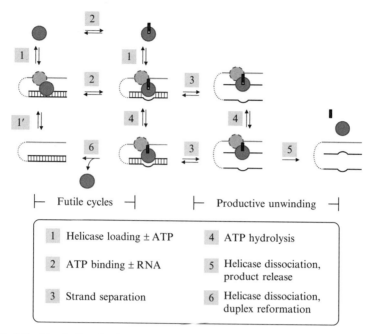

Figure 1.5 Unwinding by local strand separation. The circles represent individual helicase protomers. The closed circle is the unwinding protomer, and the light circle is the loading unit. Not all DEAD-box helicases function as oligomers (Jankowsky, 2011). The black rectangle represents ATP.

rely on pre-steady-state kinetic measurements (Jankowsky and Putnam, 2010). That is, unwinding reactions are performed with enzyme excess over the substrate. Pre-steady-state reactions minimize multiple substrate turnovers and can be described with fewer reaction parameters than steady-state reactions where the enzyme continuously turns over substrate molecules.

Pre-steady-state unwinding reactions are implemented in two different regimes, single cycle and multiple cycle. Below, we outline principles, rationales, expected insight, and protocols for performing single cycle and multiple cycle reactions, using the yeast DEAD-box helicase Ded1p as example. The approaches are generally applicable to helicases that unwind by the translocation-based mode. We highlight points where approaches and interpretations differ between the two unwinding modes.

4.2.1. The single cycle regime

The single cycle regime examines unwinding per single enzyme binding event. The helicase is bound to the substrate in the absence of ATP, and the reaction is started with scavenger and ATP (Fig. 1.6A and B). Here, the scavenger traps excess helicase and enzyme that dissociate during the course

Stopped-Flow Unwinding Measurements

Single cycle regime

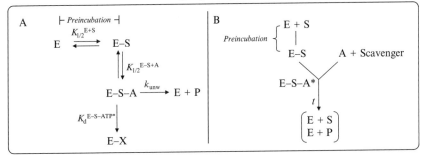

Multiple cycle regime

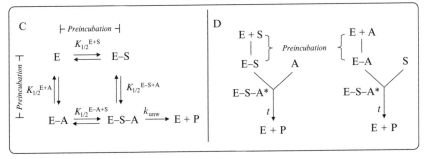

Figure 1.6 Pre-steady-state reaction regimes. (A) Single cycle regime. E, enzyme (helicase); S, substrate (RNA); A, ATP; P, product (unwound strands); X, scavenger. (B) Reaction scheme for single cycle experiment. (C) Multiple cycle regime. Abbreviations as in panel (A). (D) Reactions schemes for multiple cycle experiments, started with ATP (left) or RNA (right).

of unwinding. Rebinding of the helicase to the substrate during the reaction is thus prevented and multiple unwinding events per substrate are avoided (Fig. 1.6A). Scavenger can be RNA that does not interact with the substrate. Alternatively, heparin can be used (Paolini *et al.*, 2000; Pfeffer *et al.*, 1977). The scavenger must completely sequester the helicase. This is best tested by preincubating the scavenger with the substrate and the helicase before starting the reaction with ATP. Efficient helicase scavenging completely abrogates unwinding. Inclusion of scavenger also eliminates helicase-mediated strand annealing, as noted for several helicases tested in our laboratory (Putnam and Jankowsky, unpublished data).

For DEAD-box helicases, the unwinding rate constant measured under single cycle conditions (k'_{unw}) is the sum of the actual unwinding rate constant (k_{unw}) and the rate constant by which the enzyme dissociates from the complex that contains ATP, substrate, and enzyme ($k_d^{E-S-ATP\star}$, Fig. 1.6A).

$$k'_{\text{unw}} = k_{\text{unw}} + k_d^{\text{E-S-ATP}*} \tag{1.1}$$

This rate constant cannot be further deconvoluted by fluorescence measurements alone. However, it is possible to calculate the actual unwinding rate constant from the absolute reaction amplitude (A), which can be determined under single cycle conditions with EMSA-based methods at ATP and enzyme saturation (Jankowsky and Putnam, 2010), according to:

$$A = k_{\text{unw}}\left(k_{\text{unw}} + k_d^{\text{E-S-ATP}*}\right)^{-1} = k_{\text{unw}}(k'_{\text{unw}})^{-1} \tag{1.2}$$

This equation also yields the dissociation rate constant $k_d^{\text{E-S-ATP}*}$. Both k_{unw} and $k_d^{\text{E-S-ATP}*}$ do not depend on enzyme concentration, and the observed unwinding rate constant measured under single cycle conditions should therefore not vary with enzyme concentration (Fig. 1.6A). In contrast, the reaction amplitude under single cycle conditions depends on both enzyme and ATP concentration. It is therefore possible to determine functional dissociation constants for substrate (K_d^S) and ATP binding ($k_d^{\text{ES-ATP}}$) from plots of reaction amplitudes versus enzyme or ATP concentrations (Section 5.3). If reaction amplitudes are measured by fluorescence, it is often important to benchmark these values to absolute values obtained with EMSA-based methods.

For translocating helicases, strand separation can be examined in isolation under single cycle conditions in essentially the same manner as for DEAD-box helicases. However, as noted above (Fig. 1.4), unwinding by translocating helicases involves a series of consecutive steps (Pyle, 2008). The kinetics of strand separation under single cycle conditions are thus significantly more complex than for the nontranslocating DEAD-box helicases. Corresponding data analyses have been described in detail (e.g., Lucius et al., 2004). We refer the interested reader especially to the publication by Lucius et al. (2003) for an in-depth discussion of the topic. It is important to note that for all types of RNA helicases, single cycle conditions provide insight only if ATP binding and subsequent unwinding are at least as fast as dissociation of the enzyme from the RNA (Fig. 1.6A). If these conditions are not met, no or only a weak unwinding signal is seen under the single cycle regime.

4.2.2. Protocol for single cycle unwinding reactions

1. Prior to the stopped-flow experiments, prepare the fluorimeter by purging the equipment with nitrogen gas, warming up the lamp, and equilibrating the temperature control system. Wash the sample syringes and tubing thoroughly, followed by equilibration with the reaction buffer before each experiment and between samples.

2. For a single mixing experiment, prepare two syringes containing 750 μL each.
3. In syringe 1, mix 75 μL of 10× HRB, 75 μL of 2× helicase in helicase storage buffer, 15 μL of 5 μM 2-AP containing duplex, and 585 μL of water. The helicase concentration in syringe 1 should be double the final desired helicase concentration.
4. In syringe 2, mix 75 μL of 10× HRB, 75 μL of protein storage buffer, 15 μL of 100 μM RNA scavenger, 150 μL of 20 mM ATP/Mg, 5 μL of 20 mg/mL heparin (this concentration will need to be optimized for each helicase), 320 μL of water.
5. Inject samples into the stopped-flow syringes and equilibrate to the reaction temperature for 5 min prior to beginning the reaction.
6. Initiate the reaction, excite samples at 310 nm, and record emission record with a 350-nm high-pass filter and slit widths of 2 nm. Monitor reactions over roughly 10 half-lives.
7. The following control reactions are recommended:
    a. Include heparin in syringe 1, to ensure that the trap is 100% efficient.
    b. Omit ATP from syringe 2, to measure any ATP-independent changes in fluorescence.
    c. Omit helicase from syringe 1, to measure any helicase-independent changes in fluorescence.
8. Perform reactions by titrating helicase concentrations at saturating ATP/Mg, and titrating ATP/Mg at saturating helicase concentrations. Saturating conditions will need to be determined experimentally.

Data analysis will be discussed in Section 5.

4.2.3. The multiple cycle regime

The multiple cycle regime differs from the single cycle regime by the absence of scavenger. Enzyme is still in excess over the substrate, but multiple rounds of substrate binding can occur during the reaction (Fig. 1.6C). For multiple cycle reactions, either enzyme and substrate are preincubated without ATP, and the unwinding reaction is started by addition of ATP, or enzyme is preincubated with ATP, and the reaction is started with substrate (Fig. 1.6D). The ability to vary the order of addition is the primary advantage of the multiple cycle regime over the single cycle regime. This feature, however, comes at the cost of a significantly more complicated data analysis and interpretation, because rebinding events and possibly helicase-mediated strand annealing have to be considered, both of which are absent in the single cycle regime.

For DEAD-box helicases, the unwinding rate constant measured under multiple cycle regimes is k'_{obs}. If strand annealing is absent or has been eliminated by strand exchange (Section 2.1.2), k'_{obs} represents an apparent unwinding rate constant that includes contributions from multiple reaction steps, even at enzyme and ATP saturation (Fig. 1.6C). Only if no notable helicase dissociation occurs during the unwinding reaction (i.e., $k_{unw} >> k_d^{E-S-ATP*}$, Fig. 1.6A), and if no annealing occurs, does k'_{obs} represent the actual unwinding rate constant k_{unw} (Fig. 1.6C).

If strand annealing activity is notable, the steady state between unwinding and annealing reactions has to be considered. The observed unwinding rate constant (k''_{obs}) will then be equal to the sum of the observed unwinding rate constant k'_{obs}, discussed above, and the observed annealing rate constant k'_{ann} (Yang and Jankowsky, 2005):

$$k''_{obs} = k'_{obs} + k'_{ann} \qquad (1.3)$$

As mentioned, amplitudes measured by stopped-flow fluorescence are relative values that need to be benchmarked, most practically by EMSA-based methods. The absolute reaction amplitude is required for deconvolution of unwinding rate constants when strand annealing is not negligible. For further discussion of the calculation of unwinding and annealing rate constants under the multiple cycle regime, see Yang and Jankowsky (2005) and Yang et al. (2007b). As discussed in Section 2.1.2 (strand exchange), inclusion of excess unlabeled strand can prevent reformation of duplex with fluorescent RNA and thus eliminate problems associated with strand annealing; however, the caveats noted for such strand exchange have to be carefully considered.

4.2.4. Protocol for multiple cycle unwinding reactions

1. Prepare reactions as described in Section 4.2.2, steps 1–7, but without scavenger.
2. Exchange ATP/Mg and the RNA duplex between syringes. If the helicase has notable ATPase activity in the absence of RNA, the incubation of helicase and ATP should be limited or an ATP regeneration system should be included.
3. Perform steps 6–9 of the protocol outlined under Section 4.2.2. Representative data are shown in Fig. 1.7.

5. DATA FITTING AND ANALYSIS

The following sections describe the calculation of unwinding and annealing rate constants for DEAD-box helicases under single and multiple cycle regimes. The analysis of corresponding data for translocating helicases has been described previously by Lucius et al. (2003).

Stopped-Flow Unwinding Measurements

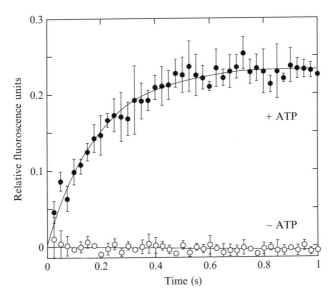

Figure 1.7 RNA unwinding measured by stopped-flow fluorescence spectroscopy. Stopped-flow measurements were conducted using a Pi-star 180 instrument (Applied Photophysics, Leatherhead, UK). Representative time courses of a 10-bp duplex with a 25-nt single-stranded region 3′ to the duplex containing an internal 2-aminopurine. Reactions were conducted at 19 °C with 50 nM RNA duplex, 1 μM 10 nt unlabeled RNA scavenger, and 0.6 μM Ded1p with (●) and without (○) 2 mM ATP/Mg as indicated. Samples were excited at 310 nm with a 2-nm bandwidth and emission was recorded through a 350-nm high-pass filter. Data points are an average of three individual measurements, and error bars represent one standard deviation.

5.1. Data fitting

The following discussion of data fitting procedures assumes that fluorescence changes are exclusively associated with the unwinding process. As noted in Section 2.2, this assumption has to be specifically verified. Only fluorescence data collected beyond the experimental dead time should be included in the data analysis. Reactions with a single phase are described by the integrated rate law for a homogeneous first-order reaction:

$$F_{obs} = A_{obs}\left(1 - e^{-k_{mes}t}\right) + F_0 \qquad (1.4)$$

F_{obs} is the fluorescence at a given time, F_0 is the initial fluorescence value ($t=0$), A_{obs} is the observed amplitude of the reaction, and k_{mes} is the measured unwinding rate constant. F_0 may vary between experiments, due to instrumental drift. Reaction amplitudes need to be benchmarked by EMSA, if they are required for additional analysis. For singe cycle reactions and multiple cycle reactions with annealing activity, the fitted value of F_0 can be subtracted from F_{obs} to display the time courses relative to each other. Calculation of rate constants and further reactions parameters from these measured data are discussed below (Sections 5.1 and 5.2).

Unwinding time courses do not always follow perfect exponentials. Such multiphasic behavior can indicate more complex reactions than shown in Fig. 1.6A and C, or heterogeneity in the RNA, the protein, or both. To determine significant deviations from a single exponential time course, residuals should analyzed. Residuals are calculated by subtraction of each actual datapoint ($F_{obs}(t)$), from the value calculated by the fit ($F_{calc}(t)$). Residuals should be evenly distributed around zero if the data are adequately fit by a single exponential. Systematic deviation of residuals over the reaction time indicates an improper model.

As described in Section 2.2, if photobleaching occurs at rate constants similar to the unwinding rate constants, photobleaching has to be considered in the data fitting. For 2-AP-labeled duplexes (d2AP), unwinding increases the fluorescence of 2-AP in the single-stranded RNA (ss2AP), that is, 2-AP is much more prone to photobleaching after unwinding than the less fluorescent 2-AP in the duplex. Thus,

$$\text{d2AP} \rightarrow \text{ss2AP} \rightarrow \text{photobleached ss2AP}$$

For this case, unwinding rate constants are calculated according to the integrated form of a consecutive two step reaction rate law:

$$F(t) = [k_{mes} A_{obs} (k_{photo} - k_{mes})^{-1}][1 - (e^{-k_{mes}t} - e^{-k_{photo}t})] + F_0 \quad (1.5)$$

Here, k_{mes} is the measured rate constant and k_{photo} is the rate constant for photobleaching. However, this data treatment is only adequate, if no significant strand annealing occurs. If both strand annealing and photobleaching are notable, the underlying reaction scheme becomes significantly more complicated, and extraction of a measured unwinding rate constant requires involved kinetic modeling. Description of the corresponding data analysis is beyond the scope of this chapter.

5.2. Determining unwinding rate constants

Having obtained the measured rate constant k_{mes}, the more instructive observed unwinding rate constants (k'_{obs} and k''_{obs}, Sections 4.2.1 and 4.2.3) have to be calculated from this value. For the single cycle regime, k_{mes} equals k'_{unw}, and functional rate constants (k_{unw} and $k_d^{E-S-ATP\star}$, Fig. 1.6A) are calculated from k'_{unw}, as outlined in Section 4.2.3. For multiple cycle reactions, k_{mes} will equal k'_{obs} in the absence of annealing, or k''_{obs} when annealing is not negligible. These calculations, however, are only correct if the unwinding reaction follows a single exponential phase; that is, if the unwinding reaction is adequately described by a single observed unwinding rate constant. Multiphasic reactions require elucidation of the cause for the multiphasicity, and corresponding kinetic modeling.

5.3. Determining functional equilibrium constants

Besides the delineation of accurate basic rate constants, pre-steady-state regimes allow the calculation of functional equilibrium constants (Fig. 1.6A and C). Functional equilibrium constants measure affinities during enzyme function and thus usually represent molecular features of the helicase more accurately than pure equilibrium binding studies that rely on the isolation of certain stages of the ATP hydrolysis cycle. For further discussion of this topic, see Chapter 2.

Single cycle experiments allow the determination of the functional affinity of the helicase for RNA without ATP ($K_{1/2}^{E+S}$), and for the affinity of the enzyme substrate complex for ATP ($K_{1/2}^{E-S+A}$) (Fig. 1.6A). To determine $K_{1/2}^{E+S}$, unwinding rate constants and reaction amplitudes are measured at saturating ATP concentrations and varying enzyme concentration. Since k'_{unw} equals $k_{unw} + k_d^{E-S-ATP\star}$ (Eq. 1.1), and thus does not depend on enzyme concentration, k'_{unw} does not vary with the enzyme concentration. The unwinding amplitude (A_{obs}), however, reflects the equilibrium distribution of free enzyme (E) and enzyme RNA complex (ES) prior to the reaction start with ATP/Mg and enzyme scavenger (Fig. 1.6A and B). Fitting of plots of the amplitude A_{obs} versus the concentration of enzyme with a Hill binding isotherm will therefore yield $K_{1/2}^{E+S}$ according to:

$$A_{obs} = A_{max}[E]^n ([E]^n + (K_{1/2}^{E+S})^n)^{-1} \qquad (1.6)$$

Here, A_{max} is the maximum amplitude at enzyme saturation, and n is the Hill coefficient. For monomeric enzymes, $n = 1$. However, several helicases have been shown to oligomerize during unwinding (Jankowsky, 2011; Lohman et al., 2008), and in such cases, $n > 1$. Notably, this analysis does not require benchmarking of the relative reaction amplitude measured by fluorescence.

To determine the affinity of the enzyme substrate complex for ATP ($K_{1/2}^{E-S+A}$, Fig. 1.6A), unwinding rate constants are measured at saturating enzyme concentrations and varying ATP concentrations. The observed unwinding rate constant k'_{obs} depends on the concentration of ATP. Fitting of plots of k'_{obs} versus the concentration of enzyme to a binding isotherm yields $K_{1/2}^{E-S+A}$ according to:

$$k'_{obs} = k'_{unw}[\text{ATP}]([\text{ATP}] + K_{1/2}^{E-S+A})^{-1} \quad (1.7)$$

It is also possible to use the Hill equation for the data fit, in case more than one ATP is used for the unwinding process. As discussed, DEAD-box proteins use a single ATP for each unwinding event (Fig. 1.5).

In a similar fashion, functional equilibrium constants can be calculated from multiple cycle experiments. However, one has to keep in mind that the observed equilibrium constants may be composites of several reaction steps (Fig. 1.6C). It is possible to combine functional equilibrium constants obtained from both single and multiple cycle experiments, according to their relation in thermodynamic boxes (Fig. 1.6C). This combination of stopped-flow fluorescence approaches is a powerful way to determine instructive, accurate, and comprehensive functional parameters for the unwinding process by RNA helicases.

ACKNOWLEDGMENTS

We thank the members of our laboratory for many fruitful discussions. We are grateful to Dr. Michael Harris (Case Western Reserve University) for the opportunity to use the stopped-flow fluorimeter and to Dr. Piet deBoer (Case Western Reserve University) for the opportunity to use the steady-state fluorimeter. Research on RNA helicases in our laboratory is supported by the NIH (GM 067700 to E. J.).

REFERENCES

Anantharaman, V., Koonin, E. V., and Aravind, L. (2002). Comparative genomics and evolution of proteins involved in RNA metabolism. *Nucleic Acids Res.* **30,** 1427–1464.

Anderson, B. J., Larkin, C., Guja, K., and Schildbach, J. F. (2008). Using fluorophore-labeled oligonucleotides to measure affinities of protein-DNA interactions. *Methods Enzymol.* **450,** 253–272.

Bernstein, J., Patterson, D. N., Wilson, G. M., and Toth, E. A. (2008). Characterization of the essential activities of Saccharomyces cerevisiae Mtr4p, a 3'->5' helicase partner of the nuclear exosome. *J. Biol. Chem.* **283,** 4930–4942.

Bizebard, T., Ferlenghi, I., Iost, I., and Dreyfus, M. (2004). Studies on three E. coli DEAD-box helicases point to an unwinding mechanism different from that of model DNA helicases. *Biochemistry* **43,** 7857–7866.

Cardullo, R. A., Agrawal, S., Flores, C., Zamecnik, P. C., and Wolf, D. E. (1988). Detection of nucleic acid hybridization by nonradiative fluorescence resonance energy transfer. *Proc. Natl. Acad. Sci. USA* **85,** 8790–8794.

Chamot, D., Colvin, K. R., Kujat-Choy, S. L., and Owttrim, G. W. (2005). RNA structural rearrangement via unwinding and annealing by the cyanobacterial RNA helicase, CrhR. *J. Biol. Chem.* **280,** 2036–2044.

Chen, Y. F., Potratz, J. P., Tijerina, P., Del Campo, M., Lambowitz, A. M., and Russell, R. (2008). DEAD-box proteins can completely separate an RNA duplex using a single ATP. *Proc. Natl. Acad. Sci. USA* **105,** 20203–20208.

Clegg, R. M. (1992). Fluorescence resonance energy transfer and nucleic acids. *Methods Enzymol.* **211,** 353–388.

Fairman-Williams, M. E., Guenther, U. P., and Jankowsky, E. (2010). SF1 and SF2 helicases: Family matters. *Curr. Opin. Struct. Biol.* **20,** 313–324.

Fischer, C. J., Maluf, N. K., and Lohman, T. M. (2004). Mechanism of ATP-dependent translocation of E. coli UvrD monomers along single-stranded DNA. *J. Mol. Biol.* **344,** 1287–1309.

Gu, Q., and Kenny, J. E. (2009). Improvement of inner filter effect correction based on determination of effective geometric parameters using a conventional fluorimeter. *Anal. Chem.* **81,** 420–426.

Gu, M., and Rice, C. M. (2010). Three conformational snapshots of the hepatitis C virus NS3 helicase reveal a ratchet translocation mechanism. *Proc. Natl. Acad. Sci. USA* **107,** 521–528.

Halls, C., Mohr, S., Del Campo, M., Yang, Q., Jankowsky, E., and Lambowitz, A. M. (2007). Involvement of DEAD-box proteins in group I and group II intron splicing. Biochemical characterization of Mss116p, ATP hydrolysis-dependent and -independent mechanisms, and general RNA chaperone activity. *J. Mol. Biol.* **365,** 835–855.

Henn, A., Cao, W., Licciardello, N., Heitkamp, S. E., Hackney, D. D., and De La Cruz, E. M. (2010). Pathway of ATP utilization and duplex rRNA unwinding by the DEAD-box helicase, DbpA. *Proc. Natl. Acad. Sci. USA* **107,** 4046–4050.

Hilliker, A., Gao, Z., Jankowsky, E., and Parker, R. (2011). The DEAD-box protein Ded1 modulates translation by the formation and resolution of an eIF4F-mRNA complex. *Mol. Cell* **43,** 962–972.

Hubner, C. G., Renn, A., Renge, I., and Wild, U. P. (2001). Direct observation of the triplet lifetime quenching of single dye molecules by molecular oxygen. *J. Chem. Phys.* **115,** 9619–9622.

Iqbal, A., Arslan, S., Okumus, B., Wilson, T. J., Giraud, G., Norman, D. G., Ha, T., and Lilley, D. M. (2008a). Orientation dependence in fluorescent energy transfer between Cy3 and Cy5 terminally attached to double-stranded nucleic acids. *Proc. Natl. Acad. Sci. USA* **105,** 11176–11181.

Iqbal, A., Wang, L., Thompson, K. C., Lilley, D. M., and Norman, D. G. (2008b). The structure of cyanine 5 terminally attached to double-stranded DNA: Implications for FRET studies. *Biochemistry* **47,** 7857–7862.

Iyer, K. S., and Klee, W. A. (1973). Direct spectrophotometric measurement of the rate of reduction of disulfide bonds. The reactivity of the disulfide bonds of bovine-lactalbumin. *J. Biol. Chem.* **248,** 707–710.

Jankowsky, E. (2011). RNA helicases at work: Binding and rearranging. *Trends Biochem. Sci.* **36,** 19–29.

Jankowsky, E., and Fairman, M. E. (2007). RNA helicases—one fold for many functions. *Curr. Opin. Struct. Biol.* **17,** 316–324.

Jankowsky, E., and Putnam, A. (2010). Duplex unwinding with DEAD-box proteins. *Methods Mol. Biol.* **587,** 245–264.

Jean, J. M., and Hall, K. B. (2001). 2-Aminopurine fluorescence quenching and lifetimes: Role of base stacking. *Proc. Natl. Acad. Sci. USA* **98,** 37–41.

Jia, H., Wang, X., Liu, F., Guenther, U. P., Srinivasan, S., Anderson, J. T., and Jankowsky, E. (2011). The RNA helicase Mtr4p modulates polyadenylation in the TRAMP complex. *Cell* **145,** 890–901.

Jose, D., Datta, K., Johnson, N. P., and von Hippel, P. H. (2009). Spectroscopic studies of position-specific DNA "breathing" fluctuations at replication forks and primer-template junctions. *Proc. Natl. Acad. Sci. USA* **106,** 4231–4236.

Law, S. M., Eritja, R., Goodman, M. F., and Breslauer, K. J. (1996). Spectroscopic and calorimetric characterizations of DNA duplexes containing 2-aminopurine. *Biochemistry* **35,** 12329–12337.

Lilley, D. M., and Wilson, T. J. (2000). Fluorescence resonance energy transfer as a structural tool for nucleic acids. *Curr. Opin. Chem. Biol.* **4,** 507–517.

Liu, F., Putnam, A., and Jankowsky, E. (2008). ATP hydrolysis is required for DEAD-box protein recycling but not for duplex unwinding. *Proc. Natl. Acad. Sci. USA* **105,** 20209–20214.

Lohman, T. M., Tomko, E. J., and Wu, C. G. (2008). Non-hexameric DNA helicases and translocases: Mechanisms and regulation. *Nat. Rev. Mol. Cell Biol.* **9,** 391–401.

Lucius, A. L., Maluf, N. K., Fischer, C. J., and Lohman, T. M. (2003). General methods for analysis of sequential "n-step" kinetic mechanisms: Application to single turnover kinetics of helicase-catalyzed DNA unwinding. *Biophys. J.* **85,** 2224–2239.

Lucius, A. L., Wong, C. J., and Lohman, T. M. (2004). Fluorescence stopped-flow studies of single turnover kinetics of E. coli RecBCD helicase-catalyzed DNA unwinding. *J. Mol. Biol.* **339,** 731–750.

Montpetit, B., Thomsen, N. D., Helmke, K. J., Seeliger, M. A., Berger, J. M., and Weis, K. (2011). A conserved mechanism of DEAD-box ATPase activation by nucleoporins and InsP6 in mRNA export. *Nature* **472,** 238–242.

Moreira, B. G., You, Y., Behlke, M. A., and Owczarzy, R. (2005). Effects of fluorescent dyes, quenchers, and dangling ends on DNA duplex stability. *Biochem. Biophys. Res. Commun.* **327,** 473–484.

Myong, S., and Ha, T. (2010). Stepwise translocation of nucleic acid motors. *Curr. Opin. Struct. Biol.* **20,** 121–127.

Myong, S., Cui, S., Cornish, P. V., Kirchhofer, A., Gack, M. U., Jung, J. U., Hopfner, K. P., and Ha, T. (2009). Cytosolic viral sensor RIG-I is a 5′-triphosphate-dependent translocase on double-stranded RNA. *Science* **323,** 1070–1074.

Nayak, R. K., Peersen, O. B., Hall, K. B., and Van Orden, A. (2012). Millisecond time-scale folding and unfolding of DNA hairpins using rapid-mixing stopped-flow kinetics. *J. Am. Chem. Soc.* **134,** 2453–2456.

Nordlund, T. M., Andersson, S., Nilsson, L., Rigler, R., Graslund, A., and McLaughlin, L. W. (1989). Structure and dynamics of a fluorescent DNA oligomer containing the EcoRI recognition sequence: Fluorescence, molecular dynamics, and NMR studies. *Biochemistry* **28,** 9095–9103.

Norman, D. G., Grainger, R. J., Uhrin, D., and Lilley, D. M. (2000). Location of cyanine-3 on double-stranded DNA: Importance for fluorescence resonance energy transfer studies. *Biochemistry* **39,** 6317–6324.

Paolini, C., De Francesco, R., and Gallinari, P. (2000). Enzymatic properties of hepatitis C virus NS3-associated helicase. *J. Gen. Virol.* **81,** 1335–1345.

Pfeffer, S. R., Stahl, S. J., and Chamberlin, M. J. (1977). Binding of Escherichia coli RNA polymerase to T7 DNA. Displacement of holoenzyme from promoter complexes by heparin. *J. Biol. Chem.* **252,** 5403–5407.

Pyle, A. M. (2008). Translocation and unwinding mechanisms of RNA and DNA helicases. *Annu. Rev. Biophys.* **37,** 317–336.

Raney, K. D., Sowers, L. C., Millar, D. P., and Benkovic, S. J. (1994). A fluorescence-based assay for monitoring helicase activity. *Proc. Natl. Acad. Sci. USA* **91,** 6644–6648.

Rogers, G. W., Jr., Richter, N. J., and Merrick, W. C. (1999). Biochemical and kinetic characterization of the RNA helicase activity of eukaryotic initiation factor 4A. *J. Biol. Chem.* **274,** 12236–12244.

Rogers, G. W., Jr., Lima, W. F., and Merrick, W. C. (2001a). Further characterization of the helicase activity of eIF4A. Substrate specificity. *J. Biol. Chem.* **276,** 12598–12608.

Rogers, G. W., Jr., Richter, N. J., Lima, W. F., and Merrick, W. C. (2001b). Modulation of the helicase activity of eIF4A by eIF4B, eIF4H, and eIF4F. *J. Biol. Chem.* **276,** 30914–30922.

Rossler, O. G., Straka, A., and Stahl, H. (2001). Rearrangement of structured RNA via branch migration structures catalysed by the highly related DEAD-box proteins p68 and p72. *Nucleic Acids Res.* **29,** 2088–2096.

Roy, S. (2004). Fluorescence quenching methods to study protein–nucleic acid interactions. *Methods Enzymol.* **379,** 175–187.

Roy, R., Hohng, S., and Ha, T. (2008). A practical guide to single-molecule FRET. *Nat. Methods* **5,** 507–516.

Singleton, M. R., Dillingham, M. S., and Wigley, D. B. (2007). Structure and mechanism of helicases and nucleic acid translocases. *Annu. Rev. Biochem.* **76,** 23–50.

Tanaka, N., and Schwer, B. (2006). Mutations in PRP43 that uncouple RNA-dependent NTPase activity and pre-mRNA splicing function. *Biochemistry* **45,** 6510–6521.

Tanner, N. K., and Linder, P. (2001). DExD/H box RNA helicases: From generic motors to specific dissociation functions. *Mol. Cell* **8,** 251–262.

Tijerina, P., Bhaskaran, H., and Russell, R. (2006). Nonspecific binding to structured RNA and preferential unwinding of an exposed helix by the CYT-19 protein, a DEAD-box RNA chaperone. *Proc. Natl. Acad. Sci. USA* **103,** 16698–16703.

Tiller, G. E., Mueller, T. J., Dockter, M. E., and Struve, W. G. (1984). Hydrogenation of triton X-100 eliminates its fluorescence and ultraviolet light absorption while preserving its detergent properties. *Anal. Biochem.* **141,** 262–266.

Uhlmann-Schiffler, H., Jalal, C., and Stahl, H. (2006). Ddx42p—A human DEAD box protein with RNA chaperone activities. *Nucleic Acids Res.* **34,** 10–22.

Valdez, B. C. (2000). Structural domains involved in the RNA folding activity of RNA helicase II/Gu protein. *Eur. J. Biochem.* **267,** 6395–6402.

Wilson, D. S., and Szostak, J. W. (1999). In vitro selection of functional nucleic acids. *Annu. Rev. Biochem.* **68,** 611–647.

Wu, P., and Brand, L. (1994). Resonance energy transfer: Methods and applications. *Anal. Biochem.* **218,** 1–13.

Yang, Q., and Jankowsky, E. (2005). ATP- and ADP-dependent modulation of RNA unwinding and strand annealing activities by the DEAD-box protein DED1. *Biochemistry* **44,** 13591–13601.

Yang, Q., and Jankowsky, E. (2006). The DEAD-box protein Ded1 unwinds RNA duplexes by a mode distinct from translocating helicases. *Nat. Struct. Mol. Biol.* **13,** 981–986.

Yang, Q., Del Campo, M., Lambowitz, A. M., and Jankowsky, E. (2007a). DEAD-box proteins unwind duplexes by local strand separation. *Mol. Cell* **28,** 253–263.

Yang, Q., Fairman, M. E., and Jankowsky, E. (2007b). DEAD-box-protein-assisted RNA structure conversion towards and against thermodynamic equilibrium values. *J. Mol. Biol.* **368,** 1087–1100.

CHAPTER TWO

ANALYZING ATP UTILIZATION BY DEAD-BOX RNA HELICASES USING KINETIC AND EQUILIBRIUM METHODS

Michael J. Bradley *and* Enrique M. De La Cruz

Contents

1. Introduction	30
2. Reagents and Equipment	31
2.1. Solution conditions and temperature	31
2.2. RNA substrate(s)	32
2.3. DEAD-box proteins	32
2.4. ATP, ADP, and mant-labeled nucleotides	32
2.5. UV–visible spectrophotometer	33
2.6. Fluorimeter	33
2.7. Stopped-flow apparatus	33
2.8. Quench-flow apparatus	33
3. Important Considerations for Initial DBP Characterization	34
3.1. DBP oligomeric state and stability/aggregation	34
3.2. RNA substrate selection/design	34
3.3. Measuring RNA binding affinity and stoichiometry	36
4. Steady-State ATPase Measurements	39
4.1. Motivation for measuring DBP ATPase activity at steady state	39
4.2. Steady-state ATPase method: Real-time enzyme-coupled assay	41
5. Transient Kinetic Analysis of the DBP ATPase Cycle	46
5.1. RNA saturation in transient kinetics experiments	46
5.2. Fluorescent-labeled ATP and ADP: Reporting on nucleotide binding/dissociation, hydrolysis, isomerization, and product release	46
5.3. Measuring the ATP hydrolysis and P_i release rates	51
6. ATP Hydrolysis Reversibility Measured by Isotope Exchange	54
7. RNA Unwinding Assays and ATPase Coupling	55
8. Kinetic Simulations	56

Department of Molecular Biophysics and Biochemistry, Yale University, New Haven, Connecticut, USA

Methods in Enzymology, Volume 511 © 2012 Elsevier Inc.
ISSN 0076-6879, DOI: 10.1016/B978-0-12-396546-2.00002-4 All rights reserved.

9. Putting It All Together: Quantitative Analysis of the DBP ATPase
   Cycle   56
Acknowledgments   58
References   58

Abstract

DEAD-box proteins (DBPs) couple ATP utilization to conformational rearrangement of RNA. In this chapter, we outline a combination of equilibrium and kinetic methods that have been developed and applied to the analysis of ATP utilization and linked RNA remodeling by DBPs, specifically *Escherichia coli* DbpA and *Saccharomyces cerevisiae* Mss116. Several important considerations are covered, including solution conditions, DBP assembly/aggregation, and RNA substrate properties. We discuss practical experimental methods for determination of DBP-RNA-nucleotide binding affinities and stoichiometries, steady-state ATPase activity, ATP binding, hydrolysis and product release rate constants, and RNA unwinding. We present general methods to integrate and analyze this combination of experimental data to identify the preferred kinetic pathway of ATP utilization and linked dsRNA unwinding.

1. INTRODUCTION

DEAD-box proteins (DBPs) utilize ATP to perform work on RNA (Henn *et al.*, 2012). Such molecular-scale work includes double-stranded RNA (dsRNA) unwinding and rearrangement, alteration of single-stranded RNA (ssRNA) conformation, and displacement of RNA-bound proteins (RNPs) (Linder and Jankowsky, 2011; Pan and Russell, 2010). Multiple and distinct low-energy RNA conformations increase the likelihood of kinetically trapping cellular RNAs in misfolded conformations. In the cell, resolution of misfolded conformations requires RNA helicases and chaperones (Herschlag, 1995; Woodson, 2010). Disruption of DBP function or loss of regulation has been linked to a variety of human pathologies, including the development and progression of several forms of cancer and heart disease (Abdelhaleem, 2004; Akao, 2009; Chao *et al.*, 2006; Godbout *et al.*, 2007; Ramasawmy *et al.*, 2006; Sahni *et al.*, 2010). These diseases likely originate from incorrect RNA folding and RNP assembly or disassembly, all of which can severely alter RNA transcription, splicing, maturation, export, translation, and degradation (Doma and Parker, 2007; Herschlag, 1995; Schroeder *et al.*, 2004).

RNA binding stimulates the ATPase activity of many DBPs (Cordin *et al.*, 2006). Therefore, rate-limiting step(s) in the intrinsic DBP ATPase cycle are accelerated or bypassed by RNA (Scheme 2.1). Progression through the ATPase cycle is coupled to transduction of the chemical energy

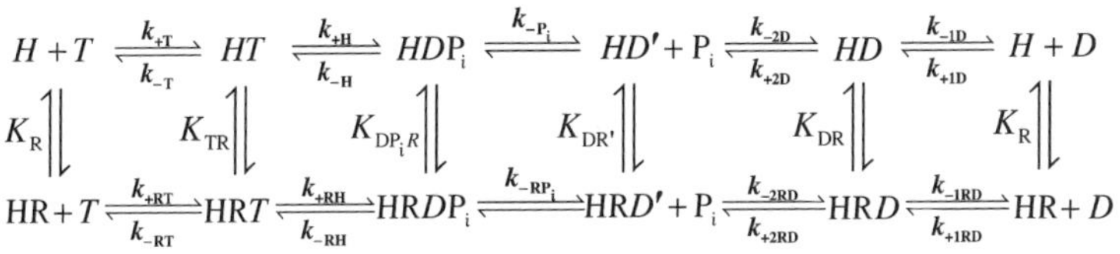

$$H+T \underset{k_{-T}}{\overset{k_{+T}}{\rightleftharpoons}} HT \underset{k_{-H}}{\overset{k_{+H}}{\rightleftharpoons}} HDP_i \underset{}{\overset{k_{-P_i}}{\rightleftharpoons}} HD'+P_i \underset{k_{+2D}}{\overset{k_{-2D}}{\rightleftharpoons}} HD \underset{k_{+1D}}{\overset{k_{-1D}}{\rightleftharpoons}} H+D$$

$$K_R \updownarrow \quad K_{TR} \updownarrow \quad K_{DP_iR} \updownarrow \quad K_{DR'} \updownarrow \quad K_{DR} \updownarrow \quad K_R \updownarrow$$

$$HR+T \underset{k_{-RT}}{\overset{k_{+RT}}{\rightleftharpoons}} HRT \underset{k_{-RH}}{\overset{k_{+RH}}{\rightleftharpoons}} HRDP_i \overset{k_{-RP_i}}{\rightleftharpoons} HRD'+P_i \underset{k_{+2RD}}{\overset{k_{-2RD}}{\rightleftharpoons}} HRD \underset{k_{+1RD}}{\overset{k_{-1RD}}{\rightleftharpoons}} HR+D$$

Scheme 2.1 Minimal ATPase cycle reaction scheme for a DBP with two ADP-bound states. H is the DBP, R is an RNA substrate, T and D are ATP and ADP, respectively, and P_i is inorganic phosphate. The k_i are rate constants and K_i are equilibrium constants of ATPase cycle transitions.

in ATP binding, hydrolysis, and/or product release to mechanical work production on RNAs and RNPs (Henn et al., 2012; Hilbert et al., 2009; Jarmoskaite and Russell, 2011). Understanding the thermodynamic coupling between RNA affinity and the chemical state (ATP, ADP-P_i, ADP) of the bound adenine nucleotide (Scheme 2.1), identifying the preferred kinetic pathway of ATP utilization and the distribution of populated biochemical and structural intermediates, and determining rate-limiting step(s) of the RNA-stimulated DBP ATPase cycle are all essential for developing mechanistic models of DBP-dependent RNA rearrangement.

In this chapter, we outline a combination of equilibrium and kinetic methods that have been developed for analysis of ATP utilization and linked RNA remodeling by DBPs (Henn et al., 2012). We discuss practical experimental methods for determination of DBP-RNA-nucleotide binding affinities and stoichiometries, steady-state ATPase activity, ATP binding, hydrolysis and product release rate constants, and RNA unwinding. These methods have been applied to *Escherichia coli* DbpA (Henn et al., 2008, 2010; Talavera and De La Cruz, 2005; Talavera et al., 2006) and *Saccharomyces cerevisiae* Mss116 (Cao et al., 2011). We refer to these examples throughout the text.

2. REAGENTS AND EQUIPMENT

2.1. Solution conditions and temperature

Quantitative kinetic and thermodynamic analyses of *E. coli* DbpA (Henn et al., 2008, 2010; Talavera and De La Cruz, 2005; Talavera et al., 2006) and *S. cerevisiae* Mss116 (Cao et al., 2011) were carried out in KMg75 buffer (20 mM Hepes, pH 7.5, 75 mM KCl, 5 mM MgCl$_2$, 1 mM DTT) or a very similar buffer (e.g., 100 mM instead of 75 mM KCl; Talavera and De La Cruz, 2005). If different solution conditions are required or desired, the ionic strength should be maintained between 50 and 200 mM with 1–5 mM Mg^{2+} in excess of [ATP] or [RNA nucleotides]. We consistently employ a temperature of 25 °C to allow direct comparison of observed reactions between experiments and different DBPs. It is critical to maintain identical solution conditions and temperature in experiments integrated for analysis. RNAase-free working conditions must be maintained in all experiments.

2.2. RNA substrate(s)

Synthesized and gel or HPLC-purified RNA substrates are commercially available (Dharmacon/Thermo-Fisher). The 2′-hydroxyl is deprotected following the manufacturer's protocol, samples desiccated in a SpeedVac concentrator, resuspended in deionized water at 200+ μM (polymer), and stored at -20 °C in 100–200 μL aliquots. RNA substrates with significant secondary structure are refolded in KMg75 by heating to 95 °C and slowly cooling to 15 °C in a thermocycler before freezing. RNA hairpins are refolded at lower concentrations, typically <40 μM polymer, to minimize oligomerization.

2.3. DEAD-box proteins

DBPs have been expressed in *E. coli* and purified to >95% purity (estimated by overloaded SDS-PAGE) (Cao et al., 2011; Henn et al., 2001, 2008). [DBP] is determined spectrophotometrically (Grimsley and Pace, 2003). DBP aliquots (100–200 μL at 20–200 μM) are flash frozen in liquid N_2 and stored at -80 °C in 20 mM K-Hepes, pH 7.5, 200 mM KCl, 1 mM DTT. Glycerol or other stabilizing agent(s) may be added for protein stability.

2.4. ATP, ADP, and mant-labeled nucleotides

ATP and ADP are prepared from >99% pure free-acid powder (Roche or Sigma). Concentrated stocks (20–100 mM) are prepared on ice in deionized water and pH adjusted to 7.0 with KOH. Aliquots (200–500 μL) are stored at -20 or -80 °C. mant-labeled nucleotides can be synthesized (Hiratsuka, 1983) or purchased (Molecular Probes/Life Technologies) and stored at -20 or -80 °C. Equimolar $MgCl_2$ is added to nucleotide stocks immediately prior to use and stored on ice while in use. mant-nucleotide stocks should be stored in opaque tubes on ice or covered with aluminum foil. Nucleotide concentrations are determined spectroscopically using absorptivities of 15,400 M^{-1} cm^{-1} at 259 nm for unlabeled (ATP and ADP), and 23,300 M^{-1} cm^{-1} at 255 nm for mant-labeled nucleotide (mantATP and mantADP, both mixed and single 3′(2′-deoxy) isomers) measured in deionized water and/or KMg75 buffer.

2.5. UV–visible spectrophotometer

A computer-controlled UV–visible spectrophotometer capable of both absorbance spectrum and kinetics measurements at multiple wave lengths and allows digital storage of data. The instrument should have high wavelength accuracy (≤ 1.0 nm), low noise (<0.001 Abs) and stray light ($<0.05\%$) in the UV–visible range, and it must be equipped with a temperature control device. We use a Perkin Elmer Lambda 20 and have used instruments made by Shimadzu, Agilent, and Jasco for satisfactory and reproducible results.

2.6. Fluorimeter

Equilibrium binding measurements utilizing changes in fluorescence, fluorescence anisotropy, or light scattering require an instrument with these capabilities. Fluorimeters equipped with relatively high-power light sources (75–200 W) in the T-format (two detectors) and monochromators on the excitation and detection channels are preferred. Instruments using optical filters on the emission channels are adequate provided changes in the emission profile upon binding and/or isomerization are significant. The instrument must be equipped with a temperature control device. We use a Photon Technology International (PTI) QuantaMaster 40.

2.7. Stopped-flow apparatus

Transient kinetics measurements of DBP ATPase cycle transitions require an instrument with millisecond time resolution to measure rate constants of several hundred per second or faster. A variety of rapid mixers with absorbance and fluorescence detection are commercially available. Those driven by compressed air or stepper motors offer the most efficient perturbation and rapid-mixing time, though manual-mixing attachments can suffice in some cases (De La Cruz and Pollard, 1994, 1995, 1996). The instrument must include a temperature control device. We have used instruments from KinTek, Applied Photophysics, Hi-Tech as well as in-house assembled instruments with satisfactory and reproducible results.

2.8. Quench-flow apparatus

Several rapid-mixing chemical-quench-flow instruments are commercially available. As with stopped-flow, millisecond time resolution is needed to measure rapid rates and rate constants. The instrument must include a temperature control device. We use the KinTek Model RQF-3 instrument.

3. Important Considerations for Initial DBP Characterization

3.1. DBP oligomeric state and stability/aggregation

The assembly state of a DBP (i.e., oligomerization) could potentially influence the affinities, rates, and internal equilibria of reaction steps that determine the overall reaction pathway and mechanism (Wyman and Gill, 1990). Protein oligomerization is strongly linked to concentration, and often to temperature, pH, solvent composition, bound ions, and ligands (e.g., nucleotide and RNA binding to DBPs). The oligomeric state of DBPs under experimental conditions must be defined if one wishes to develop mechanistic models. Ideally, self-association should be assessed at [DBP] > $10 \times$ higher than that used in other assays. If self-association occurs, it is advantageous to determine the oligomerization equilibrium constant(s), since it permits estimation of oligomeric species concentration at any [DBP], thereby identifying conditions favoring monomer or oligomer(s).

E. coli DbpA (Talavera *et al.*, 2006) and *S. cerevisiae* Mss116 (Cao *et al.*, 2011; Mallam *et al.*, 2011) behave as nonassociating monomers in solution at concentrations up to tens of micromolar. This behavior is unlikely to hold for all DBPs, given the variability in N- and C-terminal domains, which may contain oligomerization motifs (Klostermeier and Rudolph, 2009). Nonspecific aggregates, as opposed to defined oligomers (e.g., dimers, tetramers, hexamers, etc.), can also form at high [DBP], and potentially affect observed DBP activities (Cao *et al.*, 2011). As such, efforts to identify the predominant oligomeric and/or aggregation state of a DBP should not be dismissed or overlooked.

3.2. RNA substrate selection/design

3.2.1. DBP-RNA binding specificity

DBPs characterized to date display little, if any, RNA sequence specificity (Linder and Jankowsky, 2011), presumably because of a lack of direct DBP contacts with RNA nucleotide bases (Collins *et al.*, 2009; Del Campo and Lambowitz, 2009; Sengoku *et al.*, 2006). RNA recognition is achieved through sugar-phosphate backbone contacts with DBPs (Hilbert *et al.*, 2009). Accordingly, RNA-DBP binding interactions are affected by RNA length and conformation based on complementarity with the RNA binding cleft of DBPs (Del Campo and Lambowitz, 2009). In cases where DBPs display RNA sequence-specific ATPase activity, and/or dsRNA unwinding, either the DBP contains additional RNA binding domains (Kossen *et al.*, 2002; Tsu *et al.*, 2001; Wang *et al.*, 2006) or interacts with other RNA binding proteins that play a role in targeting specific cellular RNAs (Le Hir and Andersen, 2008). For DBPs that display RNA sequence specificity, observation of RNA-stimulated ATPase activity may depend on using an RNA substrate that contains a targeting sequence/structure. For example, hairpin 92 from *E. coli* rRNA is required to observe RNA stimulation of the ATPase activity of DbpA (Diges and Uhlenbeck, 2001).

3.2.2. RNA-stimulated ATPase activity
RNA-stimulated ATPase activity is a convenient diagnostic for selecting an RNA substrate for a DBP without a known target RNA. Based on current knowledge of commonalities among DBPs, it is unexpected but not impossible for a DBP to display ATP-dependent RNA remodeling of an RNA substrate without RNA-stimulated ATPase activity (Henn et al., 2012; Hilbert et al., 2009). Therefore, characterization of an RNA substrate that activates DBP ATPase can reveal valuable mechanistic information, even if it is not a physiological substrate.

3.2.3. DBP-RNA binding affinity
A key issue in RNA substrate selection is DBP-RNA affinity, particularly in the absence of bound nucleotide. DBPs possessing additional RNA binding domains bind target RNA substrates with high-affinity ($K_d < 50$ nM) (Henn et al., 2008; Polach and Uhlenbeck, 2002). Short (<20 nt) ssRNA or (<20 bp) dsRNA nonspecific DBP substrates typically bind to the nucleotide-free DBP core helicase with <100 to ~ 500 nM affinity (Cao et al., 2011; Lorsch and Herschlag, 1998). Considering that a K_d of ~ 500 nM requires 5 μM titrant for $\sim 90\%$ saturation and 50 μM titrant for $\sim 99\%$ saturation, achieving saturating DBP concentrations may not be feasible for weak DBP-RNA affinities.

3.2.4. ATP-dependent RNA unwinding
A major impetus of DBP ATPase characterization is to link ATP utilization to dsRNA unwinding. Unwinding using a strand displacement assay is readily observed for short duplexes (6–10 bp) but is slow and/or inefficient (i.e., characterized by low amplitude) for duplexes ≥ 12–16 bp (Bizebard et al., 2004; Diges and Uhlenbeck, 2001; Rogers et al., 1999; Yang and Jankowsky, 2006). It is therefore beneficial to choose RNA substrates with a short duplex region (<10 bp) (Chen et al., 2008; Henn et al., 2010) or dsRNA section interrupted by a break in the sugar-phosphate backbone (Fig. 2.1) to favor strand displacement in unwinding assays.

3.2.5. RNA substrate characteristics
DBP-RNA binding interactions depend on the RNA structure (Fig. 2.1), including the presence and length of dsRNA, hairpin loops, blunt-end dsRNA versus 5′, 3′, or both types of ssRNA overhang, and overhang length. A primary consideration is the overall length of both ss- and dsRNA regions, with the key issue being the formation of a single DBP binding site.

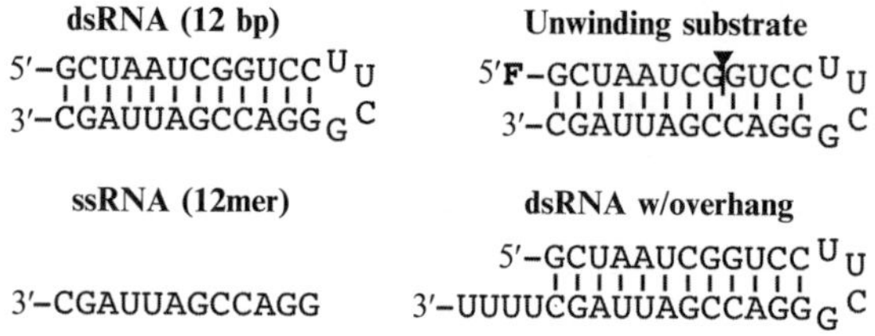

```
        dsRNA (12 bp)                    Unwinding substrate
     5'-GCUAAUCGGUCC U                5'F-GCUAAUC̸GUCC U
        ||||||||||||  U                   ||||||||||||  U
     3'-CGAUUAGCCAGG  C                3'-CGAUUAGCCAGG  C
                   G                                  G

        ssRNA (12mer)                    dsRNA w/overhang
                                      5'-GCUAAUCGGUCC U
                                         ||||||||||||  U
     3'-CGAUUAGCCAGG                  3'-UUUUCGAUUAGCCAGG  C
                                                        G
```

Figure 2.1 Model RNA substrates for use with DEAD-box proteins. Sequences illustrate important RNA substrate design principles. "unwinding substrate" has a 5′-fluorescent tag ("F"), such as fluorescein. The triangle indicates a break in the sugar-phosphate backbone, such that the labeled strand is displaced with 8 bp unwinding. 5′-fluorescent tagged dsRNA substrate(s) can be used for binding without strand displacement.

Although RNA sequence does not appear to contribute to DBP binding specificity for the helicase core, sequence significantly affects RNA structure and assembly, which could dramatically influence DBP binding. RNA substrates predicted to have a single predominant secondary topology in solution (e.g., model substrates in Fig. 2.1), predicted using Sfold (Ding et al., 2004) and RNA structure software packages (Reuter and Mathews, 2010) are desirable. Because certain RNA hairpins can potentially self-associate in solution, it is important to avoid sequence repeats within the dsRNA region, which increase the number of potential oligomerization species. Refolding of the RNA should be evaluated over a broad concentration range (e.g., 10–100 μM) to identify conditions minimizing oligomerization. Analytical ultracentrifugation should be performed to affirm the RNA substrate is predominantly monomeric following refolding (Lebowitz et al., 2002). Light scattering may provide evidence of large RNA multimers in solution (Attri and Minton, 2005; Murphy, 1997). When designing a hairpin-containing dsRNA, the loop sequence (e.g., UUCG tetraloop, Fig. 2.1) can play an important role in stabilizing the overall structure (Antao et al., 1991). An additional consideration is the potential for polypurine base-stacking to favor a helical structure (Seol et al., 2007) versus polypyrimidine sequences that adopt a more random coil in solution (Seol et al., 2004).

3.3. Measuring RNA binding affinity and stoichiometry

Structural, biochemical, and biophysical data acquired with several DBPs and short RNA substrates (≤ 20 nt or bp) reveal a 1:1 RNA:DBP binding stoichiometry (Cao et al., 2011; Polach and Uhlenbeck, 2002; Talavera et al., 2006). The combination of equilibrium and kinetic analyses presented here apply to 1:1 RNA:DBP interactions. Different expressions may be needed for other binding stoichiometries. If a DBP possesses two distinct RNA binding sites (or RNA can accommodate two DBPs), analysis of the binding data with a 1:1 binding model will, at best, give an estimate of the average affinity of the two sites and could lead to erroneous conclusions (Wyman and Gill, 1990). Therefore, quantitative determination of the binding stoichiometry is critical.

3.3.1. Measuring the equilibrium RNA binding affinity in the absence of bound nucleotide and with saturating ADP

Numerous methods for measuring RNA binding have been developed. We favor fluorescence measurements for DBP-RNA binding experiments because of their high sensitivity and the ability to measure binding under true equilibrium conditions. Changes in fluorescence intensity (quenching or enhancement) (Lakowicz, 2006) or anisotropy (Henn et al., 2010) are useful signals that can be used for real-time kinetic assays, as discussed below. Fluorescently labeled RNA and protein are amenable to fluorescence correlation spectroscopy (FCS), which assays binding from changes in diffusion (Cao et al., 2011). Fluorescent probes conjugated to RNA and/or protein can affect binding, so evaluating various labels and utilizing competition techniques to measure the "true" binding affinity of the unlabeled molecules is essential (Bujalowski and Jezewska, 2011; Thomä and Goody, 2003).

Equilibrium binding titrations are commonly performed with one reactant (DBP or RNA) at a constant concentration and varying the other "titrant" over a broad concentration range while monitoring an experimental signal that scales with bound complex. The equilibrium binding affinity is determined from least-squares fitting of the data. In an experiment where ~ 10–100 nM fluorescently labeled RNA (the smallest amount providing a reliable fluorescence signal) is titrated with a range of [protein], the following quadratic expression for the equilibrium binding density ([protein]$_{bound}$: [RNA]$_{total}$) accounts for protein depletion when [RNA]$_{total}$ is not $\ll K_d$, as is often the case (Cao et al., 2011; Henn et al., 2010):

$$\frac{[HR]}{R_{tot}} = \frac{(R_{tot} + H_{tot} + K_d) - \sqrt{(R_{tot} + H_{tot} + K_d)^2 - 4R_{tot}H_{tot}}}{2R_{tot}} \quad (2.1)$$

R_{tot} is the total [RNA], HR is the [helicase-RNA], H_{tot} is the total [helicase], and K_d is the dissociation equilibrium constant (K_R), assuming 1:1 binding. If RNA binding is associated with a change in DBP fluorescence (intrinsic or of a conjugated probe), unlabeled RNA should be titrated with a fixed [DBP]. In this case, the R_{tot} and H_{tot} terms in Eq. (2.1) should be switched.

Binding density values will range from 0.0 (no DBP) to 1.0 for stoichiometric helicase-RNA complexes. In practice, a range of helicase or RNA concentrations that span 100-fold above and below the K_d (a 10^4 range in the [titrated molecule]) will generate binding density values ranging from 0.01 to 0.99 and provide an excellent estimate of the K_d value. An iterative process is recommended, where an initial titration with $\sim$12 points spanning $\sim$1000-fold range in [helicase] (e.g., 5 nM to 5 µM) is used for a rough K_d estimate, followed by a full titration with $\sim$20 points spanning the suggested 10^4 range in [helicase] (or [RNA]).

The DBP-bound nucleotide affects the RNA binding affinity (Henn et al., 2012; Hilbert et al., 2009; Jarmoskaite and Russell, 2011; Linder and Jankowsky, 2011), such that weak and strong RNA binding states are transiently populated as a DBP progresses through its ATPase cycle (Henn et al., 2012). These nucleotide-linked changes in RNA binding are associated with work production (e.g., unwinding and rearrangement) (Howard, 2001). Quantitating the RNA binding affinity in the various nucleotide states (ATP, ADP-P_i, and ADP) is therefore of central importance.

The ADP-bound state(s) of DBPs are the only native nucleotide state(s) that can be studied under true equilibrium conditions. ATP is hydrolyzed and P_i binds too weakly to saturate without introducing secondary effects from changes in solution ionic strength. RNA binding with DBP-bound ATP analogs such as AMPpNp, ADP-BeF$_3$, ADP-AlF$_4$, etc., can also be measured as described.

The RNA binding affinity of DBP-ADP ($K_{DR,overall}$) is measured as above, except with saturating ADP. Knowledge of the DBP-ADP binding affinity (when bound to RNA, $K_{RD,overall}$) is therefore needed to ensure saturating [ADP] during the titration. The overall ADP affinity ($K_{RD,overall}$) can be estimated from the concentration-dependence of the mantADP binding amplitudes (Cao et al., 2011) and/or the rate constants and "overall binding" Eq. (2.5). However, this presents a "chicken and egg" problem in cases where the ADP and RNA affinities are strongly linked. Such a situation requires an iterative approach measuring approximate affinities of RNA and ADP binding to the DBP in the presence of each other, ultimately leading to determination of precise affinities under conditions where the other molecule is present at saturating concentration. As a first approximation, it is useful to keep in mind that the measured $K_{RD,overall}$ for both DbpA (Henn et al., 2008) and Mss116 (Cao et al., 2011) have been < 100 µM, meaning that 10 mM MgADP is saturating when bound to RNA.

The two DBP-RNA equilibrium binding experiments described provide the overall affinity of RNA binding to DBP (K_R) and DBP-ADP ($K_{DR,overall}$). Although these two states do not appear to be active unwinding intermediates of characterized DBPs (Henn et al., 2010, 2012), these two binding parameters are needed for analysis of ATP utilization by DBPs (Scheme 2.1). Knowledge of K_R combined with the suite of transient kinetic, nucleotide binding experiments described below, permits calculation of linked DBP-RNA equilibrium binding constants and provides a means to assess thermodynamic consistency in experimentally determined parameters. Adhering to the principle of detailed balance associated with the thermodynamic boxes comprising Scheme 2.1, K_{TR} and K_{DP_iR} can be calculated as follows (Cao et al., 2011; Henn et al., 2008)

$$K_{TR} = \frac{K_{RT}K_R}{K_T} \quad \text{and} \quad K_{DP_iR} = \frac{K_{RH}K_{TR}}{K_H} \quad (2.2)$$

where $K_T = k_{-T}/k_{+T}$, $K_{RT} = k_{-RT}/k_{+RT}$, $K_H = k_{-H}/k_{+H}$, and $K_{RH} = k_{-RH}/k_{+RH}$. Accordingly, K_{DR} and $K_{DR'}$ (Scheme 2.1) can be calculated from Cao et al. (2011) and Henn et al. (2008)

$$K_{DR} = \frac{K_{1RD}K_R}{K_{1D}} \quad \text{and} \quad K_{DR'} = \frac{K_{2RD}K_{DR}}{K_{2D}} \quad (2.3)$$

where $K_{1D} = k_{-1D}/k_{+1D}$, $K_{1RD} = k_{-1RD}/k_{+1RD}$, $K_{2D} = k_{-2D}/k_{+2D}$, and $K_{2RD} = k_{-2RD}/k_{+2RD}$.

The DBP-RNA equilibrium dissociation constant in the presence of saturating ADP, $K_{DR,overall}$, measured as described above, provides an internal consistency check for the combination of K_R and the fundamental ADP binding/dissociation rate constants measured by transient kinetics as follows (Cao et al., 2011; Henn et al., 2008)

$$K_{DR,overall} = \frac{K_{RD,overall}K_R}{K_{D,overall}} \quad (2.4)$$

where

$$K_{D,overall} = K_{1D}\left(\frac{K_{2D}}{1+K_{2D}}\right) \quad \text{and} \quad K_{RD,overall} = K_{1RD}\left(\frac{K_{2RD}}{1+K_{2RD}}\right) \quad (2.5)$$

4. Steady-State ATPase Measurements

4.1. Motivation for measuring DBP ATPase activity at steady state

Thorough characterization of the DBP steady-state ATPase activity (Fig. 2.2) is an essential starting point of quantitative biochemical and biophysical analysis of ATP utilization. A primary objective of steady-state ATPase analysis is to determine the maximal per-enzyme cycling rate, or k_{cat} (in units of s^{-1} DBP^{-1}), at saturating [ATP] and in the presence and absence of saturating [RNA]. Some DBPs display little or undetectable ATPase in the absence of a suitable RNA substrate (Henn et al., 2008), while others have modest intrinsic ATPase activity (Cao et al., 2011). In all

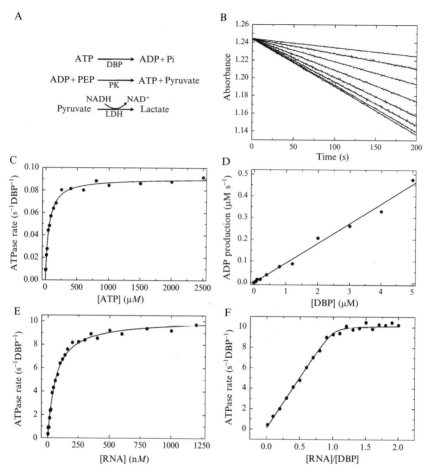

Figure 2.2 Steady-state DBP ATPase activity. (A) Schematic of the coupled enzyme reactions. DBP is DEAD-box protein, PEP is phosphoenolpyruvate, PK is pyruvate kinase, LDH is lactate dehydrogenase, NADH and NAD^+ are the reduced and oxidized forms, respectively, of nicotinamide adenine dinucleotide. (B) NADH oxidation is monitored by absorbance at 340 nm. Time courses of absorbance change for 1 μM DBP ($k_{cat} = 0.09$ s^{-1} DBP^{-1} and $K_{m,ATP} = 48$ μM) cycling in the presence of 10–1000 μM ATP (upper to lower). (C) [ATP]-dependence of the DBP steady-state ATPase rate with model fit (solid line) to Eq. (2.7) with [DBP] $\leq$ 400 nM. (D) [DBP]-dependence of the observed ATPase rate in the presence of 10 mM (saturating) ATP. The slope of the linear fit yields k_{cat}. (E) [RNA]-dependence of DBP (10 nM) steady-state ATPase in the presence of 10 mM (saturating) ATP. Continuous model curve is generated by Eq. (2.9) with $k_{cat} = 10.1$ s^{-1} DBP^{-1} and $K_{M,RNA} = 60$ nM. (F) Estimating the RNA stoichiometry of DBP ATPase activation. The [DBP] is constant (10 μ$M \gg K_{M,RNA}$) with varying [RNA] in the presence of 10 mM ATP. The continuous model curve is generated by the implicit bimolecular binding equations (Cao et al., 2011) with a stoichiometry parameter of 1.0. All data are simulated with introduced Gaussian noise.

cases studied to date, suitable RNA substrates accelerate ATPase $k_{cat} > 5$ to 2000+ fold.

The RNA-stimulated DBP ATPase cycle k_{cat} provides the minimum rate at which ATP is utilized to perform mechanical work on RNA, in the absence of "futile," nonproductive ATPase cycles. This rate can be compared with direct measurements of RNA rearrangement kinetics (Cao et al., 2011; Henn et al., 2010). If RNA rearrangement is $\ll k_{cat}$, multiple ATPase cycles are required to yield the observed rearrangement and/or nonproductive ATPase cycles occur with regularity. As part of a complete DBP ATPase kinetic analysis, k_{cat} can be directly compared with the rate constants of individual cycle steps (Scheme 2.1) to determine which biochemical transition(s) limit(s) ATP utilization, providing candidate force-producing and load-bearing state(s) during ATPase cycling.

The "Michaelis constant," or K_m, is defined as the concentration required to half-saturate the observed ATPase rate (Fig. 2.2) (Cornish-Bowden, 2004). The $K_{m,ATP}$ in the presence and absence of RNA is a true K_m, while the $K_{m,RNA}$ is an apparent K_m because RNA rearrangement is not monitored in the ATPase assay. The K_m reflects the mole-fraction weighted affinities (K_d) of DBP states populated during steady-state ATPase cycling (Henn et al., 2008). In experiments where saturating RNA or other regulatory components of the DBP ATPase activity are included, effects on the K_m of any other species (such as ATP or RNA) reveal altered binding affinities, that is, cooperativity, during one or more ATPase cycle steps. Given an ATPase cycle scheme (e.g., Scheme 2.1), relationships between K_m values for each molecular species and the rate and equilibrium constants of individual steps in the ATPase cycle can be derived (Cao et al., 2011; Henn et al., 2008, 2010). As such, the measured steady-state K_m values provide a consistency check on the combination of rate constants measured by transient kinetics. In addition, the K_m values provide constraints on difficult to measure rate constants that can instead be estimated using multiple lines of equilibrium, steady-state, and transient kinetics experimental evidence (see Section 9).

Regulatory proteins can affect k_{cat} and $K_{m,ATP}$ (with or without RNA), and $K_{m,RNA}$ of some DBPs. Examples include eIF4A (Marintchev et al., 2009; Rogers et al., 2001), eIF4AIII (Le Hir and Andersen, 2008), and Dbp5/Ddx19 (Montpetit et al., 2011; Weirich et al., 2006). For DBPs that display this additional level of ATPase regulation, it is important to determine the steady-state parameters in the presence of a saturating concentration of the regulatory molecule(s).

4.2. Steady-state ATPase method: Real-time enzyme-coupled assay

While several methods have been used to measure ATPase activity from liberation of P_i (De La Cruz and Ostap, 2009) or ADP (Charter et al., 2006), we prefer the NADH-linked enzyme-coupled assay (De La Cruz et al.,

2000; Tsu and Uhlenbeck, 1998) based on our results with DBPs (Cao et al., 2011; Henn et al., 2008, 2010) and myosins (De La Cruz et al., 2000; Henn and De La Cruz, 2005). The NADH-linked assay has several advantages over other assays, including real-time measurement, ATP regeneration, and minimal interference from P_i contamination (De La Cruz et al., 2000; Furch et al., 1998). The assay relies on absorbance detection of NADH oxidation resulting from ADP production (Fig. 2.2A). One NADH molecule is oxidized per liberated ADP, allowing for direct conversion of the change in absorbance versus time to the [ADP] produced per second using the absorptivity of NADH ($6220\ M^{-1}\ cm^{-1}$ at 340 nm).

It is helpful to confirm that the assay responds rapidly to ADP (under experimental conditions) prior to beginning an experiment. The NADH absorbance at 340 nm should decrease and reach a minimum within a few seconds, remaining flat thereafter. If the change in signal is slower, use more coupling enzymes, make fresh stocks, or both. The change in absorbance, ΔA_{340}, should be used to calculate the change in [NADH]. This value should correspond closely (within a few percentage) to the [ADP] added, up to the total final concentration of NADH. If this is not the case, use more PEP and/or a fresh PEP stock. The same is true for NADH, although the characteristic absorbance peak at 340 nm is a good indicator of the NADH integrity, and with a 1 cm path length, the starting absorbance value with 200 μM NADH should be close to 1.244 (Fig. 2.2B). Using more NADH is not recommended unless the path length is decreased, since the signal to noise is reduced at higher absorbance.

Stock solutions

1. 10× KMg75 buffer (add DTT to 5 mM upon dilution to 5× or 1 mM upon dilution to 1× prior to use) or other suitable buffer tested as above with known ADP aliquots. Store at 4 °C in a large volume (50 mL or more).
2. 100 mM PEP, prepared by dissolving pure powder (Sigma) in water and adjusting the pH to 7.0 with KOH. Aliquots (100–200 μL) can be stored at −20 or −80 °C for over 1 year as long as repeated freeze-thaw cycles are avoided. Keep on ice while in use.
3. 100 mM ATP (see Section 2) thawed on ice and vortexed to thoroughly mix upon adding MgCl$_2$ (1 M stock in deionized water, equimolar to ATP upon addition), prior to further dilution.
4. Stock aliquot of RNA substrate, thawed on ice (previously refolded if containing secondary/tertiary structure).
5. Pyruvate kinase (PK) from rabbit muscle, prepared from lyophilized powder (Sigma) in 10 mM Tris or Hepes, pH 7.0–7.5, 50% glycerol at a stock concentration of 10,000 U/mL, and stored in aliquots (200–500 μL) at −20 °C.
6. Lactate dehydrogenase (LDH) from porcine heart (recombinant), prepared from lyophilized powder (BBI enzymes) in 10 mM Tris or

Hepes, pH 7.0–7.5, 50% glycerol at a stock concentration of 4000 U/mL and stored in aliquots (200–500 µL) at −20 °C.
7. DBP of interest, aliquot thawed on ice.

Working solutions (freshly made)

1. 15 mM NADH, prepared in 1 mL deionized water from desiccated powder (Sigma) stored at 4 °C. Store the solution in an opaque tube on ice for up to 12 h. Aliquots may be frozen at −20 °C for several days, but the [NADH] should be checked spectrophotometrically.
2. 1 M DTT, prepared in deionized water from desiccated powder (Sigma or American Bioanalytical) stored at 4 °C. Store stock solution on ice.
3. 5× coupling assay "cocktail" made in 1× KMg75 buffer containing: 1 mM NADH, 100 U/mL LDH, 500 U/mL PK, 2.5 mM PEP. Make a large enough volume to mix at 1× final in the reaction volume for all replicate measurements. Note that in titrations where [ATP] will be held constant (and saturating), ATP may be included in the "cocktail" at 5× final concentration. Store on ice for up to 12 h in an opaque tube or 15 mL tube covered by aluminum foil. Check for activity at 1× with MgADP standards.
4. Diluted stocks of DBP, RNA, MgATP, and MgADP in 1× KMg75 buffer. Use the smallest stock concentration (requiring the largest volume addition to the final reaction mixture) possible to achieve the desired reaction concentration after adding all the other components. This greatly improves reproducibility and reduces experimental errors due to pipetting small volumes of highly concentrated reaction components.

Method

1. Prepare a clean (RNAase free) quartz cuvette with a 1 cm path-length that requires ≤ 100 µL to fill the observation window.
2. Prepare the thermostated UV–visible spectrophotometer according to the manufacturer's protocol and set for collection of an absorbance time course at 340 nm. Recording data points every 0.5 s for 200+ s total is usually sufficient for steady-state measurements.
3. Prepare at the reaction temperature a series of tubes ("tube A series") containing the variable component (2× final concentration in 1× buffer) in ½ the reaction volume. These tubes may also contain one or more reaction components kept at constant concentration (2× final).
4. Prepare a single tube ("tube B") containing all other components not in tube A (2× final concentration) in a large enough volume to aliquot ½ the reaction volume for each tube in the "A series." DBP and ATP must be in separate tubes prior to mixing.

5. For each measurement, mix an equal volume from tube B into tube A and gently pipette up and down to mix while avoiding introduction of bubbles. The ATPase reaction should now be "running." Quickly transfer the full reaction volume (120–150 µL) to the cuvette and check that no bubbles are visible in the cuvette window. Place the cuvette in the spectrophotometer and collect the time course absorbance data (Fig. 2.2B).

Analysis

Time courses of absorbance change should be linear (Fig. 2.2B) and fitted to obtain the slope, m (ΔA_{340} s^{-1}). Deviations from linearity at early times (<3 s) are likely due to incomplete mixing, differences in temperature between incubation tube and cuvette, contaminating ADP in the ATP, or the approach to steady state, and should not be used for analysis. Time courses must be corrected for the background oxidation of NADH by subtracting the slopes of time courses acquired in the absence of DBP from those of experimental samples.

Background-corrected time course slopes, m (in A_{340} s^{-1}), can be converted to observed ATPase rates, v_{obs}, (ADP molecules produced s^{-1} DBP^{-1}) according to

$$v_{obs} = \left(\frac{1}{l}\right)\left(\frac{m}{\varepsilon_{NADH,340} \times H_{tot}}\right) \quad (2.6)$$

where l is the path length of the cuvette (in cm), H_{tot} is the total (active) [DBP] (in M), and $\varepsilon_{NADH,340}$ is the NADH absorptivity at 340 nm (6220 M^{-1} cm^{-1}).

To obtain k_{cat} and $K_{m,ATP}$, v_{obs} is plotted as a function of [ATP] and fitted by nonlinear regression. If the [DBP] $\ll$ [ATP] for all data points (Fig. 2.2C), the hyperbolic form of the Briggs–Haldane equation is appropriate (Cornish-Bowden, 2004)

$$v_{obs} = \frac{k_{cat} T}{K_{m,ATP} + T} \quad (2.7)$$

where k_{cat} is the maximal ATP turnover rate (s^{-1} DBP^{-1}) at saturating MgATP in the absence of RNA and T is the [MgATP]$_{total}$ $\approx$ [MgATP]$_{free}$. Analysis of the raw data is identical for experiments in which the [DBP] is varied. In this case, H_{tot} is not included in Eq. (2.6) (Fig. 2.2D) since the slope of the DBP concentration dependence in the absence of RNA provides an estimate of k_{cat} (in ADP production s^{-1} DBP^{-1}), when ATP is saturating.

Since DBP ATPase is usually slow in the absence of RNA (<0.1 s^{-1} DBP^{-1}), [DBP] > [ATP]/10 may be needed to observe a measureable absorbance change above background, particularly at low [ATP]. Such conditions require the quadratic form of the Briggs–Haldane equation when fitting (Fig. 2.2C),

$$v_{obs} = k_{cat}\left(\frac{H_{tot} + T_{tot} + K_{m,ATP} - \sqrt{(H_{tot} + T_{tot} + K_{m,ATP})^2 - 4H_{tot}T_{tot}}}{2H_{tot}}\right)$$

(2.8)

where T_{tot} is the [MgATP]$_{total}$ ≠ [MgATP]$_{free}$.

RNA-stimulated ATPase activity is measured in the presence of saturating [ATP] over a broad [RNA] range (Fig. 2.2E). $K_{m,ATP}$ could vary with RNA, so an iterative process is required to estimate saturating concentrations of both RNA and MgATP, followed by precise measurements of $K_{m,RNA}$ with saturating MgATP and $K_{m,ATP}$ in the presence of saturating RNA. "Saturating" means > 20 × (ideally > 100 ×) the K_m value for either species, based on reaching > 95% (ideally > 99%) of the maximum value with either the hyperbolic or quadratic forms of the Briggs–Haldane equation (see above). The [RNA] dependence of the steady-state ATPase rate in the presence of saturating [ATP] (Fig. 2.2E) should be fitted to the following modified version of the quadratic Briggs–Haldane steady-state equation (Cao et al., 2011)

$$v_{obs} = (k_{cat,R} - k_0)\frac{H_{tot} + R_{tot} + K_{m,RNA} - \sqrt{(H_{tot} + R_{tot} + K_{m,RNA})^2 - 4H_{tot}R_{tot}}}{2H_{tot}} + k_0$$

(2.9)

where $k_{cat,R}$ is the maximal DBP ATPase rate (s^{-1} DBP^{-1}) at saturating [RNA], R_{tot} is the [RNA]$_{total}$, $K_{m,RNA}$ is the apparent Michaelis constant for RNA stimulation of the DBP ATPase activity, $k_0 = k_{cat}$ in the absence of RNA, and H_{tot} is the [DBP]$_{total}$.

The RNA:DBP stoichiometry during steady-state cycling can be estimated from the [RNA] dependence of the DBP ATPase rate saturation, fit with a system of implicit equations, when [DBP]$_{total} \gg K_{m,RNA}$ (Fig. 2.2F; Cao et al., 2011). An accurate $K_{m,RNA}$ value is best measured with [DBP] $\ll K_{m,RNA}$, whereas an accurate stoichiometry parameter, n, is best measured at [DBP] $\gg K_{m,RNA}$, while $K_{m,RNA}$ is held constant during the fitting of n.

5. Transient Kinetic Analysis of the DBP ATPase Cycle

Many important questions about DBP function require additional methods besides equilibrium binding and steady-state kinetic analysis. How does RNA stimulate DBP ATPase activity? What is the degree of thermodynamic coupling between RNA binding and transiently populated DBP intermediates (e.g., ATP and ADP-P_i bound)? When during the ATPase cycle does a DBP unwind dsRNA? What is the mechanism and preferred kinetic pathway for DBP utilization of ATP "fuel" used to rearrange RNA and RNP substrates? Addressing these questions requires quantitative analysis of the transient (pre-steady-state) kinetics of DBP binding to ATP and RNA, catalysis of ATP hydrolysis, rearrangement of the DBP-RNA complex, release of the hydrolysis products P_i and ADP, and release of RNA strand(s).

Transient kinetic measurements of biomolecules in solution usually require mixing techniques such as stopped-flow and quench-flow (Johnson, 1992). Measurement of pre-steady-state relaxations with observed rate constants $> 1\ s^{-1}$ requires the ability to mix the reaction components and acquire data in a fraction of a second. Faster relaxations with observed rate constants ~ 100–$500\ s^{-1}$ require (sub)millisecond time resolution, and demand short (<2 ms) dead times. Observed rate constants in this range or faster may occur with DBPs, especially with nucleotide and RNA association reactions (Fig. 2.3; Cao et al., 2011; Henn et al., 2008).

5.1. RNA saturation in transient kinetics experiments

The suite of transient kinetics experiments described in this review should be performed with and without saturating RNA substrate. For experiments conducted with ATP (labeled and unlabeled), a good estimate for RNA saturation is $>20 \times\ K_{m,RNA}$ measured using the steady-state ATPase method or $>20 \times\ K_R$ measured using equilibrium DBP-RNA binding, whichever is greater. For experiments conducted with added ADP (labeled or unlabeled), a good estimate for RNA saturation is $>20 \times\ K_{DR,overall}$ measured with equilibrium binding in the presence of saturating [ADP], or $>20 \times\ K_R$, whichever is greater. These conditions ensure that nucleotide binding/dissociation and other ATPase cycle transitions measured in the presence of RNA arise only from the DBP-RNA complex.

5.2. Fluorescent-labeled ATP and ADP: Reporting on nucleotide binding/dissociation, hydrolysis, isomerization, and product release

Fluorescent-labeled nucleotide analogs allow measurement of nucleotide binding, dissociation, and often other ATPase cycle transitions. ATP and ADP labeled with methylanthraniloyl (mant) labels covalently attached to

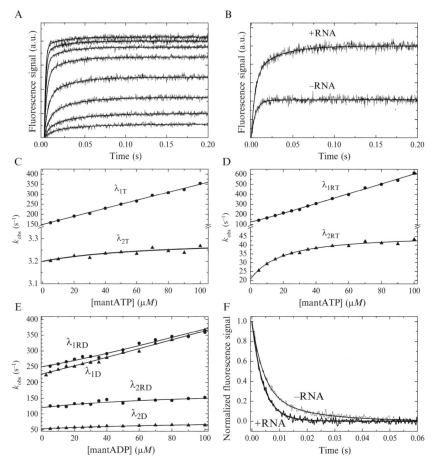

Figure 2.3 Transient kinetics of mant-nucleotide binding interactions. (A) Time courses of fluorescence change after mixing mantATP (5–100 μM, lower to upper) with DBP (500 nM) and saturating RNA (10 μM). Smooth lines are model curves with observed rate constants plotted in (D). (B) Comparison of fluorescence time courses after mixing DBP ± saturating RNA with 50 μM mantATP. (C, D) Concentration-dependence of the fast and slow relaxations from (A) and (B) of mantATP binding to DBP in the absence (C) and presence (D) of saturating RNA. Solid lines are model curves generated with Eq. (2.11). (E) Representative mantADP data. The solid lines are model curves generated with Eq. (2.14) (F) Normalized fluorescence transients of mantADP (60 μM) release from pre-equilibrated DBP (1 μM) or DBP with saturating RNA (40 μM) upon mixing with excess ADP (10 mM). The solid lines are model curves showing the sum of two exponentials with rate constants according to Eq. (2.15). All data are simulated with introduced Gaussian noise.

either the 2′ or 3′ hydroxyl of the nucleotide ribose ring have proven to be useful analogs that behave similarly to unlabeled parent nucleotides, displaying small (less than twofold) differences in some kinetic transitions (Cao et al., 2011; Henn and De La Cruz, 2005; Henn et al., 2008; Talavera and De La Cruz, 2005). The photophysical characteristics of the mant moiety allow for relatively strong FRET between mant and nearby tryptophan and/or tyrosine residues within the protein. Using tyrosine/tryptophan excitation (~278 to ~295 nm), FRET signal is observed through a 400 nm long-pass filter to selectively detect mant-nucleotide. Other labeled nucleotides can provide useful fluorescent probes for helicase/motor protein investigations (Jezewska et al., 2005; Lucius et al., 2006).

Method

Excellent FRET signals have been observed using final [DBP] of 500 nM–2 μM. Reactions must achieve pseudo-first-order conditions for the expressions presented to be valid (Gutfreund, 1995; Johnson, 1992). As a general rule, maintain [nucleotide] $\geq 10 \times$ [DBP].

mantATP/ADP association experiments

1. Prepare the stopped-flow for fluorescence measurements, including the desired excitation wavelength, emission filter, PMT voltage, etc.
2. Wash both sets of drive syringes and flow lines and the observation cell with several milliliters of RNAse-free ddH_2O followed by $1 \times$ buffer. Empty the drive syringes to prepare for sample loading.
3. Starting with the lowest [mantATP], load one drive syringe at $2 \times$ final concentration. Load the other drive syringe with DBP alone at $2 \times$ final concentration. Both syringes should be loaded with enough volume, typically 500–1000 μL total, to "prime" the flow lines when applicable, and acquire several transients ("shots").
4. Acquire multiple (3+) replicate shots at each [mantATP]. Time courses can be averaged for subsequent fitting and analysis.
5. Wash the drive syringe and flow line containing mantATP as above. Load with the lowest [mantADP] and proceed with steps 3 and 4 above using mantADP.
6. Wash both drive syringes and flow lines as above.
7. Repeat steps 3–5 except using DBP with saturating RNA substrate and/or other regulatory molecules.

mantADP dissociation

1. Prepare DBP and DBP-RNA complex samples at the concentrations used in the association experiments above, with 10–50 μM mantADP, or the smallest [mantADP] that gave a strong FRET signal change in the association experiment.
2. Prepare a stock of 10–50 mM unlabeled MgADP ($1000 \times$ the [mantADP] above) in $1 \times$ buffer in enough volume for several repeated measurements with DBP alone and DBP-RNA.

3. Using the stopped-flow prepared as above, load the equilibrated mantADP-DBP sample in one drive syringe and the excess unlabeled ADP sample in the other.
4. Record several replicate shots. The FRET signal is expected to decrease (Fig. 2.3F) due to irreversible mantADP release in the presence of excess unlabeled ADP competitor.

Analysis

FRET data from mantATP-DBP interactions (±RNA) have provided information about ATP binding/dissociation (k_{+T}, k_{-T}, k_{+RT}, k_{-RT}), as well as fundamental rate constants associated with ATP hydrolysis (k_{+H}, k_{-H}, k_{+RH}, k_{-RH}) and phosphate (P_i) release (k_{-P_i}, k_{-RP_i}; Scheme 2.1; Cao et al., 2011; Henn et al., 2008). Time courses of changes in FRET should be fit to a sum of exponentials according to

$$F = C + \sum_{i=1}^{n} A_i \left(1 - e^{-k_{i,\text{obs}}t}\right) \tag{2.10}$$

where t is time (in s), F is the fluorescence signal as a function of time, C is an arbitrary signal offset at $t=0$, A_i is the amplitude of the ith exponential component, $k_{i,\text{obs}}$ is the observed rate constant (in s^{-1}) of the ith exponential component, and n is the number of exponential components required to remove systematic trends in the best-fit residuals (typically $n=1$–4). Small amplitude exponential phases are often observed using mant nucleotides; however, their $k_{i,\text{obs}}$ do not show concentration dependence and/or are too slow ($<k_{\text{cat}}$) to be part of the ATPase cycle (Cao et al., 2011; Henn et al., 2008). Such phases could reflect isomerization of the mant moiety.

In cases where mantATP FRET time courses follow a double exponential (or two exponentials $>k_{\text{cat}}$), the fast phase $k_{i,\text{obs}}$ is λ_{1T} (or λ_{1RT} on RNA), the slow phase $k_{i,\text{obs}}$ is λ_{2T} (or λ_{2RT}), and the [mantATP]-dependence of the time courses (Fig. 2.3C and D) should be fitted to the following expression (Cao et al., 2011)

$$\lambda_{1RT,2RT} = \frac{1}{2}\left(k_{+RT}[T] + k_{-RT} + \lambda_{2RT,\infty} \pm \sqrt{\left(k_{+RT}[T] + k_{-RT} + \lambda_{2RT,\infty}\right)^2 - 4\left(k_{+RT}[T]\lambda_{2RT,\infty} + k_{-RT}\lambda_{2RT,0}\right)}\right) \tag{2.11}$$

with

$$\lambda_{2RT,\infty} = k_{+RH} + k_{-RH} + k_{-RP_i} \tag{2.12}$$

and

$$\lambda_{2RT,0} = \frac{k_{-RT}k_{-RH} + k_{-RT}k_{-RP_i} + k_{+RH}k_{-RP_i}}{k_{-RT}} \quad (2.13)$$

where $\lambda_{2RT,\infty}$ is the maximum value of λ_{2RT} at (infinite) saturating [mantATP], $\lambda_{2RT,0}$ is approximately equal to the value of λ_{2RT} at zero [mantATP], [T] is the total [mantATP], and the individual rate constants are defined in Scheme 2.1 (Henn et al., 2008). The same expressions apply in the absence of RNA.

The global fit for the [mantATP] dependence has four free parameters, k_{+RT}, k_{-RT}, $\lambda_{2RT,\infty}$, and $\lambda_{2RT,0}$ (k_{+T}, k_{-T}, $\lambda_{2T,\infty}$, and $\lambda_{2T,0}$ are fit independently for DBP alone) and Eq. (2.11) should be solved globally and simultaneously, using the two expressions (add the square root term for λ_{1RT}, subtract it for λ_{2RT}) and the [mantATP] dependence of both relaxations (Fig. 2.3C and D; Cao et al., 2011). The best-fit values for $\lambda_{2RT,\infty}$ and $\lambda_{2RT,0}$ (or $\lambda_{2T,\infty}$ and $\lambda_{2T,0}$) are used in conjunction with the P_c-value from ^{18}O-exchange measurements to solve for k_{+RH}, k_{-RH}, and k_{-RP_i} (or k_{+H}, k_{-H}, and k_{-P_i}) (Section 9).

As noted in Henn et al. (2008), Eqs. (2.11)–(2.13) are only valid under the following conditions:
(a) substrate binding and dissociation are more rapid than hydrolysis ($k_{+RT}[T] + k_{-RT} \gg k_{+RH}$),
(b) substrate binding is more rapid than substrate resynthesis ($k_{+RT}[T] > k_{-RH}$), and
(c) substrate dissociation is more rapid than product release ($k_{-RT} > k_{-RP_i}$).

Biphasic time courses of mantADP association contain information about ADP binding/dissociation (k_{+D1}, k_{-D1}, k_{+RD1}, k_{-RD1}) and ADP-bound state isomerization (k_{+D2}, k_{-D2}, k_{+RD2}, k_{-RD2}; Scheme 2.1). The time courses are fitted to a sum of exponentials as above, but the concentration-dependence of the fast and slow relaxations, λ_{1D} (or λ_{1RD} on RNA) and λ_{2D} (or λ_{2RD} on RNA) are globally/simultaneously fitted with a different function

$$\lambda_{1D,2D} = \frac{1}{2}\left(k_{+1D}[D] + k_{-1D} + k_{+2D} + k_{-2D} \pm \sqrt{(k_{+1D}[D] + k_{-1D} + k_{+2D} + k_{-2D})^2 - 4(k_{-2D}k_{-1D} + k_{-2D}k_{+1D}[D] + k_{+2D}k_{+1D}[D])}\right)$$

(2.14)

where [D] is the total [mantADP] and the fundamental rate constants are defined in Scheme 2.1. The same expressions (with different subscripts) apply for DBP-RNA. The global fitting in each case (DBP alone and DBP-RNA) involves four free parameters, k_{+1D}, k_{-1D}, k_{+2D}, k_{-2D} or k_{+1RD}, k_{-1RD}, k_{+2RD}, k_{-2RD}.

In cases where time courses follow single exponentials, k_{obs} could vary linearly or hyperbolically with [mant nucleotide]. If linear, the association rate constant is obtained from the slope and dissociation rate constant from the intercept. If hyperbolic, binding occurs following (minimally) a two-step binding mechanism (Robblee et al., 2005).

The irreversible mantADP dissociation experiment (Fig. 2.3F) provides additional data to support or refute the occurrence of a DBP-ADP (or DBP-ADP-RNA) isomerization. Time courses of mantADP dissociation should be fitted to a sum of exponentials. The expectation is that DBP-ADP (or DBP-ADP-RNA) isomerization will yield biphasic dissociation time courses. The relaxation rate constants for these two observed exponential processes (λ_{diss1}, λ_{diss2}) should be predicted by the fundamental rate constants of the underlying ADP dissociation and isomerization steps (Scheme 2.1) from

$$\lambda_{diss1,diss2} = \frac{1}{2}\left(k_{-1D} + k_{+2D} + k_{-2D} \pm \sqrt{(k_{-1D} + k_{+2D} + k_{-2D})^2 - 4k_{-1D}k_{-2D}}\right)$$

(2.15)

The same expressions are used (with different subscripts) for DBP-RNA. Note that Eq. (2.15) is a special case of Eq. (2.14) when [mantADP] = 0 for the association step, because no rebinding occurs. If dissociation time courses follow single exponentials, k_{obs} directly yields the ADP dissociation rate constant, because $\lambda_{diss1} = k_{-1D}$.

5.3. Measuring the ATP hydrolysis and P_i release rates

5.3.1. ATP hydrolysis measured using quench-flow

It is possible to measure the ATP hydrolysis/resynthesis rate constants (k_{+H}, k_{-H}, or k_{+RH}, k_{-RH}) of some ATPase enzymes using radiolabeled ATP and quench-flow methods (Johnson, 1995). This approach quantifies the total amount of ATP hydrolyzed as a function of time, including DBP-bound ADP-P_i, with millisecond resolution (Johnson, 1995). When hydrolysis (or conformational isomerization that precedes rapid hydrolysis) is either too fast to measure (completed within the mixing/measurement dead time) or (partially) rate limiting, time courses will be linear, reflecting steady-state turnover (Henn et al., 2008). However, if a step following hydrolysis such as P_i release, ADP release, or a conformational isomerization limits ATP turnover, then it may be possible to observe a pre-steady-state hydrolysis "burst" with amplitude that is proportional to [DBP]. In the presence of saturating [ATP], both amplitude and observed exponential rate constant contain information about k_{+H}, k_{-H}, and possibly k_{-P_i} (or the DBP-RNA rate constants) (Johnson, 1986, 1992). DBPs characterized to date at this level of detail do not display a detectable "burst" because hydrolysis is partially rate limiting (Henn et al., 2008). The values of k_{+H}, k_{-H}, k_{+RH}, k_{-RH}, k_{-P_i}, and k_{-RP_i} were determined by ^{18}O-exchange and mantATP/P_i-release (Section 9; Cao et al., 2011).

5.3.2. P_i release measured using stopped-flow

The P_i release experiment utilizing the A197C mutant of *E. coli* phosphate binding protein (PBP) labeled with the fluorescent dye *N*-[2-(1-maleimidyl)ethyl]-7-(diethylamino)coumarin-3-carboxamide (MDCC) has been extensively described (Brune et al., 1994; Webb, 2003). We typically use 5–10 µM MDCC-PBP and 500 nM–1 µM DBP when measuring the [MgATP] dependence of the P_i release time courses. Note that the [MDCC-PBP] should be >4 µM to saturate the K_d for P_i binding (~200 nM in KMg75) (Henn et al., 2008) and 5–10 × the [DBP] in order to observe both pre-steady-state behavior and several turnovers of the steady-state ATPase prior to P_i saturation of MDCC-PBP (Cao et al., 2011; Henn et al., 2008). It is crucial to remove contaminating P_i from the instrument, flow-lines, and drive syringes by overnight incubation with the "P_i mop" (purine nucleoside phosphorylase with excess 7-methylguanosine) (Webb, 2003). We also include the "P_i mop" in the reaction components, which is kept at a low enough concentration that it does not interfere with P_i binding to MDCC-PBP on the 10+ s timescale (Webb, 2003). The "P_i mop" is incubated with the reaction components for at least 30 min on ice prior to mixing in the stopped-flow to reduce background P_i.

5.3.3. P_i release data fitting and analysis

Time courses of P_i release from *E. coli* DbpA-RNA (Henn et al., 2008) and *S. cerevisiae* Mss116-RNA (Cao et al., 2011) display pre-steady-state lag phases (see Fig. 2.4). The lag phases are indicative of at least two consecutive ATPase cycle transitions (Scheme 2.1) with comparable rate constants (Johnson, 2003; Robblee et al., 2005). These lag phases persist in the presence of saturating ATP for DbpA and Mss116, where the calculated ATP binding rates with bound RNA are >2000 s^{-1} (Cao et al., 2011; Henn et al., 2008). Given that ADP isomerization and release do not limit the observed 2–4 s^{-1} DBP^{-1} k_{cat} values in the presence of RNA (Cao et al., 2011; Henn et al., 2008), the two or more steps with similar rate constants producing the observed lags must include both ATP hydrolysis (or an isomerization that precedes rapid hydrolysis) and P_i release (Scheme 2.1). This is apparent in the simulated data in Fig. 2.4, where k_{-RP_i} and k_{+RH} are 20 and 25 s^{-1}, respectively.

P_i release time courses that exhibit pre-steady-state lag phases (Fig. 2.4A) can be fit with a function that combines a negative amplitude exponential with a straight line arising from the steady-state ATPase (Cao et al., 2011; Henn et al., 2008):

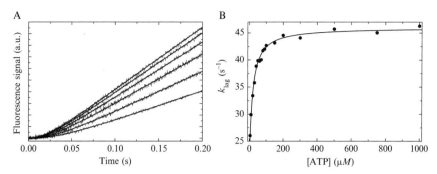

Figure 2.4 Transient phosphate (P$_i$) release from DBP-RNA assayed with MDCC-labeled phosphate binding protein (PBP). (A) Time courses of fluorescence change after mixing RNA-saturated DBP (500 nM) with a range of ATP (10 μM–1 mM, lower to upper). Smooth lines are model curves generated with Eq. (2.16). (B) [ATP]-dependence of the observed lag phase rate constant from (A). The smooth line is the model curve generated with Eq. (2.11) where $\lambda_{2RT} = \lambda_{lag}$. All data are simulated with introduced Gaussian noise.

$$[P_i] = \beta [E]_{tot} \left(\frac{1}{\lambda_{lag}} \left(e^{-\lambda_{lag} t} - 1 \right) + t \right) \quad (2.16)$$

where [P$_i$] is the phosphate concentration (as a function of time), [E]$_{tot}$ is the total [DBP], β is the steady-state ATPase rate (ATP hydrolyzed s^{-1} DBP^{-1} at a given [ATP], ±saturating RNA), and λ_{lag} is the observed rate constant of the pre-steady-state lag phase fit with a single exponential. In practice, the product $\beta [E]_{tot}$ can substituted with a constant (C) that includes the MDCC-PBP molar incremental fluorescence increase upon binding P$_i$ ($C = \beta [E]_{tot} f_{PBP-P_i}$) to fit the raw fluorescence time courses. This relationship holds as long as [P$_i$] < [MDCC − PBP]$_{active}$, where "active" refers to labeled PBP with a strong fluorescence increase upon binding P$_i$ (Brune et al., 1998). Additional caveats concerning conditions where more than one exponential may be required to fit the lag, where a lag might not occur, or where a "burst" could be observed with this assay are discussed in Henn et al. (2008).

Based on the derivation from Henn et al. (2008), the [ATP] dependence of λ_{lag} is expected to match λ_{2RT} from the mantATP experiments (Eq. 2.11). Therefore, the best-fit values of $\lambda_{lag,0}$ and $\lambda_{lag,\infty}$ (Fig. 2.4B) should be comparable to $\lambda_{2RT,0}$ and $\lambda_{2RT,\infty}$. Note that much higher [ATP] can be used in fitting $\lambda_{lag,\infty}$ versus the highest [mantATP] used with $\lambda_{2RT,\infty}$, which provides more confidence in the value from the P$_i$ release assay (compare Fig. 2.3D with Fig. 2.4B). However in the absence of RNA, or any case with a very slow steady-state ATPase activity, it may be very difficult to reliably fit the lag with an exponential due to the lack of observed signal change on the 0.01−1 s timescale. In contrast, the fluorescence signal in the mantATP assay does not suffer from this limitation, and therefore $\lambda_{2T,0}$ and $\lambda_{2T,\infty}$ may provide more reliable data in the absence of RNA. The [ATP] dependence of C (steady-state fluorescence increase) can be fit with either Eq. (2.7) or (2.8) to estimate $K_{m,ATP}$ in the presence and absence of saturating RNA and corroborate the values measured using the enzyme-coupled assay (Cao et al., 2011).

6. ATP Hydrolysis Reversibility Measured by Isotope Exchange

We refer the reader to the oxygen isotope exchange methodology chapter in this volume (Hackney, 2012). Briefly, we measure ^{18}O isotope incorporation into phosphate ions free in solution that originated in ATP molecules mixed with the DBP ($\pm$saturating RNA) under steady-state cycling conditions in the presence of $\sim$5 mM PEP and the PK enzyme (see Section 4) to regenerate ATP. The reaction proceeds until a large fraction ($\sim$80%) of the PEP is consumed, generating $\sim$4 mM free phosphate in 100–200 μL reaction volume. When ATP is hydrolyzed, an oxygen atom from a water molecule is incorporated into the phosphate product. The distribution of ^{18}O incorporated into P_i from ^{18}O—H_2O (Fig. 2.5) depends on the percentage of ^{18}O—H_2O and the probability of ATP resynthesis from bound ADP and P_i (k_{-H}, or k_{-RH}) relative to P_i release (k_{-P_i} or k_{-RP_i}), indicated by the partition coefficient, P_c, calculated as (Henn et al., 2008),

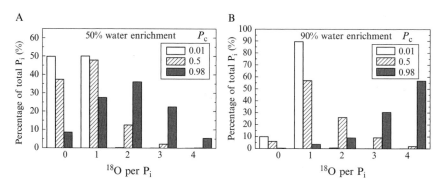

Figure 2.5 Distributions of ^{18}O incorporation into phosphate (P_i) during steady-state DBP ATPase cycling in ^{18}O-water. Up to four ^{18}O/P_i can be incorporated with multiple hydrolysis and resynthesis events per ATP turnover. The observed distribution depends on both the percentage of ^{18}O-water enrichment and the partition coefficient (P_c value) associated with the hydrolysis reversal (k_{-H} or k_{-RH}) and P_i release (k_{-P_i} or k_{-RP_i}) rate constants, see Eq. (2.17).

$$P_c = \frac{k_{-H}}{k_{-H} + k_{-P_i}} \quad (2.17)$$

where the fundamental rate constants are defined above and the same expression is used (with different subscripts) for DBP-RNA. The P_c value is calculated from the observed distribution of ^{18}O incorporation into free phosphate, detected by mass spectrometry given the percentage of ^{18}O water (Hackney, 2012). The best-fit P_c value and Eq. (2.17) are used in conjunction with either the mantATP data ($\lambda_{2T,0}$ and $\lambda_{2T,\infty}$) or the P_i release data ($\lambda_{lag,0}$ and $\lambda_{lag,\infty}$) to calculate k_{+H}, k_{-H}, and k_{-P_i} (see Section 9).

7. RNA Unwinding Assays and ATPase Coupling

A variety of RNA unwinding assays have been reported using DBPs (Chen et al., 2008; Diges and Uhlenbeck, 2001; Henn et al., 2010; Iost et al., 1999; Liu et al., 2008; Yang and Jankowsky, 2006). Most have relied on a discontinuous assay that requires separation and detection of unwinding products using gel electrophoresis and autoradiography, and often required a large excess of [DBP] to observe strand displacement. However, stopped-flow fluorescence methods using dye-labeled RNA substrates (Fig. 2.1) have shown great utility in measuring real-time dsRNA unwinding (Cao et al., 2011; Henn et al., 2010) and have the potential for even greater insights as has been demonstrated with DNA helicases (Lohman et al., 2008). The approach is discussed in detail in Chapter 1.

In general, the observed RNA unwinding amplitude depends on the [DBP]. However, an effort must be made to distinguish between mechanisms where one DBP per RNA molecule is required to observe unwinding versus multiple DBP molecules. Furthermore, an estimate of the number of ATP molecules required per unwinding event is desired, and requires comparison between the observed unwinding rate constant and the DBP ATPase k_{cat} in the presence of saturating RNA and MgATP (Cao et al., 2011; Henn et al., 2010). One good approach is to observe unwinding using fast-mixing techniques (stopped-flow or quench-flow) under conditions where the DBP-RNA complex is both (1) formed at a concentration $> 20 \times K_{m,RNA}$ and/or K_R, whichever is greater (to ensure $> 95\%$ DBP-RNA binding during ATPase cycling) and (2) formed at $\sim 1{:}1$ stoichiometry. With these starting conditions, the observed unwinding rate and amplitude should be compared between no nucleotide cofactor and in the presence of MgATP, MgADP, or ATP analogs, and as a function of [nucleotide]. It may be necessary to increase the DBP:RNA stoichiometry for unwinding, indicating multiple-bound DBP molecules per RNA unwinding event. Because DBPs are thought to be nonprocessive (Linder and Jankowsky, 2011; Pyle, 2008), observed dsRNA unwinding that requires a stoichiometry > 1DBP:1RNA may be a consequence of RNA substrate length, including both dsRNA base pairs and any ssRNA overhang. Conclusions drawn from RNA unwinding assays must consider all of the above factors.

Determining if dsRNA and ssRNA substrates differentially affect the ATP utilization mechanism, and the degree to which substrate affects the DBP-RNA affinity at each DBP-bound nucleotide intermediate, are important aspects of quantitative DBP studies. The inclusion of RNA strand displacement adds another level of complexity to the complete reaction scheme (Henn *et al.*, 2010). As a consequence, the equilibrium binding, steady-state ATPase, and transient kinetic analyses should be repeated with both dsRNA (with hairpins, eliminating strand dissociation) and ssRNA (unwinding product) substrates. The unwinding substrate should resemble the dsRNA prior to strand displacement.

8. Kinetic Simulations

Numerical integration of the differential equations describing the changes in [reactants], [intermediates], and [products] given a kinetic scheme (e.g., Scheme 2.1) have been employed to study complex enzyme reaction mechanisms for ~30 years (Barshop *et al.*, 1983). In cases with many free parameters (e.g., Scheme 2.1), multiple different experiments are required to obtain one or a few of the reaction constants, thereby allowing for their constraint during optimization of the remaining parameters. Kinetic simulations provide an excellent tool for estimation of some of the hard-to-measure parameters, including feasible upper/lower bounds (Henn *et al.*, 2010).

Given a kinetic scheme (e.g., Scheme 2.1), some simulation programs have the capability to fit the full time courses of one or more experiments and extract best-fit parameter estimates for one or more parameters while holding others constant (Johnson *et al.*, 2009b; Zimmerle and Frieden, 1989). We regularly use two program packages, KinTek Global Kinetic Explorer (Johnson, 2009; Johnson *et al.*, 2009a,b) and Tenua, which is freely downloadable (provided by Dr. D. Wachsstock, available at http://bililite.com/tenua/).

9. Putting It All Together: Quantitative Analysis of the DBP ATPase Cycle

The combination of equilibrium, steady-state, and transient kinetics experiments described above require additional analysis to assign values or ranges to the parameters in Scheme 2.1. As noted above, the P_c values ($\pm$ saturating RNA) determined by ^{18}O-isotope exchange can be combined with either $\lambda_{2T,0}$ and $\lambda_{2T,\infty}$ from mantATP experiments or $\lambda_{lag,0}$ and $\lambda_{lag,\infty}$ from P_i release experiments (±saturating RNA) using Eqs. (2.12), (2.13), and (2.17). In both cases, the best-fit value of k_{-T} (and k_{-RT}) from mantATP experiments must be known. The values of the three fundamental rate constants, k_{+H}, k_{-H}, and k_{-P_i} (or k_{+RH}, k_{-RH}, and k_{-RP_i}) can be obtained by solving these three equations with three unknowns using Mathematica (Wolfram Research).

With the entire set of fundamental rate constants assigned, calculated values for the steady-state kinetic constants can be compared with those measured experimentally. Using the derivation of $K_{m,RNA}$ (Cao et al., 2011), an additional approximation is required. For the example DBP used in this review, two ADP-bound DBP (and DBP-RNA) states are observed (Scheme 2.1; Fig. 2.3E and F). While the overall DBP-RNA binding affinity in the presence of saturating ADP, $K_{DR,overall}$, can be measured using equilibrium binding, the overall ADP release rate, $k_{-D,overall}$ (or $k_{-RD,overall}$) is estimated as follows: During vectorial cycling through the ATPase kinetic pathways on and off of the RNA (Scheme 2.1), the overall rate of ADP release is limited by isomerization of the ADP-bound DBP state, k_{-2D} (or k_{-2RD}). Therefore, $k_{-D,overall} \approx k_{-2D}$.

The following expressions account for the steady-state kinetic constants in terms of fundamental ATPase cycle rate constants (Henn et al., 2008):

$$k_{cat} = \frac{k_{+H} k_{-P_i} k_{-D,overall}}{k_{-D,overall}(k_{+H} + k_{-H} + k_{-P_i}) + k_{+H} k_{-P_i}} \quad (2.18)$$

and

$$K_{m,ATP} = \frac{k_{-T} k_{-H} + k_{-T} k_{-P_i} + k_{+H} k_{-P_i}}{k_{+T}(k_{-H} + k_{-P_i} + k_{+H})} \quad (2.19)$$

where the same expressions with different subscripts are used for the RNA-stimulated values of k_{cat} and $K_{m,ATP}$. Equation (2.19) for $K_{m,ATP}$ holds in the absence of significant [ADP], which is the case using the enzyme-coupled assay or very early in the steady-state part of the P_i release time courses. An additional expression for $K_{m,ATP}$ that varies with [ADP] and discussion of caveats are presented in (Henn et al., 2008). Similarly, the following expression yields the $K_{m,RNA}$ (Cao et al., 2011):

$$K_{m,RNA} = \frac{K_{TR} k_{-RD,overall}(k_{-RH} + k_{-RP_i}) + K_{DP_iR} k_{+RH} k_{-RD,overall} + K_{DR,overall} k_{+RH} k_{-RP_i}}{k_{-RD,overall}(k_{+RH} + k_{-RH} + k_{-RP_i}) + k_{+RH} k_{-RP_i}} \quad (2.20)$$

These expressions for the steady-state ATPase kinetic parameters in the presence and absence of RNA bring together the full combination of equilibrium and kinetic experiments presented in this review. Their derivations are based on the minimal ATPase cycle schemes, given the available DBP data, such as the existence of two ADP-bound DBP states (Cao et al., 2011; Henn et al., 2008). Qualitative agreement between the calculated and measured steady-state parameters provides additional confidence in measurements of the underlying fundamental rate and equilibrium constants and the veracity of the minimal ATPase cycle scheme. There are several potential explanations if qualitative agreement is not observed. Chief among these are changes in the DBP oligomeric state during ATPase cycling, ATPase cycles that include additional states, additional cofactors or regulatory molecules whose concentrations are not accounted for, and ATPase cycle steps that do not conform to the conditions and caveats stated in the derivations presented in Henn et al. (2008) and Cao et al. (2011).

ACKNOWLEDGMENTS

We thank Dr. Wenxiang Cao for helpful discussions on combining experiments and help with the simulated data used in this review, and Dr. Arnon Henn for valuable discussions regarding experimental design and execution. The authors thank the National Science Foundation for supporting DEAD-box protein related research activities under NSF-CAREER Award MCB-0546353 (awarded to E. M. D. L. C.) and the National Institutes of Health for supporting research activities investigating molecular motor proteins of the cytoskeleton under award GM097348. E. M. D. L. C. is an American Heart Association Established Investigator (0940075N) and Hellman Family Fellow.

REFERENCES

Abdelhaleem, M. (2004). Do human RNA helicases have a role in cancer? *Biochim. Biophys. Acta* **1704,** 37–46.

Akao, Y. (2009). A role of DEAD-Box RNA helicase rck/p54 in cancer cells. *Curr. Drug Ther.* **4,** 29–37.

Antao, V. P., Lai, S. Y., and Tinoco, I. (1991). A thermodynamic study of unusually stable RNA and DNA hairpins. *Nucleic Acids Res.* **19,** 5901–5905.

Attri, A. K., and Minton, A. P. (2005). Composition gradient static light scattering: A new technique for rapid detection and quantitative characterization of reversible macromolecular hetero-associations in solution. *Anal. Biochem.* **346,** 132–138.

Barshop, B. A., Wrenn, R. F., and Frieden, C. (1983). Analysis of numerical methods for computer simulation of kinetic processes: Development of KINSIM—A flexible, portable system. *Anal. Biochem.* **130,** 134–145.

Bizebard, T., Ferlenghi, I., Iost, I., and Dreyfus, M. (2004). Studies on three E. coli DEAD-box helicases point to an unwinding mechanism different from that of model DNA helicases. *Biochemistry* **43,** 7857–7866.

Brune, M., Hunter, J. L., Corrie, J. E. T., and Webb, M. R. (1994). Direct, real-time measurement of rapid inorganic phosphate release using a novel fluorescent probe and its application to actomyosin subfragment 1 ATPase. *Biochemistry* **33**, 8262–8271.

Brune, M., Hunter, J. L., Howell, S. A., Martin, S. R., Hazlett, T. L., Corrie, J. E. T., and Webb, M. R. (1998). Mechanism of inorganic phosphate interaction with phosphate binding protein from *Escherichia coli*. *Biochemistry* **37**, 10370–10380.

Bujalowski, W., and Jezewska, M. J. (2011). Macromolecular competition titration method: Accessing thermodynamics of the unmodified macromolecule-ligand interactions through spectroscopic titrations of fluorescent analogs. *Methods Enzymol.* **488**, 17–57.

Cao, W., Coman, M. M., Ding, S., Henn, A., Middleton, E. R., Bradley, M. J., Rhoades, E., Hackney, D. D., Pyle, A. M., and De La Cruz, E. M. (2011). Mechanism of Mss116 ATPase reveals functional diversity of DEAD-box proteins. *J. Mol. Biol.* **409**, 399–414.

Chao, C. H., Chen, C. M., Cheng, P. L., Shih, J. W., Tsou, A. P., and Lee, Y. H. (2006). DDX3, a DEAD box RNA helicase with tumor growth-suppressive property and transcriptional regulation activity of the p21waf1/cip1 promoter, is a candidate tumor suppressor. *Cancer Res.* **66**, 6579–6588.

Charter, N. W., Kauffman, L., Singh, R., and Eglen, R. M. (2006). A generic, homogenous method for measuring kinase and inhibitor activity via adenosine 5′-diphosphate accumulation. *J. Biomol. Screen.* **11**, 390–399.

Chen, Y., Potratz, J. P., Tijerina, P., Del Campo, M., Lambowitz, A. M., and Russell, R. (2008). DEAD-box proteins can completely separate an RNA duplex using a single ATP. *Proc. Natl. Acad. Sci. U. S. A.* **105**, 20203–20208.

Collins, R., Karlberg, T., Lehtiö, L., Schütz, P., van den Berg, S., Dahlgren, L. G., Hammarström, M., Weigelt, J., and Schüler, H. (2009). The DEXD/H-box RNA helicase DDX19 is regulated by an alpha-helical switch. *J. Biol. Chem.* **284**, 10296–10300.

Cordin, O., Banroques, J., Tanner, N. K., and Linder, P. (2006). The DEAD-box protein family of RNA helicases. *Gene* **367**, 17–37.

Cornish-Bowden, A. (2004). Fundamentals of Enzyme Kinetics. 3rd edn. Portland Press, London.

De La Cruz, E. M., and Ostap, M. E. (2009). Kinetic and equilibrium analysis of the myosin ATPase. *Methods Enzymol.* **455**, 157–192.

De La Cruz, E. M., and Pollard, T. D. (1994). Transient kinetic analysis of rhodamine phalloidin binding to actin filaments. *Biochemistry* **33**, 14387–14392.

De La Cruz, E. M., and Pollard, T. D. (1995). Nucleotide-free actin: Stabilization by sucrose and nucleotide binding kinetics. *Biochemistry* **34**, 5452–5461.

De La Cruz, E., and Pollard, T. D. (1996). Kinetics and thermodynamics of phalloidin binding to actin filaments from three divergent species. *Biochemistry* **35**, 14054.

De La Cruz, E. M., Sweeney, H. L., and Ostap, E. M. (2000). ADP inhibition of myosin V ATPase activity. *Biophys. J.* **79**, 1524–1529.

Del Campo, M., and Lambowitz, A. M. (2009). Structure of the Yeast DEAD box protein Mss116p reveals two wedges that crimp RNA. *Mol. Cell* **35**, 598–609.

Diges, C. M., and Uhlenbeck, O. C. (2001). *Escherichia coli* DbpA is an RNA helicase that requires hairpin 92 of 23S rRNA. *EMBO J.* **20**, 5503–5512.

Ding, Y., Chan, C. Y., and Lawrence, C. E. (2004). Sfold web server for statistical folding and rational design of nucleic acids. *Nucleic Acids Res.* **32**, W135–W141.

Doma, M. K., and Parker, R. (2007). RNA quality control in eukaryotes. *Cell* **131**, 660–668.

Furch, M., Geeves, M. A., and Manstein, D. J. (1998). Modulation of actin affinity and actomyosin adenosine triphosphatase by charge changes in the myosin motor domain. *Biochemistry* **37**, 6317–6326.

Godbout, R., Li, L., Liu, R. Z., and Roy, K. (2007). Role of DEAD box 1 in retinoblastoma and neuroblastoma. *Future Oncol.* **3,** 575–587.

Grimsley, G. R., and Pace, C. N. (2003). Spectrophotometric determination of protein concentration. *Curr. Protoc. Protein Sci.* **33,** 3.1.1–3.1.9.

Gutfreund, H. (1995). Kinetics for the Life Sciences: Receptors, Transmitters and Catalysts. 1st edn. Cambridge University Press, Cambridge.

Hackney, D. D. (2012). Oxygen isotopic exchange probes of ATP hydrolysis by RNA helicases. *Methods Enzymol.* **511,** 65–73.

Henn, A., and De La Cruz, E. M. (2005). Vertebrate myosin VIIb is a high duty ratio motor adapted for generating and maintaining tension. *J. Biol. Chem.* **280,** 39665–39676.

Henn, A., Medalia, O., Shi, S. P., Steinberg, M., Franceschi, F., and Sagi, I. (2001). Visualization of unwinding activity of duplex RNA by DbpA, a DEAD box helicase, at single-molecule resolution by atomic force microscopy. *Proc. Natl. Acad. Sci. U. S. A.* **98,** 5007–5012.

Henn, A., Cao, W., Hackney, D. D., and De La Cruz, E. M. (2008). The ATPase cycle mechanism of the DEAD-box rRNA helicase, DbpA. *J. Mol. Biol.* **377,** 193–205.

Henn, A., Cao, W., Licciardello, N., Heitkamp, S. E., Hackney, D. D., and De La Cruz, E. M. (2010). Pathway of ATP utilization and duplex rRNA unwinding by the DEAD-box helicase, DbpA. *Proc. Natl. Acad. Sci. U. S. A.* **107,** 4046–4050.

Henn, A., Bradley, M. J., and De La Cruz, E. M. (2012). ATP utilization and RNA conformational rearrangement by DEAD-box proteins. *Annu. Rev. Biophys.* **41,** 11.1–11.21.

Herschlag, D. (1995). RNA chaperones and the RNA folding problem. *J. Biol. Chem.* **270,** 20871–20874.

Hilbert, M., Karow, A. R., and Klostermeier, D. (2009). The mechanism of ATP-dependent RNA unwinding by DEAD box proteins. *Biol. Chem.* **390,** 1237–1250.

Hiratsuka, T. (1983). New ribose-modified fluorescent analogs of adenine and guanine nucleotides available as subtrates for various enzymes. *Biochim. Biophys. Acta* **742,** 496–508.

Howard, J. (2001). Mechanics of Motor Proteins and the Cytoskeleton. Sinauer Associates, Sunderland, MA.

Iost, I., Dreyfus, M., and Linder, P. (1999). Ded1p, a DEAD-box protein required for translation initiation in Saccharomyces cerevisiae, is an RNA helicase. *J. Biol. Chem.* **274,** 17677–17683.

Jarmoskaite, I., and Russell, R. (2011). DEAD box proteins as RNA helicases and chaperones. *WIREs: RNA* **2,** 135–152.

Jezewska, M. J., Lucius, A. L., and Bujalowski, W. (2005). Binding of six nucleotide cofactors to the hexameric helicase RepA protein of plasmid RSF1010. 1. Direct evidence of cooperative interactions between the nucleotide-binding sites of a hexameric helicase. *Biochemistry* **44,** 3865–3876.

Johnson, K. A. (1986). Rapid kinetic analysis of mechanochemical adenosinetriphosphatases. *Methods Enzymol.* **134,** 677–705.

Johnson, K. A. (1992). Transient-state kinetic analysis of enzyme reaction pathways. *In* "The Enzymes," (D. S. Sigman, ed.), pp. 1–61. Academic Press, New York.

Johnson, K. A. (1995). Rapid quench kinetic analysis of polymerases, adenosinetriphosphatases, and enzyme intermediates. *Methods Enzymol.* **249,** 38–61.

Johnson, K. A. (2003). Introduction to kinetic analysis of enzyme systems. *In* "Kinetic Analysis of Macromolecules. A Practical Approach," (K. A. Johnson, ed.), pp. 1–18. Oxford University Press, New York.

Johnson, K. A. (2009). Fitting enzyme kinetic data with KinTek Global Kinetic Explorer. *Methods Enzymol.* **467,** 601–626.

Johnson, K. A., Simpson, Z. B., and Blom, T. (2009a). FitSpace Explorer: An algorithm to evaluate multidimensional parameter space in fitting kinetic data. *Anal. Biochem.* **387,** 30–41.

Johnson, K. A., Simpson, Z. B., and Blom, T. (2009b). Global Kinetic Explorer: A new computer program for dynamic simulation and fitting of kinetic data. *Anal. Biochem.* **387**, 20–29.

Klostermeier, D., and Rudolph, M. G. (2009). A novel dimerization motif in the C-terminal domain of the *Thermus thermophilus* DEAD box helicase Hera confers substantial flexibility. *Nucleic Acids Res.* **37**, 421–430.

Kossen, K., Karginov, F. V., and Uhlenbeck, O. C. (2002). The carboxy-terminal domain of the DExDH protein YxiN is sufficient to confer specificity for 23S rRNA. *J. Mol. Biol.* **324**, 625–636.

Lakowicz, J. R. (2006). Principles of Fluorescence Spectroscopy. 3rd edn. Springer, New York.

Le Hir, H., and Andersen, G. R. (2008). Structural insights into the exon junction complex. *Curr. Opin. Struct. Biol.* **18**, 112–119.

Lebowitz, J., Lewis, M. S., and Schuck, P. (2002). Modern analytical ultracentrifugation in protein science: A tutorial review. *Protein Sci.* **11**, 2067–2079.

Linder, P., and Jankowsky, E. (2011). From unwinding to clamping—The DEAD box RNA helicase family. *Nat. Rev. Mol. Cell Biol.* **12**, 505–516.

Liu, F., Putnam, A., and Jankowsky, E. (2008). ATP hydrolysis is required for DEAD-box protein recycling but not for duplex unwinding. *Proc. Natl. Acad. Sci. U. S. A.* **105**, 20209–20214.

Lohman, T. M., Tomko, E. J., and Wu, C. G. (2008). Non-hexameric DNA helicases and translocases: Mechanisms and regulation. *Nat. Rev. Mol. Cell Biol.* **9**, 391–401.

Lorsch, J. R., and Herschlag, D. (1998). The DEAD box protein eIF4A. 1. A minimal kinetic and thermodynamic framework reveals coupled binding of RNA and nucleotide. *Biochemistry* **37**, 2180–2193.

Lucius, A. L., Jezewska, M. J., and Bujalowski, W. (2006). The Escherichia coli PriA helicase has two nucleotide-binding sites differing dramatically in their affinities for nucleotide cofactors. 1. Intrinsic affinities, cooperativities, and base specificity of nucleotide cofactor binding. *Biochemistry* **45**, 7202–7216.

Mallam, A. L., Jarmoskaite, I., Tijerina, P., Del Campo, M., Seifert, S., Guo, L., Russell, R., and Lambowitz, A. M. (2011). Solution structures of DEAD-box RNA chaperones reveal conformational changes and nucleic acid tethering by a basic tail. *Proc. Natl. Acad. Sci. U. S. A.* **108**, 12254–12259.

Marintchev, A., Edmonds, K. A., Marintcheva, B., Hendrickson, E., Oberer, M., Suzuki, C., Herdy, B., Sonenberg, N., and Wagner, G. (2009). Topology and regulation of the human eIF4A/4G/4H helicase complex in translation initiation. *Cell* **136**, 447–460.

Montpetit, B., Thomsen, N. D., Helmke, K. J., Seeliger, M. A., Berger, J. M., and Weis, K. (2011). A conserved mechanism of DEAD-box ATPase activation by nucleoporins and InsP6 in mRNA export. *Nature* **472**, 238–242.

Murphy, R. M. (1997). Static and dynamic light scattering of biological macromolecules: What can we learn? *Curr. Opin. Biotechnol.* **8**, 25–30.

Pan, C., and Russell, R. (2010). Roles of DEAD-box proteins in RNA and RNP Folding. *RNA Biol.* **7**, 667.

Polach, K. J., and Uhlenbeck, O. C. (2002). Cooperative binding of ATP and RNA substrates to the DEAD/H protein DbpA. *Biochemistry* **41**, 3693–3702.

Pyle, A. M. (2008). Translocation and unwinding mechanisms of RNA and DNA helicases. *Annu. Rev. Biophys.* **37**, 317–336.

Ramasawmy, R., Cunha-Neto, E., Faé, K. C., Müller, N. G., Cavalcanti, V. L., Drigo, S. A., Ianni, B., Mady, C., Kalil, J., and Goldberg, A. C. (2006). BAT1, a putative anti-inflammatory gene, is associated with chronic Chagas cardiomyopathy. *J. Infect. Dis.* **193**, 1394–1399.

Reuter, J., and Mathews, D. (2010). RNAstructure: Software for RNA secondary structure prediction and analysis. *BMC Bioinformatics* **11**, 129.

Robblee, J. P., Cao, W., Henn, A., Hannemann, D. E., and De La Cruz, E. M. (2005). Thermodynamics of nucleotide binding to actomyosin V and VI: A positive heat capacity change accompanies strong ADP binding. *Biochemistry* **44**, 10238–10249.

Rogers, G. W., Jr., Richter, N. J., and Merrick, W. C. (1999). Biochemical and kinetic characterization of the RNA helicase activity of eukaryotic initiation factor 4A. *J. Biol. Chem.* **274**, 12236–12244.

Rogers, G. W., Jr., Richter, N. J., Lima, W. F., and Merrick, W. C. (2001). Modulation of the helicase activity of eIF4A by eIF4B, eIF4H, and eIF4F. *J. Biol. Chem.* **19**, 30914–30922.

Sahni, A., Wang, N., and Alexis, J. D. (2010). UAP56 is an important regulator of protein synthesis and growth in cardiomyocytes. *Biochem. Biophys. Res. Commun.* **393**, 106–110.

Schroeder, R., Barta, A., and Semrad, K. (2004). Strategies for RNA folding and assembly. *Nat. Rev. Mol. Cell Biol.* **5**, 908–919.

Sengoku, T., Nureki, O., Nakamura, A., Kobayashi, S., and Yokoyama, S. (2006). Structural basis for RNA unwinding by the DEAD-box protein Drosophila Vasa. *Cell* **125**, 287–300.

Seol, Y., Skinner, G. M., and Visscher, K. (2004). Elastic properties of a single-stranded charged homopolymeric ribonucleotide. *Phys. Rev. Lett.* **93**, 118102.

Seol, Y., Skinner, G. M., Visscher, K., Buhot, A., and Halperin, A. (2007). Stretching of homopolymeric RNA reveals single-stranded helices and base-stacking. *Phys. Rev. Lett.* **98**, 158103.

Talavera, M. A., and De La Cruz, E. M. (2005). Equilibrium and kinetic analysis of nucleotide binding to the DEAD-box RNA helicase DbpA. *Biochemistry* **44**, 959–970.

Talavera, M. A., Matthews, E. E., Eliason, W. K., Sagi, I., Wang, J., Henn, A., and De La Cruz, E. M. (2006). Hydrodynamic characterization of the DEAD-box RNA helicase DbpA. *J. Mol. Biol.* **355**, 697–707.

Thomä, N., and Goody, R. S. (2003). What to do if there is no signal: Using competition experiments to determine binding parameters. *In* "Kinetic Analysis of Macromolecules. A Practical Approach," (K. A. Johnson, ed.), pp. 153–170. Oxford University Press, New York.

Tsu, C. A., and Uhlenbeck, O. C. (1998). Kinetic analysis of the RNA-dependent adenosinetriphosphatase activity of DbpA, an *Escherichia coli* DEAD protein specific for 23S ribosomal RNA. *Biochemistry* **37**, 16989–16996.

Tsu, C. A., Kossen, K., and Uhlenbeck, O. C. (2001). The *Escherichia coli* DEAD protein DbpA recognizes a small RNA hairpin in 23S rRNA. *RNA* **7**, 702–709.

Wang, S., Hu, Y., Overgaard, M. T., Karginov, F. V., Uhlenbeck, O. C., and McKay, D. B. (2006). The domain of the Bacillus subtilis DEAD-box helicase YxiN that is responsible for specific binding of 23S rRNA has an RNA recognition motif fold. *RNA* **12**, 959–967.

Webb, M. R. (2003). A fluorescent sensor to assay inorganic phosphate. *In* "Kinetic Analysis of Macromolecules. A Practical Approach," (K. A. Johnson, ed.), pp. 131–152. Oxford University Press, New York.

Weirich, C. S., Erzberger, J. P., Flick, J. S., Berger, J. M., Thorner, J., and Weis, K. (2006). Activation of the DExD/H-box protein Dbp5 by the nuclear-pore protein Gle1 and its coactivator InsP6 is required for mRNA export. *Nat. Cell Biol.* **8**, 668–676.

Woodson, S. A. (2010). Taming free energy landscapes with RNA chaperones. *RNA Biol.* **7**, 677.

Wyman, J., and Gill, S. J. (1990). Binding and Linkage: Functional Chemistry of Biological Macromolecules. University Science Books, Mill Valley.

Yang, Q., and Jankowsky, E. (2006). The DEAD-box protein Ded1 unwinds RNA duplexes by a mode distinct from translocating helicases. *Nat. Struct. Mol. Biol.* **13,** 981–986.

Zimmerle, C. T., and Frieden, C. (1989). Analysis of progress curves by simulations generated by numerical integration. *Biochem. J.* **258,** 381–387.

CHAPTER THREE

Oxygen Isotopic Exchange Probes of ATP Hydrolysis by RNA Helicases

David D. Hackney

Contents

1. Introduction	66
2. Determination of the k_{-2}/k_3 Ratio by Intermediate Exchange During Net ATP Hydrolysis with Pyruvate Kinase and PEP to Regenerate ATP	67
3. Derivatization of the Pi to Volatile Triethyl Phosphate	68
4. Isotopic Analysis by GC–MS	70
5. Fitting Intermediate Exchange Data to Obtain $R = k_{-2}/k_3$	70
6. Medium Pi = HOH Medium Exchange	72
7. Complication of Concurrent Medium Pi = HOH Exchange During Net Hydrolysis	72
Acknowledgment	73
References	73

Abstract

It is often possible to obtain a detailed understanding of the forward steps in ATP hydrolysis because they are thermodynamically favored and usually occur rapidly. However, it is difficult to obtain the reverse rates for ATP resynthesis because they are thermodynamically disfavored and little of their product, ATP, accumulates. Isotopic exchange reactions provide access to these reverse reactions because isotopic changes accumulate over time due to multiple reversals of hydrolysis, even in the absence of net resynthesis of significant amounts of ATP. Knowledge of both the forward and reverse rates allows calculation of the free energy changes at each step and how it changes when coupled to an energy-requiring conformational step such as unwinding of an RNA helix. This chapter describes the principal types of oxygen isotopic exchange reactions that are applicable to ATPases, in general, and helicases, in particular, their application and their interpretation.

Department of Biological Sciences, The Center for Nucleic Acid Science and Technology, Carnegie Mellon University, Pittsburgh, Pennsylvania, USA

1. Introduction

Enzymatic hydrolysis of ATP proceeds through attack of water on the γ-phosphoryl group with incorporation of the water-derived oxygen into the Pi that is produced. If the bound Pi is released rapidly from the enzyme, only this one water-derived oxygen will be incorporated, as indicated by the top line of Scheme 3.1. However, if resynthesis of ATP can occur before Pi release, then one of the original ATP oxygens can potentially be released as water and the ATP will contain one water-derived oxygen. Subsequent hydrolysis of this ATP will produce Pi with two water-derived oxygens. Additional water-derived oxygens will be incorporated if further reversals occur. Thus analysis of the amount of incorporation of water-derived oxygens into the Pi that is finally released provides an indication of the partitioning of the E·ADP·Pi between Pi release and resynthesis of ATP. This analysis assumes that the lifetime of the E·ADP·Pi intermediate is sufficiently longer than the time required for the Pi to rotate and scramble its oxygens before resynthesis of ATP (see Dale and Hackney, 1987; Sleep et al., 1980). In solution when many enzymes hydrolyze multiple ATPs, there will be a distribution of the number of reversals per turnover. It can be shown by statistical methods that the average number of reversals per net turnover is given by $R = k_{-2}/k_3$ and that the full distribution of the five isotopic species in the released Pi ($^{18}O_0^{16}O_4$, $^{18}O_1^{16}O_3$, $^{18}O_2^{16}O_2$, $^{18}O_3^{16}O_1$, $^{18}O_4^{16}O_0$) can be calculated as a function of the k_{-2}/k_3 ratio and the ^{18}O enrichment in the water and ATP (Hackney, 1980; Hackney et al., 1980). The extent of exchange can also be reported as the partition coefficient $P_c = k_{-2}/(k_{-2} + k_3) = 1/(R+1)$ that is equal to the fraction of E·ADP·Pi that reverts back to E·ATP. In an "intermediate" exchange experiment, the distribution of species produced by net ATP hydrolysis is determined at some fixed enrichment and then the distribution

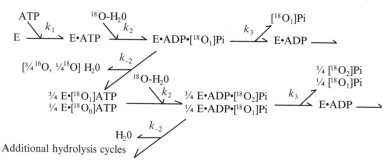

Scheme 3.1 Hydrolysis of unenriched ATP in 100% ^{18}O-enriched water. Modified from Hackney (2005).

is fit to the best value of R or to a more complex model. In a "medium" $Pi = HOH$ exchange reaction, the rate of reversal of $E \cdot ADP + Pi \rightarrow E \cdot ATP + HOH$ is determined at equilibrium from the rate of incorporation of water-derived oxygens into the Pi.

These exchange approaches have so far contributed to the detailed determination of the ATPase mechanism of the RNA helicases DpbA (Henn et al., 2008, 2010) and Mss116 (Cao et al., 2011). The approach has also been applied to a wide range of ATPases including a DNA topoisomerase (Baird et al., 2001), members of the myosin superfaminly (Hackney and Clark, 1984; Olivares et al., 2006; Sleep et al., 1980), kinesin (Hackney, 2005), dynein (Holzbaur and Johnson, 1986), and the mitochondrial and photosynthetic F1 ATPases (Hackney et al., 1979; Hutton and Boyer, 1979).

2. Determination of the K_{-2}/K_3 Ratio by Intermediate Exchange During Net ATP Hydrolysis with Pyruvate Kinase and PEP to Regenerate ATP

This exchange reaction can be performed with the ^{18}O label either in the water or in the substrate (ATP or PEP). Using labeled water, as described here, has the advantage that there is no need to synthesize or purchase labeled ATP. Regeneration of ATP allows generation of high concentrations of Pi while maintaining low fixed ATP concentrations and additionally suppresses medium exchange (see Section 7).

2.1. Prepare a $20 \times$ concentrated buffer mix: 400 mM Mops, 170 mM NaOH, 40 mM MgCl$_2$, 80 mM potassium PEP (from a 200-mM stock of monopotassium PEP adjusted to pH 7 with KOH), and 0.4 mg/ml pyruvate kinase (Sigma, Type III). The buffer, salts, and pH can be adjusted as appropriate for the particular enzyme under study. Because pyruvate kinase requires potassium, all the potassium cannot be replaced with sodium. The $20 \times$ stock can also be supplemented with ATP if all of the experiments will be at the same ATP concentration.

2.2. Prepare a 100 μL reaction mix at 50% enrichment of [^{18}O]water and 1 mM MgATP by combining: 51.5 μL 97% ^{18}O-water; 5 μL $20 \times$ buffer mix; 1 μL 100 mM MgATP (from a stock of ATP prepared by adding one equivalent of MgCl$_2$ and adjusting to pH 7); and 42.5 μL total for enzyme, RNA, and unenriched water. An enrichment of 50% is a good compromise between ease of preparation and the increased distinction between models that higher enrichment provides. Higher enrichment can be obtained if some of the volume for enzyme and

RNA can be substituted for enriched water. Alternatively, a complete reaction mix minus enzymes can be lyophilized and reconstituted with highly enriched water.

2.3. Monitor the progress of the reaction by assay of the amount of remaining PEP. This provides an evaluation of the ATPase rate and is used to determine when to withdraw samples for isotopic analysis. Add a 3-μL sample of the reaction to 200 μL of 1× reaction buffer containing 0.3 mM NADH and pyruvate kinase (rabbit muscle, Sigma Type III) and lactate dehydrogenase (bovine heart, Sigma). Any pyruvate that had been produced by coupled hydrolysis of ATP will be converted to lactate with oxidation of NADH and produce a drop in absorbance at 340 nm. Subsequent addition of excess ADP will result in an additional transient as PEP is converted to pyruvate and then lactate. A 3-μL sample of the starting 4 mM PEP in a 200-μL reaction will give a delta absorbance of 0.37 for an extinction coefficient of 6220 M^{-1} cm^{-1} at 340 nm for NADH oxidation. Reaction progress can also be monitored by assaying for Pi, but this method is not useful for locating the exact time that the end point of total PEP consumption is reached because only small differences in a high Pi concentration determine whether the end point has been passed.

2.4. Take samples of 40 μL for isotopic analysis after approximately 50% and 90% of the PEP have been consumed. After all the PEP is consumed, continued ATP hydrolysis will result in accumulation of ADP, with increased product inhibition and accelerated medium isotopic exchange that complicates the interpretation of the intermediate exchange results (see below).

2.5. The 40 μL samples are stopped by mixing with 40 μL cold 1 N HCl, placed on ice and then neutralized after approximately 5 min with 40 μL of cold 2 M Tris base. Transient exposure to very low pH is sufficient to irreversibly inactivate many enzymes, but this should be confirmed in each case. Samples at this stage can be held for hours on ice, or frozen at −80 °C for long-term storage.

3. Derivatization of the Pi to Volatile Triethyl Phosphate

3.1. Isolate the Pi by anion exchange chromatography. This procedure can be performed at room temperature and should be conducted in an exhaust hood to remove fumes of HCl.
   1. Pack BioRad Poly-Prep columns with a 0.6-ml bed volume of AG MP-1 resin (BioRad). To accelerate the flow rate, add a 25-cm

extension of 1/32-inch ID tubing. Multiple columns can be processed in parallel if mounted in a rack.
2. Wash the columns with 1 N HCl and then extensively with high-quality deionized water until the pH of the eluent is >4.
3. Dilute the neutralized samples with 11 ml of water and load onto the column. This dilution lowers the ionic strength sufficiently for the Pi to stick to the column, while keeping the volume low enough that it can all be loaded in a single batch.
4. Wash the column extensively with water taking care to rinse the walls of the column as well as the bed. Then wash with four batches of 1 ml each of 7 mM HCl. Spot check the last drop of eluent to pH <4 with a pH strip. This wash removed AMP if present.
5. Then elute the Pi with 20 mM HCl (one fraction of 0.5 ml and two fractions of 1 ml). Collect the samples in 12 × 75-mm disposable glass tubes that can also be used for the derivatization steps but check each batch of tubes for Pi contamination using the malachite green assay. The columns can be stripped with 1 N HCl and reused repeatedly.
6. Spot check the fractions for Pi using the malachite green assay to confirm that the bulk of the Pi is in the first full 1 ml fraction of 20 mM HCl. Prepare a fresh reagent mixture of 0.5 ml 0.8 M HCl, 0.5 ml 1% ammonium molybdate in 0.8 M HCl, and 50 μL of 0.5% malachite green. Add 10 μL of each column fraction to 100 μL of reagent mix. Fractions that change from pale yellow to noticeably blue by eye have sufficient Pi for GC–MS analysis.
7. Spin dry the first 1 ml fraction of 20 mM HCl (combine with the second 1 ml fraction if it also contains significant Pi) under vacuum with gentle heating (Savant SpeedVac with vacuum pump and cold trap). Turn the heat off when most, but not all, of the fluid is gone to avoid excessive heating of the H_3PO_4 film and possible scrambling of the isotopic distribution. Take the samples to complete dryness. The sample should be almost invisible as a thin clear film.
8. Prepare a solution of diazoethane in ether essentially as described by Stempel and Boyer (1986), now conveniently done in the Mini Diazald apparatus (Sigma-Aldrich). Diazoethane is potentially explosive. Both it and its precursor *N*-ethyl-*N*-nitroso-*p*-toluenesulfoamide are toxic and should be handled with care. Although the precautions needed with ethyl derivative are not well documented, they should be similar to those of the methyl derivative (see Sigma-Aldrich technical bulletin AL-180 for precautions on generation of diazomethane from *N*-methyl-*N*-nitroso-*p*-toluenesulfoamide).
9. Derivatize the samples by adding 50 μL methanol to dissolve the H_3PO_4 and then 100 μL of the diazoethane solution. The reaction

mixture will initially be pale yellow but will decolorize within a few seconds if Pi is in excess and all the diazoethane is consumed. Continue adding diazoethane until the reaction remains pale yellow indicating that all of the Pi has reacted. Leave the tubes open in an exhaust hood to concentrate the sample by evaporation of most of the ether. Then transfer the remaining sample into an autoinjector vial using a Teflon-surfaced septum and a tapered glass insert for loading of small volumes.

4. Isotopic Analysis by GC–MS

Analyze the samples on a HP6890 GC–MS using a HP-5 capillary column (Hewlett Packard) and helium carrier gas. The triethyl phosphate is eluted with a temperature ramp of 120–180 °C. Heat the column to 240 °C between samples to prevent carryover. Samples are run in duplicate. MS analysis is performed using electron impact ionization in SIM mode (selected ion monitoring) with monitoring of m/e ratios of 99, 101, 103, 105, and 107 for the five ^{18}O-enriched $P(OH)_4^+$ species. The series starting at m/e 155 can also be used (Stempel and Boyer, 1986). In either case, there are minor fragmentation products that also fall in this m/e range and corrections need to be applied as described in detail by Dale and Hackney (1987) for the 99 m/e series. The chromatographic peaks for all five isotopic species should rise and fall in parallel at fixed relative ratios. Deviations from this pattern indicate that contaminants are coeluting with the triethyl phosphate. The peak heights, after correction for baseline, are used to determine the fraction of each of the five isotopic species.

5. Fitting Intermediate Exchange Data to Obtain $R = K_{-2}/K_3$

In the absence of any reversal ($k_{-2}/k_3 \sim 0$), hydrolysis of unenriched ATP in 50% ^{18}O-water will produce Pi containing only one water-derived oxygen. This will result in approximately equal amounts of the $^{18}O_0$ and $^{18}O_1$ species as indicated in Fig. 3.1A. There will also be 0.3% of the $^{18}O_2$ species produced by the natural abundance of 0.2% in the three ATP-derived oxygens. A low level of reversal will produce additional incorporation of water-derived oxygens to yield 1.2% of the $^{18}O_2$ species, as also indicated in Fig. 3.1A for an R value of 0.05 (20 times more likely for E·ADP·Pi to release Pi rather than resynthesize E·ATP). In this case, fitting of experimental data to yield an average R value would be determined

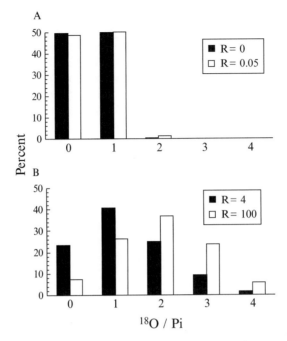

Figure 3.1 Theoretical distributions for hydrolysis of unenriched ATP in 50% 18O-enriched water at different values of $R = k_{-2}/k_3$.

almost solely by the magnitude of the $^{18}O_2$ species. A $^{18}O_0:^{18}O_1$ ratio > 1 indicates contamination with unenriched Pi. Increased reversal produces increased incorporation of water-derived oxygens until the limit of the binomial distribution for 50% ^{18}O is reached as indicated in Fig. 3.1B. For all but very low extents of exchange, fitting of the whole distribution by nonlinear regression is required to obtain the best average value of R. An additional factor in some cases is heterogeneity caused by ATP hydrolysis occurring by two different pathways with different values of R. This produces characteristic deviations from those of a single pathway with increased contribution of the high- and low enriched species at the expense of the middle of the distribution (see Sleep *et al.*, 1978, 1980).

A sample size of 50–200 nmol of Pi may seem excessive considering the high sensitivity of mass spectroscopy. It is certainly possible to perform the mass spectroscopic analysis with much smaller samples, but larger samples of this size represent a good compromise between higher consumption of valuable labeled water, enzymes, and nucleic acids and obtaining sufficient accuracy in the analysis for rigorous interpretation. One important consideration is the high dynamic range that is required. A sample size sufficient to obtain a desired S/N ratio with unenriched Pi in a single peak (as usually discussed when describing sensitivity) will require over 10,000 times more sample to obtain the same S/N ratio in quantifying the small change in enrichment of the $^{18}O_2$ needed to accurately determine the R value of 0.05 in Fig. 3.1A. An equally important complication is contamination. Unenriched Pi is everywhere. When dummy reactions are spiked with 99% Pi containing no unenriched Pi and carried through the analysis, there is always some contamination by unenriched Pi from the reaction reagents or during purification, even with extreme care. An additional concern is that derivatization with diazoethane results in side products that are difficult to totally separate from triethyl phosphate by gas chromatography. Tailing of a contaminant that elutes before Pi makes accurate estimation of base lines problematic for small Pi samples where the signal is only slightly increased over baseline. Also different contaminants make different contributions to m/e 99–107 region and so perturb the calculated contribution of each species differently. One advantage of using a high PEP concentration and going to 90% consumption of PEP is that this increased concentration of enriched Pi in the sample dilutes out the contribution of contaminants.

6. Medium Pi = HOH Medium Exchange

Incubation of the E·ADP complex of a helicase with Pi will produce no net chemical reaction, after an initial rapid transient that generates E·ADP·Pi, E·ATP, and trace amounts of free ATP. However, a water-derived oxygen will still be incorporated into the Pi at equilibrium every time an ATP is reversibly synthesized via reversal of steps 3 and 2 of Scheme 3.1. Furthermore, if the rate of ATP resynthesis saturates at high [Pi], then the values k_{-2} and the Pi dissociation constant K_3 determined separately. This exchange reaction can be performed with the label in either water or Pi and the sample derivatization and analysis are identical to that for intermediate exchange described above. The data for the change in the isotopic species with time can be fit to the theoretical profiles to obtain both the net rate of ATP resynthesis and the R value. The reaction should be free of adenylate kinase or other enzyme that can react with ADP or Pi (for controls, see Dale and Hackney, 1987).

7. Complication of Concurrent Medium Pi = HOH Exchange During Net Hydrolysis

Care must be taken that results from intermediate exchange reactions during net ATP hydrolysis are not compromised by concurrent medium exchange. Once Pi starts to accumulate by hydrolysis, it can undergo medium exchange at a rate that will vary depending on the fraction of the enzyme present as E·ADP at steady state and the rate of Pi rebinding relate to replacement of bound ADP with ATP. In the absence of concurrent medium exchange, the Pi produced by hydrolysis will have the same level of water-derived oxygens at either 50% or 90% consumption of PEP. However, concurrent medium exchange will result in a higher incorporation at 90% because the released Pi has undergone additional medium exchange. This is particularly problematic if PEP is exhausted because the steady-state level of E·ADP rises as the ATP/ADP ratio falls.

ACKNOWLEDGMENT

This work was supported by NSF Grant MCB-0615549.

REFERENCES

Baird, C. L., Gordon, M. S., Andrenyak, D. M., Marecek, J. F., and Lindsley, J. E. (2001). The ATPase reaction cycle of yeast DNA topoisomerase II. Slow rates of ATP resynthesis and P(i) release. *J. Biol. Chem.* **276,** 27893–27898.
Cao, W., Coman, M. M., Ding, S., Henn, A., Middleton, E. R., Bradley, M. J., Rhoades, E., Hackney, D. D., Pyle, A. M., and De La Cruz, E. M. (2011). Mechanism of Mss116 ATPase reveals functional diversity of DEAD-Box proteins. *J. Mol. Biol.* **409,** 399–414.
Dale, M. P., and Hackney, D. D. (1987). Analysis of positional isotope exchange in ATP by cleavage of the b P-O g P bond. Demonstration of negligible positional isotope exchange by myosin. *Biochemistry* **26,** 8365–8372.
Hackney, D. D. (1980). Theoretical analysis of distribution of [^{18}O]Pi species during exchange with water. *J. Biol. Chem.* **255,** 5320–5328.
Hackney, D. D. (2005). The tethered motor domain of a kinesin-microtubule complex catalyzes reversible synthesis of bound ATP. *Proc. Natl. Acad. Sci. USA* **102,** 18338–18343.
Hackney, D. D., and Clark, P. K. (1984). Catalytic consequences of oligomeric organization: Kinetic evidence for 'tethered' acto-heavy meromyosin at low ATP concentration. *Proc. Natl. Acad. Sci. USA* **81,** 5345–5349.
Hackney, D. D., Rosen, G., and Boyer, P. D. (1979). Subunit interaction during catalysis: Alternating site cooperativity in photophosphorylation shown by substrate modulation of [18O]ATP species formation. *Proc. Natl. Acad. Sci. USA* **76,** 3646–3650.
Hackney, D. D., Stempel, K. E., and Boyer, P. D. (1980). Oxygen-18 probes of enzymic reactions of phosphate compounds. *Methods Enzymol.* **64,** 60–83.
Henn, A., Cao, W., Hackney, D. D., and De La Cruz, E. M. (2008). The ATPase cycle mechanism of the DEAD-box rRNA helicase, DbpA. *J. Mol. Biol.* **377,** 193–205.
Henn, A., Cao, W., Licciardello, N., Heitkamp, S. E., Hackney, D. D., and De La Cruz, E. M. (2010). Pathway of ATP utilization and duplex rRNA unwinding by the DEAD-box helicase, DbpA. *Proc. Natl. Acad. Sci. USA* **107,** 4046–4050.
Holzbaur, E. L. F., and Johnson, K. A. (1986). Rate of ATP synthesis by dynein. *Biochemistry* **25,** 428–434.
Hutton, R. L., and Boyer, P. D. (1979). Subunit interaction during catalysis. Alternating site cooperativity of mitochondrial adenosine triphosphatase. *J. Biol. Chem.* **254,** 9990–9993.

Olivares, A. O., Chang, W., Mooseker, M. S., Hackney, D. D., and De La Cruz, E. M. (2006). The tail domain of myosin Va modulates actin binding to one head. *J. Biol. Chem.* **281,** 31326–31336.

Sleep, J. A., Hackney, D. D., and Boyer, P. D. (1978). Characterization of phosphate oxygen exchange reactions catalyzed by myosin through measurement of the distribution of ^{18}O-labeled species. *J. Biol. Chem.* **253,** 5235–5238.

Sleep, J. A., Hackney, D. D., and Boyer, P. D. (1980). The equivalence of phosphate oxygens for exchange and the hydrolysis characteristics revealed by distribution of [^{18}O] Pi species formed by myosin and actomyosin ATPase. *J. Biol. Chem.* **255,** 4097–4099.

Stempel, K. E., and Boyer, P. D. (1986). Refinements in oxygen-18 methodology for the study of phosphorylation mechanisms. *Methods Enzymol.* **126,** 618–639.

CHAPTER FOUR

Conformational Changes of DEAD-Box Helicases Monitored by Single Molecule Fluorescence Resonance Energy Transfer

Alexandra Z. Andreou *and* Dagmar Klostermeier

Contents

1. Introduction	76
1.1. The helicase core of DEAD-box proteins	76
1.2. C- and N-terminal domains flanking the DEAD-box helicase core	78
1.3. Helicase modules as part of larger proteins and in complex with other proteins	80
2. Theoretical Background of smFRET	80
3. Preparing DEAD-box Proteins for smFRET Studies	83
4. smFRET Experiments and the Calculation of Corrected smFRET Histograms	87
5. Extracting Distance Information from smFRET Measurements: Distance Histograms	91
6. The Conformational Cycle of DEAD-Box Proteins	94
6.1. Conformation of the YxiN helicase core in solution	94
6.2. Dissecting the catalytic cycle of the YxiN helicase core	96
6.3. Using smFRET to determine a structural model for full-length YxiN in solution	98
6.4. Regulating the helicase core: eIF4G activates eIF4A through a conformational guidance mechanism	99
6.5. Variations of a theme: Helicase modules	102
7. Perspective	103
Acknowledgments	104
References	104

Institute for Physical Chemistry, University of Muenster, Muenster, Germany

Abstract

DEAD-box proteins catalyze the ATP-dependent unwinding of RNA duplexes. The common unit of these enzymes is a helicase core of two flexibly linked RecA domains. ATP binding and phosphate release control opening and closing of the cleft in the helicase core. This movement coordinates RNA-binding and ATPase activity and is thus central to the function of DEAD-box helicases. In most DEAD box proteins, the helicase core is flanked by ancillary N-and C-terminal domains. Here, we describe single molecule fluorescence resonance energy transfer (smFRET) approaches to directly monitor conformational changes associated with opening and closing of the helicase core. We further outline smFRET strategies to determine the orientation of flanking N- and C-terminal domains of DEAD-box helicases and to assess the effects of regulatory proteins on DEAD-box helicase conformation.

1. INTRODUCTION

1.1. The helicase core of DEAD-box proteins

DEAD-box helicases (reviewed in Hilbert *et al.*, 2009; Linder and Jankowsky, 2011) are the largest family within the helicase superfamily 2. They owe their name to a conserved motif with the sequence D-E-A-D, in single letter code. All proteins of the DEAD-box helicase family share a region of 350–400 amino acids, the so-called helicase core. This core folds into two flexibly linked RecA domains (Fig. 4.1A). Notably, the core alone can display RNA-helicase activity (Rogers *et al.*, 1999). The core contains the conserved helicase signature motifs. Motifs Q, I, II, and VI contribute to nucleotide binding and hydrolysis. RNA binding is mediated by motifs Ia, Ib, GG, IV, QxxR, V, and VI. Motif III has been implicated in coupling of the energy derived from ATP hydrolysis to RNA unwinding (Pause and Sonenberg, 1992). More recently, numerous structures of the active form of DEAD-box proteins have shown that an intricate network of interactions links almost all of the helicase signature motifs to each other and to bound RNA and nucleotide (Andersen *et al.*, 2006; Bono *et al.*, 2006; Del Campo and Lambowitz, 2009; Montpetit *et al.*, 2011; Sengoku *et al.*, 2006). This interaction network suggests that the sequence motifs are functionally highly interconnected, which makes it difficult to unambiguously assign specific functions to individual side chains and motifs. Despite a wealth of functional and structural information, it is not clear how DEAD-box helicases couple the nucleotide cycle to RNA unwinding.

DEAD-box helicases unwind RNA duplexes in a nonprocessive fashion, by locally destabilizing RNA duplexes. In addition, they have also been shown to catalyze RNA structure remodeling and association or dissociation of RNA/protein complexes (Jankowsky and Bowers, 2006; Yang *et al.*, 2007). A number of crystal structures of helicase cores with and

smFRET Analysis of DEAD-Box Helicase Conformations 77

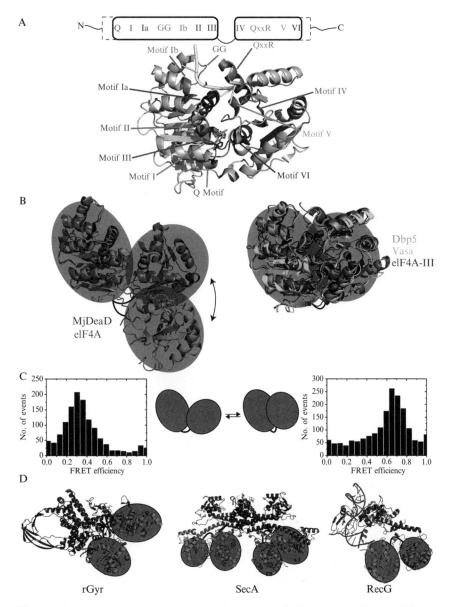

Figure 4.1 Conformations of the DEAD-box protein helicase core. (A) Top: Scheme of DEAD-box proteins, composed of a helicase core formed by two RecA domains that carry the conserved sequence motifs. Bottom: Structure of eIF4A-III (PDB-ID 2j0s) in complex with single-stranded RNA (yellow) and ADPNP (green). Motifs involved in ATP binding and hydrolysis are depicted in red tones, motifs involved in RNA binding in blue tones, and motifs involved in coupling of ATP hydrolysis to duplex separation in purple. (B) Structures of DEAD-box proteins, superimposed on the N-terminal RecA

without nucleotides and RNA have been determined. Among them are yeast translation initiation factor eIF4A (Caruthers *et al.*, 2000), DeaD from *Methanococcus jannaschii* (Story *et al.*, 2001), human eIF4A-III (Bono *et al.*, 2006), *Drosophila* Vasa (Sengoku *et al.*, 2006), the human mRNA exporter DDX19 (Collins *et al.*, 2009; Napetschnig *et al.*, 2009; von Moeller *et al.*, 2009), its yeast homologue Dbp5p (Fan *et al.*, 2009; Montpetit *et al.*, 2011), and the general RNA chaperone Mss116p (Lambowitz and Del Campo, 2009; Mallam *et al.*, 2011).

All structures in the absence of ligands show the two RecA domains separated by a wide interdomain cleft. The relative orientation of the domains varies significantly in different structures (Fig. 4.1B). In complex with the ATP analogs, ADPNP (5′–adenylyl–β,γ–imidodiphosphate), ADP·BeF$_x$ (ADP beryllium fluoride), or ADP·AlF$_x$ (ADP aluminium fluoride), and RNA, the helicase core adopts a closed state where the cleft between the two domains narrows (Fig. 4.1B) (Andersen *et al.*, 2006; Bono *et al.*, 2006; Lambowitz and Del Campo, 2009; Montpetit *et al.*, 2011; Sengoku *et al.*, 2006). In the closed state, the active site for ATP binding and hydrolysis forms, and a continuous RNA-binding site that extends over both RecA domains is created. The RNA bound to the helicase core is bent into a conformation incompatible with double strand geometry, and a role of RNA deformation in the unwinding process has been proposed (Sengoku *et al.*, 2006).

1.2. C- and N-terminal domains flanking the DEAD-box helicase core

While the helicase core is sufficient for a basic RNA-helicase activity, the majority of DEAD-box proteins also contain N- or C-terminal domains flanking the helicase core. These additional domains may enhance RNA binding or confer substrate specificity, but structural information on them is still limited. *Escherichia coli* DbpA and its *Bacillus subtilis* homolog YxiN contain a C-terminal RNA-binding domain (RBD). This domain folds into a classical RRM motif (Wang *et al.*, 2006) and binds to hairpin 92 of the

---

domain. Left: unliganded, open conformations of eIF4A (red, PDB-ID 1fuu) and mjDeaD (blue, PDB-ID 1hv8); right: closed conformation of eIF4A-III (violet, PDB-ID 2j0s), Vasa (blue, PDB-ID 2db3), and Dbp5p (green, PDB-ID 3pew) in complex with nucleotide and RNA (not shown). (C) FRET histograms for donor/acceptor-labeled YxiN_C61/267A_A115/S229C in the absence (left) and in the presence of 153mer RNA and ADPNP (right) (Karow and Klostermeier, 2009). The FRET efficiency in the open conformation is much higher than expected from the wide-open conformation observed in the crystal structure of eIF4A (B, left), suggesting that the two RecA domains are closer in solution. The FRET efficiency for the closed conformation is in good agreement with the structures (B, right). (D) Helicase modules (red) are part of larger enzymes, such as reverse gyrase (rGyr, left, PDB-ID 2gku), SecA (middle, PDB-ID 3jv2), and RecG (right, PDB-ID 1gm5). (See the Color Insert.)

ribosomal 23S rRNA with high affinity and specificity (Diges and Uhlenbeck, 2001; Fuller-Pace et al., 1993; Kossen et al., 2002). The helicase core then destabilizes the adjacent helix 91 (Kossen et al., 2002), thereby assisting in ribosome assembly (Sharpe Elles et al., 2009). A similar RBD is present in the *Thermus thermophilus* DEAD-box protein Hera (Morlang et al., 1999). 23S rRNA and RNase P RNA bind with high affinity to the Hera RBD and activate the helicase core (Linden et al., 2008). In addition, Hera contains a dimerization domain between the helicase core and the RBD (Klostermeier and Rudolph, 2009; Rudolph and Klostermeier, 2009), rendering Hera the only known dimeric DEAD-box protein (Klostermeier and Rudolph, 2009).

A different type of C-terminal extension exists in Mss116p (*Saccharomyces cerevisiae*) and Cyt-19 (*Neurospora crassa*). These DEAD-box proteins act as general RNA chaperones and are involved in splicing. Here, the helicase core is followed by an α-helical extension and a positively charged tail rich in glycine, arginine, and serine residues. In contrast to the RBD of DbpA and YxiN, the C-terminal extension of Mss116p does not constitute a separate entity but extends the RNA-binding site and engages in nonspecific contacts with the RNA (Grohman et al., 2007; Lambowitz and Del Campo, 2009; Mallam et al., 2011). Interestingly, the RNA in the Mss116p/RNA/ADPNP complex is not only bent by the helicase core, but a second bend is imposed by the C-terminal extension, suggesting a role of this domain in RNA unwinding (Del Campo and Lambowitz, 2009). A similar positively charged extension is present in Ded1p, a general RNA chaperone involved in translation (Iost et al., 1999).

An extension at the N-terminus of the helicase core is found in Dbp5p (human DDX19), a DEAD-box helicase involved in mRNA export from the nucleus (Collins et al., 2009; Napetschnig et al., 2009). This extension forms an α-helix that inserts between the two RecA domains of the helicase core (Collins et al., 2009) but interacts with the surface of the N-terminal RecA domain (RecA_N) in the closed conformation when ADPNP and ssRNA are bound (Napetschnig et al., 2009). Thus, the extension appears to regulate Dbp5p ATPase activity.

Structural models of entire DEAD-box proteins including C-terminal flanking domains have been reported for *E. coli* DbpA (Talavera et al., 2006) based on its hydrodynamic properties, for *B. subtilis* YxiN based on small-angle X-ray scattering (SAXS) (Wang et al., 2008) and from mapping by single molecule fluorescence resonance energy transfer (smFRET; Karow and Klostermeier, 2010; see below), for the yeast RNA chaperone Mss116p by X-ray crystallography (Del Campo and Lambowitz, 2009) and SAXS (Mallam et al., 2011), and for *T. thermophilus* Hera based on structures of overlapping constructs (Klostermeier and Rudolph, 2009; Rudolph and Klostermeier, 2009). More complete helicase models are required for understanding the conformational cycle of DEAD-box proteins.

1.3. Helicase modules as part of larger proteins and in complex with other proteins

Helicase modules are frequently part of larger multidomain proteins or components of larger complexes. Several enzymes involved in nucleic acid processing, such as restriction enzymes, chromatin remodelers, topoisomerases, or Dicer, rely on helicase modules (Fig. 4.1D). In SecA, the motor subunit of the bacterial polypeptide transporter, a helicase module is regulated by a C-terminal domain that mediates protein–protein and protein–lipid interactions (Breukink et al., 1995). A prominent example for a DEAD-box protein as part of a larger multicomponent functional network is the translation initiation factor eIF4A, a minimal DEAD-box protein that essentially consists only of a helicase core (Rogers et al., 2002). eIF4A interacts with a variety of other factors during translation initiation (Jackson et al., 2010), including eIF4B, eIF4G, and eIF4H. The partner proteins most likely function analogously to the N- and C-terminal extensions of other DEAD-box helicases. In the crystal structure of yeast eIF4A in complex with the middle domain of the multidomain scaffolding protein eIF4G (Schutz et al., 2008), eIF4G contacts both domains of eIF4A and restrains its helicase core in a more compact conformation than in the absence of ligands (Caruthers et al., 2000). NMR models of the mammalian eIF4A/eIF4G/eIF4H complex indicate that both factors differentially affect eIF4A activity by stabilizing different conformations (Marintchev et al., 2009).

In this chapter, we describe the use of smFRET to monitor conformations and conformational changes of DEAD-box helicases connected to the opening and closing of the helicase core, to the orientation of flanking N-andC-terminal domains and to the effect of binding proteins. smFRET is a powerful technique to measure intermolecular distances. It has no limitation regarding the size and complexity of the molecules studied and is thus an invaluable tool particularly for studying larger, flexible, multidomain, or multicomponent entities that are not readily amenable to NMR or X-ray crystallography. smFRET is well suited for monitoring conformational changes in the catalytic cycle of DEAD-box proteins. We describe the technical aspects of the application of smFRET to the analysis of DEAD-box helicase conformations and highlight the insight gained on the mechanism and regulation of these enzymes from smFRET studies.

2. Theoretical Background of smFRET

FRET is a nonradiative transfer of energy from a donor fluorophore to an acceptor fluorophore via a through-space interaction of their transition dipoles (Förster, 1948; Fig. 4.2A). The rate constant k_t for energy transfer

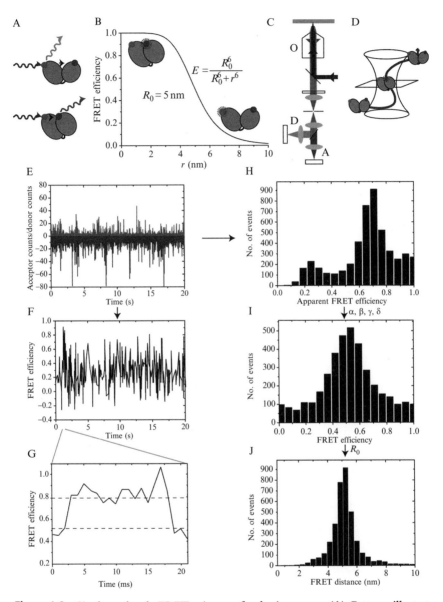

Figure 4.2 Single-molecule FRET using confocal microscopy. (A) Cartoon illustrating the principle of FRET with a two-domain protein carrying a donor fluorophore (green) on one domain and an acceptor (red) on the second domain. When the protein adopts an open conformation, the interdye distance is large, FRET will be inefficient, and donor fluorescence will be detected. When the protein adopts a closed conformation, the FRET efficiency is high and acceptor fluorescence will be detected. (B) Dependency of the FRET efficiency on the interdye distance r according to the Förster theory (Eq. 4.3) for $R_0 = 5$ nm. (C) Confocal microscopy. Laser light (blue) is focused into the confocal volume (D) through a high numerical aperture objective (O).

according to the Förster mechanism depends on the inverse 6th power of the distance separating the donor/acceptor pair, r_{DA}

$$k_t = \left(\frac{R_o}{r_{DA}}\right)^6 \tau_D \quad (4.1)$$

where R_0 is the so-called Förster distance that depends on the spectral properties of the donor and acceptor fluorophores (see Chapter 5), r_{DA} is the donor–acceptor distance, and τ_D is the fluorescence lifetime of the donor. The FRET efficiency (E) is the result of a competition between energy transfer and other radiative (k_r) and nonradiative (k_{nr}) processes deactivating the excited state of the donor.

$$E = k_t/(k_r + k_{nr} + k_t) \quad (4.2)$$

E is dependent on r_{DA} according to

$$E = \frac{R_o^6}{R_o^6 + r_{DA}^6} \quad (4.3)$$

FRET with commonly used donor–acceptor pairs allows monitoring of distances between the dyes in the range of 2–8 nm (Fig. 4.2B).

The transfer efficiency (E) reflects the number of photons transferred from the donor to the acceptor fluorophore, relative to the total number of

---

The fluorescence emission from the confocal volume is collected by the same objective and split into donor (D) and acceptor (A) spectral regions. (D) Single molecules passing the confocal volume due to free diffusion will cause bursts of fluorescence. (E) Time trace of donor (green) and acceptor (red) fluorescence from donor/acceptor-labeled YxiN. The acceptor signal is inverted for clarity. (F) Time trace of the FRET efficiency calculated from the intensities in (E). (G) Enlarged time-dependent changes in FRET efficiency during a millisecond burst. The switch in FRET efficiency from $\sim$0.35 to $\sim$0.8 and back to $\sim$0.35 represents a conformational change from the open to the closed conformation and reopening of the cleft. The broken lines indicate the mean FRET efficiencies of the open and closed conformations from FRET histograms. Reprinted from Theissen et al. (2008), with permission. © 2008 National Academy of Sciences, USA (H) Histogram of apparent FRET efficiencies, calculated burst-wise for individual molecules from donor and acceptor fluorescence intensities in (E) according to Eq. (4.4). (I) Histogram of FRET efficiencies corrected according to Eq. (4.9). (J) Distance histogram calculated from the burst-wise conversion of corrected FRET efficiencies according to Eq. (4.3). Data are for Yxin_C61/267S_S108/S229C in the presence of nucleotide and RNA ((G)–(I)) and in the absence of ligands ((H)–(J)). (For interpretation of the references to color in this figure legend, the reader is referred to the online version of this chapter.)

photons absorbed by the donor. E can be determined from the donor and acceptor fluorescence intensities I_D and I_A upon donor excitation:

$$E = \frac{I_A}{I_D + I_A} \quad (4.4)$$

Due to different quantum yields and instrument nonideality (e.g., different detection sensitivities in the spectral regions of donor and acceptor emission), this value is an apparent FRET efficiency that reports relative distance differences, not true intermolecular distances. The apparent E is therefore often called a proximity ratio.

Fluorescence intensities from ensembles of molecules naturally reflect the averaged intensities from all molecules present. FRET efficiencies determined from these intensities thus yield population-weighted averages of individual FRET efficiencies in all conformers present. Consequently, the distances obtained from the conversion of these FRET efficiencies according to Eq. (4.4) do not reflect intermolecular distances within individual molecules. This limitation can be overcome by smFRET studies that provide information about FRET efficiencies of individual molecules, and thereby reveal sub-populations. Fluorescence intensities of donor/acceptor-labeled single molecules diffusing freely in solution can be measured by confocal microscopy (Fig. 4.2C). In confocal smFRET experiments, each single molecule produces a burst of fluorescence while it diffuses through the confocal volume (Fig. 4.2D and E) with a typical duration of a few milliseconds. The measured fluorescence intensities allow the burst-wise calculation of the FRET efficiency for each individual molecule. Histograms of FRET efficiencies for a large number of molecules (typically >2000, Fig. 4.2H) measured one at a time directly provide information on subpopulations and report conformational changes. Several FRET efficiencies provide distance restraints that define global conformations and relative domain rearrangements. The time-dependent FRET efficiency within fluorescence bursts directly reports conformational fluctuations on the millisecond timescale (Fig. 4.2G). To follow conformational changes on longer timescales, an array of single molecules immobilized on a surface is imaged via scanning confocal, wide-field, or TIRF microscopy. Dwell times from FRET time traces then provide kinetic constants for conformational changes.

3. Preparing DEAD-box Proteins for smFRET Studies

Conformational changes in the catalytic cycles of DEAD-box proteins have been suggested based on differences in the orientation of the two RecA domains in different structures (Fig. 4.1B). The conformation of the

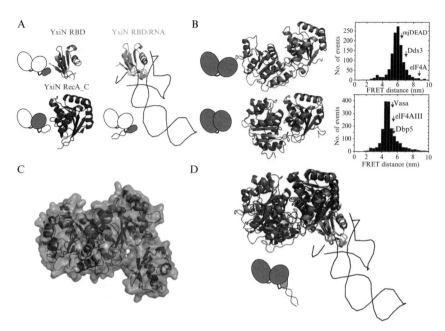

Figure 4.3 The DEAD-box protein YxiN. (A) *B. subtilis* YxiN consists of a helicase core and a C-terminal RBD (cartoon). Structures of the isolated RBD (green, PDB-ID 2g0c), RBD in complex with RNA (light blue/gray PDB-ID 3moj), RecA_C (magenta, PDB-ID 2hjv). (B) Top: homology model of YxiN created using the structure of *M. jannaschii* DeaD (open, orange) as a template, and distance histogram from smFRET measurements of YxiN_C61/267A_A115/S229C in the absence of ligands, bottom: homology model of YxiN according to eIF4A-III (closed, violet) and distance histogram from smFRET measurements of YxiN_C61/267A_A115/S229C in the presence of RNA and ADPNP. Homology models were created using Geno3D (Combet *et al.*, 2002) and Swiss-Model (Schwede *et al.*, 2003). (C) Model for full-length YxiN in solution derived by manually positioning the YxiN RBD (green) relative to the helicase core (orange) based on distance restraints from smFRET experiments with YxiN carrying one fluorophore in the N- or C-terminal RecA domain and a second fluorophor on the RBD (Karow and Klostermeier, 2010). Reprinted from Karow and Klostermeier (2010), © 2010, with permission from Elsevier. (D) Superposition of the YxiN model (orange/green, (C)) and crystal structures of the YxiN C-terminal RecA domain (violet) and the RBD (light blue) bound to RNA (gray). The binding mode of the RNA is consistent with the model derived from smFRET. (For interpretation of the references to color in this figure legend, the reader is referred to the online version of this chapter.)

helicase core of YxiN from *B. subtilis* (Fig. 4.3) has been delineated throughout the catalytic cycle using smFRET (Aregger and Klostermeier, 2009; Theissen *et al.*, 2008).

Before a protein is amenable to smFRET studies, preparative steps are required (Fig. 4.4). Attachment sites for fluorophores have to be introduced by mutagenesis, and fluorophores have to be introduced. Subsequent to

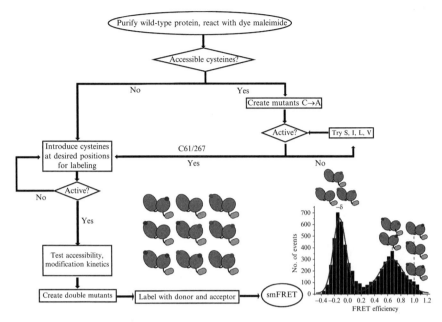

Figure 4.4 Flow chart of the procedure for preparing proteins for smFRET studies. Mutagenesis and labeling steps for smFRET experiments. Species carrying one or two donor fluorophores will contribute to the "zero" peak that appears at $-\delta$ in corrected FRET histograms due to corrections for an acceptor that is not present. Species carrying one or two acceptor fluorophores will not be excited efficiently and will only appear as a "one" peak if present in large excess. The peak at intermediate FRET efficiencies presents intermolecular FRET for donor/acceptor-labeled species. (For the color version of this figure, the reader is referred to the online version of this chapter.)

labeling, wild-type-like activity needs to be confirmed. Fluorophores can be introduced at the N-terminus, at internal cysteine or lysine residues, at the C-terminus, or at nonnatural amino acids incorporated *in vivo* through the suppression of stop codons (Ellman *et al.*, 1991; Lemke, 2011; Schuler and Pannell, 2002; Volkmann and Liu, 2009). Fluorescent double labeling is commonly achieved through the modification of thiol groups with maleimide derivatives of dyes. As a first step, solvent accessible intrinsic cysteines are replaced by mutagenesis. For YxiN, two of the four native cysteines readily reacted with cysteine-reactive reagents such as Ellman's reagent and maleimides and were therefore removed (Theissen *et al.*, 2008). A common replacement for cysteines is the almost isosteric serine residue. However, exchange of cysteines 61 and 267 of YxiN with serines resulted in a construct that was still an active, ATP-dependent RNA helicase, but showed a three- to fourfold reduced RNA-stimulated ATPase activity (Theissen *et al.*, 2008) and an increased apparent K_M value for RNA (Karow and Klostermeier, 2009). In contrast, replacement with alanines led to a construct with wild-type-like activity (Karow and Klostermeier, 2009).

In a second step, surface-exposed cysteines for fluorescent labeling can be introduced into the cysteine-free variant at desired positions. These positions should be accessible for the modification reagent, and mutations should not interfere with function. Positions on the surface outside conserved motifs are generally good candidates. Structural information can help in selecting suitable positions for the introduction of cysteines. In the absence of structural information, a homology model based on known structures can be created and used as a guide to identify potentially suitable positions, as described for the YxiN helicase core (Theissen et al., 2008). To monitor conformational changes of the YxiN helicase core via distances between the two RecA domains, cysteines were introduced on each of the two RecA domains on opposite sides of the interdomain cleft. Two and three positions were selected in the N- and C-terminal RecA domains (108 and 115 in RecA_N, 224, 229, and 262 in RecA_C), and five double cysteine mutants were constructed.

DEAD-box proteins exert a number of activities, such as RNA binding, nucleotide binding, RNA-dependent ATPase activity, and ATP-dependent RNA unwinding. Ideally, none of these activities should be affected by mutagenesis and fluorophore introduction. RNA-dependent ATP hydrolysis and RNA unwinding directly report on coupling of the ATPase with the RNA-binding site. All YxiN constructs were tested for RNA-dependent ATPase activity and for ATP-dependent RNA unwinding. The hydrolysis of ATP under steady-state conditions was followed in a coupled enzymatic assay monitoring the decrease of A_{340} due to the oxidation of NADH to NAD$^+$ (Adam, 1962). For YxiN mutants, k_{cat} and apparent K_M values for RNA were within the range determined for wild-type YxiN confirming that these constructs are RNA-dependent ATPases (Karow and Klostermeier, 2009; Theissen et al., 2008).

YxiN recognizes hairpin 92 of the 23S rRNA via its RBD and unwinds the adjacent helix 91 (Diges and Uhlenbeck, 2005). A minimal 32/9mer RNA substrate, annealed from a synthetic fluorescently labeled 9mer and a synthetic 32mer containing hairpin 92, was unwound by all YxiN mutants with yields comparable to the wild-type enzyme (Karow and Klostermeier, 2009; Karow et al., 2007; Theissen et al., 2008). Strand displacement was also monitored as a decrease in fluorescence anisotropy due to the release of the 9mer end-labeled with fluorescein (Karow et al., 2007). A similar assay has been used to measure unwinding by DbpA (Henn et al., 2010). We have also used an ensemble FRET-based assay where the substrate is generated from a 32mer carrying an acceptor at the 5′-end and a 9mer labeled with an acceptor at the 3′-end (Regula Aregger and Dagmar Klostermeier, unpublished). Both of these assays are continuous, and the time-dependent signal change directly provides rate-constants and allows a quantitative comparison of the different constructs.

YxiN mutants containing two cysteines were statistically labeled with a mixture of donor (AlexaFluor 488-maleimide) and acceptor (AlexaFluor 546-maleimide) dyes. Due to the presence of two chemically identical groups, labeling with a mixture of donor and acceptor fluorophores will lead to a mixture of donor/donor-, donor/acceptor-, and acceptor/acceptor-labeled species (Fig. 4.4).

Labeled YxiN can also be generated by expressed protein ligation from separately produced N- and C-terminal RecA domains (Karow et al., 2007). This approach allows site-specific attachment of donor and acceptor, because the two protein parts can be labeled separately before ligation. However, the overall yield for the final ligation is very low. Although smFRET histograms for a YxiN-construct created by labeling and followed by ligation did not differ from histograms for statistically labeled YxiN (Anne R. Karow and Dagmar Klostermeier, unpublished), specific labeling will be beneficial if the spectral properties of one of the dyes are affected by the attachment site.

To avoid the accumulation of donor/donor species, we typically use an excess of acceptor in the statistical labeling reaction. Labeling conditions are optimized to maximize the fraction of species carrying one donor and one acceptor fluorophore and to minimize the species carrying only donor moieties. With typical labeling efficiencies of $x = 0.3$ (donor, D) and $y = 0.6$ (acceptor, A) for each cysteine, the fractions of the individual species are as follows: D/D: $x^2 = 0.3 \times 0.3 = 0.09$; D/A, A/D: $2xy = 2 \times 0.3 \times 0.6 = 0.36$; A/A: $y^2 = 0.6 \times 0.6 = 0.36$; D/−, −/D: $2x(1 − x − y) = 2 \times 0.3 (1 − 0.3 − 0.6) = 0.06$; A/−, −/A: $2y(1 − x − y) = 2 \times 0.6(1 − 0.3 − 0.6) = 0.12$; −/−: $(1 − x − y)^2 = (1 − 0.3 − 0.6)^2 = 0.01$. In smFRET histograms, donor/donor species appear at FRET efficiencies below 0, due to the correction for direct excitation of the acceptor that is not present in this species (discussed below). Acceptor/acceptor species are excited only very inefficiently but can occasionally lead to a minor peak at a FRET efficiency of 1 (Fig. 4.4). Only the molecules carrying both a donor and an acceptor will provide FRET efficiencies reflecting intermolecular donor/acceptor distances.

4. smFRET Experiments and the Calculation of Corrected smFRET Histograms

smFRET experiments to address the conformation of YxiN were performed on a homebuilt confocal microscope (Fig. 4.2C and D). Samples were prepared in Lab-Tek chambered cover slips (NUNC™, Rochester, NY, USA). Donor fluorescence was excited with the frequency-doubled output from a mode-locked, pulsed titanium sapphire laser at 475 nm. The excitation light was focused by a 60× water immersion objective (PlanApo, N.A. 1.2, Olympus) with a typical excitation power of 75 μW, corresponding to

~1.3 kW/cm^2 in the sample. Fluorescence emitted from the sample was collected by the same objective, separated into donor and acceptor contributions using a beam splitter (Q565LP, AHF), passed through a D535/40 BP (donor) or E570LP filter (acceptor), and detected by avalanche photodiodes (SPCM-14, Perkin Elmer). Excitation light was rejected by a dichroic beam splitter (DM505, Olympus) and an external optical filter (HQ490LP, AHF). Time-correlated single photon counting was performed using a SPCM-630 counting card (Becker & Hickl, Berlin, Germany) with a time resolution of 125 ns. For each detected photon, the macrotime (time since start of the experiment), the microtime (time with respect to previous excitation pulse), and the channel are recorded. From these data, fluorescence time traces, fluorescence intensity decays, and fluorescence autocorrelation functions can be calculated. Autocorrelation functions provide information on the diffusion time of the fluorescent moiety through the confocal volume. This serves as a control for binding of high-molecular-weight ligands that lead to an increase in diffusion time.

The raw data from smFRET experiments are time traces of donor and acceptor fluorescence intensities (Fig. 4.2E). The first step of data analysis requires the identification of individual single molecule events. This is usually achieved by applying a threshold criterion, either absolute ($>n$ photons, typically $50 < n < 100$) or relative ($>n$-fold above background). In both cases, the background has to be identified for the donor and acceptor channels and then subtracted from the measured intensities.

To improve the signal-to-noise ratio, background fluorescence should be minimized. Accordingly, highly pure samples are required. To remove fluorescent impurities, buffers are generally purified by treatment with active charcoal. Nucleotides, nucleic acids, or interacting proteins are usually added in large excess to the labeled protein that is present in the picomolar range to ensure saturation and complex formation. These compounds can substantially increase the fluorescence background.

In a second step, background-corrected fluorescence intensities are corrected for nonidealities of the experimental setup, including cross talk between donor and acceptor channels (α, β), different quantum yields and detection efficiencies for donor and acceptor fluorescence (γ), and the direct excitation of the acceptor (δ). These correction parameters depend on the spectral properties of the dyes and can be determined experimentally. Since the spectral properties are influenced by the local environment at the attachment site, sets of correction parameters have to be determined individually for each construct. The correction parameters may differ for the donor/acceptor compared to the acceptor/donor configuration. Further, correction parameters may change upon ligand binding and therefore have to be determined in the absence and presence of ligands. Correction parameters for donor/acceptor-labeled cysteine variants are determined using the corresponding single cysteine mutants labeled either with donor or with acceptor (generating labeled proteins reflecting the D/–, –/D; A/–, and –/A species). The absorbance of these proteins at the donor excitation wavelength is measured (A_D, A_A) and corrected for buffer contributions (A_D^{buffer}, A_A^{buffer}).

To determine the concentration of labeled proteins, possible changes in absorbance upon attachment of dyes to the protein have to be considered. To this end, the labeled protein is degraded with proteinase K, and the dye concentrations c_D and c_A are calculated from the absorbance of this solution with extinction coefficients for the free dye, as provided by the manufacturer. The correction parameter δ as a measure for the fraction of direct excitation of the acceptor is calculated from these values according to

$$\delta = \frac{A_A - A_A^{buffer}}{A_D - A_D^{buffer}} \frac{c_D}{c_A} \quad (4.5)$$

The remaining correction parameters α, β, and γ can be determined from counting rates directly measured in the confocal microscope (Eqs. 4.6–4.8). α is determined from the counting rates of the donor-labeled protein in donor and acceptor channels (I_D^{donor}, I_A^{donor}) and corrected for buffer contributions (I_D^{buffer}, I_A^{buffer}):

$$\alpha = \frac{I_A^{donor} - I_A^{buffer}}{I_D^{donor} - I_D^{buffer}} \quad (4.6)$$

β is determined analogously with the acceptor-labeled protein:

$$\beta = \frac{I_D^{acceptor} - I_D^{buffer}}{I_A^{acceptor} - I_A^{buffer}} \quad (4.7)$$

For identical concentrations of donor- and acceptor-labeled proteins, γ is calculated as

$$\gamma = \frac{I_A^{acceptor} - I_A^{buffer}}{\delta(I_D^{donor} - I_D^{buffer})} \quad (4.8)$$

The corrected FRET efficiency (E_{FRET}) can then be calculated from the background-corrected donor and acceptor fluorescence intensities I_D and I_A:

$$E_{FRET} = \frac{(1+\beta\gamma\delta)\left(I_A - \frac{a+\gamma\delta}{1+\beta\gamma\delta}I_D\right)}{(1+\beta\gamma\delta)\left(I_A - \frac{a+\gamma\delta}{1+\beta\gamma\delta}I_D\right) + (\gamma+\gamma\delta)(I_D - \beta I_A)} \quad (4.9)$$

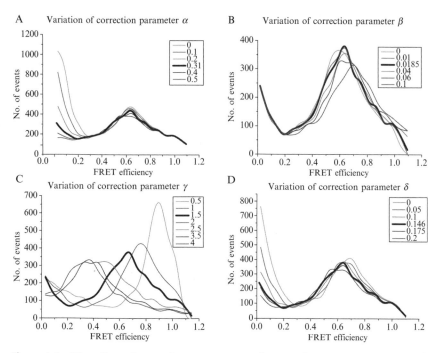

Figure 4.5 The effect of correction parameters α, β, γ, and δ on smFRET histograms. "Corrected" FRET histograms were calculated from raw data. Three correction parameters were kept constant at the experimentally determined values and the fourth was varied. The black curve represents the histogram calculated with the correct set of parameters ($\alpha = 0.31$, $\beta = 0.019$, $\gamma = 1.5$, $\delta = 0.146$). (A) Variation of correction parameter α (cross talk of donor fluorescence into the acceptor channel) between 0 and 0.5. (B) Variation of correction parameter β (cross talk of acceptor fluorescence into the donor channel) between 0 and 0.1. (C) Variation of correction parameter γ (differences in quantum yields and detection efficiencies) between 0.5 and 4. (D) Variation of correction parameter δ (direct excitation of the acceptor) between 0 and 0.2. (For the color version of this figure, the reader is referred to the online version of this chapter.)

(Theissen et al., 2008). This corrected FRET efficiency is not affected by instrumental characteristics or the environment of the dyes in a specific construct and thus reports on true interdye distances (Fig. 4.2I). The experimental error of these FRET efficiencies is <10%, taking into account the uncertainty in the determination of the correction parameters (Gubaev et al., 2009). The influence of inaccurate correction parameters on the shape of the FRET histograms is illustrated in Fig. 4.5.

smFRET experiments using confocal microscopy measure FRET efficiencies for a large number of molecules, typically >2000. A rigorous analysis to calculate the shot-noise limited accuracies of experimental FRET efficiencies has been described (Dahan et al., 1999). With typical background counts of <3000 photons/s in the donor and <7000 photons/s in the acceptor channel, a threshold for single-molecule events of >100 photons, and a value for γ of 1, differences of 0.09 between intermediate FRET efficiencies (~0.5) are significant. Even smaller differences in FRET efficiencies can be reliably determined for FRET efficiencies toward 0 and 1, and species can be reliably distinguished.

Histograms of FRET values from individual single molecules directly reflect the distribution of conformers with differences in donor/acceptor distances. FRET efficiency distributions are often uni- or bimodal but can be more complex. Due to the limited number of single-molecule events, the bin size of histograms can affect the resulting distribution. To unambiguously identify characteristic features in distributions, the bin size should be varied. In the simplest case, FRET distributions can be described by one (or more) Gaussian distributions, yielding a mean FRET efficiency and a width for each distribution. Mean FRET efficiencies of the same donor/acceptor-labeled construct, determined in independent experiments, typically do not differ by more than 0.05. The width contains contributions from shot noise due to the limited number of detected photons, from segmental flexibility of the dyes, and from conformational flexibility of the molecule. A general treatment of photon statistics in smFRET experiments can be found in Gopich and Szabo (2005). Using probability distribution analysis (PDA) (Antonik et al., 2006) or proximity ratio histogram (PRH) analysis (Nir et al., 2006), the shape of FRET histograms can be quantitatively described with high precision, either model-based or in combination with maximum entropy methods (Brochon, 1994), which do not rely on *a priori* assumptions. A more detailed discussion of these approaches is beyond the scope of this review.

5. Extracting Distance Information from smFRET Measurements: Distance Histograms

To convert corrected FRET efficiencies into interdye distances according to Eq. (4.3), the Förster distance for the donor/acceptor pair is required. The Förster distance for a specific dye pair can be calculated according to

$$R_0^6 = \frac{9000 \ln 10 \phi_D \kappa^2}{128 \pi^5 N n^4} \int_0^\infty F_D(\lambda) \varepsilon_A(\lambda) \lambda^4 d\lambda \qquad (4.10)$$

where ϕ_D is the fluorescence quantum yield of the donor in the absence of FRET, n is the refractive index of the medium between the dyes (the value for water is 1.33), κ^2 is the orientation factor describing the relative orientation of donor and acceptor dipole moments, $F_D(\lambda)$ is the normalized fluorescence of the donor-labeled protein, and $\varepsilon_A(\lambda)$ is the extinction coefficient of acceptor-labeled protein, both at wavelength λ.

The donor quantum yield ϕ_D can be determined relative to fluorescein in 0.1 M NaOH or other quantum yield standards (Magde et al., 2002; Parker and Rees, 1960), using proteins carrying only a donor. The quantum yield for the donor is sensitive to the local environment of the fluorophore and may differ for the donor/acceptor and acceptor/donor species and for different constructs. As described above for the correction parameters, it has to be determined for both configurations of each construct used. For YxiN, the quantum yield of the donor AlexaFluor 488 varied between 0.32 and 0.44, depending on the attachment site (Theissen et al., 2008). The refractive index (n) is usually set to 1.33 for water. Due to the three-dimensional through-space interaction of the dipoles, this approximation is valid even though the protein is present between the dyes. The orientation factor (κ^2) is a measure for the relative orientation of the donor and acceptor transition dipoles. For flexibly attached dyes that are freely rotating on the timescale of fluorescence emission, $\kappa^2=2/3$. When the dye motion is restricted, κ^2 can assume values between 0 and 4, depending on the angle separating the donor and acceptor transition dipoles. The rotational freedom of the attached fluorophores and the resulting range of possible κ^2 values can be estimated experimentally from the anisotropy decays of single-labeled proteins (Theissen et al., 2008). For a more detailed description, see Haas et al. (1978) and Klostermeier and Millar (2001). Fluorescence anisotropy decays usually follow double exponentials. The fast phase corresponds to the segmental mobility of the fluorophore, and the slower phase reflects the overall tumbling of the molecule. From the corresponding amplitudes α_1 (fast phase) and α_2 (slow phase), an order parameter (S) can be calculated:

$$S=\sqrt{\frac{\alpha_2}{\alpha_1+\alpha_2}} \tag{4.11}$$

S can be converted into the half-angle θ of a cone within which the fluorophore rotates relative to the attachment site according to

$$\theta = \cos^{-1}\left(0.5\left(\sqrt{1+8S}-1\right)\right) \tag{4.12}$$

For YxiN, the half-cone angles describing the rotational freedom of the donor were $\sim 30°$. Only in one case, for dyes attached to position 229, θ was reduced to 14° (Theissen et al., 2008). Hence, in all cases, the $\kappa^2=2/3$ approximation was justified. The error introduced in distances by an incorrect assumption of $\kappa^2=2/3$ is generally <26% and will not exceed 10% as long as one of the dyes rotates freely (Haas et al., 1978).

The overlap integral can be approximated from the summation of the normalized fluorescence-emission spectrum of the donor, and the absorption spectra of the acceptor, typically in 1 nm steps in the overlapping spectral region (λ_x–λ_y) as

$$\int_0^\infty F_D(\lambda)\varepsilon_A(\lambda)\lambda^4 d\lambda \approx \sum_{i=x}^{\lambda_y} F_D(\lambda_i)\varepsilon_A(\lambda_i)\lambda_i^4 \qquad (4.13)$$

With the Förster distance R_0, FRET efficiencies can be converted into interdye distances for every molecule detected (Eq. 4.3) and distance histograms can be constructed (Fig. 4.2J). Since the spectral properties may be different for donor/acceptor and acceptor/donor species, and also for different constructs, R_0 has to be calculated for each individual construct and for both dye configurations. Similarly, R_0 has to be determined in the absence and presence of ligands. For YxiN, the differences between the R_0 values for D/A and A/D configurations were typically < 0.2 nm, which is much smaller than the width of the observed distance distributions. If the difference between the configurations is extreme, an apparent bimodal distribution may arise for a homogenous sample with a single interdye distance.

It should be noted that due to the nonlinear dependence between E and r, and the resulting difference in assignment of molecules to bins, the mean distance from the distance histogram differs from the distance calculated from the mean FRET efficiency. As described above for FRET histograms, analysis with different bin sizes is recommended to identify stable features of distance histograms. The analysis of distance histograms with Gaussian distributions yields mean distances, usually with standard deviations of < 0.1 nm. The width(s) of the distance distributions contain contributions from shot noise, from the local flexibility of the dye, and from the global flexibility of the protein. The different contributions can be dissected with advanced analysis methods, such as PDA and PRH (Antonik et al., 2006; Nir et al., 2006).

In our hands, differences of the mean distances, determined from distance histograms of independently performed experiments, do not exceed 0.1 nm. For analyses with two or more distributions, the parameters for at least one of the distributions should be determined independently in control experiments. These parameters should be kept constant during the fit. Strictly speaking, Gaussian distributions are only suitable to describe symmetric distributions around medium FRET efficiencies. Model-free approaches, based on maximum entropy methods to analyze distance distributions have been described (see above). The experimental error for FRET efficiencies propagated to distances corresponds to ∼0.1–0.2 nm for FRET efficiencies < 0.8. For larger FRET efficiencies, this error increases. However, the largest uncertainty in the determination of distances arises from the length of the flexible linker by which the dyes are attached to the protein (typically < 0.5–1 nm).

To validate the experimentally determined FRET values and their conversion to distances, we calibrated the confocal microscope using a set of donor- and acceptor-labeled DNA ladders, which cover a range of FRET efficiencies. The distances calculated from FRET efficiencies were in agreement with distances expected for B-DNA duplexes and attachment geometry (Airat Gubaev & Dagmar Klostermeier, unpublished). Moreover, the good agreement of distances derived using this method with those observed in existing DEAD-box helicase structures (Hilbert et al., 2011; Theissen et al., 2008), further reinforces the validity of measurements and data analysis (Fig. 4.3B).

The data analysis described so far applies to smFRET experiments with continuous excitation of the donor. When the donor fluorescence is excited by a pulsed laser, fluorescence lifetimes for donor and acceptor can be determined. Two-dimensional plots of (apparent) FRET efficiencies and donor lifetime allow for distinction of changes in FRET due to distance changes or due to changes of the donor or acceptor spectral properties (Sisamakis et al., 2010). Alternatively, FRET efficiencies can be determined directly and without the need for correction from the burst-wise determination of the fluorescence lifetime of the donor in the absence and presence of acceptor according to:

$$E = 1 - \tau_{DA}/\tau_D \quad (4.14)$$

Due to the limited number of photons within a burst of fluorescence from a single molecule, an accurate determination of the lifetime is difficult and will become impossible if multiple lifetimes are present. This drawback becomes more pronounced for higher FRET efficiencies, as energy transfer leads to fewer photons in the donor channel. FRET efficiencies from lifetimes systematically underestimate the correct FRET value, mainly due to a cross talk of acceptor photons into the donor channel that leads to a higher apparent lifetime of the donor. Thus, the slope of a plot of E (intensities) versus E (lifetimes) is usually < 1 (Hilbert et al., 2011). Nevertheless, the equivalence of FRET efficiencies from lifetimes and intensities validates correction procedures.

6. THE CONFORMATIONAL CYCLE OF DEAD-BOX PROTEINS

6.1. Conformation of the YxiN helicase core in solution

smFRET histograms for YxiN labeled with AlexaFluor 488 (A488, donor) and AlexaFluor 546 (A546, acceptor) showed unimodal distributions. Description with Gaussian distributions yielded mean FRET efficiencies of 0.40–0.63 for all constructs, consistent with an open cleft between the RecA domains of the helicase core in the absence of RNA and nucleotide. Notably, though, the RecA domains are much closer in the open conformation of YxiN than in eIF4A (Caruthers *et al.*, 2000), where the extended interdomain linker allows for the maximum possible separation of the two domains. Such a wide-open conformation is clearly not populated in the distance histogram (Fig. 4.3B). Comparison of the interdye distances from smFRET experiments with C_β–C_β distances between the corresponding residues in the structures of other DEAD-box helicases revealed that the conformation of YxiN in solution most closely resembles the crystal structure of the DEAD-box protein DeaD from *M. jannaschii* (Story *et al.*, 2001; Fig. 4.3B). A quantitative criterion for similarity can be determined from an RMSD value between the experimental distances and the distances from the individual structures. The RMSD value over 5 distances was 9 nm for YxiN and eIF4A and 2 nm for YxiN and DeaD.

Neither the presence of nucleotide nor RNA alone induced a conformational change of the YxiN helicase core. In contrast, the cooperative binding of both RNA and ADPNP led to the appearance of a high-FRET species, corresponding to a reduced donor/acceptor distance due to a closure of the interdomain cleft (Fig. 4.3B). Such a nucleotide-driven conformational change had been suggested previously from the low affinity of DEAD-box helicases for ATP compared to ADP (Karow *et al.*, 2007; Lorsch and Herschlag, 1998a; Talavera and De La Cruz, 2005), from limited proteolysis experiments (Lorsch and Herschlag, 1998b), and, as mentioned, from structural data. The FRET increase constituted the first direct observation of such a proposed conformational change for a specific DEAD-box protein. Comparison with the crystal structures of Vasa, eIF4AIII, DDX19/Dbp5p shows that the conformation of YxiN with RNA and ADPNP in solution is in good agreement with the crystal structures of these DEAD-box proteins in the presence of RNA and ATP analogues (Andersen *et al.*, 2006; Bono *et al.*, 2006; Collins *et al.*, 2009; Montpetit *et al.*, 2011; Sengoku *et al.*, 2006) (RMSD value 2 nm). YxiN variants where intrinsic cysteines were replaced by serines exhibited reduced RNA affinity. These YxiN variants were not saturated with RNA in single molecule experiments, leading to a residual low-FRET species with an open interdomain cleft (Theissen *et al.*, 2008). In the cysteine-to-alanine variants, the RNA affinity remained high, and addition of RNA led to a complete shift from low to high FRET upon cleft closure (Karow and Klostermeier, 2009). In the presence of RNA and ATP, two populations with low- and high-FRET efficiencies were detected, reflecting the steady-state population of the open and closed conformations inter-converting during the YxiN catalytic cycle (Karow and Klostermeier, 2009).

6.2. Dissecting the catalytic cycle of the YxiN helicase core

From the initial FRET experiments, ATP and RNA binding were identified as the trigger for closure of the interdomain cleft in the helicase core at the beginning of the catalytic cycle. The sequence of conformational states of the helicase core during the nucleotide cycle of YxiN was dissected using the ATP analogues ADP·BeF$_x$ (ADP beryllium fluoride) and ADP·MgF$_x$ (ADP magnesium fluoride). ADP·BeF$_x$ mimics ATP in the prehydrolysis state (Fisher et al., 1995). While ADPNP does not support RNA unwinding, unwinding occurs in the presence of ADP·BeF$_x$ (Aregger and Klostermeier, 2009; Liu et al., 2008), suggesting that the ADP·BeF$_x$ complex constitutes an on-pathway intermediate. ADP·MgF$_x$ is thought to mimic a posthydrolysis state (Bigay et al., 1987; Chabre, 1990; Fisher et al., 1995). smFRET experiments with YxiN, RNA, and ADP·BeF$_x$ or ADP·MgF$_x$ analogues revealed that the helicase core retains the closed conformation during ATP hydrolysis. On the other hand, YxiN adopts an open conformation in the presence of ADP and RNA, implying that phosphate release triggers reopening of the interdomain cleft and resets the enzyme for a subsequent catalytic cycle (Aregger and Klostermeier, 2009) (Fig. 4.6).

Three functional mutants of YxiN, impaired in different steps of the catalytic cycle, were studied to correlate conformational changes with the catalytic cycle (Karow and Klostermeier, 2009). YxiN_K52Q, carrying a mutation in the Walker A motif, was ATPase-deficient and unable to unwind a minimal RNA substrate (Karow and Klostermeier, 2009). Interestingly, this mutant still adopted a closed conformation in the presence of RNA and ATP, identical to that of the wild-type enzyme in the presence of ADPNP and RNA, suggesting that closure of the interdomain cleft is not sufficient for RNA unwinding. This observation is consistent with the finding that ADPNP does not support unwinding but causes closure. The cleft closure was also observed for a motif V mutant, YxiN_G303A, but here the global conformation differed from the wild-type enzyme. The FRET efficiency of the closed conformation was slightly reduced, pointing toward a wider interdomain cleft. This mutant showed significantly reduced ATPase activity and was unwinding deficient. Presumably, the alanine residue introduced in the position of a conserved glycine residue sterically interferes with formation of the closed conformer and impedes assembly of the ATPase-catalytic site, leading to a reduced ATPase activity. Similarly, an incomplete closure will not assemble the RNA-binding site and may not induce a deformation of bound RNA rationalizing the lack of unwinding. Finally, a mutation in motif III (YxiN_AAA) affected neither the propensity for closure nor the global conformation of the closed state but reduced the cooperativity of ATP and RNA binding and the rate of RNA unwinding, possibly due to altered conformational dynamics.

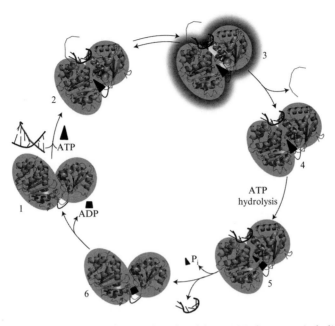

Figure 4.6 ATP-driven conformational cycle of the DEAD-box protein helicase core during RNA unwinding. Conformational states of the helicase core (orange) throughout the nucleotide cycle, as delineated from smFRET experiments with YxiN and ATP, ADPNP, ADP·BeF$_x$, and ADP·MgF$_x$ (Aregger and Klostermeier, 2009; Karow and Klostermeier, 2009; Theissen et al., 2008). Homology models of YxiN are based on *M. jannaschii* DeaD (open conformation) and eIF4A-III (closed conformation). YxiN is in the open conformation in the absence of ligands (1). The interdomain cleft closes when ATP (triangle) and RNA (black/gray) bind cooperatively, leading to local destabilization of the double-stranded RNA (2). A structural rearrangement leads to the formation of a hydrolysis- and unwinding competent state (3, highlighted by the gray aura), from which the first RNA strand (gray) can dissociate. The core remains in the closed conformation during ATP hydrolysis (4, 5), and reopening is coupled to phosphate release and RNA dissociation. (6) Nucleotide exchange then starts further catalytic cycles. (For interpretation of the references to color in this figure legend, the reader is referred to the online version of this chapter.)

Altogether, smFRET experiments have demonstrated that the helicase core of YxiN alternates between open and closed conformations during the catalytic cycle (Fig. 4.6) and have provided a molecular basis for the early notion that DEAD-box helicases may constitute nucleotide-dependent switches (Lorsch and Herschlag, 1998a,b). The helicase core closes upon cooperative binding of ATP and RNA. A subsequent activation step converts the closed conformation into a hydrolysis- and unwinding competent state. This state is not reached with ADPNP or with the K52Q mutant of YxiN, but with ATP and ADP·BeF$_x$. The existence of two ATP-bound states had already been suggested earlier from kinetic data (Henn et al., 2008) and ATPase-stimulation data (Peck and Herschlag, 1999). The molecular nature of this isomerization is unclear, as FRET efficiencies of YxiN with ADPNP and ADP·BeF$_x$ (Aregger and Klostermeier, 2009) and structures of Mss116p in complex with ADPNP and ADP·BeF$_x$ are virtually identical (Lambowitz and Del Campo, 2009). The first RNA strand is released from this unwinding-competent state, rationalizing the observation that unwinding can occur before ATP hydrolysis (Chen et al., 2008), and that ADP·BeF$_x$ can support single turnovers of RNA unwinding (Liu et al., 2008). ATP hydrolysis and phosphate release trigger reopening of the helicase core, leading to a disruption of the bipartite RNA-binding site formed by both core domains and to RNA release. ATP hydrolysis thus mainly serves to reset DEAD-box helicases for subsequent catalytic cycles. Taking into account the structural conservation observed among DEAD-box proteins, the conformational changes observed for YxiN most likely depict a common nucleotide-driven conformational cycle for DEAD-box proteins, although modifications of this common theme are possible, as discussed below.

6.3. Using smFRET to determine a structural model for full-length YxiN in solution

Only few studies have provided information about the orientation of N- or C-terminal flanking domains with regard to the helicase core, and their cooperation with the core in DEAD-box protein action. For YxiN, the structure of the RBD alone and in complex with a fragment of 23S rRNA has been determined by X-ray crystallography (Hardin et al., 2010; Wang et al., 2006) (Fig. 4.3A). Using distance restraints from smFRET experiments, the position of the RBD relative to the YxiN helicase core was determined (Karow and Klostermeier, 2010). Cysteines for labeling were introduced at six positions covering the surface of the RBD, and in the RecA_C or RecA_N domain. All six double cysteine mutants retained ATP-dependent RNA unwinding activity. smFRET histograms of A488- and A546-labeled YxiN mutants were unimodal, consistent with a defined relative orientation of the two domains. FRET efficiencies for constructs with labels on the RBD and on the C-terminal RecA domain were in the range of 0.85–0.9, corresponding to a short interdye distance of <4 nm, indicating that the RBD is very close to the C-terminal RecA domain. High FRET efficiencies are not sensitive to small changes in distances and therefore do not provide reliable restraints for mapping. Placing the second fluorophore on the N-terminal RecA domain created a more sensitive range of FRET efficiencies of 0.2–0.4 and thus more reliable distance restraints, based on which a model for full-length YxiN was built. The mean distances obtained from distance histograms calculated for both possible dye configurations (D/A, A/D) differed by 0.1–0.8 nm and were considered as limiting values of a range of possible distances for each donor/acceptor pair. In the absence of a high-resolution structure of the YxiN helicase core, model building of full-length YxiN had to rely on a homology model for the core based on *M. jannaschii* DeaD, which showed best agreement with smFRET results. Distance restraints obtained were used to manually position the RBD (Wang *et al.*, 2006) relative to the homology model of the helicase core until the distances between the corresponding C_β atoms were within the distance range determined (Fig. 4.3C). More systematic approaches have been developed that determine positions of the dyes based on molecular modeling (reviewed in Brunger *et al.*, 2011). These approaches deliver a three-dimensional probability distribution that reflects the accuracy of the dye position by taking into account experimental errors and linker dynamics (Muschielok *et al.*, 2008; Wozniak *et al.*, 2008). While these approaches recover highly accurate molecular models (Sindbert *et al.*, 2011), they require detailed knowledge of the dye attachment geometry, reliable structural information, and a large number of distance restraints.

For YxiN, two different orientations of the RBD were consistent with the experimental restraints. In one orientation (Fig. 4.3C), the RBD is in a position on the C-terminal RecA domain that is consistent with previous data on surface accessibility of several residues (Karow and Klostermeier, 2010). This orientation is also similar to a hydrodynamic model for DbpA (Talavera *et al.*, 2006). Observed FRET efficiencies between dyes on the N-terminal RecA domain and on the RBD in the closed conformation in the presence of ADPNP and RNA further support this model. In the subsequently reported structure of the RBD in complex with RNA (Hardin *et al.*, 2010), the position of the RNA is also consistent with our model (Fig. 4.3D).

In the final model (Fig. 4.3C), the RBD of YxiN lies above a slightly concave patch on the helicase core that is formed by flexible loops on the surface of the RecA_C domain. This patch constitutes a putative protein-interaction site, according to bioinformatic predictions (Karow and Klostermeier, 2010). The orientation of the YxiN RBD differs from the position of the RBD in *T. thermophilus* Hera (Klostermeier and Rudolph, 2009; Rudolph and Klostermeier, 2009) and from the position of the C-terminal extension of Mss116p (Del Campo and Lambowitz, 2009). These differences possibly reflect distinct substrate specificities and RBD functions in these RNA helicases. Strikingly, the helicase core region covered by the YxiN-RBD is involved in complex formation between eIF4A and its binding partner eIF4G (Schutz *et al.*, 2008) and between Dpb5p and Gle1p (Montpetit *et al.*, 2011). This region on the helicase core may thus constitute a commonly used, adaptable surface for interactions of DEAD-box proteins with binding partners.

6.4. Regulating the helicase core: eIF4G activates eIF4A through a conformational guidance mechanism

Changes between the open and closed state of the helicase core are central to DEAD-box protein activities. Therefore, factors that influence the transition between open and closed states have the potential to regulate

DEAD-box protein activity. An illustrative example for this type of regulation is provided by the DEAD-box protein eIF4A. Inhibition of eIF4A can be achieved by preventing formation of the closed conformer or by preventing reopening. The tumor suppressor protein PDCD4 inhibits eIF4A, and thus translation (Yang et al., 2003) by binding to both RecA domains, preventing closure (Chang et al., 2009; Fig. 4.7A and B). Possibly, an inhibitory eIF4A-binding aptamer that contacts both RecA domains also prevents closure (Oguro et al., 2003). In contrast, components of the exon junction complex inhibit the DEAD-box protein eIF4A-III by binding to the closed state (Andersen et al., 2006; Bono et al., 2006), and prevent reopening and thus product release (Nielsen et al., 2009).

The molecular basis for activation of eIF4A by eIF4G has been less clear. The crystal structure of eIF4A shows a wide-open conformation, with an extended interdomain linker and the maximum possible separation between the RecA domains (Caruthers et al., 2000). eIF4G binds to eIF4A by contacting both RecA domains and bringing them closer together, prealigning the conserved helicase motifs toward the interdomain cleft (Schutz et al., 2008). The comparison of these crystal structures has led to the proposal that eIF4G induces a "pre-closed" conformation of eIF4A, where the helicase core is predisposed for complete closure, and the separation of the RecA domains upon reopening of the interdomain cleft is prevented ("stopper" function, Schutz et al., 2008). However, mapping of the eIF4A conformation by smFRET revealed that this wide-open conformation is not the predominant state in solution (Fig. 4.7A; Hilbert et al., 2011). Seven constructs of eIF4A with donor and acceptor fluorophores attached to cysteines in different positions in each of the RecA domains showed that eIF4A adopts a well-defined conformation in the absence of ligands. However, distance restraints yielded a model for eIF4A in solution with the two domains separated by a smaller distance than observed in the crystal structure, although too large for interdomain contacts to be formed (Hilbert et al., 2011). The model was generated starting from the eIF4A conformation in complex with eIF4G by manually moving the N-terminal RecA domain away from the C-terminal RecA domain, such that the distance changes measured by FRET between the free and eIF4G-bound eIF4A were fulfilled. This approach minimized uncertainties arising from the length and flexibility of the linkers. In addition, interdye distances for the seven donor/acceptor-labeled eIF4A constructs in complex with eIF4G (Hilbert et al., 2011) were in good agreement with the eIF4A/eIF4G crystal structure (Schutz et al., 2008), rendering the structure a good starting model (Fig. 4.7A). The FRET model (Hilbert et al., 2011) for the open state demonstrated that the "stopper function" was not the primary effect of eIF4G. Instead, the relative alignment of the two domains appeared to be critical. The eIF4A/eIF4G complex is formed by two interfaces, a primary interface involving RecA_C and a secondary interface involving RecA_N

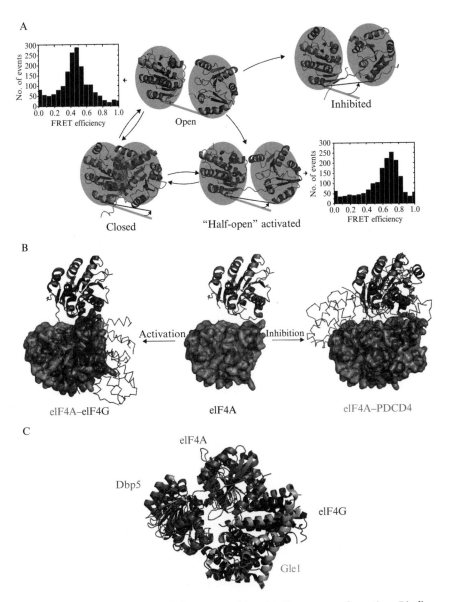

Figure 4.7 Regulation of the helicase core through effects on conformation. Binding partners of eIF4A regulate its activity by affecting the conformation of the helicase core. (A) Regulation of eIF4A activity by eIF4G and PDCD4 via conformational changes. eIF4A is in an open conformation in the absence of ligands (Hilbert et al., 2011) with the two domains closer than in the crystal structure (Schutz et al., 2008). In the presence of eIF4G, eIF4A is stabilized in a more compact, "half-open" conformation. In the presence of RNA and ATP, the helicase core closes (model based on the structure of eIF4A-III in the presence of RNA and nucleotide, PDB-ID 2hyi). Binding of the tumor suppressor programmed cell death protein 4 (PDCD4) stabilizes an eIF4A conformation with the

(Schutz et al., 2008). The secondary interface does not contribute to overall complex stability (Hilbert et al., 2011; Schutz et al., 2008). Its importance, however, is evident from smFRET experiments with eIF4A and eIF4G variants carrying mutations in the secondary interface. Despite complex formation via the primary interface, these mutants were unable to stimulate the eIF4A RNA-dependent ATPase activity and failed to induce the eIF4A conformational change (Hilbert et al., 2011). The observed changes in interdye distances for these mutants upon binding to eIF4G paralleled the degree of ATPase stimulation, clearly linking these two events. Interestingly, one eIF4A mutant bound to eIF4G with higher affinity, but its ATPase activity was not efficiently stimulated by eIF4G, indicating that a strong interaction is inhibitory. Thus, formation of the secondary interface is crucial for the stimulation of eIF4A activity by eIF4G. The transient character of the interface is essential for complex flexibility and function.

Inspection of the eIF4G-stabilized conformation of eIF4A showed that the nucleotide-binding pocket is accessible. Accordingly, nucleotide exchange and phosphate release, the rate-limiting step of the eIF4A-nucleotide cycle, are accelerated in the eIF4A/eIF4G complex (Hilbert et al., 2011). This suggests that stimulation by eIF4G is based on an increase in the rate of product release. To take this effect into account, we termed the eIF4G-bound conformation of eIF4A "half-open." Recently, it was shown that eIF4G also accelerates RNA release from eIF4A (Montpetit et al., 2011). In total, eIF4G imposes a smaller conformational change on eIF4A in solution than expected from crystal structures. In addition, eIF4G affects multiple steps of the eIF4A catalytic cycle. Collectively, these results suggest that eIF4G plays an active role in modulating eIF4A function via a conformational guidance mechanism, where eIF4G guides the transition of eIF4A between the "half-open" and closed conformation. A highly similar structure of the Dbp5p/Gle1p complex (Montpetit et al., 2011) suggests that this mechanism may be widespread for the regulation of DEAD-box proteins (Fig. 4.7C).

6.5. Variations of a theme: Helicase modules

Helicases cooperate with other proteins or occur as modules in enzymes involved in genome stability (van Brabant et al., 2000), for example, in chromatin remodeling enzymes, restriction/modification enzymes, and

---

two RecA domains closer than in the open conformation (PDB-ID 2zu6). PDCD4 binding has also been suggested to prevent closure (Chang et al., 2009). (B) Molecular models of free (green), activated (red), and inhibited eIF4A (blue), superimposed on the C-terminal RecA domain, highlighting the conformational changes. The N-terminal domain RecA domain is shown in surface representation, the C-terminal domain in cartoon representation. eIF4G (red) and PDCD4 (green) are depicted in ribbon representation. (C) Superposition of the eIF4A–eIF4G (green/red, PDB-ID 2vso) and the Dbp5p-Gle1$_{InsP6}$ (blue/orange, PDB-ID 3rrm) complexes. (See the Color Insert.)

topoisomerases. Only in selected cases has the role of the helicase module been studied in detail. In type I restriction/modification enzymes, the helicase module exhibits ATP-dependent DNA translocase activity and guides the recognition of target sites for restriction and methyl transferase activities (Stanley et al., 2006). Reverse gyrase, the only topoisomerase capable of introducing positive supercoils into DNA at the expense of ATP hydrolysis (Kikuchi and Asai, 1984), consists of an N-terminal helicase module fused to a C-terminal topoisomerase domain (Confalonieri et al., 1993). The isolated topoisomerase domain catalyzes DNA relaxation (Declais et al., 2000). The helicase domain contains insertions compared to the canonical DEAD-box helicase core that modulate its activity (Del Toro Duany et al., 2011; Ganguly et al., 2011). The isolated helicase domain undergoes a nucleotide-driven conformational cycle similar to DEAD-box proteins, but reopening of the cleft between the RecA domains occurs upon ATP hydrolysis (Del Toro Duany and Klostermeier, 2011; Del Toro Duany et al., 2011). The helicase module thus plays a central role in reverse gyrase activity by conferring a nucleotide-dependent DNA-binding site (del Toro Duany et al., 2008), locally separating the DNA strands, and presenting the single strands to the topoisomerase domain for strand passage (Del Toro Duany and Klostermeier, 2011, unpublished).

7. Perspective

Conformational changes are at the heart of DEAD-box protein function. smFRET studies of DEAD-box helicases in solution have delineated the conformational changes in the catalytic cycle and have identified the steps that trigger closure and reopening of the interdomain cleft in the helicase core. Reopening coincides with phosphate release, which has been identified as the rate-limiting step in the catalytic cycle of the DEAD-box protein DbpA (Henn et al., 2008). Possibly, the conformational change itself constitutes the rate-limiting step. The kinetics of interconversion between the open and closed states will have to be addressed in future studies using single molecule approaches on immobilized helicases. Due to the local action of DEAD-box proteins, the frequently applied strategy of indirect immobilization by binding to surface-coupled nucleic acids is only of limited value, necessitating reliable schemes for protein immobilization.

Little is known about RNA conformational changes during helicase-mediated rearrangements. Structures of DEAD-box proteins bound to small model RNAs have suggested local conformational changes in the RNA substrate that have not yet been probed experimentally. The folding of group-II introns has been studied in the absence and presence of the RNA chaperone Mss116p (Karunatilaka et al., 2010), identifying large conformational changes in the RNA and steps assisted by the helicase. The identification of *in vivo* substrates will be instrumental in rigorously studying effects of DEAD-box proteins on RNA structure.

DEAD-box proteins functionally cooperate with additional domains and other protein partners and are usually integrated into extensive functional networks, which are difficult to dissect. smFRET is an invaluable tool for studying conformational changes within large, flexible, multidomain, or multicomponent entities that are not amenable to other structural approaches. With the continuous advancement of labeling and preparation techniques, the extension to multidye experiments, and the development and refinement of analysis procedures, more insight into DEAD-box protein mechanism and the modulation of their function within networks is expected.

ACKNOWLEDGMENTS

Helicase work in the author's laboratory was funded by the Volkswagen Stiftung, the Swiss National Science Foundation (SNSF), the German Research Foundation (DFG), the Roche Research Foundation, and the Verband der Chemischen Industrie. We thank Airat Gubaev and Markus Rudolph for comments on the chapter and Airat Gubaev for help in figure preparation.

REFERENCES

Adam, H. (1962). Methoden der Enzymatischen Analyse. Bergmeyer, Weinheim, Germany.
Andersen, C. B., Ballut, L., Johansen, J. S., Chamieh, H., Nielsen, K. H., Oliveira, C. L., Pedersen, J. S., Seraphin, B., Le Hir, H., and Andersen, G. R. (2006). Structure of the exon junction core complex with a trapped DEAD-box ATPase bound to RNA. *Science* **313,** 1968–1972.
Antonik, M., Felekyan, S., Gaiduk, A., and Seidel, C. A. (2006). Separating structural heterogeneities from stochastic variations in fluorescence resonance energy transfer distributions via photon distribution analysis. *J. Phys. Chem. B* **110,** 6970–6978.
Aregger, R., and Klostermeier, D. (2009). The DEAD-box helicase YxiN maintains a closed conformation during ATP hydrolysis. *Biochemistry* **48,** 10679–10681.
Bigay, J., Deterre, P., Pfister, C., and Chabre, M. (1987). Fluoride complexes of aluminium or beryllium act on G-proteins as reversibly bound analogues of the gamma phosphate of GTP. *EMBO J.* **6,** 2907–2913.
Bono, F., Ebert, J., Lorentzen, E., and Conti, E. (2006). The crystal structure of the exon junction complex reveals how it maintains a stable grip on mRNA. *Cell* **126,** 713–725.
Breukink, E., Nouwen, N., van Raalte, A., Mizushima, S., Tommassen, J., and de Kruijff, B. (1995). The C terminus of SecA is involved in both lipid binding and SecB binding. *J. Biol. Chem.* **270,** 7902–7907.
Brochon, J. C. (1994). Maximum entropy method of data analysis in time-resolved spectroscopy. *Methods Enzymol.* **240,** 262–311.
Brunger, A. T., Strop, P., Vrljic, M., Chu, S., and Weninger, K. R. (2011). Three-dimensional molecular modeling with single molecule FRET. *J. Struct. Biol.* **173,** 497–505.

Caruthers, J. M., Johnson, E. R., and McKay, D. B. (2000). Crystal structure of yeast initiation factor 4A, a DEAD-box RNA helicase. *Proc. Natl. Acad. Sci. U. S. A.* **97,** 13080–13085.

Chabre, M. (1990). Aluminofluoride and beryllofluoride complexes: A new phosphate analogs in enzymology. *Trends Biochem. Sci.* **15,** 6–10.

Chang, J. H., Cho, Y. H., Sohn, S. Y., Choi, J. M., Kim, A., Kim, Y. C., Jang, S. K., and Cho, Y. (2009). Crystal structure of the eIF4A-PDCD4 complex. *Proc. Natl. Acad. Sci. U. S. A.* **106,** 3148–3153.

Chen, Y., Potratz, J. P., Tijerina, P., Del Campo, M., Lambowitz, A. M., and Russell, R. (2008). DEAD-box proteins can completely separate an RNA duplex using a single ATP. *Proc. Natl. Acad. Sci. U. S. A.* **105,** 20203–20208.

Collins, R., Karlberg, T., Lehtio, L., Schutz, P., van den Berg, S., Dahlgren, L. G., Hammarstrom, M., Weigelt, J., and Schuler, H. (2009). The DEXD/H-box RNA helicase DDX19 is regulated by an α-helical switch. *J. Biol. Chem.* **284,** 10296–10300.

Combet, C., Jambon, M., Deleage, G., and Geourjon, C. (2002). Geno3D: Automatic comparative molecular modelling of protein. *Bioinformatics* **18,** 213–214.

Confalonieri, F., Elie, C., Nadal, M., de La Tour, C., Forterre, P., and Duguet, M. (1993). Reverse gyrase: A helicase-like domain and a type I topoisomerase in the same polypeptide. *Proc. Natl. Acad. Sci. U. S. A.* **90,** 4753–4757.

Dahan, M., Deniz, A. A., Ha, T. J., Chemla, D. S., Schultz, P. G., and Weiss, S. (1999). Ratiometric measurement and identification of single diffusing molecules. *Chem. Phys.* **247,** 85–106.

Declais, A. C., Marsault, J., Confalonieri, F., de La Tour, C. B., and Duguet, M. (2000). Reverse gyrase, the two domains intimately cooperate to promote positive supercoiling. *J. Biol. Chem.* **275,** 19498–19504.

Del Campo, M., and Lambowitz, A. M. (2009). Structure of the Yeast DEAD-box protein Mss116p reveals two wedges that crimp RNA. *Mol. Cell* **35,** 598–609.

Del Toro Duany, Y., and Klostermeier, D. (2011). Nucleotide-driven conformational changes in the reverse gyrase helicase-like domain couple the nucleotide cycle to DNA processing. *Phys. Chem. Chem. Phys.* **13,** 10009–10019.

del Toro Duany, Y., Jungblut, S. P., Schmidt, A. S., and Klostermeier, D. (2008). The reverse gyrase helicase-like domain is a nucleotide-dependent switch that is attenuated by the topoisomerase domain. *Nucleic Acids Res.* **36,** 5882–5895.

Del Toro Duany, Y., Klostermeier, D., and Rudolph, M. G. (2011). The conformational flexibility of the helicase-like domain from Thermotoga maritima reverse gyrase is restricted by the topoisomerase domain. *Biochemistry* **50,** 5816–5823.

Diges, C. M., and Uhlenbeck, O. C. (2001). Escherichia coli DbpA is an RNA helicase that requires hairpin 92 of 23S rRNA. *EMBO J.* **20,** 5503–5512.

Diges, C. M., and Uhlenbeck, O. C. (2005). Escherichia coli DbpA is a 3′–>5′ RNA helicase. *Biochemistry* **44,** 7903–7911.

Ellman, J., Mendel, D., Anthonycahill, S., Noren, C. J., and Schultz, P. G. (1991). Biosynthetic method for introducing unnatural amino-acids site-specifically into proteins. *Methods Enzymol.* **202,** 301–336.

Fan, J. S., Cheng, Z., Zhang, J., Noble, C., Zhou, Z., Song, H., and Yang, D. (2009). Solution and crystal structures of mRNA exporter Dbp5p and its interaction with nucleotides. *J. Mol. Biol.* **388,** 1–10.

Fisher, A. J., Smith, C. A., Thoden, J. B., Smith, R., Sutoh, K., Holden, H. M., and Rayment, I. (1995). X-ray structures of the myosin motor domain of Dictyostelium discoideum complexed with MgADP·BeF$_x$ and MgADP·AlF$_4$. *Biochemistry* **34,** 8960–8972.

Förster, T. (1948). Intermolecular energy migration and fluorescence test. *Ann. Phys.* **2,** 55–75.

Fuller-Pace, F. V., Nicol, S. M., Reid, A. D., and Lane, D. P. (1993). DbpA: A DEAD-box protein specifically activated by 23s rRNA. *EMBO J.* **12**, 3619–3626.

Ganguly, A., Del Toro Duany, Y., Rudolph, M. G., and Klostermeier, D. (2011). The latch modulates nucleotide and DNA binding to the helicase-like domain of Thermotoga maritima reverse gyrase and is required for positive DNA supercoiling. *Nucleic Acids Res.* **39**, 1789–1800.

Gopich, I., and Szabo, A. (2005). Theory of photon statistics in single-molecule Forster resonance energy transfer. *J. Chem. Phys.* **122**, 14707.

Grohman, J. K., Del Campo, M., Bhaskaran, H., Tijerina, P., Lambowitz, A. M., and Russell, R. (2007). Probing the mechanisms of DEAD-box proteins as general RNA chaperones: The C-terminal domain of CYT-19 mediates general recognition of RNA. *Biochemistry* **46**, 3013–3022.

Gubaev, A., Hilbert, M., and Klostermeier, D. (2009). The DNA-gate of Bacillus subtilis gyrase is predominantly in the closed conformation during the DNA supercoiling reaction. *Proc. Natl. Acad. Sci. U. S. A.* **106**, 13278–13283.

Haas, E., Katchalskikatzir, E., and Steinberg, I. Z. (1978). Effect of orientation of donor and acceptor on probability of energy-transfer involving electronic-transitions of mixed polarization. *Biochemistry* **17**, 5064–5070.

Hardin, J. W., Hu, Y. X., and McKay, D. B. (2010). Structure of the RNA binding domain of a DEAD-box helicase bound to its ribosomal RNA target reveals a novel mode of recognition by an RNA recognition motif. *J. Mol. Biol.* **402**, 412–427.

Henn, A., Cao, W., Hackney, D. D., and De La Cruz, E. M. (2008). The ATPase cycle mechanism of the DEAD-box rRNA helicase, DbpA. *J. Mol. Biol.* **377**, 193–205.

Henn, A., Cao, W., Licciardello, N., Heitkamp, S. E., Hackney, D. D., and De La Cruz, E. M. (2010). Pathway of ATP utilization and duplex rRNA unwinding by the DEAD-box helicase, DbpA. *Proc. Natl. Acad. Sci. U. S. A.* **107**, 4046–4050.

Hilbert, M., Karow, A. R., and Klostermeier, D. (2009). The mechanism of ATP-dependent RNA unwinding by DEAD-box proteins. *Biol. Chem.* **390**, 1237–1250.

Hilbert, M., Kebbel, F., Gubaev, A., and Klostermeier, D. (2011). eIF4G stimulates the activity of the DEAD-box protein eIF4A by a conformational guidance mechanism. *Nucleic Acids Res.* **39**, 2260–2270.

Iost, I., Dreyfus, M., and Linder, P. (1999). Ded1p, a DEAD-box protein required for translation initiation in Saccharomyces cerevisiae, is an RNA helicase. *J. Biol. Chem.* **274**, 17677–17683.

Jackson, R. J., Hellen, C. U., and Pestova, T. V. (2010). The mechanism of eukaryotic translation initiation and principles of its regulation. *Nat. Rev. Mol. Cell Biol.* **11**, 113–127.

Jankowsky, E., and Bowers, H. (2006). Remodeling of ribonucleoprotein complexes with DExH/D RNA helicases. *Nucleic Acids Res.* **34**, 4181–4188.

Karow, A. R., and Klostermeier, D. (2009). A conformational change in the helicase core is necessary but not sufficient for RNA unwinding by the DEAD-box helicase YxiN. *Nucleic Acids Res.* **37**, 4464–4471.

Karow, A. R., and Klostermeier, D. (2010). A structural model for the DEAD-box helicase YxiN in solution: Localization of the RNA binding domain. *J. Mol. Biol.* **402**, 629–637.

Karow, A. R., Theissen, B., and Klostermeier, D. (2007). Authentic interdomain communication in an RNA helicase reconstituted by expressed protein ligation of two helicase domains. *FEBS J.* **274**, 463–473.

Karunatilaka, K. S., Solem, A., Pyle, A. M., and Rueda, D. (2010). Single-molecule analysis of Mss116-mediated group II intron folding. *Nature* **467**, 935–939.

Kikuchi, A., and Asai, K. (1984). Reverse gyrase—A topoisomerase which introduces positive superhelical turns into DNA. *Nature* **309,** 677–681.

Klostermeier, D., and Millar, D. P. (2001). Time-resolved fluorescence resonance energy transfer: A versatile tool for the analysis of nucleic acids. *Biopolymers* **61,** 159–179.

Klostermeier, D., and Rudolph, M. G. (2009). A novel dimerization motif in the C-terminal domain of the Thermus thermophilus DEAD-box helicase Hera confers substantial flexibility. *Nucleic Acids Res.* **37,** 421–430.

Kossen, K., Karginov, F. V., and Uhlenbeck, O. C. (2002). The carboxy-terminal domain of the DExDH protein YxiN is sufficient to confer specificity for 23S rRNA. *J. Mol. Biol.* **324,** 625–636.

Lambowitz, A. M., and Del Campo, M. (2009). Structure of the yeast DEAD-box protein Mss116p reveals two wedges that crimp RNA. *Mol. Cell* **35,** 598–609.

Lemke, E. A. (2011). Site-specific labeling of proteins for single-molecule FRET measurements using genetically encoded ketone functionalities. *Methods Mol. Biol.* **751,** 3–15.

Linden, M. H., Hartmann, R. K., and Klostermeier, D. (2008). The putative RNase P motif in the DEAD-box helicase Hera is dispensable for efficient interaction with RNA and helicase activity. *Nucleic Acids Res.* **36,** 5800–5811.

Linder, P., and Jankowsky, E. (2011). From unwinding to clamping—The DEAD-box RNA helicase family. *Nat. Rev. Mol. Cell Biol.* **12,** 505–516.

Liu, F., Putnam, A., and Jankowsky, E. (2008). ATP hydrolysis is required for DEAD-box protein recycling but not for duplex unwinding. *Proc. Natl. Acad. Sci. U. S. A.* **105,** 20209–20214.

Lorsch, J. R., and Herschlag, D. (1998a). The DEAD-box protein eIF4A. 1. A minimal kinetic and thermodynamic framework reveals coupled binding of RNA and nucleotide. *Biochemistry* **37,** 2180–2193.

Lorsch, J. R., and Herschlag, D. (1998b). The DEAD-box protein eIF4A. 2. A cycle of nucleotide and RNA-dependent conformational changes. *Biochemistry* **37,** 2194–2206.

Magde, D., Wong, R., and Seybold, P. G. (2002). Fluorescence quantum yields and their relation to lifetimes of rhodamine 6G and fluorescein in nine solvents: Improved absolute standards for quantum yields. *Photochem. Photobiol.* **75,** 327–334.

Mallam, A. L., Jarmoskaite, I., Tijerina, P., Del Campo, M., Seifert, S., Guo, L., Russell, R., and Lambowitz, A. M. (2011). Solution structures of DEAD-box RNA chaperones reveal conformational changes and nucleic acid tethering by a basic tail. *Proc. Natl. Acad. Sci. U. S. A.* **108,** 12254–12259.

Marintchev, A., Edmonds, K. A., Marintcheva, B., Hendrickson, E., Oberer, M., Suzuki, C., Herdy, B., Sonenberg, N., and Wagner, G. (2009). Topology and regulation of the human eIF4A/4G/4H helicase complex in translation initiation. *Cell* **136,** 447–460.

Montpetit, B., Thomsen, N. D., Helmke, K. J., Seeliger, M. A., Berger, J. M., and Weis, K. (2011). A conserved mechanism of DEAD-box ATPase activation by nucleoporins and InsP6 in mRNA export. *Nature* **472,** 238–242.

Morlang, S., Weglohner, W., and Franceschi, F. (1999). Hera from Thermus thermophilus: The first thermostable DEAD-box helicase with an RNase P protein motif. *J. Mol. Biol.* **294,** 795–805.

Muschielok, A., Andrecka, J., Jawhari, A., Bruckner, F., Cramer, P., and Michaelis, J. (2008). A nano-positioning system for macromolecular structural analysis. *Nat. Methods* **5,** 965–971.

Napetschnig, J., Kassube, S. A., Debler, E. W., Wong, R. W., Blobel, G., and Hoelz, A. (2009). Structural and functional analysis of the interaction between the nucleoporin Nup214 and the DEAD-box helicase Ddx19. *Proc. Natl. Acad. Sci. USA* **106,** 3089–3094.

Nielsen, K. H., Chamieh, H., Andersen, C. B., Fredslund, F., Hamborg, K., Le Hir, H., and Andersen, G. R. (2009). Mechanism of ATP turnover inhibition in the EJC. *RNA* **15**, 67–75.

Nir, E., Michalet, X., Hamadani, K. M., Laurence, T. A., Neuhauser, D., Kovchegov, Y., and Weiss, S. (2006). Shot-noise limited single-molecule FRET histograms: Comparison between theory and experiments. *J. Phys. Chem. B* **110**, 22103–22124.

Oguro, A., Ohtsu, T., Svitkin, Y. V., Sonenberg, N., and Nakamura, Y. (2003). RNA aptamers to initiation factor 4A helicase hinder cap-dependent translation by blocking ATP hydrolysis. *RNA* **9**, 394–407.

Parker, C. A., and Rees, W. T. (1960). Correction of fluorescence spectra and measurement of fluorescence quantum efficiency. *Analyst* **85**, 587–600.

Pause, A., and Sonenberg, N. (1992). Mutational analysis of a DEAD-box RNA helicase: The mammalian translation initiation factor eIF-4A. *EMBO J.* **11**, 2643–2654.

Peck, M. L., and Herschlag, D. (1999). Effects of oligonucleotide length and atomic composition on stimulation of the ATPase activity of translation initiation factor eIF4A. *RNA* **5**, 1210–1221.

Rogers, G. W., Jr., Richter, N. J., and Merrick, W. C. (1999). Biochemical and kinetic characterization of the RNA helicase activity of eukaryotic initiation factor 4A. *J. Biol. Chem.* **274**, 12236–12244.

Rogers, G. W., Jr., Komar, A. A., and Merrick, W. C. (2002). eIF4A: The godfather of the DEAD-box helicases. *Prog. Nucleic Acid Res. Mol. Biol.* **72**, 307–331.

Rudolph, M. G., and Klostermeier, D. (2009). The Thermus thermophilus DEAD-box helicase Hera contains a modified RNA recognition motif domain loosely connected to the helicase core. *RNA* **15**, 1993–2001.

Schuler, B., and Pannell, L. K. (2002). Specific labeling of polypeptides at amino-terminal cysteine residues using Cy5-benzyl thioester. *Bioconjug. Chem.* **13**, 1039–1043.

Schutz, P., Bumann, M., Oberholzer, A. E., Bieniossek, C., Trachsel, H., Altmann, M., and Baumann, U. (2008). Crystal structure of the yeast eIF4A-eIF4G complex: An RNA-helicase controlled by protein–protein interactions. *Proc. Natl. Acad. Sci. USA* **105**, 9564–9569.

Schwede, T., Kopp, J., Guex, N., and Peitsch, M. C. (2003). SWISS-MODEL: An automated protein homology-modeling server. *Nucleic Acids Res.* **31**, 3381–3385.

Sengoku, T., Nureki, O., Nakamura, A., Kobayashi, S., and Yokoyama, S. (2006). Structural basis for RNA unwinding by the DEAD-box protein Drosophila Vasa. *Cell* **125**, 287–300.

Sharpe Elles, L. M., Sykes, M. T., Williamson, J. R., and Uhlenbeck, O. C. (2009). A dominant negative mutant of the E. coli RNA helicase DbpA blocks assembly of the 50S ribosomal subunit. *Nucleic Acids Res.* **37**, 6503–6514.

Sindbert, S., Kalinin, S., Nguyen, H., Kienzler, A., Clima, L., Bannwarth, W., Appel, B., Muller, S., and Seidel, C. A. (2011). Accurate distance determination of nucleic acids via Forster resonance energy transfer: Implications of dye linker length and rigidity. *J. Am. Chem. Soc.* **133**, 2463–2480.

Sisamakis, E., Valeri, A., Kalinin, S., Rothwell, P. J., and Seidel, C. A. M. (2010). Accurate single-molecule fret studies using multiparameter fluorescence detection. *Methods Enzymol.* **475**, 455–514Single molecule tools. **474**(Pt B).

Stanley, L. K., Seidel, R., van der Scheer, C., Dekker, N. H., Szczelkun, M. D., and Dekker, C. (2006). When a helicase is not a helicase: dsDNA tracking by the motor protein EcoR124I. *EMBO J.* **25**, 2230–2239.

Story, R. M., Li, H., and Abelson, J. N. (2001). Crystal structure of a DEAD-box protein from the hyperthermophile Methanococcus jannaschii. *Proc. Natl. Acad. Sci. USA* **98**, 1465–1470.

Talavera, M. A., and De La Cruz, E. M. (2005). Equilibrium and kinetic analysis of nucleotide binding to the DEAD-box RNA helicase DbpA. *Biochemistry* **44,** 959–970.

Talavera, M. A., Matthews, E. E., Eliason, W. K., Sagi, I., Wang, J., Henn, A., and De La Cruz, E. M. (2006). Hydrodynamic characterization of the DEAD-box RNA helicase DbpA. *J. Mol. Biol.* **355,** 697–707.

Theissen, B., Karow, A. R., Kohler, J., Gubaev, A., and Klostermeier, D. (2008). Cooperative binding of ATP and RNA induces a closed conformation in a DEAD-box RNA helicase. *Proc. Natl. Acad. Sci. U. S. A.* **105,** 548–553.

van Brabant, A. J., Stan, R., and Ellis, N. A. (2000). DNA helicases, genomic instability, and human genetic disease. *Annu. Rev. Genomics Hum. Genet.* **1,** 409–459.

Volkmann, G., and Liu, X. Q. (2009). Protein C-terminal labeling and biotinylation using synthetic peptide and split-intein. *PLoS One* **4,** e8381.

von Moeller, H., Basquin, C., and Conti, E. (2009). The mRNA export protein DBP5 binds RNA and the cytoplasmic nucleoporin NUP214 in a mutually exclusive manner. *Nat. Struct. Mol. Biol.* **16,** 247–254.

Wang, S., Hu, Y., Overgaard, M. T., Karginov, F. V., Uhlenbeck, O. C., and McKay, D. B. (2006). The domain of the Bacillus subtilis DEAD-box helicase YxiN that is responsible for specific binding of 23S rRNA has an RNA recognition motif fold. *RNA* **12,** 959–967.

Wang, S., Overgaard, M. T., Hu, Y., and McKay, D. B. (2008). The Bacillus subtilis RNA helicase YxiN is distended in solution. *Biophys. J.* **94,** L01–L03.

Wozniak, A. K., Schroder, G. F., Grubmuller, H., Seidel, C. A., and Oesterhelt, F. (2008). Single-molecule FRET measures bends and kinks in DNA. *Proc. Natl. Acad. Sci. U. S. A.* **105,** 18337–18342.

Yang, H. S., Jansen, A. P., Komar, A. A., Zheng, X., Merrick, W. C., Costes, S., Lockett, S. J., Sonenberg, N., and Colburn, N. H. (2003). The transformation suppressor Pdcd4 is a novel eukaryotic translation initiation factor 4A binding protein that inhibits translation. *Mol. Cell. Biol.* **23,** 26–37.

Yang, Q., Del Campo, M., Lambowitz, A. M., and Jankowsky, E. (2007). DEAD-box proteins unwind duplexes by local strand separation. *Mol. Cell* **28,** 253–263.

CHAPTER FIVE

RNA Catalysis as a Probe for Chaperone Activity of DEAD-Box Helicases

Jeffrey P. Potratz *and* Rick Russell

Contents

1. Introduction	112
1.1. Large RNAs become trapped in nonnative conformations and fold slowly	112
1.2. DEAD-box helicase proteins function as RNA chaperones	112
2. Catalytic Activity as a Probe of RNA Folding	113
2.1. Catalytic activity distinguishes the native state from all other conformations	113
2.2. Catalytic activity can be used to study chaperone-assisted RNA folding	114
3. Self-splicing as a Readout for Native State Formation	115
3.1. Interpreting chaperone-promoted changes in observed splicing rate	117
3.2. Potential complications	118
4. Substrate Cleavage as a Readout for Native State Formation	118
4.1. Setting up a discontinuous assay: Folding and catalysis stages	120
4.2. Interpreting results from the catalysis stage	122
4.3. Using the discontinuous assay to probe chaperone-assisted RNA folding	123
5. Other Applications of the Discontinuous Assay	124
5.1. Unfolding native structure	124
5.2. Integrating results with other methods	126
Acknowledgments	126
References	127

Abstract

DEAD-box proteins are vitally important to cellular processes and make up the largest class of helicases. Many DEAD-box proteins function as RNA chaperones

Department of Chemistry and Biochemistry, Institute for Cellular and Molecular Biology, University of Texas at Austin, Austin, Texas, USA by accelerating structural transitions of RNA, which can result in the resolution of misfolded conformers or conversion between functional structures. While the biological importance of chaperone proteins is clear, their mechanisms are incompletely understood. Here, we illustrate how the catalytic activity of certain RNAs can be used to measure RNA chaperone activity. By measuring the amount of substrate converted to product, the fraction of catalytically active molecules is measured over time, providing a quantitative measure of the formation or loss of native RNA. The assays are described with references to group I and group II introns and their ribozyme derivatives, and examples are included that illustrate potential complications and indicate how catalytic activity measurements can be combined with physical approaches to gain insights into the mechanisms of DEAD-box proteins as RNA chaperones.

1. Introduction

1.1. Large RNAs become trapped in nonnative conformations and fold slowly

To carry out their cellular duties, many RNAs must fold into functional three-dimensional structures. However, the journey to the native state includes opportunities to go astray and populate inactive, misfolded structures (Russell, 2008; Treiber and Williamson, 1999). With only four standard bases, RNA is prone to forming nonnative secondary structures (Herschlag, 1995), which can persist on the time scale of hours to give "kinetically trapped" folding intermediates (Treiber and Williamson, 1999), and the same basic issue of local stability can apply to modular tertiary contacts (Russell, 2008). Slow folding to functional structure is a common problem among RNAs, as nearly every large RNA that has been studied misfolds into at least one inactive, nonnative structure (Shcherbakova *et al.*, 2008; Treiber and Williamson, 1999). In light of these intrinsic features of RNA, it is not surprising that cells can take action to rescue incorrectly folded RNAs.

1.2. DEAD-box helicase proteins function as RNA chaperones

A major way that cells promote RNA folding is through chaperone proteins, which accelerate escape from kinetically trapped intermediates (Lorsch, 2002; Mohr *et al.*, 2002; Russell, 2008; Schroeder *et al.*, 2004; Zemora and Waldsich, 2010). DEAD-box proteins are a prominent group of RNA chaperones and comprise the largest family of superfamily 2 helicases (Fairman-Williams *et al.*, 2010; Linder, 2006). They consist of a conserved helicase core made up of two RecA-like domains that may be flanked by additional domains. DEAD-box proteins bind RNA and ATP via the helicase core and are thought to promote RNA structural transitions by using the ATPase cycle to achieve tight, yet regulated RNA binding in a mode that can disrupt secondary structure (Hilbert *et al.*, 2009; Jarmoskaite and Russell, 2011; Potratz *et al.*, 2010). These disruptions are used to rescue RNAs from misfolded, nonnative structures and to convert RNAs between multiple functional structures. The prominent role played by DEAD-box proteins is highlighted by the fact that essentially every process known to be carried out by a structured RNA requires at least one DEAD-box or related protein (Fairman-Williams *et al.*, 2010; Jankowsky and Bowers, 2006; Linder, 2006, 2008). Some DEAD-box proteins have been shown to act as general RNA chaperones, interacting with multiple RNA substrates nonspecifically to promote native folding (Del Campo *et al.*, 2009; Halls *et al.*, 2007; Huang *et al.*, 2005; Mohr *et al.*, 2002).

In this chapter, we describe how to harness the power of RNA catalytic activity to study protein-assisted RNA folding. Rather than listing detailed protocols, which are likely to be different for individual proteins and RNAs, we describe the basic concepts that underlie the experimental design, and we address major issues that can complicate data analysis. For further details on basic issues that arise when using catalytic activity to measure native RNA folding, we refer readers to an earlier work (Wan *et al.*, 2010a). We also refer to examples that illustrate how catalytic activity has been used, alone and in conjunction with complementary approaches, to gain insights into the mechanisms of DEAD-box proteins as RNA chaperones.

2. Catalytic Activity as a Probe of RNA Folding

To track productive folding of an RNA, a signal must exist that allows the native state of the RNA to be distinguished from all other conformations. Functional assays provide a powerful probe because even folding intermediates with extensive native structure can be readily distinguished if they are unable to function. Catalytic RNAs are well suited for this purpose, as their native states can be readily detected by monitoring chemical conversion of a substrate to its corresponding product.

2.1. Catalytic activity distinguishes the native state from all other conformations

The central goal of using catalytic activity to monitor RNA folding is to measure the fraction of the RNA that is present in the native state unambiguously and quantitatively. This information can then be combined with other approaches to gain insights into the structural properties of folding intermediates. In the case of the ribozyme derived from a group I intron in

Tetrahymena thermophila, a misfolded conformation exists that contains the full set of native, secondary, and tertiary contacts and is difficult to distinguish from the native state by physical approaches (Russell *et al.*, 2006; Wan *et al.*, 2010b). However, it is straightforward to distinguish between the two structures in a catalytic activity assay because the natively folded ribozyme, but not the misfolded ribozyme, can cleave an oligonucleotide substrate in *trans*. Using this assay, the transition from the misfolded state to the native state has been measured (Chadee *et al.*, 2010; Russell and Herschlag, 2001; Russell *et al.*, 2006; Wan *et al.*, 2010b).

2.2. Catalytic activity can be used to study chaperone-assisted RNA folding

The ability to track native RNA folding allows the influence of chaperone proteins to be monitored. While this chapter highlights group I and group II introns and the DEAD-box helicase chaperone proteins, the general techniques can be applied to a broad range of catalytically active RNAs and chaperone proteins. Indeed, RNA catalytic activity of the hammerhead ribozyme was used nearly two decades ago to study chaperone activities of the HIV NC protein (Herschlag *et al.*, 1994; Tsuchihashi *et al.*, 1993), and group I introns have been used to monitor *in vitro* and *in vivo* chaperone activities of a variety of RNA binding proteins (Clodi *et al.*, 1999; Rajkowitsch *et al.*, 2005; Semrad and Schroeder, 1998; Semrad *et al.*, 2004).

Group I and group II introns are mobile genetic elements that catalyze their own excision from precursor RNA via two transesterification splicing reactions (Fig. 5.1A). In order for these splicing events to occur, the RNA must fold to an active three-dimensional conformation. Introns can be converted to ribozymes with constructs that lack the exons (Fig. 5.1B and C). Ribozymes of group I and group II introns cleave oligonucleotide substrates that include a 5′ splice site in reactions that mimic the first step of splicing (Kuo *et al.*, 1999; Qin and Pyle, 1997; Swisher *et al.*, 2001; Tanner and Cech, 1996; Zaug and Cech, 1986; Zaug *et al.*, 1988).

Group I and group II introns and their corresponding ribozyme constructs are well suited for studying chaperone-assisted RNA folding because they have extensive networks of secondary and tertiary structures (Fig. 5.1B and C), and RNAs from both groups have been shown to fold slowly and to accumulate intermediates (Emerick *et al.*, 1996; Fedorova *et al.*, 2007, 2010; Karunatilaka *et al.*, 2010; Russell and Herschlag, 1999; Sinan *et al.*, 2011; Steiner *et al.*, 2008; Walstrum and Uhlenbeck, 1990; Zhang *et al.*, 2005). These RNAs are sufficiently complex to include diverse sets of intermediates and corresponding kinetic barriers during folding, allowing detailed probing of the abilities of chaperones to assist in overcoming these barriers, while remaining simple enough to deeply probe the folding processes and pathways.

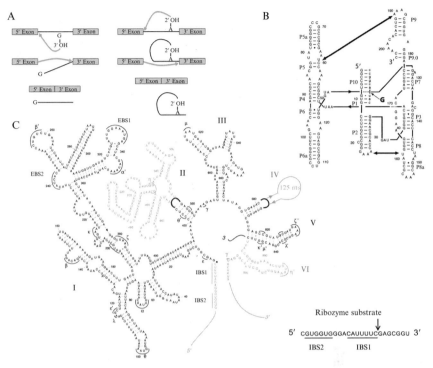

Figure 5.1 Group I and group II introns. (A) Self-splicing reactions. Left panel, group I intron splicing reaction. An exogenous guanosine attacks the 5′ splice site in the first step, and the 5′ exon attacks the 3′ splice site in the second step. Right panel, group II intron splicing reaction. A bulged adenosine near the 3′ end of the intron attacks the 5′ splice site, generating a lariat intermediate. The 5′ exon then attacks the 3′ splice site to ligate the exons together. (B) Secondary structure of the *Azoarcus* group I intron ribozyme. The nine-nucleotide substrate is lowercase. The cleavage site is indicated by a thin arrow, and two tertiary contacts are indicated by thick arrows. (C) Secondary structure of the aI5γ group II intron from *S. cerevisiae* (Swisher et al., 2001). The domains shown in black are present in the D135 ribozyme. Tertiary interactions are indicated by Greek letters. Interaction sites between exon and intron sequences are indicated by the abbreviations IBS and EBS (intron binding site and exon binding site, respectively). The 24-nt substrate is shown at the right and the cleavage site is indicated by an arrow.

3. SELF-SPLICING AS A READOUT FOR NATIVE STATE FORMATION

A straightforward way of measuring native folding is to follow self-processing of constructs that include the intron and flanking exons after folding is initiated by the addition of Mg^{2+}. The reaction can be followed in the format of a continuous assay, in which folding and the catalytic steps take place concurrently and in the same reaction. If the rate of native RNA structure formation is slower than the subsequent catalytic steps, the observed splicing rate provides a good measure of the rate of folding to the native state.

It is a common practice to body label the RNA to visualize all the products of intron splicing. This can be accomplished by including a radiolabeled nucleotide, commonly [α-^{32}P] UTP, to label nucleotides throughout the RNA (Huang et al., 2005; Mohr et al., 2006; Pyle and Green, 1994). The reaction is then monitored by denaturing polyacrylamide gel electrophoresis, which allows the unspliced precursor to be separated from the splicing products (Fig. 5.2A). The loss of precursor

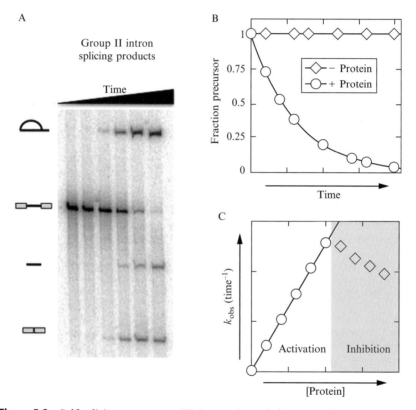

Figure 5.2 Self-splicing constructs. (A) Denaturing gel showing splicing products for a group II intron: (from top to bottom) lariat intron, unspliced precursor, linear intron, and spliced exons. (B) Simulated plot of a splicing reaction showing the fraction of precursor as a function of time. The simulated data are fit by a single exponential curve to obtain rate constants for splicing in the presence (circles) and absence (diamonds) of protein. (C) Simulated plot showing how the observed splicing rate varies with protein concentration. The rising linear portion of the data (circles) is fit with a line to obtain a second-order rate constant for chaperone-accelerated folding. The plateau and decrease in rate (diamonds in gray area) reflect inhibition by the chaperone at higher concentrations (see Section 3.1).

RNA over time is quantitated to indicate the observed rate of splicing, and in turn folding to the native state (Del Campo *et al.*, 2009; Mohr *et al.*, 2006; Potratz *et al.*, 2011) (Fig. 5.2B).

3.1. Interpreting chaperone-promoted changes in observed splicing rate

A simple reaction scheme for intron folding and splicing is shown below and is instructive for understanding how a chaperone protein affects the splicing process.

To identify whether an RNA helicase influences folding, the observed rate constant ($k_{observed}$) is compared in the absence and presence of multiple protein concentrations and in the presence and absence of ATP. If the helicase increases the rate of folding to the native state (k_{fold}), most likely by accelerating resolution of one or more intermediates (I), this increase will lead to an increase in $k_{observed}$ if the splicing rate constant (k_{splice}) is larger than the folding rate constant. For introns with robust self-splicing activity, this condition is generally met because large RNAs usually fold slowly *in vitro* (Pan and Russell, 2010; Shcherbakova *et al.*, 2008).

The observed rate constant is typically plotted as a function of protein concentration to determine the efficiency with which the chaperone promotes folding (Fig. 5.2C). The observed rate constant may increase with protein concentration at low concentrations and then level or begin to decrease at higher protein concentrations (Fig. 5.2C). Possible physical sources for this inhibition are trapping of intermediate conformations in a protein-bound state (shown in Scheme 5.1 as a single intermediate (I) for simplicity) and chaperone-induced unfolding of the native state. Although unfolding of native RNAs may be physiologically important for RNAs that must cycle between functional structures, this inhibition complicates analysis of the effects of the chaperone on the folding process toward the native state. Thus, it is most straightforward to interpret the data in terms of protein-accelerated folding at relatively low protein concentrations where these additional effects are minimized (Fig. 5.2C).

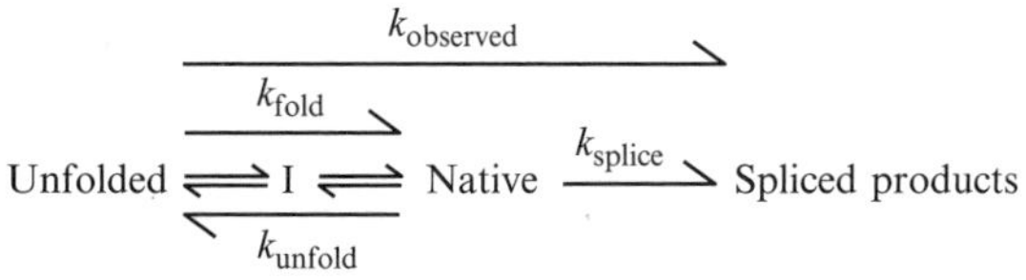

$$\text{Unfolded} \underset{k_{unfold}}{\overset{k_{fold}}{\rightleftharpoons}} \text{I} \rightleftharpoons \text{Native} \xrightarrow{k_{splice}} \text{Spliced products}$$

$$\overset{k_{observed}}{\longrightarrow}$$

Scheme 5.1 Intron folding and splicing. The intron starts from an unfolded conformation and folds through intermediates (I) to a native conformation. Once in the native state, the intron is catalytically active and splices to give products. In a continuous assay, the observed rate constant ($k_{observed}$) reflects the folding process if the rate constant for splicing (k_{splice}) is large relative to those for folding and unfolding (k_{fold} and k_{unfold}).

3.2. Potential complications

Since the chaperone is present during the splicing reaction, it could affect the catalytic steps of splicing (k_{splice}) in addition to the folding steps. If the catalytic steps (k_{splice}) are rate limiting, influencing them will alter the overall rate constant (k_{observed}) even if the protein does not affect the rate of native structure formation (k_{fold}). However, this will not be an issue provided that folding remains slower than splicing. Group I introns allow the rate-limiting step to be determined. A folding incubation is performed in the absence of the exogenous guanosine cofactor and then guanosine is added to permit splicing. If this reaction gives a larger splicing rate than the reaction in which guanosine was present continuously, it indicates that folding is rate limiting.

A second issue concerns the multistep nature of the splicing process. Although it is shown as a single step for simplicity in Scheme 5.1, some group I and group II introns display rapid, reversible first steps of splicing, giving accumulation of intermediates in a fast phase that is followed by slower completion of the second step and formation of products (Chin and Pyle, 1995; Golden and Cech, 1996; Karbstein et al., 2002; Woodson and Cech, 1989). If native folding is slower than the reversible first step but faster than the second step, the precursor will be lost with an initial rate constant that reflects folding, as in the simple case with a more rapid second step. A slow step will result in additional loss of precursor. This loss could mistakenly be assigned to an additional, slower folding pathway of the precursor, whereas it actually reflects completion of the second splicing step. This behavior can be identified by the appearance and subsequent disappearance of intermediates representing the products of the first step, that is, the free 5′ exon and the intron-3′ exon.

A final issue concerns the choice of RNA constructs. Although emphasis is typically on the intron for folding, exon length and composition can cause substantial effects (Nolte et al., 1998; Potratz et al., 2011; Zingler et al., 2010). The exons may misfold themselves and/or stabilize structure within the intron, and either of these effects may be important biologically. Care must be taken when comparing results from constructs that differ in the properties of the exons.

4. Substrate Cleavage as a Readout for Native State Formation

Many of the potential complications associated with using self-splicing constructs in continuous assays can be avoided by using ribozyme versions. Lacking exons, ribozyme constructs allow interpretation of all folding processes to reflect the intron domains. Further, although ribozyme versions can be used in continuous assays, the fact that they cleave oligonucleotide substrates in *trans* makes them well suited for discontinuous assays, in which folding and catalytic activity are separated into two discrete stages (Fig. 5.3A). The main advantage of this separation is the ability to directly assess the fraction of native ribozyme during folding. This allows productive folding to be dissected and unfolding of the native structure to be probed (Section 5.1). In addition, the discontinuous assay permits the use of folding conditions that do not support robust catalysis, an option not available using a continuous assay.

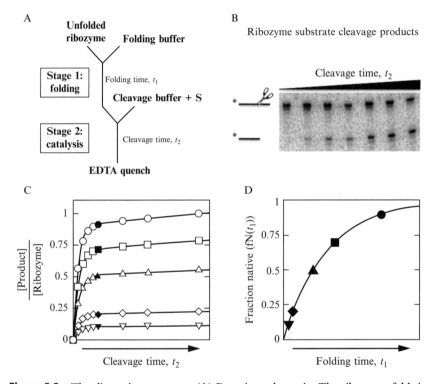

Figure 5.3 The discontinuous assay. (A) Reaction schematic. The ribozyme folds in the first stage, and then it cleaves the oligonucleotide substrate (S) in the second stage under reaction conditions that block further native folding. (B) Denaturing gel showing the results of cleavage of a 5′ labeled oligonucleotide substrate by native ribozyme. (C) Simulated plot of multiple cleavage reactions representing different folding times prior to the cleavage reaction. The curves with larger bursts represent longer folding times, giving greater accumulation of native ribozyme. The amplitude values are shown with filled symbols. (D) Simulated plot showing the fraction of native ribozyme (fN) plotted as a function of folding time (t_1). The burst amplitudes from simulated cleavage reactions in (C) are plotted as a function of folding time. In this simulated scenario, the folding progress can be fit by a single exponential function, giving a single rate constant for native state formation.

4.1. Setting up a discontinuous assay: Folding and catalysis stages

The discontinuous assay is composed of two stages, the folding stage (stage 1) and the catalysis (or cleavage) stage (stage 2) (Fig. 5.3A). The folding stage contains the ribozyme alone or with the chaperone under conditions desired for chaperone-assisted RNA folding. To prevent catalytic activity at this stage, the oligonucleotide substrate is omitted. In the catalysis stage, conditions are changed such that further folding to the native state is blocked, and radiolabeled oligonucleotide substrate is added, typically in small excess of the ribozyme (two- to threefold over the ribozyme). The cleavage reaction is allowed to proceed for various times (t_2), and the substrate and product from each time point are separated on a denaturing polyacrylamide gel (Fig. 5.3B). The fraction of product, normalized by the substrate and ribozyme concentrations, is plotted against cleavage time (t_2) (Fig. 5.3B and C). Most commonly, the cleavage stage produces a burst of product formation, with the amplitude reflecting the fraction of ribozyme in the native state (Section 4.2, below). This fraction increases as a function of time spent in the folding stage (t_1), giving a rate constant for native folding of the ribozyme (Fig. 5.3D). After initial experiments have been performed and the cleavage rate constant is known, a single time point in the cleavage reaction (stage 2) may be sufficient to determine the burst amplitude (see Fig. 5.4A). This timepoint should be chosen after completion of the burst phase, but before significant contribution from subsequent turnovers (solid symbols in Fig. 5.3C).

A key advantage of the discontinuous assay is that any set of conditions can be used for the folding stage. However, setting up the catalysis stage requires care, as conditions must support enzymatic activity while blocking further native folding. This is because ribozyme that reached the native state during stage 2 would produce cleavage products and cause the calculated fraction of native ribozyme to be artificially high.

The ability of the catalysis stage to prohibit folding is probed by comparing cleavage reactions from a ribozyme that has been prefolded to the native state and a ribozyme that is transferred directly to the catalysis stage, omitting the folding step. (In practice, testing for suitable conditions for the catalysis stage can be undertaken simultaneously with learning how to prefold ribozyme to the native state, as described below.) The burst amplitude of product from ribozyme that skipped the folding stage should be much less than that from the prefolded ribozyme, indicating little formation of native ribozyme during the cleavage reaction. It may not be possible to completely block native folding in stage 2, but a relatively small fraction of native ribozyme produced in this stage can be accounted for by normalization (Potratz et al., 2011).

Although optimal conditions for stage 2 are likely to be different for different catalytic RNAs, some general guidelines are applicable. A study

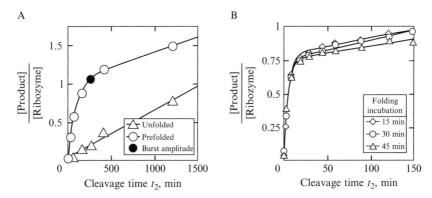

Figure 5.4 Examples of catalytic reactions with the D135 and *Azoarcus* ribozymes. (A) Identifying conditions that block folding for the catalysis stage of the discontinuous assay. Comparing burst amplitudes resulting from D135 ribozyme placed into stage 2 conditions with (circles) or without (triangles) a prior incubation in Mg^{2+}-containing buffer to allow prefolding to the native state. The burst is much smaller without the preincubation, indicating that these conditions for stage 2 (pH 8.0, 100 mM Mg^{2+}, 500 mM KCl, 15 °C) effectively block folding. The solid circle indicates how a single time point can be sufficient to determine the fraction of native ribozyme if the kinetics of cleavage are known. (B) Ribozyme prefolding to the native state. The burst amplitudes from cleavage reactions of the *Azoarcus* ribozyme are smaller than they would be for stoichiometric product formation, and further work showed that this results from an equilibrium between substrate cleavage and ligation (Sinan *et al.*, 2011). Note also that the bursts from ribozyme prefolded at 37 °C and 10 mM Mg^{2+} for 15–45 min are identical, suggesting that 15 min is sufficient for complete folding (see Section 4.1). (A) reprinted from Potratz *et al.* (2011) with permission from Elsevier. (B) adapted from Sinan *et al.* (2011). This research was originally published in *Journal of Biological Chemistry*. © The American Society for Biochemistry and Molecular Biology.

using a ribozyme engineered from the group II intron aI5γ from *Saccharomyces cerevisiae*, D135 (Fig. 5.1C), used a high pH in the catalysis stage to accelerate cleavage, and high Mg^{2+} concentration (100 mM) and low temperature (15 °C) to enhance the arrest of folding (Fig. 5.4A) (Potratz *et al.*, 2011). Cleavage is not particularly fast under these conditions (10^{-2} min^{-1}). However, slow cleavage is acceptable for the catalysis stage in a discontinuous assay as long as cleavage is significantly faster than folding. In addition, proteinase K is included at the catalysis stage to ensure that protein transferred from the folding stage has no effect on substrate cleavage.

To compare a ribozyme prefolded to the native state with an unfolded ribozyme, it must be known how to fold the ribozyme to the native state. Establishing how to fold the ribozyme to the native state can be undertaken concurrently with exploring conditions to use for the catalysis stage. Folding of a ribozyme is initiated, typically by adding Mg^{2+}, and aliquots are removed at different times from the folding stage. These aliquots are transferred to stage

2 conditions in the presence of a small excess of substrate. The product burst amplitudes are plotted against folding time. When the burst amplitude no longer increases as a function of folding time, the ribozyme has been folded to the native state as fully as it can be under that set of conditions (Fig. 5.4B). The folding time should be varied widely to determine whether there are slower folding pathways that give additional native ribozyme formation on longer time scales.

4.2. Interpreting results from the catalysis stage

To make optimal use of the discontinuous assay, it is critical to interpret the burst amplitude quantitatively in order to determine the fraction of native ribozyme. Selected examples from work involving chaperones will be covered below, and a more thorough guide to interpreting results from catalytic rate measurements was published in this series last year (Wan et al., 2010a).

The interpretation of the burst amplitudes depends on the relative rate constants of different steps in the catalytic cycle, which is shown in Scheme 5.2.

In general, the cleavage stage is performed under multiple turnover conditions, and the fraction of product is normalized by the substrate and ribozyme concentrations, giving the ratio or product to ribozyme. This ratio is plotted against cleavage time (t_2) (Fig. 5.3B and C). The reaction products are bound by base pairing, with or without additional tertiary contacts, and for many ribozymes their dissociation rate constants ($k_{P5'release}$ and $k_{P3'release}$) are small and limit the rates of subsequent turnovers. In all, there are three possible regimes: (1) both products are released quickly relative to the catalytic step ($k_{cleavage}$); (2) both are released slowly; (3) one is released quickly and the other slowly (Wan et al., 2010a). Regimes 2 and 3

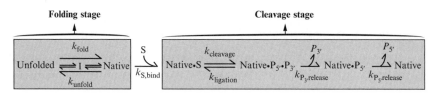

Scheme 5.2 Ribozyme folding and catalysis. In a discontinuous assay, an unfolded ribozyme is introduced to the folding stage where it can fold through intermediates (I) to a native state. Next, the ribozyme is transferred to the cleavage stage, where it can bind an oligonucleotide substrate (S). Because the conditions of the cleavage stage are chosen to block further native folding, only the ribozyme molecules that reached the native state during the folding stage are able to cleave the substrate, leaving two smaller products (native·$P_{5'}$·$P_{3'}$). These products then dissociate from the ribozyme and allow it to bind and react with another substrate molecule.

result in bursts of product formation, and the amplitude of this burst can be used to calculate the fraction of native ribozyme, as described in two examples.

The D135 ribozyme was used in a catalytic reaction with threefold excess substrate. When performed with the ribozyme prefolded to the native state, the cleavage stage resulted in a burst of product with an amplitude approximately equal to one turnover of the ribozyme (Fig. 5.4A) (Potratz et al., 2011). This is consistent with regime 3, in which there is a burst of product formation with a rate constant equal to the cleavage rate ($k_{cleavage}$) and a subsequent linear phase dictated by the slow release of one of the products. Because the 5' product of the oligonucleotide substrate forms 12 Watson–Crick base pairs with the ribozyme, it is likely to be the product that is released slowly. Under this reaction regime, the amplitude of the burst phase is equal to the fraction of D135 ribozyme folded to the native state.

The ribozyme engineered from a group I intron from *Azoarcus evansii* (Fig. 5.1B) was used in a catalytic reaction with twofold excess substrate (Sinan et al., 2011). When the ribozyme was prefolded to the native state, the cleavage stage resulted in a burst of product with an amplitude of half to three-fourths of the ribozyme concentration, depending on solution conditions (Fig. 5.4B). This is consistent with regime 2, in which the slow release of both products allows equilibrium between the cleavage ($k_{cleavage}$) and the ligation ($k_{ligation}$) to be reached. The burst amplitude is smaller than the fraction of native ribozyme and must be corrected by the value of the internal equilibrium to reveal the fraction of native ribozyme.

4.3. Using the discontinuous assay to probe chaperone-assisted RNA folding

When optimal conditions for stage 2 are established and the relationship of the burst amplitude to the fraction of native ribozyme is understood, the discontinuous assay can provide important insights into the chaperone-assisted RNA folding. The most straightforward experiment is to compare the folding reaction in the presence of various concentrations of chaperone, plotting the fraction of native ribozyme against time (t_1) (Fig. 5.3D). A chaperone may increase the rate of native state formation (Sinan et al., 2011; Tijerina et al., 2006), an effect that most likely arises from accelerated resolution of one or more kinetically trapped intermediates. This interpretation is particularly clear for RNAs that are known to misfold (Pan and Woodson, 1998; Russell and Herschlag, 1999; Sinan et al., 2011; Tijerina et al., 2006). Further confirmation can be obtained by allowing the RNA to misfold first and then adding the chaperone (Sinan et al., 2011; Tijerina et al., 2006).

It is possible for a chaperone to increase the fraction of ribozyme that reaches the native state rapidly without having a significant impact on the observed rate constant (Fig. 5.5A and B) (Potratz et al., 2011). This effect may arise from an influence of the chaperone early in folding, which decreases the probability of misfolding at later folding steps. Alternatively, resolution of an intermediate can become sufficiently fast that this pathway is then indistinguishable from pathways that avoid the intermediate.

Analogous to effects on self-splicing constructs, higher protein concentrations can inhibit folding of ribozymes to the native state (Fig. 5.5C). The physical processes responsible for inhibition are presumably the same for ribozymes and self-splicing constructs, but the discontinuous assay allows the origin of the inhibition to be distinguished. For self-splicing constructs, inhibition by trapping protein-bound nonfunctional intermediates and by unfolding native RNA both lead to a decrease in the observed splicing rate (Fig. 5.2C). In a discontinuous assay, unfolding of native ribozyme decreases the endpoint of the folding curve (Fig. 5.5C). Trapping of folding intermediates, without unfolding of natively folded RNA, would result in a decrease in the rate of native ribozyme formation but not a decrease in the endpoint.

5. OTHER APPLICATIONS OF THE DISCONTINUOUS ASSAY

The discontinuous assay is amenable to a diverse set of experiments. It can be used to monitor a decrease in the native ribozyme, and an additional stage can be included to probe the role of ATP in chaperone-mediated folding (Halls et al., 2007; Potratz et al., 2011). Further, the progress of native ribozyme formation obtained from the assay can also complement insight from other powerful physical approaches, and thus provide a more complete understanding of the action of RNA chaperones.

5.1. Unfolding native structure

Many chaperones function nonspecifically and are capable of disrupting the native states of RNAs as well as misfolded states (Bhaskaran and Russell, 2007). For probing the mechanisms of chaperone activity in structure disruptions, it can be very useful to monitor the native state because it is relatively homogeneous and the structure may be known. In contrast, folding intermediates may be heterogeneous and their structures poorly defined.

To monitor loss of the native ribozyme, the ribozyme is first prefolded to the native state, and then the chaperone protein is added. Even for general chaperones, the level of activity may be reduced for the native structure

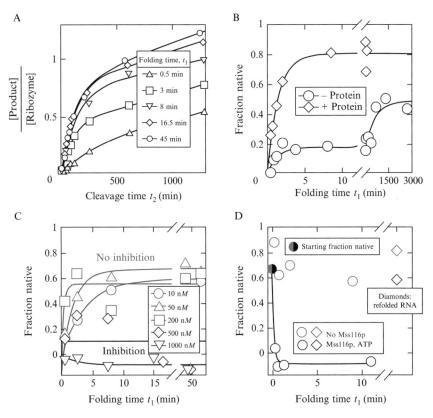

Figure 5.5 The discontinuous assay with the D135 ribozyme and the DEAD-box helicase Mss116p. (A) Cleavage time courses initiated with ribozyme folded in the presence of Mss116p for the times indicated. The increase of the burst amplitude with folding time (t_1) indicates productive folding to the native state. (B) Comparison of ribozyme folding in the presence (diamonds) and absence (circles) of Mss116p. The protein changes the folding profile from multiphasic to a single exponential phase. (C) High Mss116p concentrations inhibit accumulation of the native conformation. At 500 nM (diamonds) and 1000 nM (inverted triangles) Mss116p, the fraction of native ribozyme at steady state is lower than in the presence of lower protein concentrations (gray symbols). (D) The native conformation of D135 can be unfolded by Mss116p. Ribozyme prefolded to the native state (black/gray solid circle) is divided into reaction tubes containing a high concentration of Mss116p (black circles) or no Mss116p (gray circles). After ~10 min, the protein is proteolyzed and the ribozyme is treated to allow it to refold to the native state (black and gray diamonds), demonstrating that the loss of native ribozyme arises from unfolding and not an irreversible process. (A)–(D) adapted from Potratz et al. (2011) with permission from Elsevier.

because it is typically highly stable, so it may be necessary to lower the Mg^{2+} concentration and/or to use relatively high protein concentrations to detect net unfolding. A decrease in the fraction of native ribozyme over time indicates that the chaperone protein has mediated at least partial unfolding of the native ribozyme, giving intermediates that do not readily refold to the native state upon transfer to the stage 2 conditions. To ensure that the loss of native ribozyme reflects reversible unfolding, the protein should be proteolyzed, and the ribozyme again folded to the native state (Fig. 5.5D) (Potratz et al., 2011).

5.2. Integrating results with other methods

The materials and methods required for RNA catalytic activity assays are standard for many laboratories, and the ease of implementing this method makes it a convenient tool for obtaining a kinetic view of the fraction of native ribozyme. Results can be highly complementary to those from methods that provide physical information on folding intermediates.

Three physical probes that have been used extensively for RNA folding studies are chemical footprinting, small angle X-ray scattering (SAXS), and single-molecule FRET. Time-resolved chemical footprinting can be performed with several different probes and provides a highly specific view of nucleotides that are engaged in secondary or tertiary contacts during a folding process (King et al., 1993; Mortimer and Weeks, 2007, 2008, 2009; Sclavi et al., 1997; Tijerina et al., 2007). Hydroxyl radical footprinting has been particularly valuable for probing structured intermediates and elucidating folding pathways (Laederach et al., 2006, 2007; Sclavi et al., 1998). The orthogonal information provided by catalytic activity measurements—how much of the ribozyme is in the native state—is tremendously valuable because it can be used to place constraints on the folding pathways modeled from footprinting data (Laederach et al., 2007; Mitra et al., 2011). SAXS provides rich information on the overall size and shape of RNA as it folds, which is highly complementary to footprinting (Kwok et al., 2006; Schlatterer and Brenowitz, 2009), and again coordinated activity measurements under the same conditions can assist greatly in constraining physical descriptions of intermediates (Kwok et al., 2006; Laederach et al., 2007; Mitra et al., 2011; Roh et al., 2010; Sinan et al., 2011). Last, single-molecule FRET experiments are uniquely powerful for detecting and characterizing intermediates that do not accumulate in bulk experiments (Bokinsky and Zhuang, 2005; Bokinsky et al., 2003; Karunatilaka et al., 2010; Rueda et al., 2004; Zhuang, 2005; Zhuang et al., 2000), and the concurrent detection of catalytic activity can be used to tremendous advantage for distinguishing the native state from folding intermediates (Bokinsky et al., 2003; Liu et al., 2007; Rueda et al., 2004; Zhuang et al., 2000).

ACKNOWLEDGMENTS

Research in the Russell lab is funded by grants from the NIH (GM 070456) and the Welch Foundation (F-1563).

REFERENCES

Bhaskaran, H., and Russell, R. (2007). Kinetic redistribution of native and misfolded RNAs by a DEAD-box chaperone. *Nature* **449**, 1014–1018.
Bokinsky, G., and Zhuang, X. (2005). Single-molecule RNA folding. *Acc. Chem. Res.* **38**, 566–573.
Bokinsky, G., Rueda, D., Misra, V. K., Rhodes, M. M., Gordus, A., Babcock, H. P., Walter, N. G., and Zhuang, X. (2003). Single-molecule transition-state analysis of RNA folding. *Proc. Natl. Acad. Sci. U. S. A.* **100**, 9302–9307.
Chadee, A. B., Bhaskaran, H., and Russell, R. (2010). Protein roles in group I intron RNA folding: The tyrosyl-tRNA synthetase CYT-18 stabilizes the native state relative to a long-lived misfolded structure without compromising folding kinetics. *J. Mol. Biol.* **395**, 656–670.
Chin, K., and Pyle, A. M. (1995). Branch-point attack in group II introns is a highly reversible transesterification, providing a potential proofreading mechanism for 5′-splice site selection. *RNA* **1**, 391–406.
Clodi, E., Semrad, K., and Schroeder, R. (1999). Assaying RNA chaperone activity in vivo using a novel RNA folding trap. *EMBO J.* **18**, 3776–3782.
Del Campo, M., Mohr, S., Jiang, Y., Jia, H., Jankowsky, E., and Lambowitz, A. M. (2009). Unwinding by local strand separation is critical for the function of DEAD-box proteins as RNA chaperones. *J. Mol. Biol.* **389**, 674–693.
Emerick, V. L., Pan, J., and Woodson, S. A. (1996). Analysis of rate-determining conformational changes during self-splicing of the Tetrahymena intron. *Biochemistry* **35**, 13469–13477.
Fairman-Williams, M. E., Guenther, U. P., and Jankowsky, E. (2010). SF1 and SF2 helicases: Family matters. *Curr. Opin. Struct. Biol.* **20**, 313–324.
Fedorova, O., Waldsich, C., and Pyle, A. M. (2007). Group II intron folding under near-physiological conditions: Collapsing to the near-native state. *J. Mol. Biol.* **366**, 1099–1114.
Fedorova, O., Solem, A., and Pyle, A. M. (2010). Protein-facilitated folding of group II intron ribozymes. *J. Mol. Biol.* **397**, 799–813.
Golden, B. L., and Cech, T. R. (1996). Conformational switches involved in orchestrating the successive steps of group I RNA splicing. *Biochemistry* **35**, 3754–3763.
Halls, C., Mohr, S., Del Campo, M., Yang, Q., Jankowsky, E., and Lambowitz, A. M. (2007). Involvement of DEAD-box proteins in group I and group II intron splicing. Biochemical characterization of Mss116p, ATP hydrolysis-dependent and -independent mechanisms, and general RNA chaperone activity. *J. Mol. Biol.* **365**, 835–855.
Herschlag, D. (1995). RNA chaperones and the RNA folding problem. *J. Biol. Chem.* **270**, 20871–20874.
Herschlag, D., Khosla, M., Tsuchihashi, Z., and Karpel, R. L. (1994). An RNA chaperone activity of non-specific RNA binding proteins in hammerhead ribozyme catalysis. *EMBO J.* **13**, 2913–2924.
Hilbert, M., Karow, A. R., and Klostermeier, D. (2009). The mechanism of ATP-dependent RNA unwinding by DEAD box proteins. *Biol. Chem.* **390**, 1237–1250.
Huang, H. R., Rowe, C. E., Mohr, S., Jiang, Y., Lambowitz, A. M., and Perlman, P. S. (2005). The splicing of yeast mitochondrial group I and group II introns requires a DEAD-box protein with RNA chaperone function. *Proc. Natl. Acad. Sci. U. S. A.* **102**, 163–168.
Jankowsky, E., and Bowers, H. (2006). Remodeling of ribonucleoprotein complexes with DExH/D RNA helicases. *Nucleic Acids Res.* **34**, 4181–4188.
Jarmoskaite, I., and Russell, R. (2011). DEAD-box proteins as RNA helicases and chaperones. *Wiley Interdiscip. Rev. RNA* **2**, 135–152.

Karbstein, K., Carroll, K. S., and Herschlag, D. (2002). Probing the Tetrahymena group I ribozyme reaction in both directions. *Biochemistry* **41**, 11171–11183.
Karunatilaka, K. S., Solem, A., Pyle, A. M., and Rueda, D. (2010). Single-molecule analysis of Mss116-mediated group II intron folding. *Nature* **467**, 935–939.
King, P. A., Jamison, E., Strahs, D., Anderson, V. E., and Brenowitz, M. (1993). 'Footprinting' proteins on DNA with peroxonitrous acid. *Nucleic Acids Res.* **21**, 2473–2478.
Kuo, L. Y., Davidson, L. A., and Pico, S. (1999). Characterization of the Azoarcus ribozyme: Tight binding to guanosine and substrate by an unusually small group I ribozyme. *Biochim. Biophys. Acta* **1489**, 281–292.
Kwok, L. W., Shcherbakova, I., Lamb, J. S., Park, H. Y., Andresen, K., Smith, H., Brenowitz, M., and Pollack, L. (2006). Concordant exploration of the kinetics of RNA folding from global and local perspectives. *J. Mol. Biol.* **355**, 282–293.
Laederach, A., Shcherbakova, I., Liang, M. P., Brenowitz, M., and Altman, R. B. (2006). Local kinetic measures of macromolecular structure reveal partitioning among multiple parallel pathways from the earliest steps in the folding of a large RNA molecule. *J. Mol. Biol.* **358**, 1179–1190.
Laederach, A., Shcherbakova, I., Jonikas, M. A., Altman, R. B., and Brenowitz, M. (2007). Distinct contribution of electrostatics, initial conformational ensemble, and macromolecular stability in RNA folding. *Proc. Natl. Acad. Sci. U. S. A.* **104**, 7045–7050.
Linder, P. (2006). Dead-box proteins: A family affair—Active and passive players in RNP-remodeling. *Nucleic Acids Res.* **34**, 4168–4180.
Linder, P. (2008). mRNA export: RNP remodeling by DEAD-box proteins. *Curr. Biol.* **18**, R297–R299.
Liu, S., Bokinsky, G., Walter, N. G., and Zhuang, X. (2007). Dissecting the multistep reaction pathway of an RNA enzyme by single-molecule kinetic "fingerprinting" *Proc. Natl. Acad. Sci. U. S. A.* **104**, 12634–12639.
Lorsch, J. R. (2002). RNA chaperones exist and DEAD box proteins get a life. *Cell* **109**, 797–800.
Mitra, S., Laederach, A., Golden, B. L., Altman, R. B., and Brenowitz, M. (2011). RNA molecules with conserved catalytic cores but variable peripheries fold along unique energetically optimized pathways. *RNA* **17**, 1589–1603.
Mohr, S., Stryker, J. M., and Lambowitz, A. M. (2002). A DEAD-box protein functions as an ATP-dependent RNA chaperone in group I intron splicing. *Cell* **109**, 769–779.
Mohr, S., Matsuura, M., Perlman, P. S., and Lambowitz, A. M. (2006). A DEAD-box protein alone promotes group II intron splicing and reverse splicing by acting as an RNA chaperone. *Proc. Natl. Acad. Sci. U. S. A.* **103**, 3569–3574.
Mortimer, S. A., and Weeks, K. M. (2007). A fast-acting reagent for accurate analysis of RNA secondary and tertiary structure by SHAPE chemistry. *J. Am. Chem. Soc.* **129**, 4144–4145.
Mortimer, S. A., and Weeks, K. M. (2008). Time-resolved RNA SHAPE chemistry. *J. Am. Chem. Soc.* **130**, 16178–16180.
Mortimer, S. A., and Weeks, K. M. (2009). Time-resolved RNA SHAPE chemistry: Quantitative RNA structure analysis in one-second snapshots and at single-nucleotide resolution. *Nat. Protoc.* **4**, 1413–1421.
Nolte, A., Chanfreau, G., and Jacquier, A. (1998). Influence of substrate structure on in vitro ribozyme activity of a group II intron. *RNA* **4**, 694–708.
Pan, C., and Russell, R. (2010). Roles of DEAD-box proteins in RNA and RNP Folding. *RNA Biol.* **7**, 667–676.
Pan, J., and Woodson, S. A. (1998). Folding intermediates of a self-splicing RNA: Mispairing of the catalytic core. *J. Mol. Biol.* **280**, 597–609.
Potratz, J. P., Tijerina, P., and Russell, R. (2010). Mechanisms of DEAD-box proteins in ATP-dependent processes. *In* "RNA Helicases," (E. Jankowsky, ed.), pp. 61–98. Royal Society of Chemistry, Cambridge, UK.

Potratz, J. P., Del Campo, M., Wolf, R. Z., Lambowitz, A. M., and Russell, R. (2011). ATP-dependent roles of the DEAD-box protein Mss116p in group II intron splicing in vitro and in vivo. *J. Mol. Biol.* **411**, 661–679.

Pyle, A. M., and Green, J. B. (1994). Building a kinetic framework for group II intron ribozyme activity: Quantitation of interdomain binding and reaction rate. *Biochemistry* **33**, 2716–2725.

Qin, P. Z., and Pyle, A. M. (1997). Stopped-flow fluorescence spectroscopy of a group II intron ribozyme reveals that domain 1 is an independent folding unit with a requirement for specific Mg^{2+} ions in the tertiary structure. *Biochemistry* **36**, 4718–4730.

Rajkowitsch, L., Semrad, K., Mayer, O., and Schroeder, R. (2005). Assays for the RNA chaperone activity of proteins. *Biochem. Soc. Trans.* **33**, 450–456.

Roh, J. H., Guo, L., Kilburn, J. D., Briber, R. M., Irving, T., and Woodson, S. A. (2010). Multistage collapse of a bacterial ribozyme observed by time-resolved small-angle x-ray scattering. *J. Am. Chem. Soc.* **132**, 10148–10154.

Rueda, D., Bokinsky, G., Rhodes, M. M., Rust, M. J., Zhuang, X., and Walter, N. G. (2004). Single-molecule enzymology of RNA: Essential functional groups impact catalysis from a distance. *Proc. Natl. Acad. Sci. U. S. A.* **101**, 10066–10071.

Russell, R. (2008). RNA misfolding and the action of chaperones. *Front. Biosci.* **13**, 1–20.

Russell, R., and Herschlag, D. (1999). New pathways in folding of the Tetrahymena group I RNA enzyme. *J. Mol. Biol.* **291**, 1155–1167.

Russell, R., and Herschlag, D. (2001). Probing the folding landscape of the Tetrahymena ribozyme: Commitment to form the native conformation is late in the folding pathway. *J. Mol. Biol.* **308**, 839–851.

Russell, R., Das, R., Suh, H., Travers, K. J., Laederach, A., Engelhardt, M. A., and Herschlag, D. (2006). The paradoxical behavior of a highly structured misfolded intermediate in RNA folding. *J. Mol. Biol.* **363**, 531–544.

Schlatterer, J. C., and Brenowitz, M. (2009). Complementing global measures of RNA folding with local reports of backbone solvent accessibility by time resolved hydroxyl radical footprinting. *Methods* **49**, 142–147.

Schroeder, R., Barta, A., and Semrad, K. (2004). Strategies for RNA folding and assembly. *Nat. Rev. Mol. Cell Biol.* **5**, 908–919.

Sclavi, B., Woodson, S., Sullivan, M., Chance, M. R., and Brenowitz, M. (1997). Time-resolved synchrotron X-ray "footprinting", a new approach to the study of nucleic acid structure and function: Application to protein–DNA interactions and RNA folding. *J. Mol. Biol.* **266**, 144–159.

Sclavi, B., Sullivan, M., Chance, M. R., Brenowitz, M., and Woodson, S. A. (1998). RNA folding at millisecond intervals by synchrotron hydroxyl radical footprinting. *Science* **279**, 1940–1943.

Semrad, K., and Schroeder, R. (1998). A ribosomal function is necessary for efficient splicing of the T4 phage thymidylate synthase intron in vivo. *Genes Dev.* **12**, 1327–1337.

Semrad, K., Green, R., and Schroeder, R. (2004). RNA chaperone activity of large ribosomal subunit proteins from *Escherichia coli*. *RNA* **10**, 1855–1860.

Shcherbakova, I., Mitra, S., Laederach, A., and Brenowitz, M. (2008). Energy barriers, pathways, and dynamics during folding of large, multidomain RNAs. *Curr. Opin. Chem. Biol.* **12**, 655–666.

Sinan, S., Yuan, X., and Russell, R. (2011). The Azoarcus group I intron ribozyme misfolds and is accelerated for refolding by ATP-dependent RNA chaperone proteins. *J. Biol. Chem.* **286**, 37304–37312.

Steiner, M., Karunatilaka, K. S., Sigel, R. K., and Rueda, D. (2008). Single-molecule studies of group II intron ribozymes. *Proc. Natl. Acad. Sci. U. S. A.* **105**, 13853–13858.

Swisher, J., Duarte, C. M., Su, L. J., and Pyle, A. M. (2001). Visualizing the solvent-inaccessible core of a group II intron ribozyme. *EMBO J.* **20**, 2051–2061.

Tanner, M., and Cech, T. (1996). Activity and thermostability of the small self-splicing group I intron in the pre-tRNA(Ile) of the purple bacterium Azoarcus. *RNA* **2,** 74–83.
Tijerina, P., Bhaskaran, H., and Russell, R. (2006). Nonspecific binding to structured RNA and preferential unwinding of an exposed helix by the CYT-19 protein, a DEAD-box RNA chaperone. *Proc. Natl. Acad. Sci. U. S. A.* **103,** 16698–16703.
Tijerina, P., Mohr, S., and Russell, R. (2007). DMS footprinting of structured RNAs and RNA-protein complexes. *Nat. Protoc.* **2,** 2608–2623.
Treiber, D. K., and Williamson, J. R. (1999). Exposing the kinetic traps in RNA folding. *Curr. Opin. Struct. Biol.* **9,** 339–345.
Tsuchihashi, Z., Khosla, M., and Herschlag, D. (1993). Protein enhancement of hammerhead ribozyme catalysis. *Science* **262,** 99–102.
Walstrum, S. A., and Uhlenbeck, O. C. (1990). The self-splicing RNA of Tetrahymena is trapped in a less active conformation by gel purification. *Biochemistry* **29,** 10573–10576.
Wan, Y., Mitchell, D., and Russell, R. (2010a). Catalytic activity as a probe of native RNA folding. *Methods Enzymol.* **468,** 195–218.
Wan, Y., Suh, H., Russell, R., and Herschlag, D. (2010b). Multiple unfolding events during native folding of the Tetrahymena group I ribozyme. *J. Mol. Biol.* **400,** 1067–1077.
Woodson, S. A., and Cech, T. R. (1989). Reverse self-splicing of the tetrahymena group I intron: Implication for the directionality of splicing and for intron transposition. *Cell* **57,** 335–345.
Zaug, A. J., and Cech, T. R. (1986). The intervening sequence RNA of Tetrahymena is an enzyme. *Science* **231,** 470–475.
Zaug, A. J., Grosshans, C. A., and Cech, T. R. (1988). Sequence-specific endoribonuclease activity of the Tetrahymena ribozyme: Enhanced cleavage of certain oligonucleotide substrates that form mismatched ribozyme-substrate complexes. *Biochemistry* **27,** 8924–8931.
Zemora, G., and Waldsich, C. (2010). RNA folding in living cells. *RNA Biol.* **7,** 634–641.
Zhang, L., Xiao, M., Lu, C., and Zhang, Y. (2005). Fast formation of the P3-P7 pseudoknot: A strategy for efficient folding of the catalytically active ribozyme. *RNA* **11,** 59–69.
Zhuang, X. (2005). Single-molecule RNA science. *Annu. Rev. Biophys. Biomol. Struct.* **34,** 399–414.
Zhuang, X., Bartley, L. E., Babcock, H. P., Russell, R., Ha, T., Herschlag, D., and Chu, S. (2000). A single-molecule study of RNA catalysis and folding. *Science* **288,** 2048–2051.
Zingler, N., Solem, A., and Pyle, A. M. (2010). Dual roles for the Mss116 cofactor during splicing of the ai5gamma group II intron. *Nucleic Acids Res.* **38,** 6602–6609.

CHAPTER SIX

MOLECULAR MECHANICS OF RNA TRANSLOCASES

Steve C. Ding* *and* Anna Marie Pyle[†,‡]

Contents

1. Introduction	132
1.1. The protein domains of an RNA translocase: The RecA motor fold and companion domains	133
2. Major Examples of Monomeric RNA Translocases	134
2.1. The NS3/NPH-II family of SF2 proteins: $3' \rightarrow 5'$ translocation enzymes	134
2.2. The UPf1 family of SF1 proteins: $5' \rightarrow 3'$ translocation enzymes	138
2.3. Double-stranded translocation by monomeric RNA helicases: The RIG-I/Dicer family	140
3. Concluding Remarks	144
Acknowledgments	145
References	145

Abstract

Historically, research on RNA helicase and translocation enzymes has seemed like a footnote to the extraordinary progress in studies on DNA-remodeling enzymes. However, during the past decade, the rising wave of activity in RNA science has engendered intense interest in the behaviors of specialized motor enzymes that remodel RNA molecules. Functional, mechanistic, and structural investigations of these RNA enzymes have begun to reveal the molecular basis for their key roles in RNA metabolism and signaling. In this chapter, we highlight the structural and mechanistic similarities among monomeric RNA translocase enzymes, while emphasizing the many divergent characteristics that have caused this enzyme family to become one of the most important in metabolism and gene expression.

* Infectious and Inflammatory Disease Center, Sanford-Burnham Medical Research Institute, La Jolla, California, USA
† Department of Molecular, Cellular, and Developmental Biology, Yale University, New Haven, Connecticut, USA
‡ Howard Hughes Medical Institute, Yale University, New Haven, Connecticut, USA

1. Introduction

Progress in the field of RNA remodeling has largely resulted from the multidisciplinary nature of experimentation in this area. A combination of genetics, biochemistry, and biophysical studies in many different laboratories has established the biological functions of many RNA remodeling enzymes, the unique structural units upon which the enzymes are built, and the specific chemical and structural determinants by which RNA translocases and helicases identify their nucleic acid targets. As a result, several unifying themes and structural classes have emerged. We have limited the scope of this review to focus on monomeric RNA translocases that are phylogenetically classified as belonging to "helicase superfamily 1" (SF1) or "helicase superfamily 2" (SF2), which are nonetheless quite similar in sequence and morphology.

To date, enzymes in these families translocate unidirectionally along RNA in a processive manner. In order to exemplify the three major types of translocation behaviors that are observed for these enzymes, we focus on three different enzymes that share structural similarities while displaying divergent functional behavior (Fig. 6.1). Specifically, we discuss the following: Nonstructural protein 3 from hepatitis C virus (HCV NS3), a $3' \rightarrow 5'$

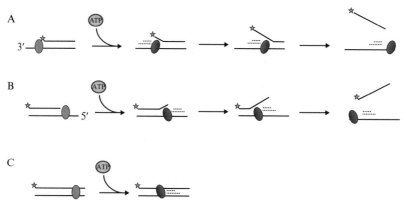

Figure 6.1 Cartoon representation of translocase activities. Depictions of diverse translocase activities are shown. These activities include (A) $3' \rightarrow 5'$ translocation on ssRNA, (B) $5' \rightarrow 3'$ translocation on ssRNA, and (C) translocation on dsRNA. In the absence of ATP, the helicases are static molecules (red ovals). Upon the addition of ATP, helicases begin to exert mechanical and dynamic movement (green ovals). Translocative behavior can lead to unwinding of RNA duplexes ((A) and (B)). However, it is worth noting that there are occasions in which translocation can occur in the absence of unwinding (C). (For interpretation of the references to color in this figure legend, the reader is referred to the online version of this chapter.)

ssRNA translocase; Upf1, a $5' \rightarrow 3'$ ssRNA translocase; and retinoic acid inducible gene I (RIG-I), a dsRNA translocase. For each enzyme, we describe structural features of the translocative motor domains, and the contributions of additional domains that aid directional movement. We also discuss the molecular determinants for substrate recognition and translocation. These findings will be interpreted in light of the fact that most of these proteins function as components of larger macromolecular machines rather than in isolation (i.e., in the exon junction complex or within replication complexes). We highlight the examples where the presence or absence of additional cofactors alters enzymatic behavior, thereby providing insight into how the cell may use different strategies to regulate helicase activity *in vivo*.

1.1. The protein domains of an RNA translocase: The RecA motor fold and companion domains

Like their DNA-remodeling cousins, the core of RNA translocase enzymes is composed of two RecA-like domains (α/β domain; a.k.a. Rossman fold) that are linked in tandem (Fairman-Williams *et al.*, 2010; Pyle, 2011; Singleton *et al.*, 2007). ATP binds in the cleft formed between these two domains and conserved amino acid motifs responsible for ATP binding and hydrolysis line up along each RecA-like domain facing inward toward the cleft. Atop these two domains is the platform where RNA binds. Depending upon the structural features of additional domains outside the two RecA-like domains, the geometry of this platform can accommodate either ssRNA or dsRNA (Fig. 6.2).

Many helicase proteins belonging to the nonprocessive DEAD-box classification, which also hydrolyze ATP upon RNA binding, only have two RecA-like domains (Linder and Jankowsky, 2011). However, most translocative and processive enzymes have additional domains that potentiate important roles in substrate recognition and helicase function. Some of these domains play positive roles in conferring processive behavior (Domain 3 of NS3), some play autoinhibitory or regulatory roles to prevent processive behavior (Domains CH and 1B in Upf1), and some mediate complex functions such as the transmission of mechanical information over long distances (pincer domain of RIG-I) (Fig. 6.2). Further, some helicases, such as *E. coli* DbpA, use these extra domains to recognize a specific RNA tertiary structure in order to activate its ATPase activity (Tsu *et al.*, 2001), whereas other helicases lacking these additional domains, such as *S. cerevisiae* Dbp5, require the binding of auxiliary cofactors (Gle1) and small molecules (inositol hexakisphosphate) to stimulate RNA release (Montpetit *et al.*, 2011). Functional dissection of these additional domains or binding cofactors represents an active area of investigation by many laboratories.

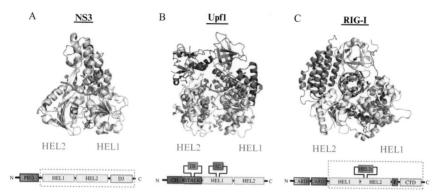

Figure 6.2 Structural attributes of translocating helicases. Structural snapshots of helicases on their respective RNA substrates are shown. These helicases are (A) HCV NS3, a 3′→5′ ssRNA translocase, (B) Upf1, a 5′→3′ ssRNA translocase, and (C) RIG-I, a dsRNA translocase. Boxed regions indicate the illustrated portion of the helicase. Domains outside the boxed regions were omitted for clarity. The two RecA-like domains are colored in yellow, and additional domains are colored according to the diagrams beneath each structure. ATP binds in the cleft between the two RecA-like domains. Regardless of the translocation direction, the ssRNAs (orange) lie in the same orientation atop the helicase motor: the 5′ end rests above HEL2 and the 3′ end rests above HEL1. This strand is called the tracking strand. For RIG-I, the tracking strand of the duplex is also colored in orange, and the activating strand is colored in purple. (See the Color Insert.)

2. Major Examples of Monomeric RNA Translocases

2.1. The NS3/NPH-II family of SF2 proteins: 3′→5′ translocation enzymes

Most of our knowledge about these enzymes stems from pioneering work on SF2 helicases of viral origins, primarily from work on the NS3 proteins from the *Flaviviridae* viruses, Dengue Virus and HCV, and NPH-II from Vaccinia Virus. The NS3 helicases from HCV and Dengue are believed to be important during replication, where they may play a role in unraveling the highly structured single-stranded genomes of these viruses. However, HCV NS3 is now known to play other roles in the life of the virus, including a major function in virus particle packaging and assembly (Jones et al., 2011; Ma et al., 2008). Thus, despite the apparent helicase activity of NS3 enzymes, their actual roles in the viral lifecycles are not firmly established. Similarly, the NPH-II helicases of pox viruses are thought to be required for efficient transcription and inhibition of R-loop formation (Gross and Shuman, 1996). However, there is increasing evidence that they may play a role in release of the polymerase.

nucleotides (Fig. 6.3A). Upon ATP hydrolysis, the products ADP and inorganic phosphate dissociate from the ATP binding cavity, and HEL2 relaxes to move away from HEL1, thus effectively advancing the helicase by one nucleotide toward the 5′ direction of the ssRNA.

Cycles of ATP binding and hydrolysis directly influence the number of hydrogen bonds formed between the helicase and RNA substrate, thus resulting in alternating high- and low-affinity states. Most interactions are observed in the ATP-free state, which represents the high-affinity state. Upon binding ATP, the number of contacts between helicase and the RNA decreases (notably, Thr416 disengages from the RNA backbone) and resulting in a low-affinity state, which is consistent with previous direct binding studies demonstrating the helicase has a lower affinity for RNA in the presence of nucleotide.

The 3′ → 5′ directionality of movement is achieved in part through conserved hydrophobic residue Trp501, which extends from Domain 3 (Fig. 6.3B). This residue stacks upon the nucleobase at the 3′ end of the ssRNA and effectively serves as an anchor point for the helicase to move unidirectionally toward the 5′ end of the ssRNA rather than slipping backward. Interestingly, through the cycles of ATP binding and hydrolysis, the orientation of Trp501 remains stacked with the nucleobase (Appleby et al., 2011). Maintaining this interaction with the nucleobase is functionally significant for the translocation and possibly for unwinding mechanisms of NS3. Indeed, single-molecule experiments have suggested that Trp501 acts as a plowshare, dragging behind the translocating helicase (which moves along the tracking strand in one nucleotide increments as described above) (Myong et al., 2007). Specifically, the data supports a model where Trp501 maintains its interactions with the same nucleobase, while the helicase undergoes two forward cycles of ATP binding and hydrolysis; during the third cycle, the helicase accumulates enough tension in the system and causes Trp501 to spring forward along the tracking strand by three nucleotides concomitant unwinding of three base pairs. Importantly, no sequence–specific interactions are formed between the nucleobases on the ssRNA strand and residues from Domain 3. This lack of specificity likely contributes to the ability of the helicase to unwind substrates in a sequence-independent fashion.

Currently, there are no crystal structures available of *bona fide* helicases in complex with duplex RNA substrates (there are structures of RIG-I with RNA duplexes, but it is not a helicase, *vide infra*). However, unwinding is likely achieved by the steric pressure from a β-hairpin that projects from HEL2. This conserved structural element is present in other related helicases and is proposed to function like a wedge to splay the two composite strands apart (Buttner et al., 2007; Luo et al., 2008b). While the tracking strand is fed through the ssRNA channel described above, the displaced strand is directed away. However, the trajectory of the displaced strand is currently unknown and represents a potential area of research.

2.2. The UPf1 family of SF1 proteins: $5' \rightarrow 3'$ translocation enzymes

Much less is known about the translocation and unwinding mechanisms of RNA helicases that move with 5'–3' directionality. This is unfortunate because members of this class include helicases of viral origin such as SARS coronavirus, which had been shown to be unusually processive. The SARS helicase can efficiently unwind RNA duplexes of several hundred base pairs under single-cycle conditions (Ivanov and Ziebuhr, 2004). However, the most structurally and functionally characterized $5' \rightarrow 3'$ translocation enzymes of the SF1 protein Upf1. Upf1 is a core component of the nonsense-mediated decay (NMD) machinery that detects and rapidly degrades aberrant mRNA transcripts (Conti and Izaurralde, 2005). Intriguingly, Upf1 is phylogenetically related to well-characterized DNA helicases such as Rep and PcrA, rather than to other SF2 RNA helicases.

Upf1 exhibits low ATPase and unwinding activities in isolation and requires the binding of additional NMD factors Upf2 and Upf3 to stimulate these activities (Chamieh et al., 2008). The structural basis for these functional observations was recently illustrated in a series of crystal structures of Upf1 in isolation and with various cofactors that alter its enzymatic activities (Chakrabarti et al., 2011; Cheng et al., 2007; Clerici et al., 2009). Like the NS3 helicase, Upf1 contains a core motor domain composed of two RecA-like domains (HEL1 and HEL2) (Fig. 6.2B). The N-terminus of Upf1 contains a zinc-knuckle domain (CH domain) that exerts an inhibitory effect *in cis* on the enzymatic activities of the protein. Between the CH domain and HEL1 is a domain composed of six antiparallel β-strands (Domain 1B), and a stalk domain composed of two α-helices. Projecting from HEL1 is an additional domain also composed of two α-helices (Domain 1C). Such modular domains embedded within the RecA-like domains are unique features that typify members belonging to SF1. Interestingly, ssRNA binds in an orientation atop the HEL1 and HEL2 platform of Upf1 similar to that seen for NS3. This is a surprising finding, given that one would expect that a helicase that translocates and unwinds with an opposite polarity as NS3 to position itself on the RNA in a reversed orientation. However, this similarity in binding polarity, despite opposite translocation directionality, has also been observed in comparisons of the structures of DNA helicases, RecD2 and PcrA (Saikrishnan et al., 2009; Velankar et al., 1999), and in the hexameric helicases, Rho and E1 (Enemark and Joshua-Tor, 2006; Thomsen and Berger, 2009). Results with Upf1 are therefore consistent with the notion that translocases do not "turn around" to run backward. Rather, they reverse their gears (Pyle, 2009).

At a molecular level, Upf1 contacts the ssRNA using main-chain amide groups and side-chain interactions that project from HEL1, HEL2, the stalk,

Molecular Mechanics of RNA Translocases 139

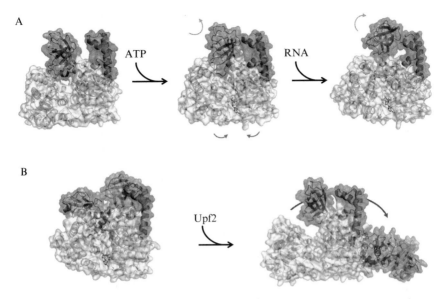

Figure 6.4 Dynamic motions of Upf1, a 5′→3′ ssRNA translocase. (A) Similar to HCV NS3, the two RecA-like domains of Upf1 close upon binding ATP. Domain 1B undergoes a small conformational change and rotates away from the interface of the two RecA-like domains. Upon binding ssRNA, Domain1B makes another conformational change to position itself away from the 3′ end of the nucleic acid. This rearrangement allows Upf1 to translocate processively with 5′→3′ directionality. Domain motions were interpreted based upon aligning all structures to HEL1. (B) In the presence of the CH domain (green), Upf1 adopts a compact, globular conformation that prevents processive translocation activity. Inhibition can be partially relieved by the binding of the cofactor Upf2 (orange). Binding leads to a large conformational rearrangement of the CH domain. Whereas the CH domain was once positioned above HEL2, binding of Upf2 repositions the CH domain behind HEL1. (For the color version of this figure, the reader is referred to the online version of this chapter.)

Domain 1B, and Domain 1C. The CH domain does not directly interact with the RNA but rather exerts an allosteric effect on RNA binding by impinging upon Domain 1B to contact two additional nucleotides at the 3′ end of the RNA (Fig. 6.4B). Impinging upon Domain 1B likely prevents forward translocative movement and may account for the intrinsically low ATPase and unwinding activities of the full-length protein. Interestingly, the crystal structure of a Upf1 construct lacking its CH domain shows that Domain 1B rotates away from the 3′ end of the RNA such that Domain 1B is no longer locked in a restricted position (Fig. 6.4A). By allowing Domain 1B to disengage from the RNA, Upf1 displays higher ATPase and unwinding activities. Therefore, the presence or absence of Domain 1B interacting with the 3′ end of the RNA may represent an autoinhibition mechanism occurring *in cis* to modulate the helicase activities of Upf1.

In a cellular context, preventing the CH domain from impinging on Domain 1B is accomplished by the addition of Upf2 (Fig. 6.4B). The C-terminus of Upf2 binds directly to the CH domain of Upf1, and this interaction is sufficient to enhance the ATPase and unwinding activities of Upf1. From a structural viewpoint, the binding of Upf2 to Upf1 induces a large conformational change in the CH domain of Upf1. Whereas the CH domain was originally positioned atop the HEL2 domain, binding of Upf2 causes the CH domain to occupy a new position distal to HEL1. By occupying this new position, the CH domain can no longer impinge upon Domain 1B to interact with the 3′ end of the ssRNA. The repositioning of the CH domain is likely the structural basis for the functional observation that Upf2 enhances the biochemical properties of Upf1.

For future studies, it will be interesting to understand the coordinated domain motions of Upf1 in the presence of full-length Upf2 and Upf3 proteins or with an RNA unwinding substrate. Having such a global view may help uncover the biological significance of translocation and unwinding by Upf1 in the context of the NMD pathway. Further, it would be valuable to conduct experiments similar to those performed using NPH-II and NS3 proteins probing mechanistic features of substrate recognition. These studies would provide a basis to compare the similarities and differences between helicases that operate on the similar nucleic acid substrates but move in different directions.

2.3. Double-stranded translocation by monomeric RNA helicases: The RIG-I/Dicer family

While members of this class have been characterized biochemically, a structural view of any member of this class had long been elusive until recently. This was especially unfortunate given that members constituting this class of enzymes are involved in important functions such as innate immune signaling (RIG-I, and related proteins MDA5 and LGP2) and RNA interference (Dicer) (Bernstein et al., 2001; Yoneyama et al., 2004, 2005). In the absence of a crystal structure, it was difficult to understand how a helicase recognizes dsRNA, which remained a lingering question in the field. For example, does the dsRNA rest on top of the two RecA-like domains in an orientation similar to those seen for helicases that bind ssRNA? If so, is the binding groove sufficiently deep to discriminate dsRNA from ssRNA? For recognizing and binding a dsRNA substrate, is one strand preferred over the other strand? These questions were especially important to answer given that the RIG-I and Dicer proteins are phylogenetically related SF2 helicases, but function in strikingly different pathways: RIG-I translocates along dsRNA and functions in innate immune signaling, whereas Dicer binds and cleaves long double-stranded pre-microRNAs into short, double-stranded siRNAs. Though both proteins are powered by the same helicase core motor, their divergent functions are conferred by accessory domains flanking the N- and C-termini of the respective protein. Fortunately, the structural basis for recognizing and binding dsRNA by RIG-I in the presence and absence of bound nucleotide was recently shown in a series of publications from numerous groups (Civril et al., 2011; Jiang et al., 2011; Kowalinski et al., 2011; Luo et al., 2011).

RIG-I recognizes and binds dsRNA using its central SF2 helicase motor core and displays dsRNA-dependent ATPase activity (Gack et al., 2008). At the N-terminus, RIG-I contains two tandem caspase activation and recruitment domains (CARDs; CARD1 and CARD2) for signaling, and at the C-terminus, it contains a domain important for recognizing 5'-triphosphate moieties (the CTD) (Cui et al., 2008) (Fig. 6.2C). The CARD fold is typically found in proteins that function in apoptosis and inflammatory signaling pathways (Hofmann et al., 1997). Therefore, it was highly unusual for a helicase to have such an accessory domain that functions outside the boundaries of the Central Dogma. In the context of RIG-I, it was later determined that the CARDs forms a platform to interact with other CARD-containing proteins to propagate a signaling cascade (Gack et al., 2008). The CTD recognizes the 5' end of either ssRNA or dsRNA, though it binds with the tightest affinity to duplexed substrates bearing a triphosphate group (Wang et al., 2010).

The recognition of triphosphorylated substrates is a defining feature of RIG-I (Hornung et al., 2006; Pichlmair et al., 2006). As most cellular RNAs are capped or modified at their 5' ends immediately following transcription, the chemical nature of free 5' triphosphate groups are immediately recognized by RIG-I as "nonself" RNA. It is generally accepted that such "nonself" RNAs are generated when viruses hijack the cellular machinery to create new strands of viral genomic RNA *de novo* using its viral polymerase. This recognition event activates RIG-I to initiate a signaling cascade through the CARDs to ward off the invading virus to maintain an antiviral cellular state.

Translocation along RNA duplex by the helicase domain was demonstrated using a single-molecule fluorescence approach (Myong et al., 2009). Using a wide spectrum of protein and RNA constructs, the authors demonstrated that RIG-I exhibits robust and repetitive translocation along duplex RNA substrates. Notably, translocation activity occurred in the absence of unwinding. This was consistent with previous studies showing that neither RIG-I, nor its cousin MDA5 display strong helicase activity. Both of these enzymes display ATPase activity that is specifically activated by duplex RNA rather than single strands (Gee et al., 2008; Kang et al., 2002). Functional dissection of the role for the CARD domains indicates that they play an inhibitory role by negatively regulating translocation activity (Myong et al., 2009) (Fig. 6.5B). Binding of duplex RNA partially alleviates this inhibitory role to allow translocation. Interestingly, maximal translocation

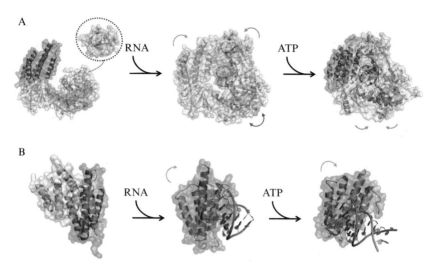

Figure 6.5 Dynamic motions of RIG-I, a dsRNA translocase. (A) As determined by SAXS measurements, RIG-I adopts an open conformation in the absence of dsRNA substrate. For clarity, we omitted the CARDs from the first structure. We model in a proposed position of the CTD in this figure (dotted circle) to account for the open conformation needed to allow dsRNA binding. HEL2i and the CTD surround the dsRNA substrate upon binding, which is likely facilitated by the V-shaped pincer domain. Additionally, the RecA-like domains close upon binding ATP, and HEL2 establishes additional contacts with the dsRNA backbone. Domain motions were interpreted based upon aligning all structures to HEL1 and dsRNA. (B) Major domain motions by HEL2i in response to dsRNA and ATP. In the *apo* conformation, the CARDs form a large hydrophobic interface with HEL2i and prevent HEL2i from interacting with the dsRNA. Likewise, the presence of the CARDs is inhibitory for processive translocative behavior. Upon binding dsRNA, HEL2i establishes a single, weak interaction with the backbone of the substrate using the face of a specific α-helix (green). Upon binding ATP, HEL2i moves further toward the dsRNA and establishes more contacts using this α-helix. This alternation between strong and weak interactions in response to ATP is likely the structural basis for dsRNA translocation. (See the Color Insert.)

activity was observed using triphosphorylated duplex RNA, thereby highlighting the specific structural and chemical features of the nucleic acid substrate requirement for RIG-I. Taken together, RIG-I uses its helicase domain to power translocation along dsRNA, potentially disrupting the viral replication machinery or assisting in the loading of additional RIG-I molecules.

From the collection of available recent RIG-I structures, we can now directly observe the coordinated motions of its multiple domains, which has led to new and unexpected structural findings. In addition to the CARDs, the canonical motor domains HEL1 and HEL2, and the CTD, RIG-I contains two domains that had never been appreciated previously. The first domain is composed of five antiparallel α-helices and is termed as the insertion domain (HEL2i). HEL2i exists as an independent structure projecting above from HEL2 and forms direct contacts with the backbone dsRNA. The second domain is composed of two α-helices that form a V-shape, which we refer to as the pincer domain, which wraps around multiple domains in order to transduce mechanical signals throughout the protein.

In the absence of dsRNA, RIG-I adopts an open, extended conformation. CARD1 and CARD2 stack upon one another, and CARD2 forms extensive interactions with HEL2i that are stabilized by salt bridges and hydrophobic contacts (Kowalinski *et al.*, 2011) (Fig. 6.5B). The interactions between CARD2 and HEL2i may partially explain the inhibitory role of the CARDs observed in the single-molecule fluorescence studies. In the presence of dsRNA, however, RIG-I collapses upon and encircles the dsRNA (Jiang *et al.*, 2011; Luo *et al.*, 2011). This compaction event is consistent with small-angle X-ray scattering (SAXS) data measuring the dimensions of these molecules in solution in the presence and absence of dsRNA. Though a structure of the full-length RIG-I protein with dsRNA is unavailable, SAXS measurements indicate that such a complex would place the CARDs away in solution and distal to the dsRNA binding interface. In fact, such a conformation might even be optimal for the CARDs to interact with other protein cofactors.

When binding an RNA duplex, it is important to distinguish one strand from the other. We refer to the RNA strand that lies across the HEL1 and HEL2 motor domains as the tracking strand (Fig. 6.2C). Interestingly, the tracking strand is in the same polarity to those seen for NS3 and Upf1, whereby the 5′ end of the tracking strand is located above HEL2, while the 3′ end is located above HEL1. We refer to the complementary sister strand as the activating strand since the 5′ end of this strand directly interacts with the CTD and would carry the triphosphate group that activates RIG-I translocation activity. The duplex RNA is oriented perpendicular to the plane of HEL1 and HEL2, and numerous polar interactions are formed between the 2′-hydroxyl groups of the ribose sugars and the phosphodiester backbone using main-chain amides and side-chain residues; hydrophobic interactions are only observed between the CTD and the 5′ end of the activating strand. The lack of base specificity likely allows RIG-I to translocate on diverse dsRNAs without sequence bias.

The bound nucleotide state directly influences the number of interactions made between RIG-I and the dsRNA. When ATP occupies the cleft formed by HEL1 and HEL2, these two domains move in close proximity to one another in order to prime the catalytic residues lining the cleft for ATP binding and hydrolysis. In this closed conformation, HEL1 and HEL2 form contacts with both the tracking and activating strands. However, in the absence of nucleotide, RIG-I adopts a more open conformation where

HEL2 moves away from HEL1 and disengages from contacting the dsRNA. Intriguingly, HEL1 maintains its grip with the dsRNA regardless of the nucleotide bound state. Similar domain movements are also observed with HEL2i. Using the face of one of its helices, HEL2i forms extensive contacts with both strands of the dsRNA in the presence of nucleotide; these contacts are lost, except one, in the absence of bound nucleotide (Fig. 6.5B).

The act of engaging and then disengaging from the dsRNA by HEL2 and HEL2i in response to the bound nucleotide state may directly reflect the RIG-I translocation mechanism. By maintaining a constant point of contact with the dsRNA through HEL1, RIG-I may use HEL2 and HEL2i to inch the protein forward along one strand of the dsRNA. Upon nucleotide binding, RIG-I maintains a tight grip on the dsRNA by forming direct contacts through HEL1, HEL2, and HEL2i. Following, ATP hydrolysis may permit RIG-I to loosen its grip on the dsRNA by allowing HEL2 and HEL2i to disengage from the dsRNA. Each cycle of ATP binding and hydrolysis would then propel RIG-I forward by one base-pair toward the $5'$ end of the tracking strand. Indeed, such directional bias along the tracking strand had been observed in a single-molecule setting. In a biological context, it is tempting to speculate that directional translocation along the tracking strand may serve to dislodge the viral replication machinery that would be actively synthesizing new viral RNA (i.e., the activating strand). However, whether RIG-I can remodel protein–RNA complexes, as had been shown for other helicases previously, remains the subject of future inquiry.

Despite having the numerous structural views of RIG-I now available, several outstanding questions remain. For example, given that the CTD has a tight affinity for the triphosphorylated duplex substrates, how does RIG-I disengage from that position to translocate along dsRNA? What are the roles of phosphorylation, ubiquitination, or tetraubiquitin binding in altering the biochemical activities of RIG-I? How do the CARDs interact with other proteins that have CARD domains, and what are the functional outcomes? Is there a role for ATP hydrolysis during the presentation of the CARD domains and in signaling?

3. Concluding Remarks

We now have complete structural views of three RNA translocases that move in different directions and function on different types of RNA. Despite the diversity in the directionalities and substrate specificities of these molecular motors, these enzymes share a common theme: they bind and hydrolyze ATP in the presence of a stimulating nucleic acid substrate. While a main research question is to decipher what ATP binding and hydrolysis accomplish, an emerging theme appears to be that this orchestrated motion advances the protein forward by one nucleotide. It is thus tempting to speculate that perhaps any helicase containing these two RecA-like modules may be able to move along nucleic acid strands, regardless of the number of additional flanking domains. Nevertheless, there is still much to uncover what is underneath the hoods of Nature's molecular engines, and future research will undoubtedly shed light on their diverse mechanical properties.

ACKNOWLEDGMENTS

This work was funded by the Howard Hughes Medical Institute and NIH grant AI089826. A. M. P. is an Investigator with the Howard Hughes Medical Institute.

REFERENCES

Appleby, T. C., Anderson, R., Fedorova, O., Pyle, A. M., Wang, R., Liu, X., Brendza, K. M., and Somoza, J. R. (2011). Visualizing ATP-dependent RNA translocation by the NS3 helicase from HCV. *J. Mol. Biol.* **405,** 1139–1153.

Beran, R. K., and Pyle, A. M. (2008). Hepatitis C viral NS3-4A protease activity is enhanced by the NS3 helicase. *J. Biol. Chem.* **283,** 29929–29937.

Beran, R. K., Bruno, M. M., Bowers, H. A., Jankowsky, E., and Pyle, A. M. (2006). Robust translocation along a molecular monorail: the NS3 helicase from hepatitis C virus traverses unusually large disruptions in its track. *J. Mol. Biol.* **358,** 974–982.

Beran, R. K., Serebrov, V., and Pyle, A. M. (2007). The serine protease domain of hepatitis C viral NS3 activates RNA helicase activity by promoting the binding of RNA substrate. *J. Biol. Chem.* **282,** 34913–34920.

Bernstein, E., Caudy, A. A., Hammond, S. M., and Hannon, G. J. (2001). Role for a bidentate ribonuclease in the initiation step of RNA interference. *Nature* **409,** 363–366.

Buttner, K., Nehring, S., and Hopfner, K. P. (2007). Structural basis for DNA duplex separation by a superfamily-2 helicase. *Nat. Struct. Mol. Biol.* **14,** 647–652.

Chakrabarti, S., Jayachandran, U., Bonneau, F., Fiorini, F., Basquin, C., Domcke, S., Le Hir, H., and Conti, E. (2011). Molecular mechanisms for the RNA-dependent ATPase activity of Upf1 and its regulation by Upf2. *Mol. Cell* **41,** 693–703.

Chamieh, H., Ballut, L., Bonneau, F., and Le Hir, H. (2008). NMD factors UPF2 and UPF3 bridge UPF1 to the exon junction complex and stimulate its RNA helicase activity. *Nat. Struct. Mol. Biol.* **15,** 85–93.

Cheng, Z., Muhlrad, D., Lim, M. K., Parker, R., and Song, H. (2007). Structural and functional insights into the human Upf1 helicase core. *EMBO J.* **26,** 253–264.

Civril, F., Bennett, M., Moldt, M., Deimling, T., Witte, G., Schiesser, S., Carell, T., and Hopfner, K. P. (2011). The RIG-I ATPase domain structure reveals insights into ATP-dependent antiviral signalling. *EMBO Rep.* **12,** 1127–1134.

Clerici, M., Mourao, A., Gutsche, I., Gehring, N. H., Hentze, M. W., Kulozik, A., Kadlec, J., Sattler, M., and Cusack, S. (2009). Unusual bipartite mode of interaction between the nonsense-mediated decay factors, UPF1 and UPF2. *EMBO J.* **28,** 2293–2306.

Conti, E., and Izaurralde, E. (2005). Nonsense-mediated mRNA decay: molecular insights and mechanistic variations across species. *Curr. Opin. Cell Biol.* **17,** 316–325.

Cui, S., Eisenacher, K., Kirchhofer, A., Brzozka, K., Lammens, A., Lammens, K., Fujita, T., Conzelmann, K. K., Krug, A., and Hopfner, K. P. (2008). The C-terminal regulatory domain is the RNA 5′-triphosphate sensor of RIG-I. *Mol. Cell* **29**, 169–179.

Dumont, S., Cheng, W., Serebrov, V., Beran, R. K., Tinoco, I., Jr., Pyle, A. M., and Bustamante, C. (2006). RNA translocation and unwinding mechanism of HCV NS3 helicase and its coordination by ATP. *Nature* **439**, 105–108.

Enemark, E. J., and Joshua-Tor, L. (2006). Mechanism of DNA translocation in a replicative hexameric helicase. *Nature* **442**, 270–275.

Fairman-Williams, M. E., Guenther, U. P., and Jankowsky, E. (2010). SF1 and SF2 helicases: family matters. *Curr. Opin. Struct. Biol.* **20**, 313–324.

Gack, M. U., Kirchhofer, A., Shin, Y. C., Inn, K. S., Liang, C., Cui, S., Myong, S., Ha, T., Hopfner, K. P., and Jung, J. U. (2008). Roles of RIG-I N-terminal tandem CARD and splice variant in TRIM25-mediated antiviral signal transduction. *Proc. Natl. Acad. Sci. U. S. A.* **105**, 16743–16748.

Gee, P., Chua, P. K., Gevorkyan, J., Klumpp, K., Najera, I., Swinney, D. C., and Deval, J. (2008). Essential role of the N-terminal domain in the regulation of RIG-I ATPase activity. *J. Biol. Chem.* **283**, 9488–9496.

Gross, C. H., and Shuman, S. (1996). Vaccinia virions lacking the RNA helicase nucleoside triphosphate phosphohydrolase II are defective in early transcription. *J. Virol.* **70**, 8549–8557.

Hofmann, K., Bucher, P., and Tschopp, J. (1997). The CARD domain: a new apoptotic signalling motif. *Trends Biochem. Sci.* **22**, 155–156.

Hornung, V., Ellegast, J., Kim, S., Brzozka, K., Jung, A., Kato, H., Poeck, H., Akira, S., Conzelmann, K. K., Schlee, M., *et al.* (2006). 5′-Triphosphate RNA is the ligand for RIG-I. *Science* **314**, 994–997.

Ivanov, K. A., and Ziebuhr, J. (2004). Human coronavirus 229E nonstructural protein 13: Characterization of duplex-unwinding, nucleoside triphosphatase, and RNA 5′-triphosphatase activities. *J. Virol.* **78**, 7833–7838.

Jankowsky, E., Gross, C. H., Shuman, S., and Pyle, A. M. (2000). The DExH protein NPH-II is a processive and directional motor for unwinding RNA. *Nature* **403**, 447–451.

Jankowsky, E., Gross, C. H., Shuman, S., and Pyle, A. M. (2001). Active disruption of an RNA-protein interaction by a DExH/D RNA helicase. *Science* **291**, 121–125.

Jiang, F., Ramanathan, A., Miller, M. T., Tang, G. Q., Gale, M., Jr., Patel, S. S., and Marcotrigiano, J. (2011). Structural basis of RNA recognition and activation by innate immune receptor RIG-I. *Nature* **479**, 423–427.

Jones, D. M., Atoom, A. M., Zhang, X., Kottilil, S., and Russell, R. S. (2011). A genetic interaction between the core and NS3 proteins of hepatitis C virus is essential for production of infectious virus. *J. Virol.* **85**, 12351–12361.

Kang, D. C., Gopalkrishnan, R. V., Wu, Q., Jankowsky, E., Pyle, A. M., and Fisher, P. B. (2002). mda-5: An interferon-inducible putative RNA helicase with double-stranded RNA-dependent ATPase activity and melanoma growth-suppressive properties. *Proc. Natl. Acad. Sci. U. S. A.* **99**, 637–642.

Kawaoka, J., Jankowsky, E., and Pyle, A. M. (2004). Backbone tracking by the SF2 helicase NPH-II. *Nat. Struct. Mol. Biol.* **11**, 526–530.

Kowalinski, E., Lunardi, T., McCarthy, A. A., Louber, J., Brunel, J., Grigorov, B., Gerlier, D., and Cusack, S. (2011). Structural basis for the activation of innate immune pattern-recognition receptor RIG-I by viral RNA. *Cell* **147**, 423–435.

Linder, P., and Jankowsky, E. (2011). From unwinding to clamping—The DEAD box RNA helicase family. *Nat. Rev. Mol. Cell Biol.* **12**, 505–516.

Luo, D., Xu, T., Hunke, C., Gruber, G., Vasudevan, S. G., and Lescar, J. (2008a). Crystal structure of the NS3 protease-helicase from dengue virus. *J. Virol.* **82**, 173–183.

Luo, D., Xu, T., Watson, R. P., Scherer-Becker, D., Sampath, A., Jahnke, W., Yeong, S. S., Wang, C. H., Lim, S. P., Strongin, A., *et al.* (2008b). Insights into RNA unwinding and ATP hydrolysis by the flavivirus NS3 protein. *EMBO J.* **27**, 3209–3219.

Luo, D., Ding, S. C., Vela, A., Kohlway, A., Lindenbach, B. D., and Pyle, A. M. (2011). Structural insights into RNA recognition by RIG-I. *Cell* **147**, 409–422.

Ma, Y., Yates, J., Liang, Y., Lemon, S. M., and Yi, M. (2008). NS3 helicase domains involved in infectious intracellular hepatitis C virus particle assembly. *J. Virol.* **82**, 7624–7639.

Montpetit, B., Thomsen, N. D., Helmke, K. J., Seeliger, M. A., Berger, J. M., and Weis, K. (2011). A conserved mechanism of DEAD-box ATPase activation by nucleoporins and InsP6 in mRNA export. *Nature* **472**, 238–242.

Myong, S., Bruno, M. M., Pyle, A. M., and Ha, T. (2007). Spring-loaded mechanism of DNA unwinding by hepatitis C virus NS3 helicase. *Science* **317**, 513–516.

Myong, S., Cui, S., Cornish, P. V., Kirchhofer, A., Gack, M. U., Jung, J. U., Hopfner, K. P., and Ha, T. (2009). Cytosolic viral sensor RIG-I is a 5′-triphosphate-dependent translocase on double-stranded RNA. *Science* **323**, 1070–1074.

Pichlmair, A., Schulz, O., Tan, C. P., Naslund, T. I., Liljestrom, P., Weber, F., and Reis e Sousa, C. (2006). RIG-I-mediated antiviral responses to single-stranded RNA bearing 5′-phosphates. *Science* **314**, 997–1001.

Pyle, A. M. (2009). How to drive your helicase in a straight line. *Cell* **139**, 458–459.

Pyle, A. M. (2011). RNA helicases and remodeling proteins. *Curr. Opin. Chem. Biol.* **15**, 636–642.

Raney, K. D., Sharma, S. D., Moustafa, I. M., and Cameron, C. E. (2010). Hepatitis C virus non-structural protein 3 (HCV NS3): A multifunctional antiviral target. *J. Biol. Chem.* **285**, 22725–22731.

Saikrishnan, K., Powell, B., Cook, N. J., Webb, M. R., and Wigley, D. B. (2009). Mechanistic basis of 5′–3′ translocation in SF1B helicases. *Cell* **137**, 849–859.

Singleton, M. R., Dillingham, M. S., and Wigley, D. B. (2007). Structure and mechanism of helicases and nucleic acid translocases. *Annu. Rev. Biochem.* **76**, 23–50.

Tai, C. L., Chi, W. K., Chen, D. S., and Hwang, L. H. (1996). The helicase activity associated with hepatitis C virus nonstructural protein 3 (NS3). *J. Virol.* **70**, 8477–8484.

Thomsen, N. D., and Berger, J. M. (2009). Running in reverse: the structural basis for translocation polarity in hexameric helicases. *Cell* **139**, 523–534.

Tsu, C. A., Kossen, K., and Uhlenbeck, O. C. (2001). The Escherichia coli DEAD protein DbpA recognizes a small RNA hairpin in 23S rRNA. *RNA* **7**, 702–709.

Velankar, S. S., Soultanas, P., Dillingham, M. S., Subramanya, H. S., and Wigley, D. B. (1999). Crystal structures of complexes of PcrA DNA helicase with a DNA substrate indicate an inchworm mechanism. *Cell* **97**, 75–84.

Wang, Y., Ludwig, J., Schuberth, C., Goldeck, M., Schlee, M., Li, H., Juranek, S., Sheng, G., Micura, R., Tuschl, T., et al. (2010). Structural and functional insights into 5′-ppp RNA pattern recognition by the innate immune receptor RIG-I. *Nat. Struct. Mol. Biol.* **17**, 781–787.

Yao, N., Hesson, T., Cable, M., Hong, Z., Kwong, A. D., Le, H. V., and Weber, P. C. (1997). Structure of the hepatitis C virus RNA helicase domain. *Nat. Struct. Biol.* **4**, 463–467.

Yoneyama, M., Kikuchi, M., Natsukawa, T., Shinobu, N., Imaizumi, T., Miyagishi, M., Taira, K., Akira, S., and Fujita, T. (2004). The RNA helicase RIG-I has an essential function in double-stranded RNA-induced innate antiviral responses. *Nat. Immunol.* **5**, 730–737.

Yoneyama, M., Kikuchi, M., Matsumoto, K., Imaizumi, T., Miyagishi, M., Taira, K., Foy, E., Loo, Y. M., Gale, M., Jr., Akira, S., et al. (2005). Shared and unique functions of the DExD/H-box helicases RIG-I, MDA5, and LGP2 in antiviral innate immunity. *J. Immunol.* **175**, 2851–2858.

CHAPTER SEVEN

Analysis of Helicase–RNA Interactions Using Nucleotide Analog Interference Mapping

Annie Schwartz,* Makhlouf Rabhi,*,† Emmanuel Margeat,‡,§,¶ and Marc Boudvillain*

Contents

1. Introduction	150
2. Materials and Reagents	152
2.1. Chemicals	152
2.2. Equipment	152
2.3. Enzymes	153
2.4. Buffers	153
3. Methods	154
3.1. Preparation of NαS-modified RNA strands by *in vitro* transcription	154
3.2. Design and preparation of the helicase substrates	155
3.3. Helicase activity selection step	158
3.4. Sequencing of the selected populations of transcripts	161
3.5. Quantitation and normalization of NAIM signals	161
3.6. Accounting for potential sequence effects	163
3.7. Global analysis and interpretation of NAIM signals	164
3.8. Analysis of the periodicity of the intereference signals	165
4. Conclusions	166
Acknowledgments	167
References	167

Abstract

Nucleotide analog interference mapping (NAIM) is a combinatorial approach that probes individual atoms and functional groups in an RNA molecule and identifies those that are important for a specific biochemical function. Here, we

* CNRS UPR4301, Centre de Biophysique Moléculaire, Orléans cedex 2, France
† Ecole doctorale Sciences et Technologies, Université d'Orléans, Orléans, France
‡ CNRS UMR5048, Centre de Biochimie Structurale, Montpellier, France
§ INSERM, Montpellier, France
¶ Universités Montpellier, Montpellier, France

Methods in Enzymology, Volume 511   © 2012 Elsevier Inc.
ISSN 0076-6879, DOI: 10.1016/B978-0-12-396546-2.00007-3   All rights reserved.

show how NAIM can be adapted to reveal functionally important atoms and groups on RNA substrates of helicases. We explain how NAIM can be used to investigate translocation and unwinding mechanisms of helicases and discuss the advantages and limitations of this powerful chemogenetic approach.

1. Introduction

In order to move along RNA, unwind duplexes, and/or dissociate RNP complexes, RNA helicases need to hold and contact the RNA chain, often transiently and repeatedly. Determining which chemical groups in the RNA chain are necessary and contacted productively by a RNA helicase can provide great insights into its mechanisms of action. Chemical requirements are traditionally probed through the introduction of site-specific modifications in the RNA substrate. However, this method is expensive and time consuming as it requires the individual preparation and testing of many modified substrates in order to obtain meaningful information. An alternative strategy relies on the use of nucleotide analog interference mapping (NAIM; Fig. 7.1) to probe chemical requirements and RNA–helicase interactions in a comparably quick, combinatorial fashion (Schwartz et al., 2009; Taylor et al., 2010). NAIM screens a library of randomly modified RNA substrates for chemical moieties that are important for helicase function. Following fractionation of the initial population of substrates into subsets of functional (e.g., unwound) and nonfunctional molecules, cleavage of phosphorothioate linkage tags with iodine (Gish and Eckstein, 1988) reveals the locations of interfering chemical modifications, and thus of functionally important molecular moieties (Fig. 7.1). Although NAIM is useful in identifying important RNA atoms and functional groups, it does not report directly on their roles and interactions. Therefore, NAIM information should be analyzed and interpreted with great care, preferably in the context of other experimental information for the helicase under study.

To date, NAIM has been used to probe RNA folding, the binding of proteins, small molecule metabolites, or metal ions to RNA, the catalytic activity of ribozymes, or ionized states of active-site nucleobases (reviewed in Cuzic and Hartmann, 2005; Fedorova et al., 2005; Soukup and Soukup, 2009; Suydam and Strobel, 2009; Waldsich, 2008). NAIM has also been used to study RNA interactions *in vivo* (Szewczak, 2008) and to dissect RNA features governing intrinsic termination of transcription by a RNA polymerase (RNAP; Schwartz et al., 2003). Recently, NAIM has been introduced as a powerful tool to unravel the 2′-OH dependent mechanism of Rho, a bacterial transcription termination factor. Rho is a ring-shaped RNA helicase (Rabhi et al., 2010a,b) that requires contacts with 2′-hydroxyl groups at periodic intervals (every $\sim$7 nt) during RNA translocation and duplex unwinding (Schwartz et al., 2009). NAIM also helped to identify the

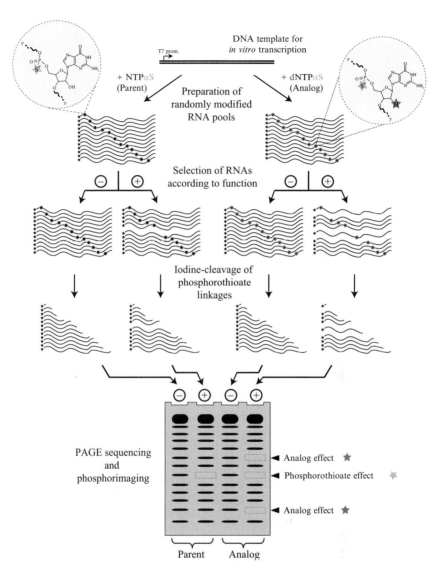

Figure 7.1 Principle of nucleotide analog interference mapping (NAIM). Nucleotide analogs are incorporated into transcripts at low (∼5%) levels. This yields chemical modifications of the RNA chain (magnifying circles) at random positions. (For color version of this figure, the reader is referred to the online version of this chapter.)

tolerance bias of the viral RNA helicase NPH-II toward purine deoxyribonucleotides (Taylor *et al.*, 2010).

The general principle (Fig. 7.1) and main technical features of NAIM experiments have been detailed thoroughly in several recent reviews (Cuzic and Hartmann, 2005; Fedorova *et al.*, 2005; Suydam and Strobel, 2009;

Waldsich, 2008). Here, we briefly discuss these aspects and rather concentrate on the adaptation of NAIM to investigate helicase mechanisms.

2. MATERIALS AND REAGENTS

2.1. Chemicals

- *Nucleotide triphosphates and analogs.* High-purity rNTP solutions (100 mM) can be obtained from GE Healthcare (#27-2025-01 for the complete rNTP set). A total of 36 distinct 5′-O-(1-thio)nucleoside triphosphates, including parental rNTPαS analogs, can be purchased from Glen Research (USA), Trilink Biotechnologies (USA), and IBA GmbH (Germany). Additional 5′-O-(1-thio)nucleoside triphosphates can be prepared from commercial nucleosides following a simple one-pot procedure developed by Arabshahi and Frey (1994) and refined by Strobel and coworkers. The procedure as well as utility of the various analogs for NAIM is thoroughly described in volume 468 of this series (Suydam and Strobel, 2009) and is thus not detailed here. The resulting preparations of the 5′-O-(1-thio) nucleoside triphosphates as well as most commercial preparations are racemic mixtures of Sp and Rp phosphorothioate diastereoisomers. Although T7 RNAP only incorporates the Sp diastereoisomers of the nucleotide analogs into transcripts (Griffiths *et al.*, 1987), the mixtures are usually used directly without further purification of the isomers (Suydam and Strobel, 2009). It is assumed that the Rp diastereoisomers do not poison T7 RNAP at the concentrations of the analogs used in the *in vitro* transcriptions (see Section 3.1) although this assumption has not been verified rigorously for most phosphorothioate-tagged nucleoside triphosphates. The purified Sp diastereoisomers of a selected set of analogs can be purchased from Biolog LSI (Germany).
- [γ-^{32}P]-ATP (6000 Ci/mmol, Perking-Elmer #NEG-035C).
- 40% Acrylamide:bis-acrylamide (19:1 ratio) solution (Interchim #UP86489B).
- Phenol:Chloroform:Isoamyl alcohol (PCA) mixture (Interchim #UP873359).
- Other chemicals and reagents are from Sigma–Aldrich.

2.2. Equipment

- Vertical, height-adjustable electrophoresis units for nucleic acid sequencing. We use Owl ADJ2 systems that are no longer sold but may be replaced by similar systems (e.g., SG-600-20 adjustable sequencing kit from CBS Scientific).

- Ludlum Model 3 survey meter equipped with a Model 44-7 detector.
- Typhoon Trio imager equipped with 35×43 cm storage phosphor screens and ImageQuant TL software (GE healthcare). Other systems for high-resolution (≤ 200 μm) storage phosphor detection from GE healthcare (e.g., Storm-820) or Bio-Rad (e.g., PharosFX Plus) are also adequate.

2.3. Enzymes

- The wild-type (WT) form (#T7905K) and permissive Y639F mutant (#D7P9205K) of T7 RNAP can be purchased from Epicentre Biotechnologies or prepared from overproducing strains obtained from Dr. Rui Sousa (University of Texas), following published procedures (Huang et al., 1997; Sousa and Padilla, 1995).
- Rho is overexpressed in BL21(DE3)pLysS cells harboring the pET28b–Rho plasmid (kindly provided by Dr. J.M. Berger, University of California at Berkeley). The procedures for the overexpression and purification of Rho are detailed elsewhere (Boudvillain et al., 2010a; Nowatzke et al., 1996).
- Alkaline phosphatase (Roche Applied Science #10713023001).
- T4 polynucleotide kinase (New England Biolabs #201 S).
- SUPERase-In (Ambion #AM2694).
- Q1 DNase (Promega #M6101).

2.4. Buffers

- Transcription buffer ($5 \times$): $0.12\ M$ $MgCl_2$, $0.4\ M$ HEPES, pH 7.5, $0.1\ M$ DTT, 0.1 % Triton X-100, and 5 mM Spermidine.
- Elution buffer: $0.3\ M$ sodium acetate, 10 mM MOPS, pH 6, and 1 mM EDTA.
- $M_{10}E_1$ buffer: 10 mM MOPS, pH 6, and 1 mM EDTA.
- Hybridization buffer ($10 \times$): 200 mM HEPES, pH 7.5, 1 mM EDTA, and 1.5 M potassium acetate.
- Helicase buffer ($10 \times$): 200 mM HEPES, pH 7.5, 1 mM EDTA, 5 mM DTT, and 1.5 M sodium glutamate.
- Denaturing loading buffer: 95% formamide, 5 mM EDTA, 0.01% (w/v) xylene cyanol, and 0.01% (w/v) bromophenol blue.
- Native loading buffer: 25% (w/v) Ficoll-400 (Sigma–Aldrich), 25 mM EDTA, 0.05% (w/v) xylene cyanol, and 0.05% (w/v) bromophenol blue.
- Quench buffer: 100 mM EDTA, 4% (w/v) SDS, and 16% (w/v) Ficoll-400.

3. Methods

3.1. Preparation of NαS-modified RNA strands by *in vitro* transcription

The RNA strands that we use for assembly of the helicase substrates never exceed 200 nt. The RNAs are easily prepared by *in vitro* transcription with T7 RNAP and synthetic DNA templates. The easiest way to obtain the DNA templates is by enzymatic fill-in of synthetic DNA oligonucleotides that overlap and base pair partially (Fig. 7.2), using standard PCR procedures (Schwartz *et al.*, 2007). In most cases, the templates do not need to be gel purified. Elimination of salts, dNTPs, and DNA polymerase from the PCR mixtures is performed by simple standard procedures such as PCA extraction followed by desalting on a spin disposable column (e.g., Sephadex G25 column) and ethanol precipitation in the presence of 3 M ammonium acetate.

The random incorporation of a NαS or 2′-deoxyNαS (dNαS) analog into RNA strands is achieved by doping transcription mixtures with the corresponding 5′-O-(1-thio)nucleoside triphosphate. The T7 RNAP variant (WT or Y639F) and rNTP and NTPαS (or dNTPαS) concentrations are chosen to yield ~5% incorporations per transcript (Table 7.1), although conditions may need reoptimization for particular transcripts (Suydam and Strobel, 2009). This low level of analog incorporation usually provides

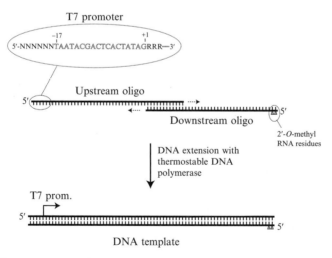

Figure 7.2 Preparation of DNA templates for T7 transcription. The gray triangles indicate the presence of 2′-O-methyl RNA residues at the 5′-end of the downstream oligonucleotide which reduces the level of nontemplated addition of extra nucleotides at the 3′-end of transcripts (Kao *et al.*, 2001).

Table 7.1 Conditions required for ~5% incorporation per transcript of the most common NαS and dNαS analogs[a]

Incorporated analog	[NTPαS] (mM)	[parent rNTP] (mM)	RNAP
AαS, CαS, GαS, UαS	0.05	1	WT
dAαS, dCαS	1.5	1	Y639F
dGαS, dUαS	0.5	1	Y639F
IαS	0.4	1	WT

[a] Adapted from Fedorova *et al.* (2005) and Suydam and Strobel (2009).

sufficient experimental signal for NAIM while minimizing the probability that a given transcript bears multiple inhibitory modifications (Suydam and Strobel, 2009). Typically, transcription mixtures contain 10 nM DNA template, 0.1 U/μL SUPERase-In, 2 U/μL T7 RNAP, and 1 mM rNTPs (excepted for the analogs and their parent rNTPs; Table 7.1) in transcription buffer and are incubated for 2–4 h at 37 °C. The DNA templates are then digested by the addition of 0.1 U/μL RQ1 DNase to the mixture followed by incubation for 20 min at 37 °C. Subsequently, the mixture is PCA extracted, ethanol precipitated, and redissolved in $M_{10}E_1$ buffer. Transcripts are purified by denaturing polyacrylamide gel electrophoresis (PAGE), visualized by UV shadowing with a 254 nm lamp, and excised and eluted from the gel (Boudvillain *et al.*, 2010a). The eluate is PCA extracted, ethanol precipitated, and redissolved in $M_{10}E_1$ buffer before being stored at −20 °C. Typical yields range from 1 to 4 nmol of purified transcripts for a 250 μL transcription. Because addition of nucleotides to the RNA chain by T7 RNAP proceeds with inversion of the configuration at the phosphorus center (S_N2-like mechanism), the modified transcripts contain only Rp phosphorothioate linkages (Griffiths *et al.*, 1987).

3.2. Design and preparation of the helicase substrates

Design of the RNA:DNA or RNA·RNA duplex constructs used in NAIM experiments is a critical step. Typically, substrates used in unwinding assays are formed by base pairing two complementary nucleic acid strands. One of the two strands is usually longer so that the "reporter" duplex region is flanked by a single-stranded overhang which is used by the helicase for loading on the substrate. The position of the overhang with respect to the duplex region is chosen based on the polarity of translocation of the helicase (*Note*: exception for nontranslocating DEAD-box helicases; see Potratz *et al.* (2010) and references therein).

When designing substrates for NAIM, three points should be considered:

(i) *General architecture of the substrate.* RNAs should be tailored to limit conformational heterogeneity and fluctuations in the starting population of helicase–substrate complexes. If the helicase is a translocase, it is important to ensure that the helicase will not start translocating from multiple nucleotide positions. Multiple start sites will result in dispersion of interference effects along the RNA, which decreases signal:noise ratios. This may complicate detection of periodic trends in interference signals that reveal translocation/unwinding step sizes (see Section 3.7).

(ii) *Length of the duplex region.* Duplex length should be sufficient to detect periodic interference signals (e.g., multiple translocation steps). However, the number of distinct species within the initial pool of substrates (e.g., substrates with a NαS or dNαS modification at a given position) usually increases with the substrate length. This lowers abundances of any given substrate species in the fractionated populations (unreacted and unwound substrates). Therefore, the signal-to-noise ratios for NAIM signals decreases with the substrate length, if this correlates with a larger number of substrate positions to be probed.

(iii) *Lack of redundant/stabilizing components.* Stringent selection conditions usually increase the number and quality of NAIM signals (Boudvillain and Pyle, 1998; Fedorova et al., 2005). For the unwinding reaction, selection stringency may be achieved by removing redundant or ancillary parts of the substrate that are not critical for helicase function. This trimming prevents multiple helicases from binding and cooperating on the same substrate (Byrd and Raney, 2004). In addition, a minimal substrate may increase the possibility that stability of the helicase:RNA complex and/or unwinding activity become critically affected by single chemical modifications. Stringency can also be improved by nonpermissive conditions for the unwinding selection step (see Section 3.3).

For Rho, the design of RNA:DNA constructs for NAIM analysis was facilitated by prior knowledge of the catalytically competent Rho:RNA configuration (Schwartz et al., 2009). It was especially helpful to know that distinct RNA regions associate with primary binding site (PBS) and secondary binding site (SBS) on the Rho hexamer (Fig. 7.3A). While PBS:RNA contacts are important for forming the Rho:RNA complex, SBS:RNA contacts govern the ATP-dependent translocation of Rho along RNA (reviewed in Rabhi et al., 2010a,b).

In our minimal tripartite RNA:RNA:DNA substrates (Fig. 7.3B; Schwartz et al., 2007), individual nucleic acid strands interact with distinct functional parts of Rho. An "anchoring" RNA arm contacts the PBS on top of the hexamer, a "tracking" arm binds the SBS in the inner Rho channel. The reporter DNA oligonucleotide contacts and is excluded sterically by Rho's bottom surface upon RNA translocation (Walmacq

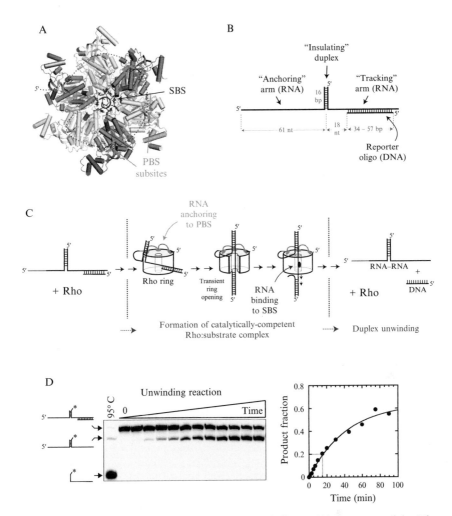

Figure 7.3 Transcription termination factor Rho helicase. (A) Structure of the Rho hexamer (PDB #3ICE) showing components of the primary binding site (PBS, in green) and secondary binding site (SBS, in blue) as well as the helicoidally shaped rU_6 fragment (in red) occupying the inner channel of the hexamer (Thomsen and Berger, 2009). Although no PBS-bound RNA is visible in this particular structure, PBS pockets on Rho's subunits are occupied by 5'-UC dinucleotides in other structures (Skordalakes and Berger, 2003, 2006). The red dotted line thus depicts the putative RNA trajectory on the crown-like PBS. (B) Schematic depicting general architecture of the minimal RNA:RNA:DNA constructs used to probe Rho mechanisms by NAIM. (C) Probable reaction pathway leading to Rho anchoring to and unwinding of the RNA:RNA:DNA constructs (Rabhi et al., 2010a,b). (D) A representative gel showing the specific Rho-directed unwinding of the RNA:DNA hybrid region of a minimal RNA:RNA:DNA construct (adapted from Rabhi et al., 2011). Red dotted lines on the graph identify the incubation time required for the NAIM selection (20% reaction extent). (For interpretation of the references to color in this figure legend, the reader is referred to the online version of this chapter.)

et al., 2006; Fig. 7.3B). The configuration of the RNA:RNA:DNA constructs facilitates the assignment of residues revealed by NAIM to interactions with either the PBS or SBS (Fig. 7.3C). The presence of an "insulating" RNA duplex upstream from the reporter RNA:DNA helix (Fig. 7.3B) limits the conformational dispersion of reactive Rho:RNA species (Fig. 7.3C) while the lack of a canonical, C-rich *Rut* (Rho utilization) loading site in the "anchoring" RNA arm reduces the number of potential stabilizing contacts with Rho (Schwartz et al., 2007).

3.2.1. Preparation and purification of the tripartite RNA:RNA:DNA substrates

In each NAIM experiment, only one of the two RNA strands is modified with the NαS (or dNαS) analogs and labeled with ^{32}P. Analog-containing transcripts are first dephosphorylated with alkaline phosphatase, labeled with [γ-^{32}P]-ATP and T4 polynucleotide kinase, and then purified by PAGE following standard protocols (Boudvillain et al., 2010a; Gimple and Schön, 2005). Then, approximately 10 pmol of the ^{32}P-labeled RNA strand are mixed with 1.1–1.2 molar equivalents of the other, unlabeled strands (second RNA arm and reporter DNA oligonucleotide; Fig. 7.3B) in 10 μL of hybridization buffer. The mixture is heated for 2 min at 95 °C and then slowly cooled (3 °C/min) to 20 °C in a thermocycler equipped with a heated lid to avoid sample evaporation. The mixture is then combined with 0.2 volumes of native loading buffer and resolved by native PAGE at room temperature. The RNA:RNA:DNA substrates are visualized by autoradiography and excised and eluted from the gel (Boudvillain et al., 2010a). Elution is usually performed at 15 °C for 3–5 h in a Peltier incubator (e.g., Thermomixer, Eppendorf) under gentle shaking. The eluate is PCA extracted, ethanol precipitated, and redissolved in helicase buffer before being stored at −20 °C. Yields generally do not exceed 20% (≤2 pmol of RNA:RNA:DNA substrates).

3.3. Helicase activity selection step

As noted above, selection stringency is critical to ensure optimal NAIM conditions (Boudvillain and Pyle, 1998; Fedorova et al., 2005; Suydam and Strobel, 2009). Various parameters can be tested to reach nonpermissive unwinding conditions for a given helicase. For instance, NAIM trials with the NPH-II helicase revealed that interference signals were highly dependent on salt concentration, reaction extent, and enzyme turnover. Strong signals required relatively high-salt concentrations (4 mM Mg(OAc)$_2$ and ≥70 mM NaCl), less than 20% duplex unwinding, and single-cycle unwinding conditions (obtained by scavenging of unbound helicases with a trap RNA molecule; Fedorova et al., 2005; Taylor et al., 2010).

For Rho, wide variations of reaction conditions were precluded by the minimal tripartite RNA:RNA:DNA substrates (Fig. 7.3B). Unwinding of these substrates requires glutamate ions in the reaction and unwinding under single-cycle conditions is not efficient (Schwartz et al., 2007). Multiple-cycle unwinding conditions introduce additional steps in the reaction pathway due to multiple helicase binding events. As discussed above, this situation potentially complicates interpretation of interference signals. The choice of the minimal tripartite RNA:RNA:DNA substrates was thus a tradeoff between advantageous topological features of the substrates and stringent reaction parameters. Adequate selection conditions for NAIM with these substrates included high-salt (150 mM sodium glutamate), a 4:1 Rho:substrate ratio, and an extent of duplex unwinding of roughly 20% (Schwartz et al., 2009).

Before performing selection reactions with a helicase, it is necessary to evaluate the rate of duplex unwinding. To this end, a scaled-down unwinding experiment is performed with ~0.1 pmol of the substrates under the conditions used for the selection. Reaction aliquots are withdrawn, quenched at various times, and resolved by PAGE as described below. Gel bands are visualized and quantified by phosphorimaging (Typhoon Trio Imager, GE). Rate constants are calculated from the rate of appearance of the product band (Fig. 7.4) according to

$$F_P = A\left(1 - e^{-k\exp^t}\right) + k_{\text{lin}}t$$

F_p is the fraction of product formed, A is the amplitude of the fast reaction phase, and k_{exp} and k_{lin} are the rate constants of the exponential and linear phases of the reaction, respectively (k_{lin} is zero if the reaction exhibits first-order kinetics; Rabhi et al., 2011; Walmacq et al., 2004).

To perform unwinding selections, the ^{32}P-labeled RNA:RNA:DNA substrates (~10^5 cpm; 5 nM final concentration) are mixed with Rho hexamers (20 nM final concentration) in helicase buffer and incubated for 3 min at 30 °C. Then, an initiation mixture containing Mg(OAc)$_2$ and ATP (1 mM, final concentrations) and a DNA trap (400 nM, final concentration; the DNA trap is complementary to the reporter oligonucleotide released upon duplex unwinding) is added before further incubation at 30 °C. The reaction is stopped at 20% duplex unwinding by addition of 0.3 volumes of quench buffer and transfer to ice (Fig. 7.3D). Reaction products are separated on a 7.5% (w/v) polyacrylamide gel (1× TBE, 0.5% (w/v) SDS), visualized by autoradiography, and excised and eluted from the gel (Boudvillain et al., 2010a). The eluates are PCA extracted, ethanol precipitated, and redissolved in M$_{10}$E$_1$ buffer. Amounts of unwound and unreacted species recovered at this stage should represent ~10^4 and ~5×10^4 cpm of radioactive decay, respectively. To facilitate subsequent stages, it is

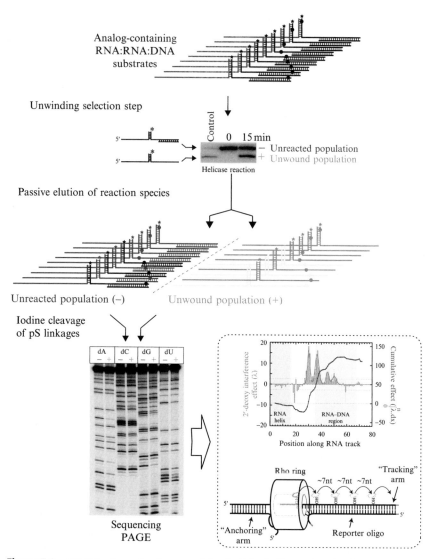

Figure 7.4 NAIM scheme implemented for the Rho helicase. Randomly incorporated nucleotide analogs are depicted by black circles. Black stars represent ^{32}P-labelings of the RNA transcripts probed by NAIM. The graph shows the periodic distribution of $2'$-dNαS interference effects along the tracking RNA arm. This type of NAIM data was used to deduce that the 2-hydroxyl-dependent translocation of Rho has an $\sim$7 nt "activation" step size (Schwartz et al., 2009). (For color version of this figure, the reader is referred to the online version of this chapter.)

recommended to equal amounts of unwound and unreacted species (e.g., 10^3 cpm/μL). Note that the cpm counts indicated above were determined at the bench with a Ludlum 3.44-7 survey meter; they do not necessarily correspond to accurate measurements performed with a scintillation counter.

3.4. Sequencing of the selected populations of transcripts

To cleave the phosphorothioate linkages in the transcripts (Fig. 7.1), one half of each solution of purified unwound and unreacted species is mixed with 0.1 volume of a 10 mM ethanolic solution of iodine that has been freshly prepared. Each of the other halves is mixed with 0.1 volume of ethanol to provide uncleaved control samples. The mixtures are incubated for 1 min at 52 °C and mixed with 1 volume of denaturing loading buffer. Then, 8 μL of the mixtures are heated for 2 min at 95 °C before being loaded on a prewarmed, denaturing PAGE gel containing 7 M urea. Gels are run at a high wattage (e.g., 45 W for a 40 × 20 cm gel). Typically, different acrylamide percentages (4–20%) are used to resolve different RNA sizes. Gels containing less than 15% acrylamide are vacuum-dried while gels containing higher percentages are phosphorimaged directly. An example of NAIM sequencing gel is presented in Fig. 7.4.

3.5. Quantitation and normalization of NAIM signals

Storage phosphor screens are exposed overnight to the sequencing gels before imaging with the Typhoon Trio imager using standard, built-in phosphorimaging settings. Then, the ImageQuantTL software is used to quantify the intensity of each band in duplex (Dupl) and unwound (Unw) product lanes for both parental (rNαS) and modified (e.g., dNαS) nucleoside phosphorothioate-containing RNAs. We use the automated 1D gel analysis routine of ImageQuantTL (lane creation, "rubber-band" background correction, peak detection, and integration). Nonetheless, visual inspection and manual adjustments are usually required to account for distortions in the shapes of gel lanes and bands and for detection of bands with low intensities. Additional useful advices for optimal quantitation of gel bands can be found in Suydam and Strobel (2009).

The NAIM signals are determined from the intensities of individual bands after normalization for differences in loading of the gel lanes. The interference (κ) at a given position is the ratio of normalized band intensities (Ortoleva-Donnelly et al., 1998).

For the parental NαS modifications,

$$\kappa = I_{N\alpha S}(\text{Dupl})/I_{N\alpha S}(\text{Unw})$$

For other modifications, such as dNαS modifications if raw interference signals are desired:

$$\kappa = I_{dN\alpha S}(\text{Dupl})/I_{dN\alpha S}(\text{Unw})$$

To discount effects due to phosphorothioate tags (normalized interference signals),

$$\kappa = [I_{dN\alpha S}(\text{Dupl})/I_{dN\alpha S}(\text{Unw})]/[I_{N\alpha S}(\text{unreacted})/I_{N\alpha S}(\text{unwound})]$$

Raw κ values are used to quantify interference effects due to dNαS substitutions as whole entities, a procedure that is sometimes useful to increase experimental output (Schwartz et al., 2009). Raw κ values are also used in conservative quantitation approaches that assume effects from the phosphorothioate tag and from the accompanying nucleoside modification are not necessarily independent (Waldsich, 2008).

A $\kappa > 1$ indicates a nucleotide modification that is detrimental to helicase-directed unwinding of the RNA:DNA substrate. A $\kappa < 1$ reveals a modification that favors strand separation. Although different thresholds for interference effects have been used in NAIM studies, conservative analyzes usually posit that only κ values below 0.5 or above 2 reflect significant effects (Suydam and Strobel, 2009). In any case, it is important to ensure that the interference patterns are reproducible and are not biased by iodine-independent RNA cleavage (e.g., nonspecific degradation; Suydam and Strobel, 2009). Typically, errors on κ values derived from 2 to 3 independent experiments should not exceed 20%.

To facilitate data analysis and comparisons, we convert interference effects into normalized λ discrimination factors (Schwartz et al., 2003). For each homogeneous set of κ values (for instance, 2′-dAαS signals observed for a specific RNA:DNA substrate under given unwinding conditions), we exclude κ values smaller than 0.5 and larger than 2 and calculate the standard deviation (SD) on the reduced data set. Then, we define a discrimination factor λ for every κ value of the original data set:

$$\lambda = (\kappa - 1)/\text{SD}, \quad \text{for} \kappa > 1$$
$$\lambda = 1 - (1/\kappa)/\text{SD}, \quad \text{for} \kappa < 1$$

NAIM signals are thus normalized for varying experimental quality and composition of the data sets (Schwartz et al., 2003). The λ discrimination factors also provide identical intensity scales for favorable ($\lambda < 0$) and detrimental ($\lambda > 0$) effects (Schwartz et al., 2003). We traditionally use conservative confidence intervals of >98.8% and >99.9% to define moderate

($2.5 < |\lambda| < 3.5$) and strong ($|\lambda| > 3.5$) interference effects, respectively (Schwartz et al., 2003, 2009).

3.6. Accounting for potential sequence effects

The nucleotide sequence of the substrate can have various effects on the activity of the helicase under study. These effects may stem from sequence-dependent helicase:substrate interactions and/or strengths of base-pairings. For instance, the RNA helicase NPH-II displays a strong purine bias and, as a result, can efficiently proceed along a purine-rich DNA track (Taylor et al., 2010). This bias results in 2′-deoxy interference effects that are generally much stronger at pyrimidines than at purine positions (Taylor et al., 2010). However, the nucleotide sequence also impacts the efficiency of incorporation of NαS analogs into transcripts, which may prevent NAIM experiments with analogs that are not easily incorporated by T7 RNAP (Suydam and Strobel, 2009; Waldsich, 2008).

It is generally not practical to probe each of the 4^z combinations of sequences corresponding to a RNA substrate with z nucleotides, given that z is usually larger than 20. However, NAIM experiments with a small set of substrates of well-chosen compositions may already reveal sequence bias. While only four distinct substrates are required to systematically vary nucleotide identity at each position, sequence combination effects may escape notice upon testing only few specimens.

For Rho, NAIM experiments were conducted with six different substrates where only the identity of base pairs within the most informative RNA:DNA hybrid region were systematically varied (Schwartz et al., 2009). The substrates were designed to level, as best as possible, the representation of the nearest-neighbor base pair combinations within the RNA:DNA hybrid while maintaining sufficient sequence variations for adequate PAGE sequencing (see Section 3.4) and signal quantitation (see Section 3.5). Moreover, sequence variations at helix termini were limited to doublets of G/C pairs in order to limit spontaneous opening fluctuations. The impact of hybrid sequence on NAIM signals was then examined by searching for correlations between base pairing strengths and interference signals with a quantitative structure activity relationships (QSAR) algorithm (see Section 3.7). We also compared, for each position of the hybrid, the average interference effect obtained as a function of nucleotide identity.

The average interference $\lambda_{A(p)}$, for an A residue at the p position was calculated according to

$$\lambda_{A(p)} = 1/i \cdot \sum_{0 \to i} [\lambda_A]$$

where λ_A is the discrimination factor calculated for the p position for each of the i hybrids containing an A residue at this position (Schwartz et al., 2009).

It may be also useful to generate sequence-weighted NAIM signals using the following formula:

$$\lambda_{\text{weighted}(p)} = 1/4i \cdot \sum_{0 \to i}[\lambda_A] + 1/4j \cdot \sum_{0 \to j}[\lambda_C] + 1/4m \cdot \sum_{0 \to m}[\lambda_G] + 1/4n \cdot \sum_{0 \to n}[\lambda_U]$$

Here, λ_A, λ_C, λ_G, and λ_U are the discrimination factors for the hybrids containing, respectively, an A, C, G, and U residue at the p position and i, j, m, and n are the numbers of hybrids containing an A, C, G, and U residue at this position (Schwartz et al., 2009).

3.7. Global analysis and interpretation of NAIM signals

Duplex unwinding by helicases is a multistep reaction. Consequently, chemical modifications at critical substrate positions can induce a variety of effects, which can be difficult to identify and to deconvolute. Deleterious NAIM signals resulting from NαS or dNαS modifications of the substrate can arise from: (i) the rate-limiting step(s) of the helicase reaction slowed down by the modification; (ii) another step becoming rate-limiting; (iii) forcing the enzyme onto an alternative reaction pathway with a new slow step (for instance, sliding from a position of detrimental 2′-deoxy modification to a position where a 2′-hydroxyl group is available for adequate function); (iv) destabilizing the helicase:substrate complex. These scenarios are not mutually exclusive. In contrast to deleterious NAIM signals, beneficial signals can be attributed to only one of two factors: (i) the rate-limiting step(s) is (are) faster; (ii) the fraction of substrate unwound per helicase run is higher because the chemical modification renders the helicase more processive or because it reduces the number of helicase translocation steps required before the duplex dissociates.

For Rho, interpretation of NAIM signals was aided by the configuration of the minimal RNA:RNA:DNA substrates, by knowledge of substrate–Rho interactions (Fig. 7.3C), and by the steric exclusion model. (Walmacq et al., 2006). These constraints made clear that interference effects within the RNA:DNA region of the substrates were not linked to helicase activation since this step precedes RNA:DNA unwinding (Fig. 7.3). Consequently, these interference effects report on stepping/unwinding of Rho (Schwartz et al., 2009).

We usually attribute the NAIM signals to perturbations of the helicase that result from disruption of interactions with functional groups (e.g., 2′-OH groups for dNαS modifications) or nonbridging pro-R oxygen atoms (phosphorothioate effects). Although a single 2′-OH → 2′-H change can affect the local structure and stability of a RNA:DNA helix, effects are generally weak (Freier and Altmann, 1997; Wang and Kool, 1995; Wyatt and Walker, 1989). These stability changes are unlikely to account alone for large interference signals, such as those observed in the RNA:DNA hybrid region with the Rho helicase (Fig. 7.4; Schwartz et al., 2009).

In general, 2′-deoxy effects are easier to interpret than effects by phosphorothioate substitutions, which induce subtle structural and electronic perturbations that may alter interaction networks, local structures, or charge distributions (Brautigam and Steitz, 1998; LeCuyer et al., 1996; Loverix et al., 1998; Smith and Nikonowicz, 2000). Phosphorothioate effects may thus not solely stem from alteration of contacts between the helicase and nonbridging pro-R oxygens of the RNA substrate. Favorable phosphorothioate effects may result from beneficial, sulfur-mediated interactions with the helicase. Given the complexity of causes for NAIM effects, the data are best interpreted in conjunction with other mechanistic information.

3.8. Analysis of the periodicity of the intereference signals

To quantify the period of the bell-shaped 2′-deoxy interference peaks, we first performed an autocorrelation analysis of the dNαS interference signals on the tracking arm (Fig. 7.4). We used a simple built-in LabVIEW (National Instruments, Austin, TX) routine (AutoCorrelation.vi). The input array was the λ values as a function of the position along the tracking arm ($\lambda(p)$), and the correlation function was calculated as:

$$A(t) = \int_0^\infty \lambda(p)\lambda(p+t)dt$$

The calculated autocorrelation curves obtained for the Rho helicase display a second maximum around position 7, indicating a periodicity of $\sim$7 bp for the interference signals (Schwartz et al., 2009). Similar results were obtained by a power spectrum analysis that extracts the frequency content of a process (Press et al., 2007; own unpublished data).

To detect correlation between the interference values at specific positions ($\lambda(p)$) and local and/or global physicochemical parameters, we used a set of global parameters, whose value does not depend on the position p, including:

- k_{exp}, the rate constant for the exponential phase of the helicase reaction measured with the corresponding hybrid (Section 3.3)
- A, the amplitude of the exponential phase of the helicase reaction (Section 3.3)
- k_{lin}, the rate constant of the linear, steady-state phase of the helicase reaction (Section 3.3).

In addition, we used local parameters with a specific value for each interference position probed, p:

- $\Delta G(p)$, the Gibbs free energy of the RNA:DNA base pair (Sugimoto et al., 1995);
- $T_m(p)$, the melting temperature of the RNA:DNA region between the interference position and the end of the hybrid;
- $Nb_{end}(p)$, the number of base pairs between the interference position and the end of the hybrid.

We then performed a numerical analysis to detect correlation between the interference value at a specific position ($\lambda(p)$), and these physicochemical descriptors. We used tools originally developed for QSAR studies (TSAR software, Accelerys, San Diego, CA). However, any software that can calculate correlation coefficients between datasets can be used. The best linear combination of descriptors gave an overall correlation coefficient of 0.54. The best individual correlation was with the helicase amplitude, A, although it was rather low ($R = 0.28$). Additionally, we performed a principal component analysis and a cluster analysis (using the built-in routines in TSAR), to search for composite correlations that could quantitatively explain amplitude variations of the interference signals. However, no significant correlation was found. The absence of significant correlations suggested that the 7 bp periodicity of the dNαS interference signals stems from an intrinsic feature of Rho chemomechanical cycle. Based on these findings and on other structural and biochemical data, we proposed a set of models for the chemomechanical cycle of Rho (Boudvillain et al., 2010b; Patel, 2009).

4. Conclusions

NAIM has been adapted to a large variety of RNA-based systems (reviewed in Cuzic and Hartmann, 2005; Fedorova et al., 2005; Suydam and Strobel, 2009). Although the method has proven robust and versatile in most cases, NAIM only identifies atoms and functional groups that are important for a particular system under the conditions of selection. To attribute precise functional roles to the identified functional groups, NAIM data are usually combined with data obtained by other methods, such as structural analysis and enzyme kinetics. Notwithstanding, sophisticated NAIM schemes make it possible to detect patterns of energetic communications or specific tertiary contacts in RNA (Boudvillain and Pyle, 1998; Boudvillain et al., 2000; Fedorova and Pyle, 2005; Jansen et al., 2006; Strobel and Ortoleva-Donnelly, 1999; Strobel et al., 1998; Szewczak et al., 1998, 1999). Energetic coupling is revealed by suppression of interference signals in NAIM profiles of mutant RNA where specific groups have been altered (nucleotide analog interference suppression (NAIS) strategy). To date, NAIS has been restricted to RNA-only systems. However, there is no conceptual obstacle to the identification of RNA–protein interactions through NAIS probing of molecular complexes containing single-point protein mutants. This type of analysis may be particularly useful to deconvolute NAIM information on highly dynamic systems such as RNA helicases. We are currently examining the use of NAIS for RNA helicases. The NAIM studies of Rho (Schwartz et al., 2009), NPH-II (Taylor et al., 2010), and intrinsic termination of transcription by T7 RNAP (Schwartz et al., 2003) demonstrate the potential of this approach to unravel transient mechanistic features of molecular motors working on RNA.

ACKNOWLEDGMENTS

Financial support from the Agence Nationale de la Recherche (PCV, 2006 and Blanc, 2010) and the Conseil Régional du Centre (AO, 2007) is gratefully acknowledged.

REFERENCES

Arabshahi, A., and Frey, P. A. (1994). A simplified procedure for synthesizing nucleoside 1-thiotriphosphates: dATP alpha S, dGTP alpha S, UTP alpha S, and dTTP alpha S. *Biochem. Biophys. Res. Commun.* **204**, 150–155.

Boudvillain, M., and Pyle, A. M. (1998). Defining functional groups, core structural features and inter-domain tertiary contacts essential for group II intron self-splicing: A NAIM analysis. *EMBO J.* **17**, 7091–7104.

Boudvillain, M., de Lencastre, A., and Pyle, A. M. (2000). A tertiary interaction that links active-site domains to the 5' splice site of a group II intron. *Nature* **406**, 315–318.

Boudvillain, M., Walmacq, C., Schwartz, A., and Jacquinot, F. (2010a). Simple enzymatic assays for the in vitro motor activity of transcription termination factor Rho from Escherichia coli. *In* "Helicases: Methods and Protocols," (M. Abdelhaleem, ed.) Vol. 587, , pp. 137–154. Humana Press Inc., Totowa.

Boudvillain, M., Nollmann, M., and Margeat, E. (2010b). Keeping up to speed with the transcription termination factor Rho motor. *Transcription* **1**, 70–75.

Brautigam, C. A., and Steitz, T. A. (1998). Structural principles for the inhibition of the 3'–5' exonuclease activity of Escherichia coli DNA polymerase I by phosphorothioates. *J. Mol. Biol.* **277**, 363–377.

Byrd, A. K., and Raney, K. D. (2004). Protein displacement by an assembly of helicase molecules aligned along single-stranded DNA. *Nat. Struct. Mol. Biol.* **11**, 531–538.

Cuzic, S., and Hartmann, R. K. (2005). Nucleotide analog interference mapping: Application to the RNase P system. *In* "Handbook of RNA Biochemistry," (R. K. Hartmann, A. Bindereif, A. Schön, and E. Westhof, eds.), Vol. 1, pp. 294–318. Wiley-VCH, Weinheim.

Fedorova, O., and Pyle, A. M. (2005). Linking the group II intron catalytic domains: Tertiary contacts and structural features of domain 3. *EMBO J.* **24**, 3906–3916.

Fedorova, O., Boudvillain, M., Kawaoka, J., and Pyle, A. M. (2005). Nucleotide analog interference mapping and suppression: Specific applications in studies of RNA tertiary structure, dynamic helicase mechanism and RNA-protein interactions. *In* "Handbook of RNA Biochemistry," (R. K. Hartmann, A. Bindereif, A. Schön, and E. Westhof, eds.), Vol. 1, pp. 259–293. Wiley-VCH, Weinheim.

Freier, S. M., and Altmann, K. H. (1997). The ups and downs of nucleic acid duplex stability: Structure-stability studies on chemically-modified DNA:RNA duplexes. *Nucleic Acids Res.* **25**, 4429–4443.

Gimple, O., and Schön, A. (2005). Direct determination of RNA sequence and modification by radiolabeling methods. *In* "Handbook of RNA Biochemistry," (R. K. Hartmann, A. Bindereif, A. Schön, and E. Westhof, eds.), Vol. 1, pp. 133–171. Wiley-VCH, Weinheim.

Gish, G., and Eckstein, F. (1988). DNA and RNA sequence determination based on phosphorothioate chemistry. *Science* **240**, 1520–1522.

Griffiths, A. D., Potter, B. V., and Eperon, I. C. (1987). Stereospecificity of nucleases towards phosphorothioate-substituted RNA: Stereochemistry of transcription by T7 RNA polymerase. *Nucleic Acids Res.* **15**, 4145–4162.

Huang, Y., Eckstein, F., Padilla, R., and Sousa, R. (1997). Mechanism of ribose $2'$-group discrimination by an RNA polymerase. *Biochemistry* **36**, 8231–8242.

Jansen, J. A., McCarthy, T. J., Soukup, G. A., and Soukup, J. K. (2006). Backbone and nucleobase contacts to glucosamine-6-phosphate in the glmS ribozyme. *Nat. Struct. Mol. Biol.* **13**, 517–523.

Kao, C., Rudisser, S., and Zheng, M. (2001). A simple and efficient method to transcribe RNAs with reduced $3'$ heterogeneity. *Methods* **23**, 201–205.

LeCuyer, K. A., Behlen, L. S., and Uhlenbeck, O. C. (1996). Mutagenesis of a stacking contact in the MS2 coat protein-RNA complex. *EMBO J.* **15**, 6847–6853.

Loverix, S., Winquist, A., Stromberg, R., and Steyaert, J. (1998). An engineered ribonuclease preferring phosphorothioate RNA. *Nat. Struct. Biol.* **5**, 365–368.

Nowatzke, W., Richardson, L., and Richardson, J. P. (1996). Purification of transcription termination factor Rho from Escherichia coli and Micrococcus luteus. *Methods Enzymol.* **274**, 353–363.

Ortoleva-Donnelly, L., Szewczak, A. A., Gutell, R. R., and Strobel, S. (1998). The chemical basis of adenosine conservation throughout the Tetrahymena ribozyme. *RNA* **4**, 498–519.

Patel, S. S. (2009). Structural biology: Steps in the right direction. *Nature* **462**, 581–583.

Potratz, J. P., Tijerina, P., and Russell, R. (2010). Mechanisms of DEAD-box proteins in ATP-dependent processes. *In* "RNA Helicases," (E. Jankowsky, ed.) Vol. 19, pp. 61–98. RSC Publishing, Cambridge (UK).

Press, W. H., Teukolsky, S. A., Vetterling, W. T., and Flannery, B. P. (2007). Numerical Recipes: The Art of Scientific Computing. Cambridge University Press, Cambridge.

Rabhi, M., Rahmouni, A. R., and Boudvillain, M. (2010a). Transcription termination factor Rho: A ring-shaped RNA helicase from bacteria. *In* "RNA Helicases," (E. Jankowsky, ed.) Vol. 19, pp. 243–271. RSC Publishing, Cambridge (UK).

Rabhi, M., Tuma, R., and Boudvillain, M. (2010b). RNA remodelling by hexameric helicases. *RNA Biol.* **7**, 16–27.

Rabhi, M., Gocheva, V., Jacquinot, F., Lee, A., Margeat, E., and Boudvillain, M. (2011). Mutagenesis-based evidence for an asymmetric configuration of the ring-shaped transcription termination factor Rho. *J. Mol. Biol.* **405**, 497–518.

Schwartz, A., Rahmouni, A. R., and Boudvillain, M. (2003). The functional anatomy of an intrinsic transcription terminator. *EMBO J.* **22**, 3385–3394.

Schwartz, A., Walmacq, C., Rahmouni, A. R., and Boudvillain, M. (2007). Noncanonical interactions in the management of RNA structural blocks by the transcription termination rho helicase. *Biochemistry* **46**, 9366–9379.

Schwartz, A., Rabhi, M., Jacquinot, F., Margeat, E., Rahmouni, A. R., and Boudvillain, M. (2009). A stepwise $2'$-hydroxyl activation mechanism for the ring-shaped transcription termination Rho helicase. *Nat. Struct. Mol. Biol.* **16**, 1309–1316.

Skordalakes, E., and Berger, J. M. (2003). Structure of the rho transcription terminator. Mechanism of mRNA recognition and helicase loading. *Cell* **114,** 135–146.

Skordalakes, E., and Berger, J. M. (2006). Structural insights into RNA-dependent ring closure and ATPase activation by the Rho termination factor. *Cell* **127,** 553–564.

Smith, J. S., and Nikonowicz, E. P. (2000). Phosphorothioate substitution can substantially alter RNA conformation. *Biochemistry* **39,** 5642–5652.

Soukup, J. K., and Soukup, G. A. (2009). Identification of metabolite-riboswitch interactions using nucleotide analog interference mapping and suppression. *Methods Mol. Biol.* **540,** 193–206.

Sousa, R., and Padilla, R. (1995). A mutant T7 RNA polymerase as a DNA polymerase. *EMBO J.* **14,** 4609–4621.

Strobel, S. A., and Ortoleva-Donnelly, L. (1999). A hydrogen-bonding triad stabilizes the chemical transition state of a group I ribozyme. *Chem. Biol.* **6,** 153–165.

Strobel, S. A., Ortoleva-Donnelly, L., Ryder, S. P., Cate, J. H., and Moncoeur, E. (1998). Complementary sets of noncanonical base pairs mediate RNA helix packing in the group I intron active site. *Nat. Struct. Biol.* **5,** 60–66.

Sugimoto, N., Nakano, S., Katoh, A., Nakamura, H., Ohmichi, T., Yoneyama, M., and Sasaki, M. (1995). Thermodynamic parameters to predict the stability of RNA/DNA hybrid duplexes. *Biochemistry* **34,** 11211–11216.

Suydam, I. T., and Strobel, S. A. (2009). Nucleotide analog interference mapping. *Methods Enzymol.* **468,** 3–30.

Szewczak, L. B. (2008). In vivo analysis of ribonucleoprotein complexes using nucleotide analog interference mapping. *Methods Mol. Biol.* **488,** 153–166.

Szewczak, A. A., Ortoleva-Donnelly, L., Ryder, S. P., Moncoeur, E., and Strobel, S. A. (1998). A minor groove RNA triple helix within the catalytic core of a group I intron. *Nat. Struct. Biol.* **5,** 1037–1042.

Szewczak, A. A., Ortoleva-Donnelly, L., Zivarts, M. V., Oyelere, A. K., Kazantsev, A. V., and Strobel, S. A. (1999). An important base triple anchors the substrate helix recognition surface within the Tetrahymena ribozyme active site. *Proc. Natl. Acad. Sci. USA* **96,** 11183–11188.

Taylor, S. D., Solem, A., Kawaoka, J., and Pyle, A. M. (2010). The NPH-II helicase displays efficient DNA × RNA helicase activity and a pronounced purine sequence bias. *J. Biol. Chem.* **285,** 11692–11703.

Thomsen, N. D., and Berger, J. M. (2009). Running in reverse: The structural basis for translocation polarity in hexameric helicases. *Cell* **139,** 523–534.

Waldsich, C. (2008). Dissecting RNA folding by nucleotide analog interference mapping (NAIM). *Nat. Protoc.* **3,** 811–823.

Walmacq, C., Rahmouni, A. R., and Boudvillain, M. (2004). Influence of substrate composition on the helicase activity of transcription termination factor Rho: Reduced processivity of Rho hexamers during unwinding of RNA-DNA hybrid regions. *J. Mol. Biol.* **342,** 403–420.

Walmacq, C., Rahmouni, A. R., and Boudvillain, M. (2006). Testing the steric exclusion model for hexameric helicases: Substrate features that alter RNA-DNA unwinding by the transcription termination factor Rho. *Biochemistry* **45,** 5885–5895.

Wang, S., and Kool, E. T. (1995). Origins of the large differences in stability of DNA and RNA helices: C-5 methyl and 2′-hydroxyl effects. *Biochemistry* **34,** 4125–4132.

Wyatt, J. R., and Walker, G. T. (1989). Deoxynucleotide-containing oligoribonucleotide duplexes: Stability and susceptibility to RNase V1 and RNase H. *Nucleic Acids Res.* **17,** 7833–7842.

CHAPTER EIGHT

Crystallization and X-Ray Structure Determination of an RNA-Dependent Hexameric Helicase

Nathan D. Thomsen[*,1] and James M. Berger[*,2]

Contents

1. Introduction	172
2. Biochemical Foundation	173
3. Protein Expression, Purification, and Storage	175
4. Ligand Preparation and Storage	176
5. Trapping Rho–RNA–Nucleotide Complexes	177
6. Design of a Substrate-Centric Crystal-Screening Strategy	177
7. Crystallization of Rho Bound to RNA and Adenosine Nucleotides	178
7.1. Crystal-form I	178
7.2. Crystal-form II	180
7.3. Crystal-form III	182
8. Conclusion	185
9. Additional Methods	186
9.1. Structure solution and refinement	186
9.2. Structural analysis	187
Acknowledgment	187
References	187

Abstract

Hexameric helicases couple the energy of ATP hydrolysis to processive movement along nucleic acids and are critical components of cells and many viruses. Molecular motion derives from ATP hydrolysis at up to six distinct catalytic centers, which is coupled to the coordinated action of translocation loops in the center of the hexamer. Due to the structural dynamics and catalytic complexity of hexameric helicases, few have been crystallized with a full complement of bound substrates, and instead tend to form crystals belonging to high-symmetry space

[*] Department of Molecular and Cell Biology and The Institute for Quantitative Biosciences, University of California, Berkeley, California, USA
[1] Current address: Department of Pharmaceutical Chemistry, University of California, San Francisco, California, USA
[2] corresponding author. E-mail: jmberger@berkeley.edy

Methods in Enzymology, Volume 511
ISSN 0076-6879, DOI: 10.1016/B978-0-12-396546-2.00008-5

© 2012 Elsevier Inc.
All rights reserved.

groups that obscure the differences among catalytic subunits. We were able to overcome these difficulties and solve an asymmetric structure of the Rho transcription termination factor from *Escherichia coli* bound to ATP mimics and RNA. Here, we present some considerations used for crystallization of this hexameric helicase, discuss the utility of substrate-centric crystal-screening strategies, and outline a crystal-aging screen that allowed us to overcome the adverse effects of nonmerohedral twinning.

1. Introduction

Hexameric helicases are ring-shaped motor proteins that encircle and move along nucleic acid strands to unwind duplex DNA and RNA substrates (Singleton et al., 2007). Enzymes of this type broadly fall into two evolutionarily related, but distinct families—RecA and AAA+—that participate in a myriad number of vital DNA- and RNA-dependent transactions in the cell (Iyer et al., 2004; Lyubimov et al., 2011). Because of their complex action and essential functions, hexameric helicases have garnered significant interest both as archetypal molecular machines and as prospective targets for therapeutic intervention.

Alongside biochemistry, structural investigations have helped generate key insights into hexameric helicase mechanism. However, while numerous members of disparate hexameric helicase families have been crystallized over the past decade, it has proven exceedingly difficult to obtain structures of these enzymes bound to both target nucleic acid substrates and the nucleotide cofactors that fuel motor movement. Studies of hexameric helicases have been complicated by many factors, including their large size and inherent flexibility, the presence of multiple ligand binding sites, and a tendency for hexamers to form mixtures of closely related symmetric and quasi-symmetric states of heterogeneous nucleotide occupancy.

This latter problem can be one of the most insidious. Hexameric helicases contain six ATPase sites (positioned between each subunit interface), along with six nucleic acid binding sites housed in the center of the ring. These sites frequently display some level of positive and negative cooperativity between each other, giving rise to both strong and weak binding sites within a single hexamer (Bujalowski and Klonowska, 1993; Geiselmann and von Hippel, 1992; Hingorani and Patel, 1996; Kim and Patel, 1999; Kim et al., 1999; Seifried et al., 1992; Stitt, 1988; Xu et al., 2003). The presence of multiple classes of ATP binding sites can lead to difficulties in determining which ATP analogs and/or nucleotide mixtures might be optimal for use in structural studies. Even if this hurdle is surmounted, the high degree of conformational similarity between subunits in the hexamer can favor crystal packing arrangements that blur out essential details, such as when a nucleic acid segment bound in the center of a helicase ring, lies on a crystallographic symmetry axis (Skordalakes and Berger, 2006). At its worst, pseudo-symmetry can produce crystals that display merohedral or nonmerohedral twinning—a situation where single crystals are actually composed of blocks of overlapping microcrystals in similar but differing orientations—preventing structure solution and refinement (Yeates and Fam, 1999).

An approach we recently used in determining the structure of a fully liganded hexameric helicase—the full-length *Escherichia coli* Rho transcription termination factor—took these problems into account. Using prior biochemical knowledge of Rho's catalytic mechanism, as well as our earlier experience with structural studies of this system, we implemented a crystallization strategy that screened a highly focused set of chemical conditions against a panel of RNA substrates and different ATP mimetics. The result was three distinct, but related crystals forms, one of which contained the desired protein–ligand complex. Next, we conducted a time-dependent crystallization screen to overcome an otherwise intractable, interleaved form of coincident nonmerohedral twinning that was present in our most promising crystals. Finally, by collecting the data on specific regions of the crystal, we avoided contributions from the unwanted twin domain resulting in a complete 2.8 Å dataset that allowed structure determination (Thomsen and Berger, 2009). Here, we present considerations for the crystallization of hexameric helicases bound to nucleic acid and nucleotide cofactors, as well as a crystal screening method that can be used to mitigate the effects of nonmerohedral twinning that can occur with pseudo-symmetric protein samples.

2. Biochemical Foundation

In designing a crystal-screening strategy for enzymes that make use of multiple substrates, there is an overwhelmingly large combination of chemical and ligand space to explore (McPherson, 1999). Consideration of the activities and features unique to the system of interest is therefore critical to the design of a successful screen. Discovered in 1969, the Rho transcription termination factor has since been the subject of more than 40 years of study (Roberts, 1969). Rho is a 47 kDa protein consisting of an N-terminal oligonucleotide binding (OB) fold and a C-terminal RecA-like domain (Allison *et al.*, 1998; Briercheck *et al.*, 1998; Burgess and Richardson, 2001; Dombroski and Platt, 1988; Miwa *et al.*, 1995; Opperman and Richardson, 1994; Skordalakes and Berger, 2003). The N-terminal region houses an RNA binding locus, termed the "primary site," that preferentially associates with cytosine-rich nucleic acid segments, and is responsible for recruiting Rho to target the mRNAs (Bear *et al.*, 1985; Bogden *et al.*, 1999; Martinez *et al.*, 1996a,b; Modrak and Richardson, 1994). The C-terminal part of Rho also binds RNA, forming the so-called "secondary site," and serves as the ATP-dependent motor element of the protein (Dolan et al., 1990; Wei and Richardson, 2001). Rho is capable of translocating 5′–3′ along RNA and can both unwind RNA/DNA duplexes and release RNA from ternary transcription elongation complexes (Brennan et al., 1987; Park and Roberts, 2006; Richardson and Conaway, 1980; Roberts, 1969; Shigesada and Wu, 1980). At the time we began this project, it was known that RNA bound to Rho's secondary site with maximal affinity only when the protein's ATP binding sites were saturated. It was also established that Rho possessed a mixture of strong and weak ATP binding sites, whose relative strengths and number changed depending on the type of nucleotide being assayed. Hence, there was strong evidence that some type of ATP analog would be required to crystallize an intact Rho hexamer with RNA productively bound to the motor domains, and that the choice of analog would play a significant role in the type of conformational state we obtained.

Studies of Rho/RNA interactions provided boundary points for the length and sequence of RNA substrates used in our crystal-screening strategy. Early experiments revealed that Rho could engage and wrap long stretches of cytosine-rich RNA through its primary RNA binding site (Lowery and Richardson, 1977; Lowery-Goldhammer and Richardson, 1974; Wang and von Hippel, 1993). By contrast, the secondary site had been shown to bind only ∼8 bases of RNA (Richardson, 1982). Furthermore, although cytosine-rich pyrimidine RNAs >22 nucleotides in length were seen to maximally stimulate Rho's ATPase activity at roughly stoichiometric concentrations, pyrimidine-based RNA substrates as short as nine nucleotides also strongly activated ATP turnover when present at saturating levels (Lowery and Richardson, 1977; Richardson, 1982). These data suggested that poly-$r(CU)_n$ or -$r(U)_n$ RNA polymers between 8 and 30 nucleotides in length could target the secondary RNA binding site in the center of the Rho ring; poly-$r(C)_n$ oligos were not chosen for study, as we had seen earlier that these substrates can have solubility problems when used at crystallographic concentrations (Bogden et al., 1999).

Extensive studies of ATP and ATP analog binding and hydrolysis by Rho further helped refine our crystallization strategy. As with ATP, Rho had been found to engage adenosine 5′-(γ-thio) triphosphate (ATPγS) using a mixture of three strong and three weak binding sites; however, these studies further showed that this cofactor could be readily hydrolyzed (Stitt and Webb, 1986; Stitt and Xu, 1998; Xu et al., 2003), indicating that it would not be suitable for stabilizing or trapping a translocation intermediate. The ATP analogs adenosine 5′-(β, γ-imido) triphosphate (AMP-PNP) and adenosine 5′-(β,γ-methylene)triphosphate (AMP-PCP) were found to bind much more weakly to the helicase and tightly occupy only one of Rho's six possible sites (Xu et al., 2003), properties shared with ADP rather than ATP. During pre-steady-state ATP hydrolysis assays, in which a single active site was filled with ATP, the addition of AMP-PCP or AMP-PNP produced burst hydrolysis rates significantly lower than those seen from the addition of ATP (Browne *et al.*, 2005), further suggesting that they would be poor substrate mimetics. In the end, only one nonhydrolyzable ATP analog, ADP•BeF$_3$, was reported to bind Rho in the presence of RNA with an affinity similar to that of ATP (Adelman *et al.*, 2006; Xu *et al.*, 2003). This observation suggested that ADP•BeF$_3$ might be the most viable candidate for a nonhydrolyzable ATP analog in cocrystallization trials.

3. Protein Expression, Purification, and Storage

Although each type of hexameric helicase will necessitate a particular purification strategy, and will further exhibit a unique set of solution properties once pure, a number of considerations proved important for the crystallization of Rho that may be applicable to other systems. One such factor was the level of soluble protein that could be obtained during expression. Over the course of a crystallization project, expression levels correlate with the number of crystallization conditions that ultimately can be explored, a factor that can have a marked outcome on the success of a project. Expression levels further define a "signal-to-noise" ratio that influences the purity of the final preparation. Particularly, high expression levels also provide more flexibility in the purification protocol, such as eliminating a need to use and remove affinity tags (see below), which can speed up purifications to produce protein of higher quality.

In looking for ways to boost expression, we found that full-length *E. coli* Rho was best overproduced from a pET-based vector in BL21 pLysS cells by inducing at an A_{600} of 0.6 at 37 °C for 3 h. In this instance, a range of IPTG concentrations produced fairly comparable results. BL21 pLysS cells provided approximately twofold greater yields of soluble Rho protein compared to BL21 Codon (+) (Agilent Technologies) or Rosetta (EMD Millipore) cells. With other Rho constructs (such as mutants), C41 (Lucigen) cells have proven even more effective for expression, underlying the importance of screening a large number of cell lines at the earliest stages of a project. Once optimized, we found that Rho accounted for ~80% of the total soluble cellular protein content, obviating a need for affinity purification tags. Lysis was performed by resuspending induced cells into buffer A (25 mM Tris, pH 7.5, 50 mM KCl, 10% glycerol, 1 mM DTT) and sonicating on ice for a total of 2 min (30 s on, 1 min off, repeated four times) on a power setting of 5.5 (Sonicator 3000, Misonix), after which cells were centrifuged at 16,000 rpm (4 °C) in an SS34 rotor (Sorvall) for 30 min. Cells were not frozen between harvesting and lysis steps.

Untagged Rho was purified to homogeneity by applying the clarified lysate to a cation-exchange column in buffer A (Poros HS, Life Technologies). Protein was eluted using a 100 ml gradient into buffer B (25 mM Tris, pH 7.5, 1 M KCl, 10% glycerol, 1 mM DTT), the dominant center peak fractions were pooled, the salt concentration was reduced to 75 mM using diluent buffer (25 mM Tris, pH 7.5, 10% glycerol, 1 mM DTT), and the sample was reapplied to the same Poros HS column for a second round of enrichment. This procedure is fast, typically requiring $\sim$2 h. Rho hexamers were next concentrated by ultrafiltration (Amicon Ultra 10,000 MWCO, Millipore) and purified away from aggregated material using size exclusion chromatography. Two S-300 columns (GE Healthcare) were run in succession using buffer C (50 mM Tris, pH 7.5, 500 mM KCl, 10% glycerol, 1 mM DTT), reconcentrating the peak hexamer fractions of protein between runs to ensure that all traces of aggregated protein or Rho subassemblies were removed. The purified protein was reconcentrated and stored in buffer C at 4 °C.

The Rho protein produced from this procedure is of very high purity and homogeneity and readily gives crystals in a number of conditions. Some of the steps, such as avoiding freezing and running two gel-filtration columns, were empirically found to improve crystal quality and diffraction data resolution in our early work on this protein (Skordalakes and Berger, 2003, 2006). Similarly, we found it important to use only freshly prepared protein (<1 week old) for crystallization. Although these types of treatments are clearly not necessary for crystallization success with all proteins, their use may be a good practice for increasing the chances of obtaining a hit in difficult cases.

4. Ligand Preparation and Storage

RNA oligonucleotides were purchased from IDT, dissolved in Ambion RNAse-free water to 4 mM final concentration and stored at -80 °C in single-use aliquots. Nucleotides such as ADP, ATP, AMP·PNP, and ATPγS were all dissolved in deionized H$_2$O to 100 mM and stored at -80 °C, again in single-use aliquots. For our initial experiments, we prepared ADP·BeF$_3$ and ADP·AlF$_x$ using ratios taken from the literature, with special attention paid to the work on the Bovine F$_1$-ATPase by Walker, Leslie, and colleagues (Braig et al., 2000; Kagawa et al., 2004; Menz et al., 2001). We used a ratio of 1ADP:3Be:15F for ADP·BeF$_3$ and a ratio of 1ADP:4Al:20F for ADP·AlF$_x$. To prepare the analogs, either 1 M AlCl$_3$ or 1 M BeCl$_2$ was mixed with 1 M NaF in the appropriate volumes to generate 1:5 ratios of Be:F and Al:F. These mixtures were then added to 100 mM stocks of ADP in order to produce a final concentration of 25 mM ADP. Finally, all nucleotides were adjusted to an approximate pH of 7.5 using pH paper and drop wise addition of 1 N HCl or NaOH.

5. Trapping Rho–RNA–Nucleotide Complexes

Protein/nucleic acid complexes can, in principle, be prepared either by mixing the constituents at the desired ratios/concentrations and setting trays directly or by mixing at lower concentrations, purifying the complex using column chromatography, and then reconcentrating prior to setting trays. Given the high cost of synthesized RNA and the toxicity of the ADP·BeF$_3$ and ADP·AlF$_x$ nucleotide preparations, along with a concern that low-affinity RNA or nucleotide binding sites might be stripped during the course of a column purification strategy, we chose to directly mix purified protein and ligands immediately prior to crystallization screening.

To prevent precipitation and produce the best protein crystals, a specific procedure for ligand addition was developed. Purified Rho at $\sim$60 mg/ml in buffer C (50 mM Tris, pH 7.5, 500 mM KCl, 10% glycerol, 1 mM DTT) was first exchanged into dialysis buffer (10 mM Tris, pH 7.5, 100 mM NaCl, 1 mM DTT) for $\sim$3 h using a 10,000 MWCO Slide-A-Lyzer dialysis cup (Thermo). We empirically found that lower quality crystals resulted if Rho was dialyzed into a low-salt buffer for longer than $\sim$6 h, suggesting that dialysis might not have gone to completion. Following dialysis, the concentration of Rho typically approached $\sim$40 mg/ml due to sample dilution. As trays were set at 20 mg/ml, this procedure provided a 2× protein stock with ample room for ligand addition. More importantly, it allowed us to predilute various ligands into dialysis buffer, thus buffering them and reducing their local concentration upon addition to the protein solution. Reasoning that flooding Rho with high concentrations of a nonhydrolyzable ATP analog alone might trap a fraction of the helicase population in nonproductive conformations incompatible with RNA binding, we first added individual RNA substrates and let the sample sit on ice for 15 min. We next added nonhydrolyzable nucleotide complexes at 2.5 mM (and spiked with 5–10 mM MgCl$_2$), mixed the solution and incubated the reagents together on ice for another 15 min prior to setting trays. Early crystal trials suggested that incubation of Rho–ligand complexes for up to 30 min produced better crystals than setting trays immediately after mixing.

6. Design of a Substrate-Centric Crystal-Screening Strategy

We next used the biochemical data to inform the design of a substrate-centered screening strategy. Because our early work with Rho had found that the protein crystallized preferentially under low-salt conditions with organic (e.g., PEG- or MPD-based) precipitants, we focused on diversifying

Table 8.1 Substrate matrix used for Rho crystal screening

	ANP	ATPγS	ADP	ADP·AlF$_4$	ADP·BeF$_3$
r(CU)$_4$	–	–	–	–	X
r(U)$_{12}$	–	–	–	–	II, III
r(CU)$_6$	–	–	–	–	II, III
r(CU)$_8$	–	–	–	–	II, III
r(CU)$_{10}$	X	I	I	I	I, II
r(CU)$_{12}$C	X	I	I	I	I, II
r(CU)$_{15}$	X	I	I	I	I, II

Key: (X)—no crystal; (I, II, III)—crystal-forms; (–)—not screened.

the number and types of substrates and cofactors screened, rather than thoroughly searching the chemical space. To maximize our exploration of ligand space, we conducted screens using r(CU)$_n$ and r(U)$_n$ nucleic acid polymers between 8 and 30 nucleotides in length, alongside a variety of adenosine nucleotides (Table 8.1). We initially focused on longer r(CU)$_n$ RNA's between 20 and 30 nucleotides in length, reasoning that they might bind first to the high affinity primary RNA binding site and then feed down into the central channel in the presence of the proper adenosine nucleotide. For the nucleotides themselves, we individually tested ADP, ATPγS, AMP-PNP, ADP·BeF$_3$, and ADP·AlF$_x$. After hits were obtained, we screened various combinations of nucleotides as part of our optimization efforts. Mixtures of ADP and AMP-PNP have proved important for the crystallization of Rho's closest homolog, the F$_1$-ATPase (Abrahams *et al.*, 1994), but such mixtures did not improve the quality of our crystals. For each substrate combination, we selected a relatively limited set of commercial crystallization screens, typically including Index HT (Qiagen), The MPD's (Qiagen), PEG-Ion (Hampton Research), and Natrix (Hampton Research). Ultimately, the screening strategy yielded three different crystal-forms (Fig. 8.1).

7. Crystallization of Rho Bound to RNA and Adenosine Nucleotides

7.1. Crystal-form I

Crystal-form I was obtained with r(CU)$_n$ polymers between 20 and 30 bases in length and either ADP, ATPγS, or ADP·AlF$_4$ (Table 8.1). These crystals were grown by mixing 1 μl of the Rho–substrate complex in dialysis buffer (see above) with 1 μl of a crystallization solution containing 100 mM HEPES (pH 7.5), 8% PEG 6000, and 5% MPD and incubating under paraffin oil at 18 °C. Crystals grew within 24 h and reached full-size in ~1 week. Since the

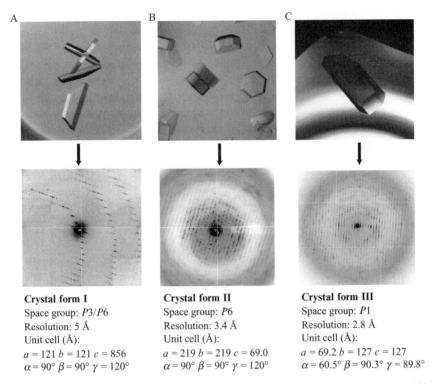

Figure 8.1 Three Rho crystal-forms obtained from substrate matrix screening. (A) Crystal-form I (top) grew as thin hexagonal plates that diffract to relatively low resolution. A closeup view of the diffraction pattern (bottom) reveals the long unit cell edge that caused severe overlaps and prevented collection of a complete dataset. (B) Crystal-form II grew as thick hexagonal "lug-nuts" (top). Diffraction patterns revealed a $P6$ packing arrangement. Usable diffraction maxima extended to 3.4 Å resolution. (C) Crystal-form III grew as large hexagonal rods free from any obvious morphological defects (top). While the diffraction pattern is highly mosaic, these crystals diffracted to the highest resolution of any of the forms. (For interpretation of the references to color in this figure legend, the reader is referred to the online version of this chapter.)

crystals that grew in three unique nucleotide conditions were all indistinguishable morphologically, we reasoned that Rho likely associated with ADP in each instance. Crystal-form I diffracted to ∼5 Å in a hexagonal space group and possessed an unusually long unit cell edge (>850 Å) (Fig. 8.1A). Matthew's analysis suggested that as many as 12 Rho hexamers were present per unit cell, while the space group ($P3$ or $P6$) implied that the hexamers were likely centered around the crystallographic symmetry axes (Kantardjieff and Rupp, 2003; Matthews, 1968). Numerous approaches were explored in order to improve the resolution of these crystals and/or alter the symmetry or unit cell dimensions, including dehydration, freeze/thaw annealing, and chemical cross-linking using glutaraldehyde; however, none of these techniques improved diffraction (Newman, 2006). Data processing was further hindered by the long unit cell, which caused severe overlaps in the diffraction patterns and ultimately prevented us from solving the structure. Given these results, we continued to explore the substrate space with the goal of obtaining crystals that diffracted to higher resolution and possessed more manageable unit cell dimensions.

7.2. Crystal-form II

Crystal-form II was obtained with $r(CU)_n$ polymers between 12 and 30 bases in length and the ATP mimic ADP·BeF$_3$ (Table 8.1). Crystals were grown by mixing 1 μl of the Rho–substrate complex in dialysis buffer with 1 μl of a crystallization solution containing 100 mM HEPES (pH 7.0), and 4% MPD and incubating under paraffin oil at 18 °C. The initial hit took ~1 month to appear, although later optimization produced crystals that would appear within 3–7 days. Crystals were cryoprotected by the addition of mother liquor supplemented with 20% MPD and flash frozen in liquid nitrogen. The hexagonal crystals resembled "lug-nuts" and proved capable of diffracting to ~3.4 Å after refinement of the initial growth conditions (Table 8.2). Molecular replacement using a Rho monomer model (Skordalakes and Berger, 2003) was used to phase the data (which processed cleanly as $P6$) and indicated that three copies were present per asymmetric unit (Fig. 8.2A). Application of crystal symmetry operators to these

Table 8.2 Data collection and partial refinement for crystal-form II

Data	
Space group	$P6$
Unit cell	$a=218.9\ b=218.9\ c=68.95$
	$\alpha=90\ \beta=90\ \gamma=120$
Wavelength (Å)	0.9796
Resolution (Å)	50–3.4
Unique reflections	50,668
Redundancy	2.3 (2.3)
Completeness (%)	99.7 (99.8)
I/σ	8.3 (2.5)
R_{merge}	0.104 (0.348)
Refinement	
R_{work} (%)	33.5[a]
R_{free} (%)	35.7[b]
RMSD bonds (Å)	0.010
RMSD angles (°)	1.174

Table 8.2 (Continued)

Data	
Ramachandran	
Preferred (%)	99.0
Allowed (%)	1.0
Outliers (%)	0
Number of atoms	9936
Protein	9840
Ligands	96

Values in parenthesis correspond to the highest resolution bin.
[a] $R_{work} = \sum\sum |F_o - F_c|/\sum F_o$.
[b] R_{free} is calculated using 5% of the data omitted from refinement.

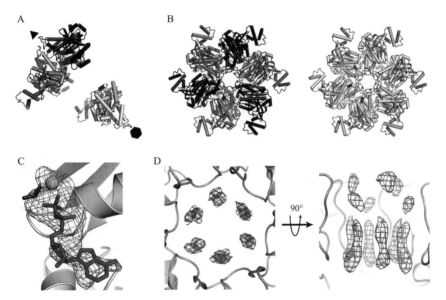

Figure 8.2 Structure solution and ligand binding in crystal-form II. (A) Molecular replacement reveals one Rho dimer (blue/dark gray and green/gray) and one monomer (pink/light gray) per asymmetric unit. Crystallographic six- and threefold symmetry axes are indicated by a hexagon and triangle, respectively. (B) Crystallographic symmetry generates one threefold symmetric trimer-of-dimers (blue/dark gray and green/gray) and one sixfold symmetric hexamer (pink/light gray). These structures represented our first crystallographic views of a full-length, closed-ring Rho hexamer. (C) Initial refinement of molecular replacement solutions produced $F_o - F_c$ difference electron density maps revealing the presence of nucleotide (modeled as ADP·BeF$_3$) at all three ATP binding sites per asymmetric unit. ADP is colored magenta/gray, BeF$_3$ is colored black, and the Mg^{2+} ion is colored yellow-green/light gray. Electron density

solutions revealed that the monomers belonged to two different Rho hexamers, one exhibiting threefold symmetry and the other sixfold (Fig. 8.2B).

Refinement of the models from these data could not proceed beyond an R_{work}/R_{free} of 33.6%/35.7% (Table 8.2). However, $F_o - F_c$ difference electron density maps obtained after rigid body fitting of the MR solution revealed clear density for a nucleotide bound at the active site of each Rho monomer (Fig. 8.2C). This density was modeled as a complex of ADP·BeF$_3$·Mg^{2+}, although the presence of the BeF$_3$ group and the Mg^{2+} ion could not be confirmed at this resolution and level of overall model accuracy. Six strong bands of $F_o - F_c$ difference density also were observed around the three- and sixfold crystal symmetry axes at the center of each hexamer (Fig. 8.2D), suggesting that RNA was bound in the center of each ring; however, the electron density was not of sufficient quality for modeling, likely due to rotational averaging of the RNA about the crystallographic symmetry axes. Unfortunately, these issues obscured any details that might have allowed us to identify the ATP and RNA binding status of individual subunits, preventing us from gaining useful insights into Rho's translocation mechanism. Nonetheless, these data revealed that short RNA substrates and the ATP analog ADP·BeF$_3$ could be used to capture a fully liganded complex with RNA bound in the center of the ring. We therefore expanded our search of chemical space around these substrate conditions to identify a crystal-form that would contain at least one complete hexamer whose conformational state was not influenced by crystal symmetry.

7.3. Crystal-form III

Our inability to fully refine crystal-form II suggested either that certain pathologies were present in the diffraction data or that some insurmountable form of model bias was carrying over from molecular replacement. Reasoning that any bias might be due to significant conformational differences between our search model (which was from an RNA-free, open-ring Rho state) and the subunits in a closed-ring, fully liganded Rho complex, we purified selenomethionine-labeled Rho for use with phasing. Crystal-form III was obtained with this protein, using short $r(CU)_6$ (and eventually $r(U)_{12}$) RNA polymers in the presence of ADP·BeF$_3$ (Table 8.1). Crystals (100 × 500 μm hexagonal rods) were grown by mixing 2 μl of protein–ligand solution at 20 mg/ml in modified dialysis buffer (10 mM Tris, pH

(green mesh) is contoured at 3σ. (D) Initial refinement of molecular replacement solutions produced $F_o - F_c$ difference electron density maps revealing the presence of RNA in the central channel of each Rho hexamer. The RNA density appeared to be averaged around the crystallographic symmetry axis, producing uninterpretable maps. Electron density (green mesh) is contoured at 2.5σ. (See the Color Insert.)

7.5, 300 mM NaCl, and 1 mM TCEP) with 2 μl of a solution of 5% MPD, 100 mM HEPES (pH 7.9), 20 mM NaCl, and 10 mM spermidine–HCl and incubating under paraffin oil at 18 °C (Fig. 8.1C). After 3 days, the crystals were cryoprotected by adding mother liquor supplemented with 25% MPD directly to the 4 μl drop in 1 μl increments separated by 1–5 min intervals. Once 4 μl of cryoprotectant was added in this way, the drop was fully exchanged into cryoprotectant by: (1) removal of ½ drop volume, (2) addition of ½ drop volume of cryoprotectant, and (3) incubation for 1 min intervals. The entire cryoprotection process was conducted under paraffin oil in the original crystallization plates to minimize crystal manipulations. Upon complete exchange, the crystals were looped and flash-frozen in liquid nitrogen (Thomsen and Berger, 2009).

Attempts to collect data from these crystals were not initially promising. Diffraction images obtained from the central region of the rods showed a mixture of lattices. However, exposures taken from different points along the length of the rods revealed that one end (the tip that nucleated from the drop or crystallization plate surface) diffracted to ∼3.5 Å resolution and indexed as space group *P*6, whereas the distal end diffracted to ∼3.0 Å and appeared to be *P*1. Interestingly, the *P*6 end possessed the same unit cell dimensions as seen in crystal-form II. Molecular replacement solutions obtained with data from the *P*1 end of the crystals revealed that a single Rho hexamer was present in the asymmetric unit and revealed clear difference density for both RNA and nucleotide (Thomsen and Berger, 2009). However, the MR-phased maps were of low quality and refinement stalled at an early stage. Close inspection of the diffraction patterns revealed that the *P*1 lattice was conflated with the overlapping hexagonal lattice at low resolution (Fig. 8.3A), suggesting that these crystals were nonmerohedrally twinned.

Although one end of the crystal was primarily *P*1, the nearly perfect crystal morphology prevented us from identifying the nature of the *P*6 contribution. Without this knowledge, isolation of the two twin domains or improvement of the crystals proved exceedingly difficult. However, one element of the two lattice arrangements that proved fortuitously variable was their stability over time. While some crystals remained intact for several months, others exhibited age dependent defects and eventually degraded or disappeared. Reasoning that the two different crystal-forms in the nonmerohedral twins might degrade over time at different relative rates, we conducted a series of crystallization screens around 10 mM Spermidine–HCl, 20 mM NaCl, 100 mM HEPES (pH values between 7.5 and 8.2), and MPD (1–12% MPD). After reaching full size, the crystals were left to slowly degrade over a period of 2 months as the drops slowly dehydrated. By 1 month, we observed that one portion of the rod was degrading, leaving behind a cone-shaped crystal at its nucleating end (Fig. 8.3B). This finding indicated that the *P*1 portion of the rods was being corrupted by the

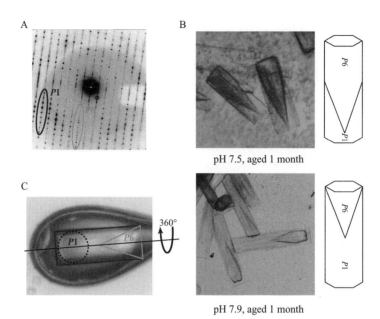

Figure 8.3 Overcoming a vexing form of coincident nonmerohedral twinning. (A) Initial diffraction patterns obtained from crystal-form III (Fig. 8.1C), revealed two distinct lattices. While the P1 diffraction dominated the pattern (blue/dark gray), a P6 diffraction pattern corresponding to crystal-form II was present at low resolution (pink/light gray). Since crystal-forms I and II shared a coincident $\sim$69 Å unit cell edge, the diffraction patterns almost perfectly overlapped and prevented accurate measurement of low resolution P1 data. (B) Aging caused the P1 portion of each crystal to degrade at a faster rate than the P6 portion. When viewed under bright-field illumination, the two twin domains are clearly distinguishable. The P6 twin domain forms at the nucleating end of the crystals and has a "cone-shaped" protrusion that interleaved with the P1 twin domain (top). An X-ray beam aimed at the P1 tip of the crystal thus passes through a small fraction of the P6 twin domain, producing P6 diffraction only at low resolution as seen in panel A. Increasing the pH of crystal growth increased the relative size of the P1 twin domain (bottom). (C) Crystals grown at a pH of 7.9 were mounted in bendable cryoloop and oriented such that the axis of crystal rotation (solid line) was parallel with the long axis of the hexagonal rods. A 100 μM collimated beam (dashed circle) was directed into the P1 tip (blue/dark gray) of the crystal. Data collection in this orientation avoided contributions from the P6 twin domain (pink/light gray). (See the Color Insert.)

invasion of a highly related P6 packing arrangement, explaining why twinning was present in our diffraction experiments.

Significantly, this "aging" screen also revealed that the size of the P1 twin domain was dependent on pH. Crystallization at pH 7.5 produced a nearly complete overlap of the two twin domains. By contrast, crystallization at pH 7.9 caused the P1 twin domain to increase in size relative to the

P6 domain and dominate the length of the crystal. Using this information, we grew fresh crystals at pH 7.9 for data collection. Data collection was carried out using bendable crystal mounting loops (Hampton Research), which together with our knowledge of the twin domain structure, allowed us to properly orient the crystals in the X-ray beam to avoid contributions from the P6 portion (Fig. 8.3C). Although the P1 datasets still displayed a relatively high degree of disorder (1.5–2° mosaicity) that caused some difficulties in data processing, and also likely contained some residual amount of pseudo-merohedral twinning, the data were of sufficiently high quality to successfully solve and refine a Rho model bound to both RNA and the ATP mimic ADP·BeF$_3$. The resultant structure shows how a homohexameric helicase adopts an asymmetric ring-shaped structure around a single-stranded RNA coil and provides a model for how ATP turnover is coupled to translocation. Together with a DNA- and ADP-bound structure of a distantly related helicase—the papillomavirus E1 protein—the Rho model further provides an explanation for how helicases can translocate with opposite polarities along nucleic acid substrates(Enemark and Joshua-Tor, 2006; Thomsen and Berger, 2009).

8. Conclusion

Our efforts with Rho yielded many useful lessons for crystallizing a substrate-bound hexameric helicase complex. First and foremost is the importance of varying the sample itself during crystal screening. Although we could have applied this principle by looking at Rho homologs from different bacterial species, we chose instead to focus on the well-characterized E. coli Rho and screen-different ligand combinations. Extensive efforts to improve crystal-form I using only our starting panel of RNA and nucleotide substrates failed: while thousands of crystallization experiments were performed, only a single relatively low quality crystal-form was obtained. However, immediately upon switching from ADP to ADP·BeF$_3$, new higher quality crystals (form II) were obtained. Efforts to improve upon these crystals also proved extremely difficult, and only after shortening the RNA oligonucleotides did crystal-form III arise. The most promising strategy with Rho, which may be useful when working with other multisubstrate enzymes, was thus to screen a relatively limited region of chemical space (200–300 conditions), but a wide range of substrate conditions (Table 8.1). This recommendation is in line with metaanalyses of protein crystallization experiments, which have suggested that most proteins crystallize from a relatively small subset of conditions (Kimber et al., 2003; Page et al., 2003). In the same vein, biochemical data is essential to help focus early screens around the most useful substrate combinations. This is particularly important when looking at closely related cofactors like ATP analogs, which can have unexpected effects on the target system. For example, AMP-PNP behaves more like ADP (and ADP·BeF$_3$ like ATP) in how it interacts with Rho; whereas, in other systems we have studied, the opposite has been true. Finally, it may be worth taking additional care in how protein samples are treated during purification, for instance, avoiding sample freezing or doubly ensuring that no aggregated material is present during crystallization.

Rho also exemplified many of the challenges that can arise when crystallizing pseudo-symmetric protein–ligand complexes, such as those formed by many homohexameric protein systems. These problems include the "averaging-out" of interpretable ligand density due to disadvantageous crystal packing, high mosaicity, and twinning. Nonetheless, even nonideal crystals can provide indispensible clues as to how to move forward. For instance, the solution and partial refinement of models from crystal-form II, while not useful for directly understanding Rho mechanism *per se*, did reveal that we had succeeded in capturing a fully liganded complex. This observation in turn allowed us to confidently focus on a set of substrate conditions that were likely to trap the desired complex, allowing us to expand our search of chemical space while screening for new crystals. Moreover, although nonmerohedral twinning with coincident cell orientations can cause severe difficulties during structure solution and refinement, this work showed that crystal-trays are the experiment that keeps on giving: just as it is worth running out old crystallization drops on gels from time to time to ascertain sample integrity, visually inspecting crystals over time enabled us to identify an insidious twin domain morphology and uncover the conditions that best mitigated its effects. Although each macromolecular system presents its own set of unique challenges in so far as purification and crystallization, these experiences have shaped our thinking and approach to future projects and may prove useful to those working on similarly complex systems.

9. ADDITIONAL METHODS

9.1. Structure solution and refinement

Data on all crystals were collected at Beamline 8.3.1 at the advanced light source (ALS) (MacDowell *et al.*, 2004) and processed with HKL-2000 (Otwinowski and Minor, 1997). Crystal-form II was solved using molecular replacement (Phaser) (McCoy *et al.*, 2007), by searching for three copies of a Rho monomer from the open-ring Rho structure (Skordalakes and Berger, 2003). Initial rigid body and translation libration screw (TLS) refinement was conducted using Phenix (Adams *et al.*, 2010). Crystal-form III was solved using molecular replacement and refined as described (Thomsen and Berger, 2009).

9.2. Structural analysis

All structural superpositions and figures were prepared using PyMol (Schrödinger) (DeLano, 2002).

ACKNOWLEDGMENT

This work was supported by the NIGMS (RO1-GM077373).

REFERENCES

Abrahams, J. P., Leslie, A. G., Lutter, R., and Walker, J. E. (1994). Structure at 2.8 Å resolution of F1-ATPase from bovine heart mitochondria. *Nature* **370,** 621–628.

Adams, P. D., Afonine, P. V., Bunkoczi, G., Chen, V. B., Davis, I. W., Echols, N., Headd, J. J., Hung, L. W., Kapral, G. J., Grosse-Kunstleve, R. W., McCoy, A. J., Moriarty, N. W., et al. (2010). PHENIX: A comprehensive Python-based system for macromolecular structure solution. *Acta Crystallogr. D Biol. Crystallogr.* **66,** 213–221.

Adelman, J. L., Jeong, Y. J., Liao, J. C., Patel, G., Kim, D. E., Oster, G., and Patel, S. S. (2006). Mechanochemistry of transcription termination factor Rho. *Mol. Cell* **22,** 611–621.

Allison, T. J., Wood, T. C., Briercheck, D. M., Rastinejad, F., Richardson, J. P., and Rule, G. S. (1998). Crystal structure of the RNA-binding domain from transcription termination factor rho. *Nat. Struct. Biol.* **5,** 352–356.

Bear, D. G., Andrews, C. L., Singer, J. D., Morgan, W. D., Grant, R. A., von Hippel, P. H., and Platt, T. (1985). *Escherichia coli* transcription termination factor rho has a two-domain structure in its activated form. *Proc. Natl. Acad. Sci. U. S. A.* **82,** 1911–1915.

Bogden, C. E., Fass, D., Bergman, N., Nichols, M. D., and Berger, J. M. (1999). The structural basis for terminator recognition by the Rho transcription termination factor. *Mol. Cell* **3,** 487–493.

Braig, K., Menz, R. I., Montgomery, M. G., Leslie, A. G., and Walker, J. E. (2000). Structure of bovine mitochondrial F(1)-ATPase inhibited by Mg(2+) ADP and aluminium fluoride. *Structure* **8,** 567–573.

Brennan, C. A., Dombroski, A. J., and Platt, T. (1987). Transcription termination factor rho is an RNA-DNA helicase. *Cell* **48,** 945–952.

Briercheck, D. M., Wood, T. C., Allison, T. J., Richardson, J. P., and Rule, G. S. (1998). The NMR structure of the RNA binding domain of *E. coli* rho factor suggests possible RNA–protein interactions. *Nat. Struct. Biol.* **5,** 393–399.

Browne, R. J., Barr, E. W., and Stitt, B. L. (2005). Catalytic cooperativity among subunits of *Escherichia coli* transcription termination factor Rho. Kinetics and substrate structural requirements. *J. Biol. Chem.* **280,** 13292–13299.

Bujalowski, W., and Klonowska, M. M. (1993). Negative cooperativity in the binding of nucleotides to *Escherichia coli* replicative helicase DnaB protein. Interactions with fluorescent nucleotide analogs. *Biochemistry* **32,** 5888–5900.

Burgess, B. R., and Richardson, J. P. (2001). RNA passes through the hole of the protein hexamer in the complex with the *Escherichia coli* Rho factor. *J. Biol. Chem.* **276,** 4182–4189.

DeLano, W. L. (2002). The PyMOL Molecular Graphics System. DeLano Scientific, San Carlos, CA, USA.

Dolan, J. W., Marshall, N. F., and Richardson, J. P. (1990). Transcription termination factor rho has three distinct structural domains. *J. Biol. Chem.* **265,** 5747–5754.

Dombroski, A. J., and Platt, T. (1988). Structure of rho factor: an RNA-binding domain and a separate region with strong similarity to proven ATP-binding domains. *Proc. Natl. Acad. Sci. U. S. A.* **85,** 2538–2542.

Enemark, E. J., and Joshua-Tor, L. (2006). Mechanism of DNA translocation in a replicative hexameric helicase. *Nature* **442,** 270–275.

Geiselmann, J., and von Hippel, P. H. (1992). Functional interactions of ligand cofactors with Escherichia coli transcription termination factor rho. I. Binding of ATP. *Protein Sci.* **1,** 850–860.

Hingorani, M. M., and Patel, S. S. (1996). Cooperative interactions of nucleotide ligands are linked to oligomerization and DNA binding in bacteriophage T7 gene 4 helicases. *Biochemistry* **35,** 2218–2228.

Iyer, L. M., Leipe, D. D., Koonin, E. V., and Aravind, L. (2004). Evolutionary history and higher order classification of AAA+ ATPases. *J. Struct. Biol.* **146,** 11–31.

Kagawa, R., Montgomery, M. G., Braig, K., Leslie, A. G., and Walker, J. E. (2004). The structure of bovine F1-ATPase inhibited by ADP and beryllium fluoride. *EMBO J.* **23,** 2734–2744.

Kantardjieff, K. A., and Rupp, B. (2003). Matthews coefficient probabilities: Improved estimates for unit cell contents of proteins, DNA, and protein–nucleic acid complex crystals. *Protein Sci.* **12,** 1865–1871.

Kim, D. E., and Patel, S. S. (1999). The mechanism of ATP hydrolysis at the noncatalytic sites of the transcription termination factor Rho. *J. Biol. Chem.* **274,** 32667–32671.

Kim, D. E., Shigesada, K., and Patel, S. S. (1999). Transcription termination factor Rho contains three noncatalytic nucleotide binding sites. *J. Biol. Chem.* **274,** 11623–11628.

Kimber, M. S., Vallee, F., Houston, S., Necakov, A., Skarina, T., Evdokimova, E., Beasley, S., Christendat, D., Savchenko, A., Arrowsmith, C. H., Vedadi, M., Gerstein, M., et al. (2003). Data mining crystallization databases: Knowledge-based approaches to optimize protein crystal screens. *Proteins* **51,** 562–568.

Lowery, C., and Richardson, J. P. (1977). Characterization of the nucleoside triphosphate phosphohydrolase (ATPase) activity of RNA synthesis termination factor p. II. Influence of synthetic RNA homopolymers and random copolymers on the reaction. *J. Biol. Chem.* **252,** 1381–1385.

Lowery-Goldhammer, C., and Richardson, J. P. (1974). An RNA-dependent nucleoside triphosphate phosphohydrolase (ATPase) associated with rho termination factor. *Proc. Natl. Acad. Sci. U. S. A.* **71,** 2003–2007.

Lyubimov, A. Y., Strycharska, M., and Berger, J. M. (2011). The nuts and bolts of ring-translocase structure and mechanism. *Curr. Opin. Struct. Biol.* **21,** 240–248.

MacDowell, A. A., Celestre, R. S., Howells, M., McKinney, W., Krupnick, J., Cambie, D., Domning, E. E., Duarte, R. M., Kelez, N., Plate, D. W., Cork, C. W., Earnest, T. N., et al. (2004). Suite of three protein crystallography beamlines with single superconducting bend magnet as the source. *J. Synchrotron Radiat.* **11,** 447–455.

Martinez, A., Burns, C. M., and Richardson, J. P. (1996a). Residues in the RNP1-like sequence motif of Rho protein are involved in RNA-binding affinity and discrimination. *J. Mol. Biol.* **257,** 909–918.

Martinez, A., Opperman, T., and Richardson, J. P. (1996b). Mutational analysis and secondary structure model of the RNP1-like sequence motif of transcription termination factor Rho. *J. Mol. Biol.* **257,** 895–908.

Matthews, B. W. (1968). Solvent content of protein crystals. *J. Mol. Biol.* **33,** 491–497.

McCoy, A. J., Grosse-Kunstleve, R. W., Adams, P. D., Winn, M. D., Storoni, L. C., and Read, R. J. (2007). Phaser crystallographic software. *J. Appl. Crystallogr.* **40,** 658–674.

McPherson, A. (1999). Crystallization of Biological Macromolecules. Cold Spring Harbor Laboratory Press, Cold Spring Harbor, NY.

Menz, R. I., Walker, J. E., and Leslie, A. G. (2001). Structure of bovine mitochondrial F(1)-ATPase with nucleotide bound to all three catalytic sites: Implications for the mechanism of rotary catalysis. *Cell* **106,** 331–341.

Miwa, Y., Horiguchi, T., and Shigesada, K. (1995). Structural and functional dissections of transcription termination factor rho by random mutagenesis. *J. Mol. Biol.* **254,** 815–837.

Modrak, D., and Richardson, J. P. (1994). The RNA-binding domain of transcription termination factor rho: Isolation, characterization, and determination of sequence limits. *Biochemistry* **33,** 8292–8299.

Newman, J. (2006). A review of techniques for maximizing diffraction from a protein crystal in stilla. *Acta Crystallogr. D Biol. Crystallogr.* **62,** 27–31.

Opperman, T., and Richardson, J. P. (1994). Phylogenetic analysis of sequences from diverse bacteria with homology to the *Escherichia coli* rho gene. *J. Bacteriol.* **176,** 5033–5043.

Otwinowski, Z., and Minor, W. (1997). Processing of X-ray diffraction data collected in oscillation mode. *In* "Methods in Enzymology," (C. W. Carter, Jr. and R. M. Sweet, eds.), pp. 307–326. Academic Press, New York.

Page, R., Grzechnik, S. K., Canaves, J. M., Spraggon, G., Kreusch, A., Kuhn, P., Stevens, R. C., and Lesley, S. A. (2003). Shotgun crystallization strategy for structural genomics: an optimized two-tiered crystallization screen against the Thermotoga maritima proteome. *Acta Crystallogr. D Biol. Crystallogr.* **59,** 1028–1037.

Park, J. S., and Roberts, J. W. (2006). Role of DNA bubble rewinding in enzymatic transcription termination. *Proc. Natl. Acad. Sci. U. S. A.* **103,** 4870–4875.

Richardson, J. P. (1982). Activation of rho protein ATPase requires simultaneous interaction at two kinds of nucleic acid-binding sites. *J. Biol. Chem.* **257,** 5760–5766.

Richardson, J. P., and Conaway, R. (1980). Ribonucleic acid release activity of transcription termination protein rho is dependent on the hydrolysis of nucleoside triphosphates. *Biochemistry* **19,** 4293–4299.

Roberts, J. W. (1969). Termination factor for RNA synthesis. *Nature* **224,** 1168–1174.

Seifried, S. E., Easton, J. B., and von Hippel, P. H. (1992). ATPase activity of transcription-termination factor rho: Functional dimer model. *Proc. Natl. Acad. Sci. U. S. A.* **89,** 10454–10458.

Shigesada, K., and Wu, C. W. (1980). Studies of RNA release reaction catalyzed by *E. coli* transcription termination factor rho using isolated ternary transcription complexes. *Nucleic Acids Res.* **8,** 3355–3369.

Singleton, M. R., Dillingham, M. S., and Wigley, D. B. (2007). Structure and mechanism of helicases and nucleic acid translocases. *Annu. Rev. Biochem.* **76,** 23–50.

Skordalakes, E., and Berger, J. M. (2003). Structure of the Rho transcription terminator: Mechanism of mRNA recognition and helicase loading. *Cell* **114,** 135–146.

Skordalakes, E., and Berger, J. M. (2006). Structural insights into RNA-dependent ring closure and ATPase activation by the Rho termination factor. *Cell* **127,** 553–564.

Stitt, B. L. (1988). *Escherichia coli* transcription termination protein rho has three hydrolytic sites for ATP. *J. Biol. Chem.* **263,** 11130–11137.

Stitt, B. L., and Webb, M. R. (1986). Absence of a phosphorylated intermediate during ATP hydrolysis by *Escherichia coli* transcription termination protein rho. *J. Biol. Chem.* **261,** 15906–15909.

Stitt, B. L., and Xu, Y. (1998). Sequential hydrolysis of ATP molecules bound in interacting catalytic sites of *Escherichia coli* transcription termination protein Rho. *J. Biol. Chem.* **273,** 26477–26486.

Thomsen, N. D., and Berger, J. M. (2009). Running in reverse: The structural basis for translocation polarity in hexameric helicases. *Cell* **139,** 523–534.

Wang, Y., and von Hippel, P. H. (1993). *Escherichia coli* transcription termination factor rho. I. ATPase activation by oligonucleotide cofactors. *J. Biol. Chem.* **268,** 13940–13946.

Wei, R. R., and Richardson, J. P. (2001). Identification of an RNA-binding Site in the ATP binding domain of *Escherichia coli* Rho by H_2O_2/Fe-EDTA cleavage protection studies. *J. Biol. Chem.* **276,** 28380–28387.

Xu, Y., Johnson, J., Kohn, H., and Widger, W. R. (2003). ATP binding to Rho transcription termination factor. Mutant F355W ATP-induced fluorescence quenching reveals dynamic ATP binding. *J. Biol. Chem.* **278,** 13719–13727.

Yeates, T. O., and Fam, B. C. (1999). Protein crystals and their evil twins. *Structure* **7,** R25–R29.

CHAPTER NINE

Structural Analysis of RNA Helicases with Small-Angle X-ray Scattering

Manja A. Behrens,[*,†] Yangzi He,[†,‡] Cristiano L. P. Oliveira,[§] Gregers R. Andersen,[†,‡] Jan Skov Pedersen,[*,†] *and* Klaus H. Nielsen[†,‡,1]

Contents

1. Introduction	192
2. Basic Principles of Solution Scattering and Initial Data Treatment	193
3. Model-Independent Analysis	194
4. Shape Determination Without Use of *A Priori* Information	198
5. Atomic Resolution Models and Solution Scattering	201
6. Using Atomic Resolution Models to Investigate Solution Structures	203
7. Analysis of Protein Complexes	206
8. Concluding Remarks	210
Acknowledgments	210
References	210

Abstract

Small-angle X-ray scattering (SAXS) is a structural characterization method applicable to biological macromolecules in solution. The great advantage of solution scattering is that the systems can be investigated in near-physiological conditions and their response to external changes can also be easily investigated. In this chapter, we discuss the application of SAXS for studying the conformation of helicases alone and in complex with other biological macromolecules. The DEAD-box helicase eIF4A and the DEAH/RHA helicase Prp43 are investigated for their solution structures, and the analysis of the collected scattering data is presented. A wide range of methods for analysis of SAXS data are presented and discussed. *Ab initio* methods can be used to yield

---

[*] Department of Chemistry and iNANO Interdisciplinary Nanoscience Centre, Aarhus University, Aarhus, Denmark
[†] Centre for mRNP Biogenesis and Metabolism, Aarhus University, Aarhus, Denmark
[‡] Department of Molecular Biology and Genetics, Aarhus University, Aarhus, Denmark
[§] Instituto de Física, Universidade de São Paulo, São Paulo, Brazil
[1] Current address: Department of Molecular Genetics and Cell Biology, University of Chicago, IL, Chicago, USA

Methods in Enzymology, Volume 511 © 2012 Elsevier Inc.
ISSN 0076-6879, DOI: 10.1016/B978-0-12-396546-2.00031-0 All rights reserved.

low-resolution solution structures, and when models with atomic resolution are available, these can be included to aid the determination of solution structures. Using such prior information relating to the systems studied and applying a variety of methods, substantial insight can be gained about solution structures and interactions of biological macromolecules through small-angle scattering.

1. Introduction

In small-angle X-ray scattering (SAXS), the particles studied are randomly oriented in solution. In contrast, in X-ray crystallography, the biological macromolecules are arranged in a lattice. This arrangement amplifies the signal, which makes it possible to obtain atomic resolution for a structural model. However, having the biological macromolecules arranged in a lattice can cause crystal packing effects which may affect the conformation of the molecule. The scattering at small angles does not yield models with atomic resolution but instead allows for the possibility of conducting structural investigations in solution without the need for crystals, making small-angle scattering a powerful low-resolution tool for the investigation of biological macromolecules as either single molecules or complexes. Nuclear magnetic resonance (NMR) can, as X-ray crystallography, yield models with atomic resolution and can be performed on solubilized macromolecules. However, because of limitations in molecular weight of the macromolecules investigated by NMR, the technique is not suited for large macromolecules or complexes of these, whereas SAXS is. The development of sophisticated tools for analyzing small-angle scattering data from biological macromolecules through the past 20 years mainly by Svergun and coworkers makes it possible to obtain detailed information from solution small-angle scattering.

Here, we discuss the application of SAXS for studying RNA helicases (Jankowsky, 2011). We focus on two different RNA helicases. The first is the abundant DEAD-box protein eIF4A that is involved in translation initiation, where it is thought to unwind small double stranded mRNA hairpins, thus allowing the 40S ribosomal subunit to bind the messenger RNA (mRNA) and subsequently scan the mRNA in search of the AUG start codon (Sonenberg and Hinnebusch, 2009). While eIF4A has been widely studied, no crystal structure of eIF4A in complex with RNA is known. Using SAXS, we describe the structure of human eIF4A alone and in complex with two of its binding partners eIF4G and eIF4B, the latter complex also containing RNA. The second is the DEAH/RHA helicase, Prp43, an essential protein in *S. cerevisiae*, involved in both pre-mRNA splicing and ribosome biogenesis (He *et al.*, 2011). We describe the solution structure of Prp43 alone and in complex with Pfa1 (Lebaron *et al.*, 2009). In the analysis, both model-independent and -dependent methods have been applied, and we describe these methods in relation to the analysis of the data presented.

2. Basic Principles of Solution Scattering and Initial Data Treatment

The total scattering intensity from a solution of monodisperse particles can be described as

$$I(q) = nP(q)S(q), \quad (9.1)$$

where n is the number density of particles, $P(q)$, the form factor, related to the size and shape of the particles giving rise to the scattering, and $S(q)$, the structure factor, describing the effects arising from the interference of scattering originating from interacting particles (Box 9.1). In dilute solutions, the structure factor effect is negligible and $S(q) = 1$.

All scattering data presented in this chapter were collected for dilute solutions, where no effect of interparticle interactions was observed. A dilute solution is normally below 5 mg/ml; however, whether it can be considered to be dilute depends on the size of the macromolecules and charge as well as ionic strength. A concentration series (c, $c/2$, and $c/4$)

BOX 9.1   BASIC PRINCIPLES OF SCATTERING

Scattering of X-rays in a sample gives rise to interference effects, which in turn results in a scattering pattern. Owing to the random orientation of particles in solution, the scattering intensity can be represented as a one-dimensional dataset that depends on the modulus of the scattering vector, q. This is given as $q = 4\pi \sin\theta/\lambda$, where λ is the wavelength of the X-rays and 2θ the angle between the incoming and scattered X-rays.

The scattering intensity is equal to that of a single particle averaged over all orientations, $I(q) = n\langle |A_1(\vec{q})|^2 \rangle$, where the scattering amplitude, $A_1(\vec{q})$, is obtained by integrating the sum of all scattering events within the particle over its volume.

$$A_1(\vec{q}) = \int_V \Delta\rho(\vec{r}) \exp(-i\vec{q}\cdot\vec{r}) \mathrm{d}\vec{r} \quad (9.2)$$

Here $\exp(-i\vec{q}\cdot\vec{r})$ is the phase factor that describes how the waves generated at each scattering event is shifted and $\Delta\rho(\vec{r})$ is the excess scattering length density difference between the object and the buffer at $\vec{r}$. The average of the excess scattering length density, $\Delta\rho = |\Delta\rho(\vec{r})|_{av}$, is the contrast of the scattering particles (Glatter and Kratky, 1982a,b; Lindner and Zemb, 2002a,b).

BOX 9.2 PRACTICAL POINTS TO CONSIDER

- Background sample★
  - *Elution buffer*: if size exclusion chromatography is the final step of purification
  - *Dialysis buffer*: if the final chromatography step contains a gradient
- Additives can have high small-angle scattering★★
  - Poly(ethylene glycol) (PEG)
  - High salt content

★The background sample used to subtract scattering from the buffer and the capillary (in order to obtain scattering solely from the biological macromolecule) must be the same as that of the sample.
★★These can completely dominate the scattering pattern, thus effectively masking the scattering from the macromolecule.

should be collected to be certain that no concentration effects are present in the solution. Additionally, two important practical points must be considered when performing a solution scattering experiment (Box 9.2).

Once the scattering patterns from the sample and the buffer alone have been collected, the initial data reduction can be performed. The sample data are subtracted for background and converted to an absolute scale using water as a primary standard (Pedersen, 2004). The final scattering dataset consists of the modulus of the scattering vector, q_i; the intensity, $I(q_i)$; and the standard error, σ_i. Here $i = 1, \ldots, N$, where N is the number of points for which the scattering intensity is measured (Pedersen, 1997). All data presented in this chapter have been collected on a laboratory-based SAXS instrument situated in the Department of Chemistry at Aarhus University in Denmark (Pedersen, 2004), and the initial data treatment was performed using the SUPERSAXS program package (Oliveira, C.L.P. and Pedersen, J.S.; http://stoa.usp.br/crislpo/files/).

3. MODEL-INDEPENDENT ANALYSIS

The first step in analyzing SAXS data is to evaluate the scattering data using model-independent methods. This also provides a check of the data quality and might also reveal concentration or aggregation effects in the data (Box 9.3).

Nonlinearity of the data in the Guinier plot is a clear sign of aggregation or interparticle interactions in the sample; however, linearity does not guarantee a solution of monomeric noninteracting particles. To ensure this, other characteristic parameters, such as the molecular mass and overall dimension of the particles in solution, must be considered.

BOX 9.3   RADIUS OF GYRATION, FORWARD SCATTERING, AND MOLECULAR MASS

- Directly obtainable parameters
  - Radius of gyration, R_g
  - Forward scattering, $I(q=0)$
- Guinier approximation
From a Guinier plot, $(\log(I(q))$ vs. $q^2)$ R_g and $I(q=0)$ can be obtained following the Guinier expression $I(q) \approx I(q=0)\exp(-R_g^2 q^2/3)$, where the slope of the curve yields R_g and the interception with the y-axis $I(q=0)$ (Guinier, 1939). Note that for globular particles, Guinier approximation is valid only for $qR_g < 1$.
- Molecular mass
Having the scattering data on an absolute scale, the average molecular mass of the particles giving rise to scattering can be calculated as $M = I(q=0)/(c\Delta\rho_m^2)$, where c is the concentration in mg/ml of the macromolecule and $\Delta\rho_m$ the scattering length density difference per unit mass. This is about 2.0×10^{10} cm/g for proteins.

The indirect Fourier transformation (IFT) of the scattering data (Box 9.4) yields the pair distance distribution function, $p(r)$, which for a dilute monodisperse solution is a histogram of distances between pairs of points inside the particle, weighted by the excess scattering length density at the points and by the distance length.

From the $p(r)$ function, the forward scattering and radius of gyration are directly obtained as,

$$I(q=0) = 4\pi \int_0^{D\max} p(r)dr, \quad R_g^2 = \frac{\int_0^{D\max} p(r)r^2\, dr}{2\int_0^{D\max} p(r)dr} \quad (9.5)$$

The values derived by this method are more reliable, compared to those obtained from the Guinier approximation, as it makes use of the entire scattering dataset (Glatter and Kratky, 1982c).

The $p(r)$ function contains information about the overall size and shape of the macromolecules giving rise to the scattering. The difference in particle shape and size is more intuitively understood from $p(r)$, as this representation is in real space, contrary to the scattering data in the "reciprocal" scattering vector space (Glatter, 1979).

The $p(r)$ function for eIF4A indicates that the macromolecule has a slightly elongated/ellipsoidal shape with a maximum diameter of 85 Å (Fig. 9.1). If the particles had been more globular, the $p(r)$ would be bell

BOX 9.4 INDIRECT FOURIER TRANSFORMATION

The intensity can be expressed as

$$I(q) = 4\pi \int_0^{D_{max}} p(r) \frac{\sin qr}{qr} dr \quad (9.3)$$

where D_{max} is the maximum diameter of the particle. The IFT provides the best fit to the scattering data in the finite measured q range without extrapolations of the data (Glatter, 1977). The $p(r)$ is expressed as a series of base functions

$$p(r) = \sum_{i=1}^{M} c_i \phi_i(r) \quad \text{for } 0 \leq r \leq D_{max} \quad (9.4)$$

where $\phi_i(r)$ are cubic b-splines, c_i are the expansion coefficients, and M is the number of b-splines and expansion coefficients. The maximum diameter, D_{max}, is selected manually. It is chosen so that the $p(r)$ function goes smoothly to zero at large distances (Hansen and Pedersen, 1991; Svergun and Pedersen, 1994). Sharp features cannot occur in the $p(r)$ function close to D_{max} due to the low resolution of the data and the fact that it is an orientationally averaged self-correlation function of the particle excess electron density. If the error bars on the $p(r)$ function are large close to D_{max}, it is usually because D_{max} is chosen too large; however, it can also be due to poor quality of the data in the low-q region. Since D_{max} is selected in this way, all known information about the system being investigated should be taken into account to make the best possible guess taking restrictions from the molecular weight into considerations. This will help the interactive fitting procedure and will also allow determining whether or not the investigated protein is in its monomeric state. Due to numerical instabilities in the fitting procedure, a smoothness constraint is imposed on the $p(r)$ function (Pedersen et al., 1994). The agreement with the experimental data is optimized by a weighted linear least-squares procedure. Having the fit functions in reciprocal (scattering) space allows smearing originating from instrumental broadening such as geometry and wavelength effects to be accounted for (Glatter and Kratky, 1982b; Lindner and Zemb, 2002b,c).

shaped with maximum close to $D_{max}/2$; however, as the maximum for eIF4A is closer to $D_{max}/3$, this shows that the particles have a nonspherical structure. The $p(r)$ function for eIF4A does not correspond to that of an elongated particle either, as this would have a maximum around the cross-section radius of the particle and a linear decay until D_{max}, as observed from the $p(r)$ obtained for eIF4G-MC (Fig. 9.1). Thus, the $p(r)$ functions suggest that eIF4A is slightly elongated, being more ellipsoidal, whereas eIF4G-MC is elongated with a D_{max} of 220 Å. The molecular masses obtained for eIF4A and eIF4G-MC correspond to those expected for monomeric

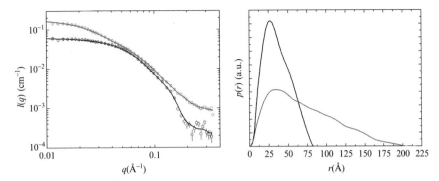

Figure 9.1 *Left*: Scattering data obtained from eIF4A (black circle) and eIF4G-MC (gray circle) with their corresponding fit from the IFT as black and gray lines, respectively. *Right*: $p(r)$ functions obtained for eIF4A (black) and eIF4G-MC (gray) through the IFT. From the $p(r)$ function, it is evident that eIF4A has a slightly elongated shape with a D_{max} of 85 Å and eIF4G-MC has an elongated shape with a D_{max} of 220 Å.

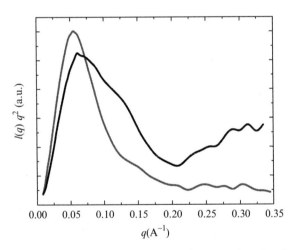

Figure 9.2 Experimental scattering data obtained for Prp43 (gray) and eIF4A (black) displayed as $I(q)q^2$ versus q (Kratky plot). The flatness of the scattering curve at intermediate and high q indicates rigid structures.

particles, and as it can be concluded that good-quality scattering data have been obtained, further modeling can be performed.

Prior to more advanced modeling, a first impression of flexibility/rigidity within the structure can be obtained by analyzing the data in a Kratky plot ($I(q)q^2$ vs. q) (Glatter and Kratky, 1982d,e). The signature of a flexible structure is more pronounced in a Kratky plot with an increase in $I(q)q^2$ at intermediate and high q, whereas a Kratky plot for compact structure will decrease steadily toward zero and appear flat at high q values. The scattering data for Prp43 and eIF4A are displayed in a Kratky plot in Fig. 9.2. The shape of the curves suggests that Prp43 has a rigid structure, as concluded from the flatness of the curve at intermediate and high q. eIF4A has a slight upturn at high q, which might indicate some degree of flexibility at small distances inside the macromolecule. However, the particle can still be considered to be relatively rigid.

The model-independent procedures are important first steps in the data analysis, as they provide the molecular mass, radius of gyration, and overall dimension of the macromolecule. The IFT procedure can also provide important information on particle oligomerization and aggregation and the interested reader can refer to several examples in the literature (Glatter and Kratky, 1982c; Oliveira *et al.*, 2009).

4. Shape Determination Without Use of *A Priori* Information

Obtaining structural information from SAXS data for a biological macromolecule without using *a priori* information is done by *ab initio* modeling (Box 9.5). This was one of the first approaches developed to analyze small-angle scattering data from biological macromolecules in solution. The development of the *ab initio* methods was started as early as in the 1970s (Stuhrmann, 1970) and continued in the early 1990s (Svergun and Stuhrmann, 1991; Svergun *et al.*, 1996). However, it was not until in the late 1990s that the approach was developed by Chacón, Svergun, and

BOX 9.5   AB INITIO MODELING

> The modeling is done by minimizing the function
>
> $$f(X) = \chi^2 + P(X) \qquad (9.6)$$
>
> This function is computed for each spatial arrangement of dummy atoms X, where χ^2 is the discrepancy between scattering calculated from the model and the experimental data, and $P(X)$ is the penalty function ensuring a physically meaningful model. Optimization of the dummy atom model is carried out by simulated annealing. One dummy atom is moved at a time and a new function is computed and compared to the old function. In most cases, the dummy atom is moved to the position, which decreases $f(X)$; however, it is also allowed to move to positions increasing $f(X)$ in order to escape local minima. The frequency of the latter is decreased during the course of minimization. The procedure is implemented in the program DAMMIN (Svergun, 1999).

coworkers (Chacón et al., 1998; Svergun et al., 1995; Walther et al., 2000) to an extent where it was useful in shape determination from solution small-angle scattering. Several examples of the implementation of this procedure are described in the literature, but in all cases, a low-resolution model of the biological macromolecule is obtained directly from the scattering curve. In the procedure developed by (Svergun, 1999), the structural model is obtained by defining a spherical search space, with sufficiently large volume with radius $R_{sphere} = D_{max}/2$. The search volume is filled with N dummy atoms on a regular lattice, with a size of $r \ll R_{sphere}$. These dummy atoms belong either to the solvent or to the macromolecule, where only the latter contributes to the scattering from the model. The scattering from the assembly of dummy atoms are calculated using spherical harmonics which allows for fast calculation of the scattering, as only the scattering amplitudes of the dummy atoms which are changed must be recalculated. The agreement between the calculated scattering data from the model and the experimental scattering data is optimized as the model is varied. The compactness and connectivity of the model are imposed by adding a penalty function to the function describing the agreement between intensities calculated from the current model and the experimental data.

Recently, a new version of DAMMIN, called DAMMIF (Franke and Svergun, 2009), has been developed in order to increase the speed of computation and to circumvent the problems in *ab initio* modeling arising when using DAMMIN (Svergun, 1999). The constraint concerning the interconnectivity of the dummy atoms is implemented differently in DAMMIF, since the model is checked for interconnectivity and discarded if the connectivity constraint is violated before computing the scattering amplitudes of the altered model. This is contrary to DAMMIN where the scattering amplitudes are calculated before the connectivity restraint is implemented. Additionally, some aspects of the penalty function differ in the two implementations. The most significant difference, apart from the increase in computation speed, is the unlimited search volume eliminating boarder effects that occur if the particle moves close to the boarder. Using DAMMIN this occasionally led to artificial bending of the particle structure. Additionally, both programs have the possibility of using knowledge of symmetry of the particles investigated.

Another program for *ab initio* modeling called GASBOR (Svergun et al., 2001) is also available, in which the number of dummy atoms is equal to the number of amino acid residues. The modeling of the structure is performed by having the dummy atoms as a gas confined in the search space instead of having the dummy atoms on a regular lattice. The restraints are also implemented differently, as the number of dummy atoms (or dummy residues) equals the number of amino acids and as one aims at obtaining a folded chain-like structure in the model. The solution is thus biased toward a model without overlap of the dummy residues and a typical distribution of the neighbors as a function of distance similar to what is found in proteins. Further, an attempt to model the hydration layer around the particle is implemented.

Due to the simulated annealing algorithm applied in the three programs, the models generated are not unique. Therefore, several models, 10 or more, must be determined and compared to obtain sufficient statistics on the reproducibility and reliability of the model. The individual models are aligned, compared, and filtered using the program package DAMAVER (Volkov and Svergun, 2003) containing several programs. The comparison and alignment of the models are performed using the program SUPCOMB (Kozin and Svergun, 2001). Here, pairs of structures are aligned by representing each structure by a set of points (atoms or beads). The alignment is performed by minimizing a similarity measure called the average normalized spatial discrepancy (NSD). An NSD value close to unity shows that two models are similar. The aligned models are averaged and filtered using the programs DAMAVER and DAMFILT, respectively (Volkov and Svergun, 2003). This procedure yields both an average model and the most representative model from the set of models, that is, the model having the lowest average NSD with the other models in the set. Through the comparison of the models, the similarity of the models is investigated, and this is used to evaluate the reproducibility and reliability of the models. A good solution is obtained when only one class of models is dominant. It should be noted that SAXS cannot distinguish between enantiomers. However, when placing high-resolution models within low-resolution envelopes determined by *ab initio* methods, one enantiomer of the envelope may fit the high-resolution structures better than the alternative, thereby overcoming this ambiguity.

Investigating the solution structure of eIF4G-MC, *ab initio* modeling was used even though partial atomic structures representing 2/3 of the protein were available. However, the structure of the middle part (residues 992–1235) is unknown, and from the *ab initio* model, it is evident that the structure has a distinct bend in middle of the molecule. The modeling was performed by DAMMIN (Svergun, 1999). The limited search volume was compensated for by having a search space 10% larger than D_{max} of the particle obtained from the $p(r)$ function (Fig. 9.1B). Ten individual runs were performed, compared, and averaged using the DAMAVER program package (Volkov and Svergun, 2003). From the comparison of the individual models, it was evident that only one species was dominant, as all the models were alike. This shows that the structure is well defined, with a distinct bend of approximately 130° (Fig. 9.3). It has been suggested that the middle part of eIF4G is unstructured; however, the SAXS analysis supports a well-structured entity and this is in agreement with circular dichroism measurements that revealed approximately half of the unknown part of eIF4G-MC is organized in β-sheets (Nielsen *et al.*, 2011).

SAXS Analysis of Helicases 201

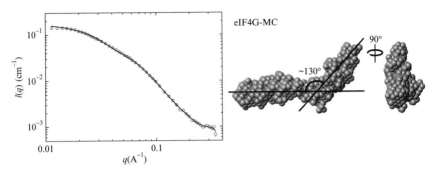

Figure 9.3 *Left*: Scattering data obtained for eIF4G-MC (open circle) and corresponding model fit (black line) from the *ab initio* modeling. *Right*: Most representative *ab initio* model for eIF4G-MC (reproduced with permission from Nielsen *et al.*, 2011).

5. ATOMIC RESOLUTION MODELS AND SOLUTION SCATTERING

A biological macromolecule in solution will associate with water, forming a hydration layer surrounding the molecule. In a small-angle scattering study, this water layer is significantly different from bulk water. If a model with atomic resolution is available for the macromolecule, the theoretical solution scattering can be computed (Box 9.6). However, Svergun and coworkers realized that the hydration layer has to be included in the calculation of the intensity (Svergun *et al.*, 1995).

To investigate the solution structure of eIF4A, models were compared to known crystal structures of an open conformation of eIF4A (Caruthers *et al.*, 2000) and a closed conformation of its homologue eIF4AIII bound to RNA and an ATP analogue (Andersen *et al.*, 2006). The amino acid sequences of eIF4A (406 residues) and eIF4AIII (411 residues) are 67% identical and 81% similar. eIF4AIII is a component of the exon junction complex where it works as an RNA clamp in a complex with MLN51 and the heterodimer Y14/Magoh (Andersen *et al.*, 2006).

The solution scattering data for eIF4A are compared to the theoretical solution scattering from the two models in Fig. 9.4. It is clear that the closed structure cannot describe the scattering data, as the model overestimates the data at intermediate q and underestimates the data at low q giving a poor agreement between the scattering data and the model intensity. It can thus be concluded that the structure of eIF4A in solution does not correspond to the closed conformation. This is in agreement with the requirement of RNA and nucleotide for eIF4A to be in the closed conformation. A better agreement between the scattering data is obtained for the scattering

BOX 9.6 COMPUTING SCATTERING FROM A STRUCTURE WITH ATOMIC RESOLUTION

> Average scattering of a single particle
>
> $$I(q) = \langle |A_p(q) - \rho_s A_s(q) + \Delta\rho_h A_h(q)|^2 \rangle \qquad (9.7)$$
>
> where the scattering amplitudes from the particle in vacuum, $A_p(q)$, the excluded volume, $A_s(q)$, that is, the volume inaccessible to the solvent, and the hydration layer, $A_h(q)$, are calculated using spherical harmonics. Further, ρ_s and ρ_h are the scattering length densities of the solvent and the hydration layer, respectively. The difference between the scattering length density of the solvent and the hydration layer, $\Delta\rho_h = \rho_h - \rho_s$, gives a significant contribution to the total scattering, due to the high density of the water in this layer. The computed scattering intensity can be compared to the experimentally obtained scattering data, to evaluate the discrepancy between model and actual solution structure. This approach is implemented in the program CRYSOL (Svergun et al., 1995).

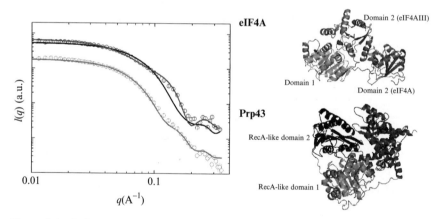

Figure 9.4 *Left*: Scattering data obtained for Prp43 (gray) and eIF4A (black) (scaled by a factor of 10) with their respective CRYSOL fits. For eIF4A, two structures are available with atomic resolution, an open conformation (PDB entry 1FUU) (magenta line) and a closed conformation (PDB entry 2HXY) (black line). For Prp43, the model obtained from the X-ray crystallography (PDB entry 3KX2) was used (solid gray line). *Right*: Crystal structure of eIF4A in the open and closed conformation. Here, domain 1 (the N-terminal RecA-like domain) is colored cyan, while domain 2 (the C-terminal RecA-like domain) in the open structure is magenta and orange in the closed structure. In the crystal structure for Prp43, the N-terminal RecA-like domain is colored green, the C-terminal RecA-like domain is colored blue, and the remaining part of the molecule is colored red. (See the Color Insert.)

calculated for the open structure of eIF4A. However, at high q, there are minor deviations between data and the model intensity. The reasonable agreement suggests that the structure of eIF4A in solution is more similar to that of the open than the closed conformation. The solution structure will be investigated further using rigid body modeling.

A crystal structure for Prp43 in complex with ADP is available (He *et al.*, 2010). Comparing this structure to the scattering data obtained for Prp43 shows that X-ray crystallography slightly underestimates the data at high q (Fig. 9.4). The scattering curve here contains information on the shorter distance in the structure, for example, relative positioning of the domains. Thus, further investigation of the model must be conducted using the information with atomic resolution.

6. Using Atomic Resolution Models to Investigate Solution Structures

As shown for Prp43, solution structures of biological macromolecules do not always agree with the structure derived from X-ray crystallography or NMR. Further, in some cases, only part of the macromolecule is available with atomic resolution. If the particles are expected to be composed of rigid structures, this can be used in rigid body refinement procedures, also developed by Svergun and coworkers in the mid-2000s (Box 9.7).

The scattering amplitudes for the individual subunits or domains are calculated using spherical harmonics, as this allows for a fast calculation. The

---

BOX 9.7   RIGID BODY MODELING

Provided atomic resolution structures of all the subunits or domains composing the macromolecule are available, the total scattering intensity can be calculated, as the orientational average of the sum of the subunit amplitudes

$$I(q) = \left\langle \left| \sum_{n=1}^{N} A_n(q) \right|^2 \right\rangle \quad (9.8)$$

where $A_n(q)$ is the scattering amplitude of the nth subunit at a given position. This amplitude is composed of the scattering amplitude at a reference position and six rotational and translational parameters describing the position and orientation of the subunit. The scattering amplitudes of the subunit at the reference position are computed using the program CRYSOL (Svergun *et al.*, 1995), as described in the previous section.

individual subunits can be fixed at the initial position or prohibited from certain rotations and/or translations, if this is justified by experimental evidence. Otherwise, the subunits are rotated and shifted to improve the agreement of the scattering intensity for the model with the experimental data. For practical reasons, only one subunit is rotated and shifted at a time. For each rotation and shift, the amplitude for the affected subunit is recalculated. The optimization of the model is performed using simulated annealing. To ensure a physically sensible model, the subunits must be interconnected and nonoverlapping. This is accomplished by the addition of a penalty function to the function describing the agreement between the model intensity and the experimental data. Further, available knowledge of interresidue distances can be employed as an additional constraint. This procedure is implemented in the program SASREF (Petoukhov and Svergun, 2005).

Svergun and coworkers also developed an approach to rigid body refinement for cases where part of the structure is unknown. Here, a combination of rigid body refinement, as described in the previous paragraph, and *ab initio* modeling is used. The parts of the macromolecule, where structures with atomic resolutions are available, are used and the remaining part is represented as dummy residues. The optimization of the model is again performed with simulated annealing. When either a part of the dummy residue chain or a domain with atomic resolution structure is reoriented, the nonoverlapping of the domains is ensured by the employment of a penalty function in the minimization. The theoretical scattering from the dummy residues are calculated by assigning a form factor for an average residue in aqueous solution and then computing the scattering from the complex as described in (Svergun *et al.*, 2001). The scattering from the subunits with atomic resolution is calculated by spherical harmonics. This procedure is implemented in the program BUNCH (Petoukhov and Svergun, 2005). Similar to the *ab initio* modeling, both approaches consider particle symmetry.

A new procedure for rigid body modeling has been recently developed. This procedure is applicable to macromolecular protein complexes, where the individual proteins consist of multiple domains. In this approach, contact constraints between the proteins and interdomain linkers represented as dummy residues are implemented. The procedure is based on the methods used in the programs SASREF and BUNCH and implemented in the program CORAL (Petoukhov and Svergun, 2005). This approach has not been used in the work described here.

The solution structure of Prp43 was modeled using rigid body refinement of the domains. The helicase was divided into three structural domains (two RecA-like domains while the remaining part of Prp43 was treated as one domain, Fig. 9.5). No residues are missing in the structure connecting the domains and therefore the method implemented in the program SASREF (Petoukhov and Svergun, 2005) was used. Additional restraints on

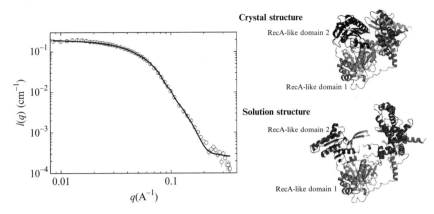

Figure 9.5 *Left*: Scattering data obtained for Prp43 (open circle) and best fit from rigid body modeling (black line). *Right*: Most representative model obtained from rigid body modeling (bottom) compared to the crystal structure model (top), where domain 1 in both structures is green and with the same orientation. Domain 2 is blue and domain 3 is red. (See the Color Insert.)

interresidue distances were imposed on the model, given that the end of one chain in one domain and beginning of the chain in another domain have to be in close proximity. To probe reproducibility, 10 individual runs were performed and these were compared and averaged using the DAMAVER program package (Volkov and Svergun, 2003). One solution was dominant and the most representative model corresponds well with the scattering data (Fig. 9.5). The solution structure of Prp43 is more open than the crystal structure, where binding of RNA is also obstructed (He *et al.*, 2010). The more open solution structure of Prp43 could potentially accommodate RNA, in agreement with RNA pull-down experiments which demonstrated binding to RNA in the apo state of Prp43 (He *et al.*, 2010).

The solution structure of eIF4A was also investigated using rigid body modeling. The molecule has a linker of 18 amino acids between the two domains, and therefore the modeling was performed using rigid body approach implemented in BUNCH (Petoukhov and Svergun, 2005), with the linker modeled as dummy residues. One dominant species was found in 10 individual runs. The solutions were compared and averaged by the DAMAVER program package (Volkov and Svergun, 2003). A somewhat open structure of eIF4A was evident from the rigid body modeling (Fig. 9.6). It is less open than in the crystal structure of yeast eIF4A (Caruthers *et al.*, 2000), where the crystal packing might be forcing a more open eIF4A conformation.

If the biological macromolecule cannot be described as a rigid structure, the structure may be too flexible. Highly flexible macromolecules can adopt many different conformations in solution. Modeling for highly flexible macromolecules therefore requires a different approach than for modeling

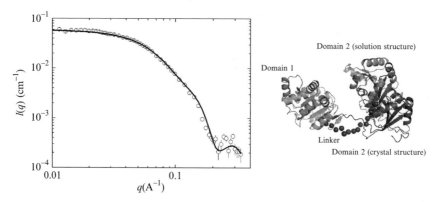

Figure 9.6 *Left*: Scattering data obtained for eIF4A (open circle) and best fit from rigid body refinement (black line). *Right*: Best model obtained from rigid body refinement, where domain 1 is colored cyan, domain 2 orange, and the linker is displayed as gray spheres, compared to the open crystal structure model, where domain 1 is colored cyan, domain 2 purple, and the linker gray. (For interpretation of the references to color in this figure legend, the reader is referred to the online version of this chapter.)

rigid macromolecules. This procedure is implemented in the Ensemble Optimization Method (the EOM program package) (Bernadó *et al.*, 2007). However, high degree of flexibility was not found in any of the structures investigated here.

7. ANALYSIS OF PROTEIN COMPLEXES

When using SAXS for structural investigation of macromolecular complexes, it is most important to first verify that the scattering data originate from the complex, and not from individual macromolecules. Figure 9.7 shows that a linear combination of the individual scattering curves (Box 9.8), obtained for eIF4A and eIF4G-MC, cannot describe the data obtained for the complex. Thus, the scattering data from the complex do not originate from a mixture of the individual macromolecules.

Investigating the solution structure of the complex between eIF4A and eIF4G-MC performed with DAMMIN (Svergun, 1999) demonstrated good agreement with the *ab initio* models. In addition, only one dominant model was found. The model displayed in Fig. 9.8 reveals the distinct bend observed for eIF4G-MC, which was conserved upon complex formation. A comparison of the *ab initio* model obtained for eIF4G-MC and the complex also shows an additional mass above the bend, which most likely represents eIF4A (Nielsen *et al.*, 2011).

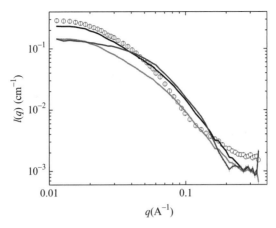

Figure 9.7 Linear combination (black line) of scattering data obtained for eIF4A (red) and eIF4G-MC (gray) to the scattering data for the sample containing both eIF4A and eIF4G-MC (open circle). (For interpretation of the references to color in this figure legend, the reader is referred to the online version of this chapter.)

BOX 9.8   LINEAR COMBINATION OF SCATTERING DATA

> Scattering data can be described as a linear combination of the components giving rise to the scattering
>
> $$I(q) = c_1 I_1(q) + \cdots + c_i I_i(q) \quad (9.9)$$
>
> where $I_i(q)$ is the scattering from the individual component and c_i is the weight of each component. A mixture of two macromolecules would yield a different scattering curve than that obtained if the two molecules form a complex. Therefore, a first check of the scattering data obtained from a complex is to investigate if it can be described by a linear combination of the scattering from the individual macromolecules.

In spite of having the entire eIF4A and 2/3 of the eIF4G-MC available with atomic resolution, it was not possible to optimize the model of the complex utilizing this information. An attempt to model the complex was carried out by representing the missing residues in eIF4G-MC as dummy residues and performing a combined *ab initio* and rigid body refinement using the program BUNCH (Petoukhov and Svergun, 2005). However, the missing part of eIF4G-MC was too large and a physically meaningful stable model could not be obtained, which was also the case for eIF4G-MC alone. However, by superimposing the parts of the complex with atomic resolution on the *ab initio* model, a likely interpretation can be obtained (Fig. 9.9).

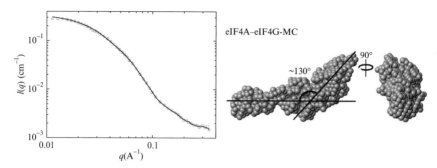

Figure 9.8 *Left*: Scattering data obtained for the complex between eIF4A and eIF4G-MC (open circle) and corresponding model fit (black line) from the *ab initio* modeling. *Right*: Most representative *ab initio* model for the complex between eIF4A and eIF4G-MC (reproduced with permission from Nielsen *et al.*, 2011).

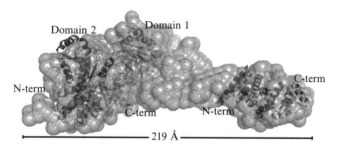

Figure 9.9 Super-positioning of the available structural models with atomic resolution for the complex between yeast eIF4A and eIF4G (PDB entry 2VSO) (Schutz *et al.*, 2008) and the first HEAT domain from the C-terminal portion of human eIF4G (PDB entry 1UG3) (Bellsolell *et al.*, 2006) on the *ab initio* model obtained from SAXS. Good agreement is evident(For the color version of this figure, the reader is referred to the online version of this chapter.) (reproduced with permission from Nielsen *et al.*, 2011).

The complex between eIF4A and eIF4BΔC (residues 1-431) has also been investigated using *ab initio* modeling. The maximum dimension and molecular mass obtained from the scattering data suggest that the scattering macromolecules are dimers of the eIF4A–eIF4BΔC complex. Therefore, $P2$ symmetry was imposed on the modeling performed using DAMMIN (Svergun, 1999) and only one class of structures was found. The result indicates that the protein complex is well defined in solution with an elongated shape. However, because more than half of the complex is composed of eIF4BΔC and the only known structures are the individual domains of eIF4A, it was not possible to retrieve any information about the solution position of eIF4BΔC, since the unknown part of the structure was too extensive. The proposed model nevertheless agrees well with the published data demonstrating that eIF4BΔC forms a dimer (Methot *et al.*, 1996), suggesting that the heterodimer between eIF4A and eIF4BΔC might also form a dimer (Fig. 9.10).

The complex for Prp43–Pfa1 was found to be a 1:1 complex based on the molecular mass estimated from the SAXS data, and the complex was investigated by *ab initio* modeling using DAMMIF (Franke and Svergun, 2009). Again, one structural model dominated the set of solutions and a reliable model was obtained. Despite the availability of an atomic model of Prp43, it was not possible to reconstruct the entire complex because the missing part of Pfa1 was too extensive. In Fig. 9.11, the solution structure obtained for Prp43 is superimposed on the *ab initio* model of the Prp43–Pfa1 complex. Prp43 was placed on the *ab initio* model by using the program

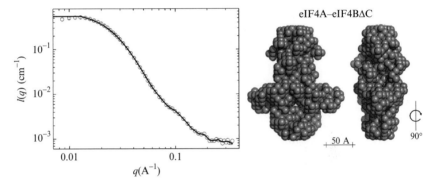

Figure 9.10 *Left*: The fit from the *ab initio* model of the scattering data. *Right*: *Ab initio* model for the dimer of the heterodimer eIF4A-eIF4BΔC.

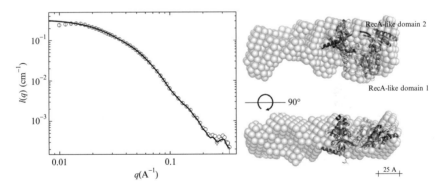

Figure 9.11 *Left*: Scattering data obtained for Prp43–Pfa1 complex (open circle) and best *ab initio* model fit obtained from DAMMIF. *Right*: Best *ab initio* model with the solution structure of Prp43 obtained by rigid body modeling superimposed. In the structure of Prp43, domain 1 is green, domain 2 blue, and domain 3 red. (For interpretation of the references to color in this figure legend, the reader is referred to the online version of this chapter.)

SUPCOMP (Kozin and Svergun, 2001), which compared two structures and gives the best overlap of the two. The Prp43 takes up a little less than half of the structure. The program gives three possible solutions, all having good agreement with the *ab initio* model. However, the best agreement was obtained when Prp43 was placed as shown in Fig. 9.11.

8. Concluding Remarks

SAXS has become a valuable tool in the investigation of the solution structure of proteins and their complexes, here exemplified by RNA helicases and their interacting partners. Important insight was gained into the solution structures of the helicases using the information available for the systems investigated combined with the information obtained from the solution scattering data. In spite of the relative low information content in the scattering data, solution SAXS is an invaluable tool for characterizing solution states of biological macromolecules, as it is essentially the only technique (apart from small-angle neutron scattering) which allows *in situ* investigation of large complexes and their response to external stimuli like changes in pH, salt, or addition of ligands.

ACKNOWLEDGMENTS

K. H. N. was supported by Alfred Benson Foundation; G. R. A. was supported by The Danish Council for Independent Research, Natural Sciences, a Hallas-Møller stipend from the Novo Nordic foundation, and the Danish National Research Foundation; M. A. B. and J. S. P. were supported by The Danish Council for Independent Research, Natural Sciences; and J. S. P. was supported by the Danish National Research Foundation.

REFERENCES

Andersen, C. B. F., Ballut, L., Johansen, J. S., Chamieh, H., Nielsen, K. H., Oliveira, C. L. P., Pedersen, J. S., Séraphin, B., Hir, H. L., and Andersen, G. R. (2006). Structure of the exon junction core complex with a trapped DEAD-box ATPase bound to RNA. *Science* **313**, 1968–1972.

Bellsolell, L., Cho-Park, P. F., Poulin, F., Sonenberg, N., and Burley, S. K. (2006). Two structurally atypical HEAT domains in the C-terminal portion of human eIF4G support binding to eIF4A and Mnk1. *Structure* **14**, 913–923.

Bernadó, P., Mylonas, E., Petoukhov, M. V., Blackledge, M., and Svergun, D. I. (2007). Structural characterization of flexible proteins using small-angle X-ray scattering. *J. Am. Chem. Soc.* **129**, 5656–5664.

Caruthers, J. M., Johnson, E. R., and McKay, D. B. (2000). Crystal structure of yeast initiation factor 4A, a DEAD-box RNA helicase. *Proc. Natl. Acad. Sci. U. S. A.* **97**, 13080–13085.

Chacón, P., Morán, F., Díaz, J. F., Pantos, E., and Andreu, J. M. (1998). Low-resolution structures of proteins in solution retrieved from X-ray scattering with a genetic algorithm. *Biophys. J.* **74**, 2760–2775.

Franke, D., and Svergun, D. I. (2009). DAMMIF, a program for rapid ab-initio shape determination in small-angle scattering. *J. Appl. Crystallogr.* **42**, 342–346.

Glatter, O. (1977). A new method for the evaluation of small-angle scattering data. *J. Appl. Crystallogr.* **10**, 415–421.

Glatter, O. (1979). The interpretation of real-space information from small-angle scattering experiments. *J. Appl. Crystallogr.* **12**, 166–175.

Glatter, O., and Kratky, O. (1982a). General theory. *Small Angle X-ray Scattering.* Academic Press, London, pp. 17–52 (Chapter 2).

Glatter, O., and Kratky, O. (1982b). Data treatment. *Small Angle X-ray Scattering.* Academic Press, London, pp. 119–166 (Chapter 4).

Glatter, O., and Kratky, O. (1982c). Interpretation. *Small Angle X-ray Scattering.* Academic Press, London, pp. 167–196 (Chapter 5).

Glatter, O., and Kratky, O. (1982d). Natural high polymers. *Small Angle X-ray Scattering.* Academic Press, London, pp. 361–386 (Chapter 11).

Glatter, O., and Kratky, O. (1982e). Synthetic polymers in solution. *Small Angle X-ray Scattering.* Academic Press, London, pp. 387–432 (Chapter 12).

Guinier, A. (1939). La diffraction des rayons X aux tres petits angles; application a l'etude de phenomenes ultramicroscopiques. *Ann. Phys. (Paris)* **16**, 161–237.

Hansen, S., and Pedersen, J. S. (1991). A comparison of three different methods for analysing small-angle scattering data. *J. Appl. Crystallogr.* **24**, 541–548.

He, Y., Andersen, G. R., and Nielsen, K. H. (2010). Structural basis for the function of DEAH helicases. *EMBO Rep.* **11**, 180–186.

He, Y., Andersen, G. R., and Nielsen, K. H. (2011). The function and architecture of DEAH/RHA helicases. *BioMol. Concepts* **2**, 315–326.

Jankowsky, E. (2011). RNA helicases at work: Binding and rearranging. *Trends Biochem. Sci.* **36**, 19–29.

Kozin, M. B., and Svergun, D. I. (2001). Automated matching of high- and low-resolution structural models. *J. Appl. Crystallogr.* **34**, 33–41.

Lebaron, S., Papin, C., Capeyrou, R., Chen, Y.-L., Froment, C., Monsarrat, B., Caizergues-Ferrer, M., Grigoriev, M., and Henry, Y. (2009). The ATPase and helicase activities of Prp43p are stimulated by the G-patch protein Pfa1p during yeast ribosome biogenesis. *EMBO J.* **28**, 3808–3819.

Lindner, P., and Zemb, T. (2002a). General theorems in small-angle scattering. *Neutrons, X-rays and Light: Scattering Methods Applied to Soft Condensed Matter.* North-Holland/Elsevier, Amsterdam, pp. 49–71 (Chapter 3).

Lindner, P., and Zemb, T. (2002b). The inverse scattering problem in small-angle scattering. *Neutrons, X-rays and Light: Scattering Methods Applied to Soft Condensed Matter.* North-Holland/Elsevier, Amsterdam, pp. 73–102 (Chapter 4).

Lindner, P., and Zemb, T. (2002c). Fourier transformation and deconvolution. *Neutrons, X-rays and Light: Scattering Methods Applied to Soft Condensed Matter.* North-Holland/Elsevier, Amsterdam, pp. 103–126 (Chapter 5).

Methot, N., Song, M., and Sonenberg, N. (1996). A region rich in aspartic acid, arginine, tyrosine, and glycine (DRYG) mediates eukaryotic initiation factor 4B (eIF4B) self-association and interaction with eIF3. *Mol. Cell. Biol.* **16**, 5328–5334.

Nielsen, K. H., Behrens, M. A., He, Y., Oliveira, C. L. P., Sottrup Jensen, L., Hoffmann, S. V., Pedersen, J. S., and Andersen, G. R. (2011). Synergistic activation of eIF4A by eIF4B and eIF4G. *Nucl. Acids Res.* **39**(7), 2678–2689.

Oliveira, C. L. P., Behrens, M. A., Pedersen, J. S., Erlacher, K., Otzen, D., and Pedersen, J. S. (2009). A SAXS study of glucagon fibrillation. *J. Mol. Biol.* **387**, 147–161.

Pedersen, J. S. (1997). Analysis of small-angle scattering data from colloids and polymer solutions: Modeling and least-squares fitting. *Adv. Colloid Interface Sci.* **70,** 171–210.
Pedersen, J. (2004). A flux- and background-optimized version of the NanoSTAR small-angle X-ray scattering camera for solution scattering. *J. Appl. Crystallogr.* **37,** 369–380.
Pedersen, J. S., Hansen, S., and Bauer, R. (1994). The aggregation behavior of zinc-free insulin studied by small-angle neutron scattering. *Eur. Biophys. J.* **22,** 379–389.
Petoukhov, M. V., and Svergun, D. I. (2005). Global rigid body modeling of macromolecular complexes against small-angle scattering data. *Biophys. J.* **89,** 1237–1250.
Schutz, P., Bumann, M., Oberholzer, A. E., Bieniossek, C., Trachsel, H., Altmann, M., and Baumann, U. (2008). Crystal structure of the yeast eIF4A-eIF4G complex: An RNA-helicase controlled by protein–protein interactions. *Proc. Natl. Acad. Sci. U.S.A.* **105,** 9564–9569.
Sonenberg, N., and Hinnebusch, A. G. (2009). Regulation of translation initiation in eukaryotes: Mechanisms and biological targets. *Cell* **136,** 731–745.
Stuhrmann, H. B. (1970). Interpretation of small-angle scattering functions of dilute solutions and gases. A representation of the structures related to a one-particle scattering function. *Acta Crystallogr. A* **26,** 297–306.
Svergun, D. I. (1999). Restoring Low Resolution Structure of Biological Macromolecules from Solution Scattering Using Simulated Annealing. *Biophys. J.* **76,** 2879–2886.
Svergun, D. I., and Pedersen, J. S. (1994). Propagating errors in small-angle scattering data treatment. *J. Appl. Crystallogr.* **27,** 241–248.
Svergun, D. I., and Stuhrmann, H. B. (1991). New developments in direct shape determination from small-angle scattering. 1. Theory and model-calculations. *Acta Crystallogr. A* **47,** 736–744.
Svergun, D., Barbcrato, C., and Koch, M. H. J. (1995). CRYSOL—A program to evaluate X-ray solution scattering of biological macromolecules from atomic coordinates. *J. Appl. Crystallogr.* **28,** 768–773.
Svergun, D. I., Volkov, V. V., Kozin, M. B., and Stuhrmann, H. B. (1996). New developments in direct shape determination from small-angle scattering. 2. Uniqueness. *Acta Crystallogr. A* **52,** 419–426.
Svergun, D. I., Petoukhov, M. V., and Koch, M. H. J. (2001). Determination of domain structure of proteins from X-ray solution scattering. *Biophys. J.* **80,** 2946–2953.
Volkov, V. V., and Svergun, D. I. (2003). Uniqueness of ab initio shape determination in small-angle scattering. *J. Appl. Crystallogr.* **36,** 860–864.
Walther, D., Cohen, F. E., and Doniach, S. (2000). Reconstruction of low-resolution three-dimensional density maps from one-dimensional small-angle X-ray solution scattering data for biomolecules. *J. Appl. Crystallogr.* **33,** 350–363.

CHAPTER TEN

Analysis of Cofactor Effects on RNA Helicases

Crystal Young[*,†] and Katrin Karbstein[†]

Contents

1. Introduction	214
1.1. Why are cofactors needed?	214
1.2. How do cofactors modulate the enzymatic activities of RNA helicases?	218
2. How to Study Cofactors	219
2.1. Identification of cofactors	219
2.2. Studying the effects of cofactors on RNA helicases	220
3. Reagents	221
3.1. Reaction buffers	221
3.2. Gels and loading dyes	221
3.3. Labeling RNAs with ^{32}P	221
3.4. Purification of nucleotides	223
3.5. Purification of RNAs	223
4. Protocols	224
4.1. RNA binding via gel-shift	224
4.2. RNA release rate constants via gel-shift	227
4.3. ATPase activity using single-turnover conditions	228
4.4. Measuring nucleotide affinities	230
4.5. Preparation of RNA duplexes	231
4.6. RNA unwinding	232
4.7. RNA annealing	233
Acknowledgments	233
References	233

Abstract

RNA helicases are involved in all aspects of RNA metabolism. Since the helicase core is conserved between all helicases, specificity for particular cellular roles must arise from interactions with specific cofactors, which can regulate RNA binding and enzymatic activity. While recent structural studies have provided

[*] Department of Chemistry, University of Michigan, Ann Arbor, Michigan, USA
[†] Department of Cancer Biology, The Scripps Research Institute, Jupiter, Florida, USA

Methods in Enzymology, Volume 511 © 2012 Elsevier Inc.
ISSN 0076-6879, DOI: 10.1016/B978-0-12-396546-2.00010-3 All rights reserved.

invaluable insight into some mechanisms of cofactor effects on RNA helicases, biochemical experiments must ultimately be conducted in order to validate these predictions. Here, we provide a guide for identifying helicase-specific cofactors and then studying their effects on helicase function. By measuring RNA binding and release, ATPase activity, nucleotide affinity, and unwinding and annealing activities, cofactor effects on an RNA helicase can be fully characterized.

1. INTRODUCTION

1.1. Why are cofactors needed?

Eukaryotic RNA helicases belong to either of the two helicase superfamilies (SF): SF1 or SF2 (Fairman-Williams et al., 2010; Jankowsky, 2011). All of these helicases share a structurally conserved helicase core that consists of two tandem RecA-like domains connected by a flexible linker. In most helicases, these domains contain sequence motifs that are conserved within the SF and are involved in ATP binding and hydrolysis, RNA binding and interdomain contacts. Since the helicase core is highly conserved between RNA helicases (Schütz et al., 2010), specificity for particular cellular roles must arise outside of these conserved regions and at least partially derive from the N- and C-terminal extensions that surround the two RecA-like domains. These additional sequences are unique to each helicase. However, the only RNA helicases known to have sequence specificity in the absence of a cofactor are bacterial orthologs of DbpA, an *Escherichia coli* DEAD-box protein that is involved in assembly of the large ribosomal subunit (Fuller-Pace et al., 1993; Iost and Dreyfus, 2006; Sharpe Elles et al., 2009). DbpA specifically recognizes hairpin 92 in 23S ribosomal RNA (Fuller-Pace et al., 1993). Yeast proteins such as Mss116 and Ded1 have a preference for structured RNA (Fairman et al., 2004; Mohr et al., 2008) but no specific target RNAs have been identified. It is therefore believed that by specifically binding individual RNA sequences, helicase cofactors may increase the specificity of RNA helicases, thereby contributing to their unique and generally nonoverlapping roles in various biological processes. Surprisingly, there is no experimental evidence for this simple hypothesis.

In addition to possibly increasing biological specificity for substrates, cofactors can also modulate enzymatic activity. RNA helicases such as Prp43 and Upf1 have very low intrinsic ATPase and helicase activities (Chakrabarti et al., 2011; Schutz et al., 2008). Their cofactors Pfa1 and Upf2, respectively, enhance ATPase and helicase activities, thereby contributing to the biological functions of these helicases (Table 10.1).

Examples in the literature indicate that cofactors can affect all steps of the RNA helicase kinetic cycle (Fig. 10.1). For DEAD-box proteins (SF2), it

Table 10.1 S. cerevisiae RNA helicases and their cofactors

RNA helicase	Family	Cofactor	Means of ID	Cofactor effects			
				Substrate binding	ATPase activity	Helicase activity	Product release
Prp43	DEAH	Ntr1/Ntr2	Genetic interaction[a]; Yeast two-hybrid[b]; Co-IP[b]; in vitro binding[a]	N/D*	No effect	Increase[a]	N/D
Prp43	DEAH	Pfa1	TAP-purification[c]; in vitro binding[c]	ATP: No effect[c]	Increase[c]	Increase[c]	N/D
Prp2	DEAH	Spp2	Genetic interaction[d]; Yeast two-hybrid[d,e]; Reconstituted complex[f]	N/D	N/D	N/D	N/D
Upf1	SF1 (Upf1-like)	Upf2	Genetic interaction[g]; Co-IP[h]; Crystal structure[h];	N/D	Increase[h]	Increase[h]	N/D
Fal1	DEAD	Sgd1	Genetic interaction[i]; Co-IP[i]	N/D	N/D	N/D	N/D
eIF4A	DEAD	eIF4G	Co-IP[j]; in vitro binding[j,k]; Crystal structure[k]	ATP and ADP: Decrease[l]	Increase[k,l]	N/D	P_i: Increase[l]
eIF4A	DEAD	eIF4B		RNA[n] and ATP: Increase	Increase[n,q,r]	Increase[p]	N/D

(Continued)

Table 10.1 (Continued)

RNA helicase	Family	Cofactor	Means of ID	Cofactor effects			
				Substrate binding	ATPase activity	Helicase activity	Product release
Dbp8	DEAD	Esf2	Genetic interaction[m]; Reconstituted complex[n,o,p]				
Dbp5	DEAD	Gle1$_{InsP6}$**	Yeast two-hybrid[s]; Co-IP[s]; *in vitro* binding[s]	N/D	Increase[s]	N/D	N/D
Dbp5	DEAD	Gle1$_{InsP6}$**	Genetic interaction[t]; Yeast two-hybrid[t,u]; *in vitro* binding[t]; Crystal structure[v]	RNA[u] and ATP[x]: Increase	Increase[u,w]	N/D	RNA: Increase[v]
Dbp5	DEAD	Nup159	Genetic interaction[y]; Crystal structure[z]	RNA: Decrease[z,aa]	Decrease[z]	N/D	ADP: Increase[bb]
Brr2	Ski2-like	Prp8	Genetic interaction[cc]; Yeast two-hybrid[cc]; *in vitro* binding[dd,ee]	ssRNA: Decrease[dd] dsRNA: Increase[ee]	Decrease[dd,***]	Increase[dd,ff]	N/D

References: [a]Tanaka et al. (2007), [b]Tsai et al. (2005), [c]Lebaron et al. (2009), [d]Last et al. (1987), [e]Silverman et al. (2004), [f]Warkocki et al. (2009), [g]He et al. (1997), [h]Chakrabarti et al. (2011), [i]Alexandrov et al. (2011), [j]Dominguez et al. (1999), [k]Schutz et al. (2008), [l]Hilbert et al. (2011), [m]Coppolecchia et al. (1993), [n]Abramson et al. (1988), [o]Bi et al. (2000), [p]Rozen et al. (1990), [q]Rogers et al. (2001), [r]Rogers et al. (1999), [s]Grannemar et al. (2006), [t]Strahm et al. (1999), [u]Weirich et al. (2006), [v]Montpetit et al. (2011), [w]Alcazar-Roman et al. (2006), [x]Noble et al. (2011), [y]Hodge et al. (1999), [z]Montpetit et al. (2011), [aa]von Moeller et al. (2009), [bb]Noble et al. (2011), [cc]van Nues and Beggs (2001), [dd]Maeder et al. (2009), [ee]Zhang et al. (2009), [ff]Pena et al. (2009).

★ N/D indicates that this cofactor effect has not been determined *in vitro*.
★★ Gle1$_{InsP6}$: Gle1 has stronger effects on Dbp5's activities when it is bound to InsP$_6$, an endogenous small molecule (Alcazar-Roman et al., 2006; Weirich et al., 2006).
★★★ The effect on ATPase activity likely arises from weakened RNA binding[dd].

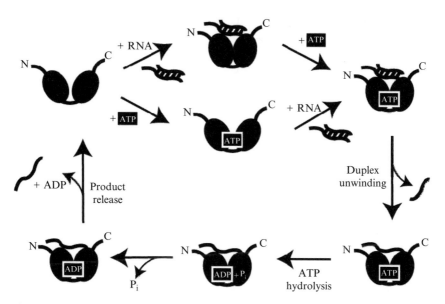

Figure 10.1 Kinetic cycle of RNA helicases. *In vitro* studies with RNA helicases have shown that cofactors can influence each step in the ATPase cycle (Table 10.1).

has recently been shown that ATP hydrolysis regulates release of the single-stranded RNA product and not duplex unwinding (Liu *et al.*, 2008). It is therefore not surprising that cofactors can also affect the rates of RNA release (Table 10.1). Further, since phosphate-release, not ATP hydrolysis, is the irreversible step in the kinetic cycle of both helicases and other ATPases (Henn *et al.*, 2008; Hilbert *et al.*, 2011), cofactors also affect this step (Table 10.1). As reported for the helicase Dbp5 and its cofactors Nup159 and Gle1, different cofactors can have opposing effects on the same helicase (Alcazar-Roman *et al.*, 2006; Montpetit *et al.*, 2011; von Moeller *et al.*, 2009; Weirich *et al.*, 2006). *In vivo*, these cofactors therefore likely regulate the progression of Dbp5 through its catalytic cycle. In some cases, it has been shown that the same cofactor can have multiple effects (Table 10.1). However, because the kinetic cycle in these examples is dominated by a single rate-limiting step, it is not clear whether many of these effects have *physiological* repercussions. Additional effects reported to arise from cofactors include an increase (Ballut *et al.*, 2005; Weirich *et al.*, 2006) or decrease (von Moeller *et al.*, 2009) in RNA affinity and a decrease (Maeder *et al.*, 2009) in ATPase activity (Table 10.1).

By modulating activity and potentially increasing specificity of RNA helicases, cofactors provide an additional level of regulation. Further, by encoding the helicase core and the cofactor on two distinct polypeptides, modularity is achieved *in vivo*. As an example, the yeast helicase Prp43 functions in both pre-mRNA splicing and ribosome assembly, where it is associated with and regulated by Ntr1/Ntr2 and Pfa1, respectively (Lebaron *et al.*, 2009; Martin *et al.*, 2002; Tanaka *et al.*, 2007; Tsai *et al.*, 2005; Walbott *et al.*, 2010). Therefore, in order to thoroughly understand their cellular functions, helicases must be studied together with the cofactors that modulate them.

1.2. How do cofactors modulate the enzymatic activities of RNA helicases?

While a mechanistic understanding of cofactor effects is still lacking for many helicases, recent biochemical and structural data have indicated a mechanism that is shared by at least two yeast RNA helicases: the DEAD-box protein Dbp5 and the SF1-like helicase Upf1 (Chakrabarti *et al.*, 2011; Hodge *et al.*, 2011; Noble *et al.*, 2011, see also Chapters 11 and 12). These helicase-cofactor complexes share structural similarity but no detectable sequence homology. The regulatory N-terminal domain of Upf1, which inhibits its ATPase activity, is displaced upon binding of its cofactor Upf2. This rearrangement results in the stimulation of Upf1's ATPase activity (Chakrabarti *et al.*, 2011). The inhibitory domain of Upf1 also extends the RNA binding site (Chakrabarti *et al.*, 2011). As a result, it is expected to increase RNA-binding affinity. The binding of the Upf2 cofactor is therefore predicted to weaken RNA binding by shortening the RNA–protein interface. Further, it was shown that nucleotide binding moderately weakens RNA binding (Chakrabarti *et al.*, 2011). In a similar mechanism, addition of the cofactor Gle1 to Dbp5 weakens RNA binding and is predicted to displace an N-terminal extension, which also has autoinhibitory effects on Dbp5's ATPase activity (Montpetit *et al.*, 2011).

Even though the crystal structures of Dbp5 and Upf1 and their respective cofactors were tremendously insightful in predicting the mechanisms of cofactor activation, biochemical experiments were required to validate the hypothesized effects. Additionally, many complexes that contain combinations of helicase, cofactor, nucleotide, and/or nucleic acid are highly unstable and therefore often poorly accessible with structural biology methods. In this case, biochemical approaches are perhaps the only available method for probing synergistic effects from multiple ligands (nucleotide, RNA, cofactor) on helicase function. For example, crystal structures also suggest that the effects from ADP and Gle1 on RNA binding should be synergistic, as both prevent RNA binding (Montpetit *et al.*, 2011). Similarly, kinetic methods are the only methods to probe transition states, which are associated with the conformational changes that modulate the helicases conformational cycle. Below, we provide a biochemical guide, both technical as well as conceptual, for identifying helicase-specific cofactors and then studying their effects on helicase function.

2. How to Study Cofactors

2.1. Identification of cofactors

In many cases, the study of RNA helicases and their biological functions has been slowed by the lack of information about their cofactors (e.g., in yeast, cofactors with known *in vitro* effects have been identified for only 8 of the 41 helicases). This has precluded a meaningful analysis, as the biologically active complex is undefined. Additionally and likely linked to this deficiency, RNA targets often remain unknown. Further dissection of the roles of yeast helicases will therefore almost certainly require a systematic approach to identify cofactors.

Yeast two-hybrid assays have many documented shortcomings (Bruckner *et al.*, 2009), yet they are a strong starting point for the identification of helicase-cofactor interactions. Especially for the analysis of the roles of the many nuclear helicases involved in pre-mRNA splicing and ribosome assembly, this technique can be fruitful, as evidenced by a complete map of protein–protein interactions within preribosomal subcomplexes obtained with this technique (Champion *et al.*, 2008; Charette and Baserga, 2010; Freed and Baserga, 2010); most of the RNA helicase cofactors in Table 10.1 were also identified by yeast two-hybrid analyses. As an untested, yet potentially interesting, extension of this technique, three-hybrid assays using the helicase and/or the identified cofactor could be used to identify RNA-binding sites.

Genetic interactions can also provide insight into potential cofactors. For example, a point mutation within a protein–protein interaction module of the RNA-binding protein Rrp5 is suppressed by overexpressing the helicase Rok1 (Torchet *et al.*, 1998), indicating that this genetic interaction could reflect on a direct physical interaction. We have verified this hypothesis using recombinant proteins *in vitro* (manuscript in preparation). Genetic interactions have also been substantiated *in vitro* for the helicases Prp43 and Upf1 and their respective cofactors Ntr1/Ntr2 and Upf2 (He *et al.*, 1997; Tanaka *et al.*, 2007). Similarly, a creative genetic strategy was recently described to isolate RNA targets for helicases (Proux *et al.*, 2011). Iost and coworkers screened a library of RNA mutants for suppressors of growth defects observed in the absence of SrmB, an *E. coli* DEAD-box protein involved in ribosome assembly (Charollais *et al.*, 2003). The isolated mutants suggest a role for SrmB in resolving an inhibitory structure between 23S rRNA and 5S rRNA, although some data are not explained by this model (Proux *et al.*, 2011).

In theory, co-immunoprecipitation assays are powerful tools to isolate protein-binding partners for any given protein; however, they have not always been fruitful for the study of RNA helicases. A likely reason is that a large majority of helicases are involved in either ribosome assembly or pre-mRNA splicing. In both cases, helicases associate with very large macromolecular complexes, which are also immunoprecipitated. As a result, dozens of associated proteins are often identified, but most are indirectly bound, and this methodology provides no means to distinguish between direct or indirect interactions.

Another method to identify helicase cofactors is cross-linking. Photoactivatable (e.g., diazirine analogs; Suchanek et al., 2005) or chemical cross-linkers such as formaldehyde can be used to covalently link a helicase to any nearby protein, and subsequent purification can be carried out under nonnative conditions (e.g., using a His-tag on the helicase of interest) to retain cofactors that are bound to the helicase via covalent interactions. Similarly, RNA cross-linking has been recently used to map the sites of interactions for several ribosome assembly factors, including the helicase Prp43 (Bohnsack et al., 2009; see also Chapter 13). Comparison of cross-links obtained from helicases to those of putative cofactors should reveal neighboring binding sites.

While these tools are of tremendous value for identifying potential cofactors, it is important to realize that they are merely a starting point. Interactions should be confirmed by combining the methods described above. For example, genetic interactions between a helicase and potential cofactor can be corroborated by yeast two-hybrid data and *vice versa*. Genetic experiments and yeast two-hybrid data can also be used in combination to map protein regions involved in interactions. In the case of Brr2 and Prp8, this approach has been used to demonstrate that both the N- and C-termini of Prp8 interact with Brr2 (van Nues and Beggs, 2001). The gold standard for a helicase–cofactor interaction is demonstration of direct protein–protein interactions using recombinant proteins and, ideally, the analysis of cofactor effects on the enzymatic activity of the RNA helicase.

2.2. Studying the effects of cofactors on RNA helicases

In order to gain a mechanistic understanding of how cofactors affect an RNA helicase and to delineate the function of the helicase-cofactor complex, *in vitro* studies should be conducted. This is especially important since helicases cycle through different steps. Thus, analyzing effects *in vivo* will only allow the analysis of one such step - effects on the entire catalytic cycle, including RNA and nucleotide binding and ATPase, helicase, or annealing activities, should be analyzed. Here, we provide detailed protocols for analyzing these RNA helicase characteristics and observing how they vary in the presence of cofactors.

3. REAGENTS

3.1. Reaction buffers

1. 10× transcription buffer: 400 mM Tris, pH 8.1, 250 mM MgCl$_2$, 20 mM spermidine, and 0.1% Triton X-100
2. 10× folding/binding buffer: 1 M KCl, 500 mM HEPES, pH 7.6
3. 50 mM and 1 M (NH$_4$)HCO$_3$ solutions, degassed until pH 8.0
4. 10× ATPase buffers:
   a. 1 M KCl, 0.5 M HEPES, pH 7.6
   b. 0.5 M KCl, 0.4 M Tris, pH 8.3, 20 mM DTT
5. Quench: 0.75 M KH$_2$PO$_4$, pH 3.3
6. TLC developing solution: 1 M LiCl, 300 mM NaH$_2$PO$_4$, pH 3.8
7. TE buffer: 10 mM Tris, pH 8.0, 1 mM EDTA
8. 10× helicase reaction buffer (HRB): 400 mM Tris, pH 8.3, 5 mM MgCl$_2$, 20 mM DTT
9. 10× duplex annealing buffer (DAB): 100 mM MOPS, pH 6.5, 10 mM EDTA, 0.5 M KCl
10. 2× helicase reaction SDS stop buffer (HRSB): 50 mM EDTA, 1% SDS, and 0.1% bromophenol blue in 20% glycerol

3.2. Gels and loading dyes

1. 2× high EDTA denaturing load dye: 0.2 M EDTA, 0.1% bromophenol blue, and 0.1% xylene cyanol in 100% formamide
2. 2× native load dye: 0.1% bromophenol blue and 0.1% xylene cyanol in 50% glycerol
3. 10× TBE: 500 mM Tris, 400 mM boric acid, and 5 mM EDTA
4. 10× THEM: 330 mM Tris, 670 mM HEPES, 10 mM EDTA, and 100 mM MgCl$_2$

3.3. Labeling RNAs with ^{32}P

3.3.1. General considerations and useful hints

1. Body-labeling works best with longer RNAs, while 5′-end labeling is preferred for shorter RNAs. 5′-end labeling requires RNAs with a 5′-OH group, such as those obtained from chemical synthesis or hammerhead/HDV ribozyme-mediated cleavage. *In vitro* transcribed RNAs contain 5′-triphosphate groups and must first be dephosphorylated using calf intestinal or shrimp alkaline phosphatase (NEB).
2. ^{32}P-α-ATP can be substituted with any other desired nucleotide.

3.3.2. Body-labeling with ^{32}P-α-ATP

1. Use a linearized DNA template that has a T7 RNA polymerase (RNAP) promoter sequence. To obtain the best yield, the first two transcribed nucleotides must be GG (Milligan *et al.*, 1987). In a 30 μL reaction, combine 0.5 mg T7 RNAP, 1 μg template, 0.2 mM ATP, 1 mM CTP, 1 mM GTP, 1 mM UTP, and 50 μCi ^{32}P-α-ATP in the presence of 1× transcription buffer and 40 mM DTT. Incubate at 37 °C for 2 h, quench with a 2× high EDTA denaturing load dye and purify on a native (for RNAs less than 30 nucleotides) or a denaturing (8 M urea; for RNAs greater than 30 nucleotides) acrylamide gel in 1× TBE. Gels should be prerun and wells must be thoroughly flushed (especially for the urea-containing gels).
2. Visualize RNA and cut out bands using razor blades (ethanol-flamed to destroy nucleases). For RNAs greater than 30 nucleotides, RNAs can be eluted via an electroelution system; this method provides almost full recovery of the RNA while ensuring that the RNA is urea-free. Shorter RNAs can be passively eluted in water overnight at 4 °C following three 10 min freeze-thaws of the RNA-containing gel bits. Upon elution, RNA may be ethanol precipitated: add 10% volume of 3 M sodium acetate and three times the volume of cold 100% ethanol. Precipitate overnight at −20 °C or 30 min at −80 °C. Spin for 20 min at 14,000 rpm in a microcentrifuge; remove supernatant and wash RNA pellet with cold 80% ethanol. Spin again for 20 min and remove supernatant. Thoroughly dry pellet and resuspend in water. Determine the concentration of the ^{32}P-labeled RNA via a scintillation counter. Equation (10.1) can be used to convert X counts per minute per microliter to Y millimolar:

$$\frac{X\,\text{cpm}}{1\,\mu L} \times \frac{3\,\text{dpm}}{1\,\text{cpm}} \times \frac{1\,\mu\text{Ci}}{2{,}220{,}000\,\text{dpm}} \times \frac{10^{-6}\,\text{Ci}}{1\,\mu\text{Ci}} \times \frac{1\,\text{mmol}}{Z\,\text{Ci}} \times \frac{10^{6}\,\mu L}{1\,L} = Y\,\text{m}M \quad (10.1)$$

where Z = the specific activity of the stock ^{32}P-ATP.

3. Store purified RNAs at −20 °C.

3.3.3. 5′-end-labeling with ^{32}P-γ-ATP

1. In a 10 μL labeling reaction, mix equimolar amounts of RNA and ^{32}P-γ-ATP with 1 unit of T4 polynucleotide kinase (PNK) in 1× PNK buffer. Incubate at 37 °C for 1 h, quench with a 2× high EDTA denaturing load dye and purify on a native (for RNAs less than ∼30 nucleotides) or an 8 M urea denaturing (for RNAs greater than 30 nucleotides) acrylamide gel in 1× TBE. Follow step 2 under Section 3.3.2 to complete RNA purification.

3.4. Purification of nucleotides

3.4.1. Unlabeled ATP

ATP from standard providers (e.g., Sigma) is usually significantly contaminated with ADP, which is most likely produced by spontaneous hydrolysis of ATP. The extent of this varies from batch to batch. Because ADP is a potent inhibitor of any ATPase, the stock ATP must be purified before use. The same considerations apply for AMPPNP and ADP (contaminated with AMP). We have not found it necessary to purify AMPPCP.

1. Resuspend 25 mg of adenosine-5'-triphosphate (ATP) in 50 mM (NH$_4$)HCO$_3$ and FPLC purify ATP over a MonoQ column in a linear gradient from 50 to 715 mM (NH$_4$)HCO$_3$; lyophilize peak fractions. ATP will elute after any AMP and ADP contaminants.
2. Resuspend purified ATP in water and determine the concentration via absorbance at 260 nm using an extinction coefficient of 15,400 M^{-1} cm^{-1}. TLC analysis can be used to verify the homogeneity of the purified stock ATP (Section 4.3.2).

3.4.2. ^{32}P-labeleled ATP

1. The ^{32}P-γ-ATP stock should be gel purified in order to remove ADP, P$_i$, or any other contaminants from the synthesis. To do so, add an equal volume of high EDTA load dye to 250 μCi of ^{32}P-γ-ATP and purify over a 24% native acrylamide/1× TBE gel (gel should be prerun and wells must be flushed). Visualize the ^{32}P-γ-ATP and cut out the band using an ethanol-flamed razor blade.
2. To allow for efficient passive elution of the labeled ATP, freeze and thaw the gel bit at −80 °C or on dry ice three times, 10 min each in a 1.5 mL microcentrifuge tube. Add water to the gel bit and passively elute overnight at 4 °C. Determine the purified ^{32}P-γ-ATP concentration via scintillation counting and store at −20 °C.

3.5. Purification of RNAs

3.5.1. Unlabeled RNA

1. Since *in vitro* transcription is less efficient for shorter RNAs, we use chemically synthesized RNA oligonucleotides. The RNA is then gel purified to ensure that it is pure and free of contaminants.
2. Resuspend 250 nmol of RNA oligo in 75 μL of TE buffer. Add equal volumes of high EDTA denaturing load dye and purify over a 15% native acrylamide gel in 1× TBE (prerun gel and thoroughly flush wells). To visualize the RNA, UV-shadow the gel using a hand-held

short-wavelength UV lamp and a fluor-coated TLC plate. Areas of the gel that contain RNA will appear purple on the TLC plate; use ethanol-flamed blades to excise the RNA bands from the gel. Submerge the gel bit in TE buffer and passively elute overnight at 4 °C. Remove buffer from the gel piece and ethanol precipitate the RNA (Section 3.3.2). Use the extinction coefficient and absorbance at 260 nm to determine the RNA concentration.

4. Protocols

4.1. RNA binding via gel-shift

4.1.1. General considerations and useful hints

1. Each RNA–protein interaction will require different buffer compositions to allow for optimal experimental conditions. Specifically, RNA–protein interactions are typically strongly dependent on both mono- and divalent salt concentrations. In the gel-shift assay described below (Section 4.1.2), both KCl and $MgCl_2$ concentrations can be varied so that the RNA–protein interaction can be fully characterized. The physiologically relevant concentration ranges that we use for KCl and $MgCl_2$ are 50–150 mM and 0.5–2 mM, respectively.
2. Other typical components of buffers for RNA–protein binding experiments include tRNA or heparin (nonspecific competitors can be used to screen against nonspecific interactions, but might not be necessary at higher salt concentrations), or low concentrations of detergent.
3. High-quality preparations of protein are essential for these experiments, as contaminating nucleases often degrade free, but not protein bound, RNA (bound RNA is protected by the protein). As a result, the binding affinity will be overestimated. An indicator for this problem is a significant loss of signal from total RNA as the protein concentration is increased. We therefore combine traditional affinity purification steps (His- or maltose binding protein (MBP)-tag) with ion exchange chromatography and gel filtration; in some cases, we also include ammonium sulfate fractionation.
4. To optimize the mobility of a protein–RNA or multi-protein–RNA complex, a variety of variables should be considered:
    a. Temperature—running the acrylamide/1 × THEM gel at 4 °C instead of room temperature results in tighter RNA bands, especially for larger protein–RNA complexes. At this lower temperature, the gel can be run at a higher voltage since the gel itself will remain cool.
    b. pH—the electrophoretic mobility of a RNA–protein complex is partially dependent upon the pI of the protein. Varying the pH of

the native acrylamide gel and running buffer (staying within the physiologically relevant pH range of 6–8) may affect the mobility of the RNA–protein complex.

4.1.2. Gel-shift RNA binding assay

1. Determine an appropriate protocol for RNA folding that includes both a heating step and the presence of divalent metal ions such as Mg^{2+}. To avoid RNA degradation, mixtures cannot be heated above ~70 °C after the addition of Mg^{2+}. Single tight bands in native gels and distinct patterns in structure probing experiments are indicative of RNA folding into distinct structures. A commonly used protocol is as follows: first, denature RNAs in the presence of 1 × folding buffer at 95 °C for 1 min. Next, add $MgCl_2$ to 10 mM and incubate at 55 °C for 15 min. Equilibrate RNA at desired binding temperature for 10 min.
2. Incubate trace amounts of the prefolded RNA (<20 nM) with increasing concentrations of protein in the presence of 1 × binding buffer and 10 mM $MgCl_2$ (KCl, HEPES, and $MgCl_2$ contributions from the folded RNA should also be considered) at the desired temperature. Incubation time will vary from protein to protein and should be optimized (our standard incubation time is 2 h). The accurate measurement of binding affinities requires equilibration of the RNA–protein interaction. Therefore, for tight interactions, caution must be taken to ensure that incubation covers several half-life times for the dissociation rate constant by varying the incubation time over at least a four- to five-fold range.
3. After incubation, add an equal volume of native load dye and load samples onto a running 6–15% (depending on RNA size) acrylamide/ 1 × THEM gel (gel should be prerun and wells must be thoroughly flushed). Dry gel and expose to a phosphor screen. Protein-bound and free RNA bands can be quantified using phosphorimager technology and analyzed.

4.1.3. Interpreting gel-shift results

1. In this gel-shift assay, the fraction of RNA that has protein bound is determined as follows from quantifying bands in phosphorimages:

$$\text{fraction}_{\text{bound}} = \frac{[\text{protein}_n \cdot \text{RNA}]}{[\text{protein}_n \cdot \text{RNA}] + [\text{RNA}]} \quad (10.2)$$

where n, the number of protein molecules bound.

The binding affinity, $K_{1/2}$, is defined by the binding equilibrium shown in Eq. (10.3):

$$K_{1/2} = \frac{[\text{protein}]_f^n \cdot [\text{RNA}]}{[\text{protein}_n \cdot \text{RNA}]} \quad (10.3)$$

where $[\text{protein}]_f$, concentration of free protein.
Assuming that $[\text{protein}]_f \sim [\text{protein}]$ added (which is true if trace RNA is present), solving Eq. (10.3) for $[\text{protein}_n \cdot \text{RNA}]$ and substituting this into Eq. (10.2) gives Eq. (10.4):

$$\text{fraction}_{\text{bound}} = \frac{\text{fraction}_{\text{bound,max}} \cdot [\text{protein}]^n}{[\text{protein}]^n + K_{1/2}} \quad (10.4)$$

2. In Eq. (10.3), n can be determined one of two ways. One approach is to not constrain the value for n during the fitting of the data. The value for n can then be extracted from the resulting fit. The second, and more exact, approach is to determine the stoichiometry of a particular RNA–protein interaction via stoichiometric titration experiments.[1] For this approach, incubate increasing concentrations of protein with saturating concentrations of unlabeled RNA spiked with trace amounts of the same ^{32}P-labeled RNA to produce varying protein:RNA ratios. The protein:RNA stoichiometry, or n-value, is the ratio of protein:RNA at which the maximum fraction of RNA bound value is first achieved (see example in Lamanna and Karbstein, 2009). This ratio is determined by the intersection point of two lines: one is drawn through the data points corresponding to full binding (high protein:RNA ratios) and is parallel to the x-axis; the other increasing line at low protein:RNA ratios is drawn through data that represent an increase in fraction of protein bound RNA as protein concentration is increased.
3. Using Eq. (10.4), the RNA–protein affinity can be determined. The affinities of both the cofactor and RNA helicase for the RNA should first be individually measured. To measure the effects of a cofactor on the binding affinity of an RNA helicase, incubate increasing concentrations of one protein in the presence of an unchanging, saturating amount of the second protein. This requires that binding of the second protein results in a super shift in order to differentiate between RNA–protein

---

[1] The term "stoichiometric titration" is used here to differentiate this experiment (which is conducted at high and saturating concentrations of RNA) from the previously described experiments in which trace and subsaturating RNA concentrations were used. The purpose is not to determine affinities (in fact, these must already be known) but instead to determine stoichiometry.

complexes containing one or both bound proteins. Experiments in which the cofactor concentration is varied should be compared to experiments that have varying RNA helicase concentrations. Variations in the binding order (e.g., preincubating both the cofactor and RNA helicase vs. preincubating cofactor and RNA vs. preincubating RNA helicase and RNA) must be optimized.

4.1.4. Controls for RNA binding specificity

1. Competition experiments should be used to test for specific RNA binding. Using trace prefolded (see Section 4.1.2) ^{32}P-labeled RNA (specific or control), form the RNA–protein complex in the presence of either the helicase or helicase/cofactor complex (at saturating concentrations) and saturating concentrations of unlabeled (competitor) RNA. One competitor RNA should be the suggested specific substrate, which is used as a positive control; this RNA should be able to compete with the protein-bound ^{32}P-labeled RNA. In addition, use unrelated RNAs as nonspecific competitors; these RNAs should not be able to compete for protein binding. Commercially available nonspecific RNAs are tRNA and homopolymers (i.e., polyU, etc.). At the same time, test a radiolabeled nonspecific RNA to demonstrate that any other RNA can compete for this interaction.
2. Incubate the protein and RNA mixture, separate free and bound RNA over an acrylamide/1× THEM gel and quantify bands via the approaches explained in Sections 4.1.2 and 4.1.3. If a helicase or helicase/cofactor complex is specifically binding an RNA, then only the addition of the same unlabeled RNA should be able to compete off the binding interaction.

4.2. RNA release rate constants via gel-shift

4.2.1. Pulse-chase experiments

1. Preform the RNA–protein complex using trace folded ^{32}P-labeled RNA and protein at concentrations that are three- to five-fold above the K_d value (according to the method used to determine the K_d value; see Sections 4.1.2 and 4.1.3 for an example). This will be the "pulse." Incubate at the desired temperature for at least 30 min; add a large excess of unlabeled, folded RNA (the "chase").
2. At varying time points, take out 3–5 µL samples, add an equal volume of 2× native load dye and immediately load onto a running native acrylamide/1× THEM gel. Dry gel and expose to a phosphor screen. Protein-bound and free RNA bands can be quantified using phosphorimager technology and analyzed.

3. Control experiments must be carried out to determine the amount of RNA–protein complex formed at time $t=0$ (before addition of chase): remove a small sample of the reaction before the addition of chase, add $2\times$ native load dye and directly load onto the gel. Controls to test the effectiveness of the chase must also be included. For this, premix both labeled and unlabeled RNA before the addition of protein, incubate as before, add $2\times$ native load dye, and load onto the gel. An effective chase has at most 20% of the RNA bound; however, if the chase is less effective, the endpoint of the dissociation reaction can be adjusted to reflect the fraction bound in the chase control.
4. If unlabeled RNA is limiting for cost or experimental reasons, a dilution chase can also be carried out. In this case, pre-form the RNA-protein complex as described above; as a chase, dilute the mixture at least 50-fold (from $\sim$5-fold above the $K_{1/2}$ to $\sim$10-fold below is a good range) and load time points directly onto the gel as described above. Note that this requires the use of a lot of unlabeled RNA in the initial complex formation in order to have enough signal on the gel for efficient visualization. Controls for time $t=0$ and chase effectiveness are carried out as normal (see above).

4.2.2. Analyzing pulse-chase experiments

Using Eq. (10.2), determine the fraction of RNA that is protein bound at each time point. Plotting fraction of RNA bound versus time (t), the data can be fit to Eq. (10.5) to determine the rate constant k_{obs} for RNA release:

$$\text{fraction}_{bound} = \text{fraction}_{bound,\, t=0} \cdot \exp(-k_{off} \cdot t) \quad (10.5)$$

4.3. ATPase activity using single-turnover conditions

The advantage of single-turnover conditions over multiple-turnover conditions is that product inhibition or a rate-limiting step after the chemical step rarely have to be considered. Product inhibition is especially a concern for helicases, as many have been shown to bind the ADP product more tightly than ATP (Karow et al., 2007; Talavera et al., 2006). Pre-steady-state ATPase assays using a different approach are discussed in detail in Chapter 2.

4.3.1. General considerations and useful hints

To verify that the ATPase activity is coming from the RNA helicase itself and not additional contaminants, we monitor copurification of protein and ATPase activities during purification. Additionally, mutants that abolish ATPase activity can be used to verify that the observed ATPase activity is coming from the RNA helicase itself. Further, active-site titrations with inhibitory nucleotides can be used. This experiment requires tight binding of inhibitors such as ADP or AMPPNP. Use helicase concentrations at which binding of the inhibitor is saturated (ideally ≥ five-fold above the K_d) and add increasing concentrations of inhibitor. As each helicase is expected to bind one ADP (or AMPPNP) molecule, complete inhibition is expected when each helicase molecule has one inhibitor molecule bound at a ratio of 1:1. Significant downward deviation from the 1:1 ratio would indicate that the ATPase activity is not coming from the helicase but from minor contaminants; therefore, a lower concentration of inhibitor than expected is required. Minor downward deviation can also arise if not all molecules are active.

4.3.2. Single-turnover ATPase assays

1. Determine an appropriate reaction condition for the ATPase assay that considers temperature, buffer composition, and salt concentrations. Our standard reaction temperatures are 20 or 30 °C and our two common reaction buffers are listed above (Section 3.1).
2. In a 10 μL reaction, mix the desired saturating concentration of RNA helicase and 10 mM MgCl$_2$ in the presence of 1× ATPase buffer. Equilibrate at reaction temperature for 10 min. Add a trace amount (<1 nM) of ^{32}P-γ-ATP to initiate the ATPase reaction. At various time points, stop the reaction by adding 1 μL of reaction mixture to an equal volume of quench.
3. Wash a PEI or PEI-F cellulose TLC plate (EMD Chemicals) once with 10% NaCl and twice with water; thoroughly dry. Using strips of TLC plates that are 2 in high, spot 1 μL of reaction mixture 1 cm from the bottom of the plate. Elute the TLC plate in developing solution until the solvent front is 1 cm from the top of the plate. Dry plate and expose to a phosphor screen. The amount of ^{32}P-γ-ATP versus free ^{32}P–P$_i$ can be quantified using phosphorimager technology and analyzed.
4. Equation (10.6) can be used to calculate the fraction of P$_i$:

$$\text{fraction } P_i = \frac{[P_i]}{[^{32}P-\gamma-ATP] + [P_i]} \qquad (10.6)$$

Plotting fraction P$_i$ versus time (t), the data can be fit to Eq. (10.7) to determine the k_{obs} for ATPase hydrolysis:

$$\text{fraction } P_i = \text{fraction}_{\text{reacted}}^{\text{max}} - \text{fraction}_{\text{reacted}}^{\text{max}} \cdot \exp(-k_{obs} \cdot t) \qquad (10.7)$$

5. The ATPase activities of both the RNA helicase and the cofactor should be individually determined so that effects from the cofactor on helicase activity can be isolated from those of minute contaminants copurifying with the cofactor. To examine the effects of a cofactor on the ATPase activity of an RNA helicase, add increasing amounts of cofactor to a constant amount of helicase. In all cases, the concentration of cofactor must exceed the amount of helicase; otherwise, there will not be enough cofactor to bind the helicase and ATPase activities from mixed complexes will be observed, therefore complicating the analysis. Equilibrate the reaction mixture at the desired temperature for 10 min before the addition of ^{32}P-γ-ATP. Conduct the ATPase experiment as described above. To determine the effects from the cofactor, subtract the rate constant observed in the presence of cofactor alone from that observed in the presence of cofactor plus helicase and divide by the rate constant observed for the helicase alone:

$$\text{fold increase} = \frac{k_{obs}^{helicase+cofactor} - k_{obs}^{cofactor}}{k_{obs}^{helicase}} \qquad (10.8)$$

4.4. Measuring nucleotide affinities

Single-turnover ATPase inhibition assays can be used to measure the affinity of a helicase for a particular nucleotide. Since many helicases have been shown to preferentially bind ADP over ATP (Karow et al., 2007; Talavera et al., 2006), comparing nucleotide affinities in the presence and absence of cofactor will indicate if the cofactor alters the nucleotide preference of the helicase. In order to accurately measure ATP binding, a nonhydrolyzable form (e.g., AMPPNP), should be used.

1. In 10 μL reactions, mix the desired saturating concentration of RNA helicase, 10 mM MgCl$_2$, and increasing concentrations of nucleotide (e.g., ADP or AMPPNP) in the presence of 1 × ATPase buffer. If the concentration of the ATP analog is 1 mM or higher, then an equal concentration of additional MgCl$_2$ should also be added to each reaction, as nucleotides titrate out Mg^{2+}. Equilibrate at the reaction temperature for 10 min.
2. Carry out reactions as explained in Section 4.3.2. Separate reaction mixtures on TLC plates, quantify the results using phosphoimager technology and calculate the fraction P$_i$ and k_{obs} values using Eqs. (10.6) and (10.7), respectively. Plot k_{obs} versus nucleotide concentration and fit the inhibition plot to Eq. (10.9):

$$k_{obs} = \frac{k_{obs}^{max}}{1 + \frac{[nucleotide]}{K_I}} \qquad (10.9)$$

where K_I, the inhibition constant, is equal to the nucleotide concentration at which the helicase is half inhibited and, consequently, the $K_{1/2}$ for the affinity of the helicase for nucleotide. To accurately determine the $K_{1/2}$, the highest nucleotide concentrations should result in saturating inhibition of the helicase. Importantly, this equation requires the helicase concentrations to be subsaturating relative to ATP (the reactant) and the inhibitor (ADP or AMPPNP). Especially for tight-binding inhibitors, the experiment needs to be repeated at different helicase concentrations to ensure this is the case. For a correct experiment, the K_I value is independent of helicase concentration (see also Section 4.3.1).

3. Compare nucleotide affinity for the helicase plus cofactor to that obtained in the presence of helicase only. Equation (10.8) can be used to quantify the cofactor effect.

4.5. Preparation of RNA duplexes

1. To determine the optimal RNA concentrations for duplex formation, saturating conditions must first be identified. Mix subsaturating concentrations of ^{32}P-end-labeled RNA with increasing concentrations of unlabeled RNA in the presence of 1× DAB. Denature mixtures at 95 °C for 1 min in an aluminum heat block. Remove block from heat source and slowly cool on the bench top to $\sim$35 °C; equilibrate at the experimental temperature.
2. Separate unduplexed and duplexed RNAs over a 15% native acrylamide/1× THEM gel; dry gel and expose to a phosphor screen. Quantify the fraction of duplexed RNA, plot against the concentration of unlabeled RNA and fit with Eq. (10.4).
3. In theory, the same result should be obtained regardless of which strand is labeled. In practice, however, we often observe that one strand self-anneals at higher concentrations (Fig. 10.2). This can have two undesired effects. First, two different duplexes with different affinities for the test protein will be mixed in the same experiment. Since both duplexes often have very similar electrophoretic mobility, it will be impossible to distinguish one duplex from the other. Further, self-annealing decreases the available free concentration of the unlabeled strand; the concentration of unlabeled strand added therefore no longer equals the concentration of free RNA (the assumption in Eq. 10.3). The self-annealing strand is therefore generally kept at trace concentrations.
4. Titration experiments can be used to test that the 1:1 duplex and not unwanted higher order complexes are formed. These are carried out as described for RNA:protein stoichiometry in Section 4.1.3.

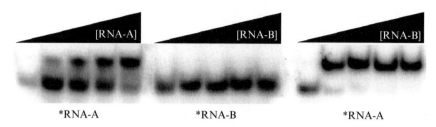

Figure 10.2 Duplex formation control. ^{32}P-labeled RNAs are indicated with an (*). Increasing the concentration of RNA-A results in self-annealing (left); RNA-B does not self-anneal to give a duplex at any concentration tested (middle). Therefore, the best combination is trace ^{32}P-labeled RNA-A and unlabeled RNA-B (right).

4.6. RNA unwinding

This protocol has been adapted from Jankowsky and Putnam (2009).

1. Following the directions in Section 4.5, form an appropriate volume of RNA duplex. In a 27 μL mixture, dilute the RNA duplex 10-fold in mixtures that contain increasing concentrations of RNA helicase in the presence of 1× HRB and 45 mM KCl (final [KCl] is 50 mM after the addition of RNA). Incubate at 20 °C for 5 min.
2. Add 3 μL of preincubated mixture to an equal volume of HRSB and place on ice ("0 s" time point). Add 3 μL of a 20 mM ATP/MgCl$_2$ mixture and pipette up and down. At various time points, quench 3 μL of the reaction in an equal volume of HRSB; place aliquots on ice.
3. Separate duplexed and free RNA on a 15% native acrylamide/THEM gel at 4 °C; dry gel and expose to a phosphor screen. Quantify the amount of free versus duplexed RNA using phosphorimager technology and fit with Eq. (10.7) from Section 4.3.2.
4. Determine the unwinding activities of both the RNA helicase and cofactor before combining both of them to observe the effects of a cofactor on the unwinding activity of the RNA helicase.
5. If RNA binding experiments reveal significant binding of the helicase to the unlabeled single-stranded RNA under the concentrations used in the unwinding experiments, then the duplex must first be gel-purified. This will ensure that unlabeled single-stranded RNA does not titrate out helicase. Unwinding effects that appear to arise from cofactors could otherwise arise from releasing (or sequestering) the helicase from single-stranded RNA. Carrying out unwinding experiments at saturating concentrations of helicase can prevent this latter problem. It will, however, not result in accurate determination of K_d values, as added helicase does not equal free helicase.

4.7. RNA annealing

This protocol has been adapted from Jankowsky and Putnam (2009).

1. Determine an appropriate reaction condition for the annealing assay that considers temperature, buffer composition, and salt concentration. Our standard reaction temperature is 20 °C and all reactions are done in 40 mM Tris buffer (pH 8.3) and 50 mM KCl.
2. In a 27 μL mixture, mix increasing concentrations of RNA helicase in the presence of 1 × HRB and 45 mM KCl (final reaction volume will be 30 μL after the addition of RNA). Incubate at 20 °C for 5 min.
3. Mix 5 nM of labeled RNA with an appropriate amount of the unlabeled complementary RNA strand (see Section 4.5) in the presence of 1 × DAB. To discourage spontaneous duplex formation, do not heat and then cool. Dilute 3 μL of this RNA into the preincubated RNA helicase mixture (reaction is now 50 mM KCl total) and pipette up and down. At various time points, quench 3 μL of the reaction in an equal volume of HRSB; place aliquots on ice.
4. Separate duplexed and free RNA on a 15% native acrylamide/1 × THEM gel at 4 °C; dry gel and expose to a phosphor screen. Quantify data and fit with Eq. (10.7). The annealing activities of both the RNA helicase and the cofactor should be individually determined. To examine the effects of a cofactor on the annealing activity of an RNA helicase, compare annealing activities at increasing concentrations of helicase in the presence and absence of cofactor. Incubate the RNA helicase and cofactor for 10 min before the addition of RNA; carry out annealing assay as normal.

ACKNOWLEDGMENTS

We would like to thank Eckhard Jankowsky and Andrea Putnam for invaluable help in setting up helicase and annealing experiments in our lab. We would also like to thank Bethany Strunk for comments on the chapter and insightful conversations. Research on RNA helicases in our lab is funded by a Beginning Investigator Grant from the AHA and R01-GM086451.

REFERENCES

Abramson, R. D., Dever, T. E., and Merrick, W. C. (1988). Biochemical evidence supporting a mechanism for cap-independent and internal initiation of eukaryotic mRNA. *J. Biol. Chem.* **263,** 6016–6019.

Alcazar-Roman, A. R., Tran, E. J., Guo, S., and Wente, S. R. (2006). Inositol hexakisphosphate and Gle1 activate the DEAD-box protein Dbp5 for nuclear mRNA export. *Nat. Cell Biol.* **8,** 711–716.

Alexandrov, A., Colognori, D., and Steitz, J. A. (2011). Human eIF4AIII interacts with an eIF4G-like partner, NOM1, revealing an evolutionarily conserved function outside the exon junction complex. *Genes Dev.* **25**, 1078–1090.

Ballut, L., Marchadier, B., Baguet, A., Tomasetto, C., Seraphin, B., and Le Hir, H. (2005). The exon junction core complex is locked onto RNA by inhibition of eIF4AIII ATPase activity. *Nat. Struct. Mol. Biol.* **12**, 861–869.

Bi, X., Ren, J., and Goss, D. J. (2000). Wheat germ translation initiation factor eIF4B affects eIF4A and eIFiso4F helicase activity by increasing the ATP binding affinity of eIF4A. *Biochemistry* **39**, 5758–5765.

Bohnsack, M. T., Martin, R., Granneman, S., Ruprecht, M., Schleiff, E., and Tollervey, D. (2009). Prp43 bound at different sites on the pre-rRNA performs distinct functions in ribosome synthesis. *Mol. Cell* **36**, 583–592.

Bruckner, A., Polge, C., Lentze, N., Auerbach, D., and Schlattner, U. (2009). Yeast two-hybrid, a powerful tool for systems biology. *Int. J. Mol. Sci.* **10**, 2763–2788.

Chakrabarti, S., Jayachandran, U., Bonneau, F., Fiorini, F., Basquin, C., Domcke, S., Le Hir, H., and Conti, E. (2011). Molecular mechanisms for the RNA-dependent ATPase activity of Upf1 and its regulation by Upf2. *Mol. Cell* **41**, 693–703.

Champion, E. A., Lane, B. H., Jackrel, M. E., Regan, L., and Baserga, S. J. (2008). A direct interaction between the Utp6 half-a-tetratricopeptide repeat domain and a specific peptide in Utp21 is essential for efficient pre-rRNA processing. *Mol. Cell. Biol.* **28**, 6547–6556.

Charette, J. M., and Baserga, S. J. (2010). The DEAD-box RNA helicase-like Utp25 is an SSU processome component. *RNA* **16**, 2156–2169.

Charollais, J., Pflieger, D., Vinh, J., Dreyfus, M., and Iost, I. (2003). The DEAD-box RNA helicase SrmB is involved in the assembly of 50S ribosomal subunits in Escherichia coli. *Mol. Microbiol.* **48**, 1253–1265.

Coppolecchia, R., Buser, P., Stotz, A., and Linder, P. (1993). A new yeast translation initiation factor suppresses a mutation in the eIF-4A RNA helicase. *EMBO J.* **12**, 4005–4011.

Dominguez, D., Altmann, M., Benz, J., Baumann, U., and Trachsel, H. (1999). Interaction of translation initiation factor eIF4G with eIF4A in the yeast Saccharomyces cerevisiae. *J. Biol. Chem.* **274**, 26720–26726.

Fairman, M. E., Maroney, P. A., Wang, W., Bowers, H. A., Gollnick, P., Nilsen, T. W., and Jankowsky, E. (2004). Protein displacement by DExH/D "RNA helicases" without duplex unwinding. *Science* **304**, 730–734.

Fairman-Williams, M. E., Guenther, U. P., and Jankowsky, E. (2010). SF1 and SF2 helicases: Family matters. *Curr. Opin. Struct. Biol.* **20**, 313–324.

Freed, E. F., and Baserga, S. J. (2010). The C-terminus of Utp4, mutated in childhood cirrhosis, is essential for ribosome biogenesis. *Nucleic Acids Res.* **38**, 4798–4806.

Fuller-Pace, F. V., Nicol, S. M., Reid, A. D., and Lane, D. P. (1993). DbpA: A DEAD box protein specifically activated by 23s rRNA. *EMBO J.* **12**, 3619–3626.

Granneman, S., Lin, C., Champion, E. A., Nandineni, M. R., Zorca, C., and Baserga, S. J. (2006). The nucleolar protein Esf2 interacts directly with the DExD/H box RNA helicase, Dbp8, to stimulate ATP hydrolysis. *Nucleic Acids Res.* **34**, 3189–3199.

He, F., Brown, A. H., and Jacobson, A. (1997). Upf1p, Nmd2p, and Upf3p are interacting components of the yeast nonsense-mediated mRNA decay pathway. *Mol. Cell. Biol.* **17**, 1580–1594.

Henn, A., Cao, W., Hackney, D. D., and De La Cruz, E. M. (2008). The ATPase cycle mechanism of the DEAD-box rRNA helicase, DbpA. *J. Mol. Biol.* **377**, 193–205.

Hilbert, M., Kebbel, F., Gubaev, A., and Klostermeier, D. (2011). eIF4G stimulates the activity of the DEAD box protein eIF4A by a conformational guidance mechanism. *Nucleic Acids Res.* **39**, 2260–2270.

Hodge, C. A., Colot, H. V., Stafford, P., and Cole, C. N. (1999). Rat8p/Dbp5p is a shuttling transport factor that interacts with Rat7p/Nup159p and Gle1p and suppresses the mRNA export defect of xpo1-1 cells. *EMBO J.* **18,** 5778–5788.

Hodge, C. A., Tran, E. J., Noble, K. N., Alcazar-Roman, A. R., Ben-Yishay, R., Scarcelli, J. J., Folkmann, A. W., Shav-Tal, Y., Wente, S. R., and Cole, C. N. (2011). The Dbp5 cycle at the nuclear pore complex during mRNA export I: dbp5 mutants with defects in RNA binding and ATP hydrolysis define key steps for Nup159 and Gle1. *Genes Dev.* **25,** 1052–1064.

Iost, I., and Dreyfus, M. (2006). DEAD-box RNA helicases in Escherichia coli. *Nucleic Acids Res.* **34,** 4189–4197.

Jankowsky, E. (2011). RNA helicases at work: Binding and rearranging. *Trends Biochem. Sci.* **36,** 19–29.

Jankowsky, E., and Putnam, A. (2009). Helicases. Springer-Verlag, New York, LLC.

Karow, A. R., Theissen, B., and Klostermeier, D. (2007). Authentic interdomain communication in an RNA helicase reconstituted by expressed protein ligation of two helicase domains. *FEBS J.* **274,** 463–473.

Lamanna, A. C., and Karbstein, K. (2009). Nob1 binds the single-stranded cleavage site D at the 3'-end of 18S rRNA with its PIN domain. *Proc. Natl. Acad. Sci. USA* **106,** 14259–14264.

Last, R. L., Maddock, J. R., and Woolford, J. L., Jr. (1987). Evidence for related functions of the RNA genes of Saccharomyces cerevisiae. *Genetics* **117,** 619–631.

Lebaron, S., Papin, C., Capeyrou, R., Chen, Y. L., Froment, C., Monsarrat, B., Caizergues-Ferrer, M., Grigoriev, M., and Henry, Y. (2009). The ATPase and helicase activities of Prp43p are stimulated by the G-patch protein Pfa1p during yeast ribosome biogenesis. *EMBO J.* **28,** 3808–3819.

Liu, F., Putnam, A., and Jankowsky, E. (2008). ATP hydrolysis is required for DEAD-box protein recycling but not for duplex unwinding. *Proc. Natl. Acad. Sci. U. S. A.* **105,** 20209–20214.

Maeder, C., Kutach, A. K., and Guthrie, C. (2009). ATP-dependent unwinding of U4/U6 snRNAs by the Brr2 helicase requires the C terminus of Prp8. *Nat. Struct. Mol. Biol.* **16,** 42–48.

Martin, A., Schneider, S., and Schwer, B. (2002). Prp43 is an essential RNA-dependent ATPase required for release of lariat-intron from the spliceosome. *J. Biol. Chem.* **277,** 17743–17750.

Milligan, J. F., Groebe, D. R., Witherell, G. W., and Uhlenbeck, O. C. (1987). Oligoribonucleotide synthesis using T7 RNA polymerase and synthetic DNA templates. *Nucleic Acids Res.* **15,** 8783–8798.

Mohr, G., Del Campo, M., Mohr, S., Yang, Q., Jia, H., Jankowsky, E., and Lambowitz, A. M. (2008). Function of the C-terminal domain of the DEAD-box protein Mss116p analyzed in vivo and in vitro. *J. Mol. Biol.* **375,** 1344–1364.

Montpetit, B., Thomsen, N. D., Helmke, K. J., Seeliger, M. A., Berger, J. M., and Weis, K. (2011). A conserved mechanism of DEAD-box ATPase activation by nucleoporins and InsP6 in mRNA export. *Nature* **472,** 238–242.

Noble, K. N., Tran, E. J., Alcazar-Roman, A. R., Hodge, C. A., Cole, C. N., and Wente, S. R. (2011). The Dbp5 cycle at the nuclear pore complex during mRNA export II: Nucleotide cycling and mRNP remodeling by Dbp5 are controlled by Nup159 and Gle1. *Genes Dev.* **25,** 1065–1077.

Pena, V., Jovin, S. M., Fabrizio, P., Orlowski, J., Bujnicki, J. M., Luhrmann, R., and Wahl, M. C. (2009). Common design principles in the spliceosomal RNA helicase Brr2 and in the Hel308 DNA helicase. *Mol. Cell* **35,** 454–466.

Proux, F., Dreyfus, M., and Iost, I. (2011). Identification of the sites of action of SrmB, a DEAD-box RNA helicase involved in Escherichia coli ribosome assembly. *Mol. Microbiol.* **82,** 300–311.

Rogers, G. W., Jr., Richter, N. J., and Merrick, W. C. (1999). Biochemical and kinetic characterization of the RNA helicase activity of eukaryotic initiation factor 4A. *J. Biol. Chem.* **274,** 12236–12244.

Rogers, G. W., Jr., Richter, N. J., Lima, W. F., and Merrick, W. C. (2001). Modulation of the helicase activity of eIF4A by eIF4B, eIF4H, and eIF4F. *J. Biol. Chem.* **276,** 30914–30922.

Rozen, F., Edery, I., Meerovitch, K., Dever, T. E., Merrick, W. C., and Sonenberg, N. (1990). Bidirectional RNA helicase activity of eukaryotic translation initiation factors 4A and 4F. *Mol. Cell. Biol.* **10,** 1134–1144.

Schutz, P., Bumann, M., Oberholzer, A. E., Bieniossek, C., Trachsel, H., Altmann, M., and Baumann, U. (2008). Crystal structure of the yeast eIF4A-eIF4G complex: An RNA-helicase controlled by protein-protein interactions. *Proc. Natl. Acad. Sci. U. S. A.* **105,** 9564–9569.

Schütz, P., Karlberg, T., van den Berg, S., Collins, R., Lehtiö, L., Högbom, M., Holmberg-Schiavone, L., Tempel, W., Park, H.W., Hammarström, M., Moche, M., Thorsell, A. G., Schüler, H. (2010). Comparative structural analysis of human DEAD-box RNA helicases. *PLoS One.* **30,** pii: e12791.

Sharpe Elles, L. M., Sykes, M. T., Williamson, J. R., and Uhlenbeck, O. C. (2009). A dominant negative mutant of the E. coli RNA helicase DbpA blocks assembly of the 50S ribosomal subunit. *Nucleic Acids Res.* **37,** 6503–6514.

Silverman, E. J., Maeda, A., Wei, J., Smith, P., Beggs, J. D., and Lin, R. J. (2004). Interaction between a G-patch protein and a spliceosomal DEXD/H-box ATPase that is critical for splicing. *Mol. Cell. Biol.* **24,** 10101–10110.

Strahm, Y., Fahrenkrog, B., Zenklusen, D., Rychner, E., Kantor, J., Rosbach, M., and Stutz, F. (1999). The RNA export factor Gle1p is located on the cytoplasmic fibrils of the NPC and physically interacts with the FG-nucleoporin Rip1p, the DEAD-box protein Rat8p/Dbp5p and a new protein Ymr 255p. *EMBO J.* **18,** 5761–5777.

Suchanek, M., Radzikowska, A., and Thiele, C. (2005). Photo-leucine and photo-methionine allow identification of protein-protein interactions in living cells. *Nat. Methods* **2,** 261–267.

Talavera, M. A., Matthews, E. E., Eliason, W. K., Sagi, I., Wang, J., Henn, A., and De La Cruz, E. M. (2006). Hydrodynamic characterization of the DEAD-box RNA helicase DbpA. *J. Mol. Biol.* **355,** 697–707.

Tanaka, N., Aronova, A., and Schwer, B. (2007). Ntr1 activates the Prp43 helicase to trigger release of lariat-intron from the spliceosome. *Genes Dev.* **21,** 2312–2325.

Torchet, C., Jacq, C., and Hermann-Le Denmat, S. (1998). Two mutant forms of the S1/TPR-containing protein Rrp5p affect the 18S rRNA synthesis in Saccharomyces cerevisiae. *RNA* **4,** 1636–1652.

Tsai, R. T., Fu, R. H., Yeh, F. L., Tseng, C. K., Lin, Y. C., Huang, Y. H., and Cheng, S. C. (2005). Spliceosome disassembly catalyzed by Prp43 and its associated components Ntr1 and Ntr2. *Genes Dev.* **19,** 2991–3003.

van Nues, R. W., and Beggs, J. D. (2001). Functional contacts with a range of splicing proteins suggest a central role for Brr2p in the dynamic control of the order of events in spliceosomes of Saccharomyces cerevisiae. *Genetics* **157,** 1451–1467.

von Moeller, H., Basquin, C., and Conti, E. (2009). The mRNA export protein DBP5 binds RNA and the cytoplasmic nucleoporin NUP214 in a mutually exclusive manner. *Nat. Struct. Mol. Biol.* **16,** 247–254.

Walbott, H., Mouffok, S., Capeyrou, R., Lebaron, S., Humbert, O., van Tilbeurgh, H., Henry, Y., and Leulliot, N. (2010). Prp43p contains a processive helicase structural architecture with a specific regulatory domain. *EMBO J.* **29,** 2194–2204.

Warkocki, Z., Odenwalder, P., Schmitzova, J., Platzmann, F., Stark, H., Urlaub, H., Ficner, R., Fabrizio, P., and Luhrmann, R. (2009). Reconstitution of both steps of Saccharomyces cerevisiae splicing with purified spliceosomal components. *Nat. Struct. Mol. Biol.* **16,** 1237–1243.

Weirich, C. S., Erzberger, J. P., Flick, J. S., Berger, J. M., Thorner, J., and Weis, K. (2006). Activation of the DExD/H-box protein Dbp5 by the nuclear-pore protein Gle1 and its coactivator InsP6 is required for mRNA export. *Nat. Cell Biol.* **8,** 668–676.

Zhang, L., Xu, T., Maeder, C., Bud, L. O., Shanks, J., Nix, J., Guthrie, C., Pleiss, J. A., and Zhao, R. (2009). Structural evidence for consecutive Hel308-like modules in the spliceosomal ATPase Brr2. *Nat. Struct. Mol. Biol.* **16,** 731–739.

CHAPTER ELEVEN

Analysis of DEAD-Box Proteins in mRNA Export

Ben Montpetit,* Markus A. Seeliger,[†] and Karsten Weis*

Contents

1. Introduction	240
2. Purification of Dbp5, Gle1, and Nup159	242
3. Steady-State ATPase Assay	244
3.1. Part 1: Preparing reaction mixtures	244
3.2. Part 2: Performing the ATPase assay	245
3.3. Part 3: Calculating ATPase rates	246
4. Use of Fluorescence Polarization to Monitor RNA or Nucleotide Binding and Release	246
4.1. Reagents and instrumentation setup	248
4.2. Determining steady-state binding affinities for RNA and adenosine nucleotide	250
4.3. Measuring RNA and nucleotide release rates from Dbp5	251
5. Concluding Remarks	252
Acknowledgments	252
References	252

Abstract

DEAD-box ATPases/helicases are a large family of enzymes (>35 in humans) involved in almost all aspects of RNA metabolism including ribosome biogenesis, RNA splicing, export, translation, and decay. Many members of this family are ATP-dependent RNA-binding proteins that interact with the RNA phosphodiester backbone and promote structural remodeling of target complexes through ATP binding and hydrolysis. Here, we describe the methods used in our laboratory to characterize the DEAD-box ATPase Dbp5 of *Saccharomyces cerevisiae*. Dbp5 is essential for the process of mRNA export in budding yeast and highly conserved orthologs can be found in all eukaryotes. Specifically, we describe enzyme assays to measure the catalytic activity of Dbp5 in association with RNA and known binding partners, as well as assays developed to measure

* Department of Molecular and Cell Biology, University of California, Berkeley, California, USA
[†] Department of Pharmacological Sciences, State University of New York at Stony Brook, Stony Brook, New York, USA

Methods in Enzymology, Volume 511   © 2012 Elsevier Inc.
ISSN 0076-6879, DOI: 10.1016/B978-0-12-396546-2.00011-5   All rights reserved.

the binding affinities and release kinetics of RNA and adenosine nucleotides from Dbp5. These assays have provided important information that has shaped our current models of Dbp5 function in mRNA export and should be useful for the characterization of other DEAD-box family members.

1. Introduction

mRNA export refers to the directional transport of an mRNA from the nuclear to the cytoplasmic compartment. This is an essential step in the expression of every eukaryotic gene since the translation of mRNA into protein occurs exclusively in the cytoplasm. The spatial separation of mRNA production and translation allows for multiple layers of regulation to coordinate the production, maturation, and eventual export of each mRNA species and to ensure the appropriate production of a given gene product. Key to the regulation is the packaging of mRNA into messenger ribonucleoprotein (mRNP) complexes, with the protein composition of an mRNP being important to determine the RNA's specific cellular fate (e.g., translation or decay) (Moore, 2005). Indeed, mRNP composition changes during nuclear mRNA maturation to promote capping, splicing, and polyadenylation in an efficient and orderly manner (Iglesias and Stutz, 2008). Thus, only those mRNPs that contain properly processed transcripts and the appropriate protein composition are competent for export to the cytoplasm and ultimately gain access to the translational machinery (Stutz and Izaurralde, 2003).

To escape the confines of the nucleus, an mRNP must transit through a highly selective transport channel that spans the nuclear envelope termed the nuclear pore complex (NPC). NPCs form channels with eightfold rotational symmetry and are inserted into the nuclear envelope at fusion sites between the inner and outer nuclear membrane (Fahrenkrog and Aebi, 2003; Hetzer and Wente, 2009; Strambio-De-Castillia et al., 2010). Cargo (e.g., protein and RNA) is directionally transported via interactions with soluble transport factors that permit passage through the pore (Stewart, 2007b; Tetenbaum-Novatt and Rout, 2010). With respect to mRNA, work in different model systems has provided convincing evidence that Mex67 in yeast and its metazoan ortholog NXF1/TAP is the major export receptor facilitating mRNA transport through the NPC by interacting with both the mRNA and components of the NPC inner channel (Katahira et al., 1999; Segref et al., 1997; Stewart, 2007a). However, a full mechanistic understanding of how mRNP export occurs remains to be described. For example, we do not understand how an mRNP is assembled for transport and what complement of proteins make up a mature transport substrate, nor do we know how the mRNP is altered during or after the transport step to enforce directionality. Immunoelectron microscopy studies have revealed that mRNPs are remodeled during export and that specific proteins are displaced from the mRNP at various stages of export (Daneholt, 2001). This remodeling of the mRNP structure and protein composition during export is likely key to mediating directional mRNA export since removal of critical export factors from the mRNA in the cytoplasm will prevent the retrograde entry into the NPC and thus transit back to the nucleus.

Dbp5, a conserved DEAD-box ATPase, has been suggested to play a critical role in mRNP remodeling during export on the cytoplasmic face of the NPC (Hodge et al., 1999; Lund and Guthrie, 2005; Schmitt et al., 1999; Strahm et al., 1999; Tran et al., 2007; Tseng et al., 1998; Weirich et al., 2004, 2006). In general, DEAD-box ATPases are considered to be RNA chaperones that can use the energy of ATP binding and hydrolysis to alter RNA–RNA or RNA–protein interactions (Cordin et al., 2006; Jankowsky, 2011; Rocak and Linder, 2004). However, the precise cellular function of most DEAD-box enzymes remains poorly understood and often the exact nature of their mRNP substrates remain ill defined. The role of Dbp5 in mRNA export was originally identified in budding yeast, and conditional mutations in *DBP5* result in the specific accumulation of poly (A) RNA inside the nucleus (Snay-Hodge et al., 1998; Tseng et al., 1998). *In vivo*, Dbp5 has been shown to associate with mRNPs early in their biogenesis and to shuttle between the nuclear and cytoplasmic compartments, but localizes most prominently to the cytoplasmic face of NPCs at steady state (Hodge et al., 1999; Zhao et al., 2002). Dbp5 is recruited to the NPC by the cytoplasmic-orientated NPC protein Nup159, which is critical for efficient mRNA export (Hodge et al., 1999; Schmitt et al., 1999; Weirich et al., 2004). At the NPC, Dbp5 can be stimulated by the cytoplasmic-orientated NPC component Gle1 together with the small molecule inositol hexakisphosphate (IP_6), which synergistically promotes the hydrolysis cycle of Dbp5 (Alcazar-Roman et al., 2006, 2010; Montpetit et al., 2011; Weirich et al., 2006). Together, these data have led to a model of localized mRNP remodeling at the cytoplasmic face of the NPC by Dbp5 where its regulators, Gle1 and Nup159, are anchored.

Recent biochemical and structural work from our lab has provided further understanding of this system with respect to the function of Gle1-IP_6 and Nup159 in promoting the ATPase activity of Dbp5 and mRNA export. Specifically, we have found that Gle1-IP_6 promotes recycling of the enzyme by promoting conformational changes in the DEAD-box protein leading to release of the RNA substrate (Montpetit et al., 2011). Our structural work also suggested that a ternary complex of Dbp5, Gle1-IP_6, and Nup159 may form to promote adenosine nucleotide release, which is supported by recent biochemical studies showing that Nup159 can accelerate the release of ADP from Dbp5 (Noble et al., 2011). These findings are consistent with other works performed on diverse DEAD-box ATPases from *Escherichia coli* (DbpA) and *Saccharomyces cerevisiae* (Mss116) demonstrating that enzyme recycling and product release are critical rate-limiting steps in the ATPase cycle of this enzyme family (Cao et al., 2011; Henn et al., 2008, 2010). To aid in these studies, we have developed detailed protocols to express and characterize the function of Dbp5 with its multiple binding partners (e.g., RNA, Gle1, Nup159, and IP$_6$) as described below.

2. Purification of Dbp5, Gle1, and Nup159

Expression constructs used for biochemical characterization encode *S. cerevisiae* proteins spanning residues 2–387 of Nup159 (Nup159 NTD), residues 1–482 of *S. cerevisiae* Dbp5 (Dbp5), and residues 244–538 of Gle1 (Gle1 CTD). Nup159 (pKW1459) and Dbp5 (pKW1329) expression constructs were made in the bacterial expression vector pSV271, generating an N-terminal 6× His fusion construct with a linker encoding the TEV-protease cleavage sequence (Weirich et al., 2004, 2006). Gle1 (pKW1716) was expressed from the bacterial expression vector pSV272 that encodes an N-terminal 6× His-MBP fusion protein and a TEV-protease cleavage site (Weirich et al., 2006). Addition of the MBP tag was found to be necessary to achieve high levels of expression and maintain Gle1 solubility. The following protocol can be used to purify Dbp5, Nup159, or Gle1 with the alterations detailed below for each protein.

Day 1: Culture growth and induction

1. Dilute an overnight culture of BL21-CodonPlus (DE3)-RILP cells (Novagen) transformed with a pSV271/2 construct to an OD$_{600}$ of 0.025 and grow in LB + kanamycin (25 µg/mL) + chloramphenicol (35 µg/mL) at 25 °C until an OD$_{600}$ of 0.4 is reached. Shift cells to 16 °C and induce with 0.15-mM IPTG for 16–18 h. Culture volumes of 1.5 (Dbp5 and Nup159) and 3.0 L (Gle1) are recommended but can be adjusted according to the amount of protein required.

Day 2: Protein purification

2. After centrifugation, wash cells once with water, collect by centrifugation, and then resuspend in lysis buffer (30 mM HEPES [pH 7.5], 400 mM NaCl, 10 mM imidazole, 2 mM MgCl$_2$, 1 µM pepstatin A, 15 M leupeptin, 1 mM PMSF, and 2 mM mercaptoethanol (β-ME)) and lyse by two passes through a French Press.
3. Following clarification by centrifugation for 30 min at 20,000 × g, incubate the lysate with 1 mL of a Nickel Affinity Gel (prewashed with lysis buffer) for 30 min at 4 °C with mixing.

4. Load the Nickel Affinity Gel onto a gravity flow chromatography column and allow the column to drain. Wash with 10 bed volumes of lysis buffer adjusted to 20 mM imidazole. Repeat wash three times.
5. Wash column twice with 5 bed volumes of wash buffer. (For Nup159, use 30 mM HEPES [pH 7.5], 100 mM NaCl, 2 mM MgCl$_2$. For Dbp5 and Gle1, use 25 mM sodium phosphate [pH 6.5], 2 mM MgCl$_2$, and 100 mM (Dbp5) or 200 mM (Gle1) NaCl.) The salt concentrations are optimized to maintain protein solubility and maximize binding to the ion exchange column used in step 7.
6. Elute protein with 5 bed volumes of elution buffer (wash buffer plus 400 mM imidazole). Monitor each fraction for the eluted protein using a suitable assay.
7. Combine protein-containing fractions and dilute 1:5 with wash buffer to reduce the concentration of imidazole.
8. Using an FPLC setup, load the partially purified protein onto the appropriate ion exchange column (for Nup159, use Pharmacia Q Sepharose HP, and for Dbp5/Gle1, use Pharmacia SP Sepharose HP) using the wash buffer described above. After loading and washing, elute protein from the ion exchange column using a linear gradient of NaCl up to 1 M over 20 column volumes at 4 °C.
9. Collect protein-containing fractions and dialyze at 4 °C (using a membrane with an 8 kDa MW cutoff) into storage buffer (30 mM HEPES [pH 7.5], 150 mM (Dbp5 and Nup159) or 400 mM (Gle1) NaCl, 2 mM MgCl$_2$, 1 mM DTT, and 10% glycerol). Dialysis is performed overnight in a volume of 500 mL with at least one change of buffer.

Day 3: Concentration and storage

10. After dialysis is complete, add protease inhibitors to the protein solution.
11. Concentrate the purified protein stocks using a centrifugal concentration device. Recommended concentrations are 100 μM for Dbp5 or Gle1 and 200 μM for Nup159 with each liter of culture typically yielding 1.5, 0.75, and 6 mg of protein.
12. Store Dbp5 and Gle1 protein preparations at 4 °C or alternatively add glycerol to 50% and store at −20 °C for long-term storage. Aliquot and flash freeze Nup159 in liquid nitrogen and store at −80 °C.

Note on storage conditions: Dbp5 and Gle1 are susceptible to loss of activity during freezing and thawing. Therefore, we recommend storing these proteins at 4 °C and using the protein stocks within 1 month or at −20 °C in 50% glycerol for long-term storage. At 4 °C, Dbp5 is particularly vulnerable to proteolytic cleavage within the N-terminus upstream of the first RecA-like domain. Given the role of the N-terminus in

RNA-mediated ATPase stimulation and autoinhibition of Dbp5 (Collins et al., 2009; Montpetit et al., 2011), it is important to monitor the integrity of the Dbp5 protein and periodically add fresh protease inhibitors.

3. STEADY-STATE ATPASE ASSAY

The following protocol is a modified version of an ATPase assay originally developed in the 1950s based on the work of Warburg (Beisenherz et al., 1955; Bücher and Pfleiderer, 1955; Warburg, 1948). The NADH-coupled ATPase assay is very sensitive, allowing real-time monitoring of the reaction, and regeneration of ATP from liberated ADP to maintain ATP concentrations and prevent buildup of reaction products that may alter reaction kinetics (e.g., ADP). The assay monitors the change in absorbance at 340 nm that follows the oxidation of NADH through a series of coupled enzymatic reactions. ADP molecules generated by the ATPase activity of Dbp5 are converted back to ATP by the actions of pyruvate kinase (PK) and lactate dehydrogenase (LDH) with phospho(enol) pyruvate (PEP) in a reaction that is coupled to the oxidation of NADH to NAD^+. This results in the loss of NADH absorbance at 340 nm, which can be easily monitored by a suitable, equipped plate reader. Importantly, the production of one ADP molecule by Dbp5 and subsequent regeneration of ATP is linked to the consumption of one NADH molecule. This allows for the activity of Dbp5 to be directly monitored through the decrease in NADH absorbance at 340 nm. Adaptation of this assay for continuous monitoring of ATPase activity is described in Chapter 2.

3.1. Part 1: Preparing reaction mixtures

The assay is performed in 384-well plates (NUNC, Rochester, NY) in a final volume of 40 μL. At a minimum, reactions are performed in triplicate. All chemicals and reagents obtained were purchased from Sigma Aldrich (St. Louis, MO).

Mixture #1 can contain the following variables: Dbp5, Gle1, Nup159, IP_6, RNA, and ATPase buffer (30 mM HEPES [pH 7.5], 100 mM NaCl, and 2 mM $MgCl_2$) in a total volume of 10 μL. Prepare a mix for each condition being tested (e.g., with or without RNA or at varying protein and RNA concentrations). Incubate the reactions for 10 min at room temperature prior to pipetting into the 384-well plate. It is also important to perform all reactions in the absence of Dbp5 to account for contaminating ATPases in the various protein preparations. An example reaction setup is shown below:

	Stock	Final concentration	Volume/reaction (μL)
Dbp5 (or buffer)	100 μM	0.5 μM	0.2
Gle1 (or buffer)	100 μM	1.0 μM	0.4
Nup159 (or buffer)	200 μM	1.0 μM	0.2
Poly(A) RNA (or H$_2$O)	10 mg/mL	0.1 mg/mL	0.4
IP$_6$ (or H$_2$O)	100 μM	1.0 μM	0.4
ATPase buffer	10×	1×	0.9
H$_2$O	–	–	7.5

Mixture #2 contains the following: ATP, DTT, PEP, NADH, PK/LDH, and ATPase buffer in a volume of 30 μL. Make a mixture large enough for all the conditions being tested. NADH is susceptible to oxidation and this solution should be prepared fresh. An example reaction setup is shown below:

	Stock	Final concentration	Volume/reaction (μL)
ATP	100 mM	2.5 mM	1.0
ATPase buffer	10 × buffer	1×	3
DTT	100 mM	1 mM	0.4
PEP	30 mM	6 mM	8.0
NADH	12 mM	1.2 mM	4.0
PK/LDH	600–1000 units/mL	125–250 units/mL	0.8[a]
	900–1400 units/mL	180–280 units/mL	
ddH$_2$O	–	–	12.80

[a] PK/LDH is purchased as a mixture from Sigma-Aldrich with the specific activity varying from batch to batch.

3.2. Part 2: Performing the ATPase assay

The ATPase assay is performed in a plate reader capable of reading Abs$_{340}$ over a kinetic interval. Reactions are routinely performed at 30 °C. To set up the reactions,

1. Aliquot 10 μL of mixture #1 for all conditions being tested into a 384-well plate.
2. Add 30 μL of mixture #2 to each of the wells. Be sure that no bubbles are present as they can lead to errors in the absorbance reading.
3. Mix the reactions on the plate reader by shaking to ensure proper mixing.
4. Measure the path length for each well being read (this feature is available on some plate readers; otherwise, the path length for 40 μL of liquid in

the 384-well plate being used must be determined). The advantage of reading the path length of each individual well is that it corrects for slight variations in well volumes due to pipetting errors.

5. Read Abs_{340} at intervals to record loss of NADH absorbance over time. A 30-min time window is routinely used for Dbp5 and can be extended if needed while measuring low ATPase rates. However, longtime periods may result in sample evaporation and altered Abs_{340} readings and reaction conditions.

3.3. Part 3: Calculating ATPase rates

1. Correct the Abs_{340} reading for a 1-cm path length (Abs_{340}/path length in cm).
2. Plot the corrected Abs_{340} versus time to obtain the rate of change in Abs_{340}.
3. Use the extinction coefficient of NADH ($Abs_{340} = 6220\ M^{-1}\ cm^{-1}$) to convert the rate of absorbance change into the rate of NADH consumed. This is equal to the rate of ATP use as described above.
4. Calculate the specific activity by normalizing the ATP consumption rate to the concentration of ATPase added to obtain the steady-state ATPase rate. This is expressed as the number of ATP molecules hydrolyzed per unit time (usually per second) by a single Dbp5 protein.

By performing these assays under varying concentrations of nucleotide, RNA, or accessory proteins (e.g., Gle1 or Nup159), various reaction parameters can be obtained (e.g., K_M of Dbp5 for ATP) (Weirich et al., 2006). This is illustrated in Fig. 11.1 where the addition of Gle1-IP_6 can be seen to increase the overall k_{cat} and decrease K_M for ATP by Dbp5 in the presence of RNA. Thus, this technique has the ability to provide essential insight into the activity of the ATPase being tested and how this activity is regulated by other proteins, RNA, and small molecule cofactors.

4. Use of Fluorescence Polarization to Monitor RNA or Nucleotide Binding and Release

The use of fluorescence spectroscopy to study molecular interaction offers the advantages of high sensitivity, small sample volumes, and an ability to work under equilibrium conditions in solution. Furthermore, it can monitor time-resolved events from nano seconds to minutes. This is in contrast to many other techniques that are relatively slow and cannot be easily used to resolve rapid kinetic reactions. Moreover, these techniques often rely upon the separation of bound and unbound materials, such as

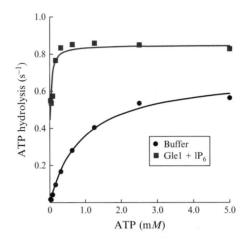

Figure 11.1 Example ATPase measurements of Dbp5 with RNA in the presence (squares) or absence (circles) of Gle1-IP$_6$. This experiment demonstrates that Gle1-IP$_6$ increases overall k_{cat} of the enzyme from 0.7 to 0.85 ATP s^{-1} and decreases the K_M for ATP from 0.93 to 0.018 mM.

electrophoresis, chromatography, or sedimentation, which allow new equilibria to form during the separation process.

In the case of polarization/anisotropy methods, binding interactions between two sets of molecules are measured by monitoring the change in one of the molecule's tumbling rate in solution. This change in rotational diffusion can be detected through the change in fluorescence polarization over the lifetime of the excited fluorescent state (Cantor and Schimmel, 1980; Lakowicz, 2006). These measurements are typically performed on a fluorimeter with two rotatable polarizers. The polarizer in the excitation light path provides linearly polarized light to the sample chamber. Only fluorophores with the appropriate orientation can absorb this light whereas most fluorophores will be in an unfavorable orientation that prevents them from absorbing the light. The fluorophore diffuses and rotates during the lifetime of the excited fluorescent state, which depends on the fluorophores, but is typically on the order of 1–10 ns. When the fluorophore emits its photon, some fluorophores will be in an orientation parallel to the orientation at photon absorption and some will be perpendicular to it. The slower the fluorophore tumbles during the 1–10 ns when it is in the excited state, the more fluorophores will be in the original orientation. The faster the fluorophores tumble, the more fluorophores will be randomly oriented relative to the starting orientation and emit their light perpendicular to the excitation polarization plane. To detect the ratio of emitted light in the planes parallel and perpendicular to the excitation plane, the polarizer in the emission light path can be rotated to be parallel or perpendicular to the orientation of the excitation polarizer. In fluorimeters, the orientation of the polarizers typically refers to some external reference as horizontal (H) or vertical (V) and the relative orientation of excitation and emission monochromators is referred to by reporting the orientation of the excitation monochromator first and the emission monochromator second. The two parallel orientations of the monochromators would be described as HH and VV and the two perpendicular orientations as HV and VH. While it would be sufficient in theory to record only one perpendicular and one horizontal orientation to calculate the anisotropy of the sample, in practice, all four measurements need to be taken because of effects in the light path of the instruments that systematically change the intensity of polarized light. The readings in two orientations control for these imperfections. Because the lifetime of the fluorophore in the excited state acts as the constant time increment during which rotational diffusion is measured and because of the photoselection with the excitation polarizer, this method does not require a flash lamp or very fast time measurements. Importantly, it can be performed under steady-state conditions to measure equilibrium binding or binding kinetics, which are much slower than the life time of the fluorophore.

We have applied this technique to determine the steady-state binding affinity of Dbp5 for both RNA and adenosine nucleotide (Fig. 11.2A). By combining this technique with stopped flow methodologies, we have also been able to determine RNA and nucleotide release rates for Dbp5 (Montpetit et al., 2011) (Fig. 11.2B).

4.1. Reagents and instrumentation setup

The RNA substrate used for the binding and release assays is a 29 nt ssRNA of the sequence 5′-GGGUAAAAAAAAAAAAAAAAAAAAAAAAA-3′ carrying a fluorescein-conjugated UTP made by *in vitro* transcription (MEGAshortscript Kit, Ambion). MANT–ADP labeled at the 2′ or 3′ position of the base (Invitrogen) is used for the nucleotide binding and release assays. To measure fluorescent anisotropy, samples are excited with vertically polarized light at 492 nm for FITC-labeled samples and 370 nm for MANT-labeled samples. The emission intensity of parallel (I_{VV}) and horizontally (I_{VH}) polarized light is measured at 521 nm for FITC-labeled samples and at 445 nm for MANT-labeled samples. Anisotropy (r) is calculated using Eq. (11.1):

$$r = \frac{(I_{VV} - (gI_{VH}))}{(I_{VV} + (2gI_{VH}))}, \text{ where } g = \frac{I_{HV}}{I_{HH}} \quad (11.1)$$

Anisotropy is measured using a Fluoromax-3 fluorimeter (Horiba Jobin Yvon) that is also equipped with a stopped flow apparatus (RX2000;

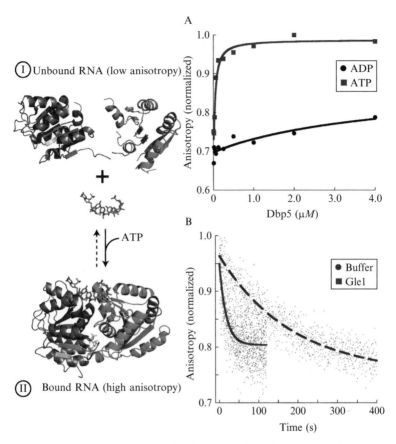

Figure 11.2 Measurement of RNA binding and release by Dbp5 E240Q, a catalytically inactive version of Dbp5 that binds both ATP and RNA but is unable to hydrolyze ATP. (A) In the presence of ATP (squares), but not ADP (circles), Dbp5 is able to bind fluorescein-labeled RNA causing a measured change in anisotropy (complex I→II). (B) Stopped flow experiment measuring the release rate of RNA (complex II→I), which is accelerated by the presence of Gle1 (solid line) in comparison to reactions containing Dbp5 alone (dashed line). Crystal structure model in state I is PDB ID: 3FHO (Fan et al., 2009) and state II is PDB: 3PEY (Montpetit et al., 2011). (See the Color Insert.)

Applied Photophysics) at 30 °C. Gle1, IP_6, Nup159, and RNA can be easily added to the reactions as needed to test the impact of these various factors on aspects of the binding and release reactions (Fig. 11.2B). Prior to use, both the cuvettes and stopped flow device are cleaned with a solution of 100 mM NaOH and 2 mM EDTA (to inactivate RNAses), rinsed with ddH_2O, and then the buffer being used in the experiment.

Fluorescence anisotropy measurements are very sensitive to errors from light scattering and it is important that the sample and buffers being used are free of dust, precipitation, and air bubbles. Filtration of samples through 0.22 μM filters or centrifugation for 10 min at 13,000 rpm are typically sufficient to remove dust and large aggregates. Because scattered light has the same wavelength (Raleigh scattering) as the excitation light or is slightly shifted to higher wavelengths (Stokes scattering), it is important that excitation and emission wavelengths are chosen sufficiently far enough from each other and that the bandwidth for the monochromators are small enough to prevent detection of scattered excitation light. If extra sensitivity is needed, set the emission monochromator to a higher wavelength and the excitation monochromator to lower wavelength than the respective excitation and emission maxima and increase the detection bandwidth.

4.2. Determining steady-state binding affinities for RNA and adenosine nucleotide

1. Mix 20 nM 5′-fluorescein-labeled RNA or 100 nM of MANT-nucleotide in a solution containing 50 mM HEPES [pH 7.5], 140 mM KCl, 5 mM MgCl$_2$, 1 mM DTT, 20% glycerol, and 0.1 mg/mL BSA in a quartz cuvette (500 μL reaction volumes) and incubate at the appropriate temperature for 5 min. ATP (to 1 mM) must also be added to the reaction buffer in RNA binding reactions.
2. Measure and record anisotropy values for the unbound RNA or nucleotide (use the average of three consecutive measurements). The sample needs to be thermostated because the tumbling rates are dependent on the temperature of the sample.
3. Add Dbp5 to a final concentration of 100 nM and mix. Allow to equilibrate for ∼1 min before recording anisotropy values (use the average of three consecutive measurements).
4. Repeat addition of Dbp5 until a change in anisotropy is no longer observed (i.e., at concentrations about 10-fold above the expected K_D).
5. To obtain the K_D value, plot anisotropy values (y-axis) versus [Dbp5] (x-axis) and fit to Eq. (11.2), which considers the dilution of the RNA or nucleotide as more protein is added. Note that Eq. (11.2) is used to fit single-binding site data. Data composed of multiple independent- or dependent-binding sites require different types of analysis (e.g., refer to Connors, 1987).

$$A\big([\text{Dbp5}]_T\big) = A_{\min} - \frac{[\text{Dbp5}]_T + K_D + [\text{RNA}]_T - \sqrt{([\text{Dbp5}]_T + K_D + [\text{RNA}]_T)^2 - 4[\text{RNA}]_T[\text{Dbp5}]_T}}{2[\text{RNA}]_T}(A_{\min} - A_{\max})$$

(11.2)

where $[Dbp5]_T$ is the total concentration of Dbp5, A_{min} is the minimum anisotropy reading, K_D is the dissociation constant, $[Dbp5]_T$ is the total Dbp5 concentration, and A_{Max} is the maximum anisotropy value.

Note: If you are titrating increasing amounts of Dbp5 to the cuvette, the volume of the reaction changes and therefore the concentration of RNA changes upon addition of Dbp5. To take this dilution factor into account, replace

$$[RNA]_T = \frac{[RNA]_0}{1 + \left(\frac{[Dbp5]_T}{[Dbp5]_{Stock} - [Dbp5]_T}\right)} \quad (11.3)$$

where $[RNA]_0$ is the concentration of RNA at the beginning of the titration (here the RNA is fluorescently labeled); $[Dbp5]_{Stock}$ is the stock concentration of the titrant, here Dbp5, and $[Dbp5]_T$ is the calculated total concentration of Dbp5 at which the anisotropy data is recorded.

When measuring MANT-nucleotide binding, the same formulae are used with [RNA] being replaced with [MANT-Nuc].

4.3. Measuring RNA and nucleotide release rates from Dbp5

1. Mix Dbp5 (1.0 μM) with 40 nM 5′-fluorescein-labeled RNA and 2.0 mM ATP or 200 nM of MANT-nucleotide in a solution containing 50 mM HEPES [pH 7.5], 140 mM KCl, 5 mM MgCl$_2$, 1 mM DTT, 20% glycerol, and 0.1 mg/mL BSA. Incubate for 5 min at room temperature.
2. In a second vial, make a mixture that contains 50 mM HEPES [pH 7.5], 140 mM KCl, 5 mM MgCl$_2$, 1 mM DTT, 20% glycerol, and 0.1 mg/mL BSA with 1.0 mg/mL polyA RNA or 2.5 mM ATP as competitors in the RNA and nucleotide binding reactions, respectively.
3. After rapidly mixing 400 μL of each of the two mixtures using the stopped flow apparatus, fluorescence intensities are measured every 0.01–0.2 s (depending on the release rate being measured) following excitation with polarized light. As a general guide, measure at least a 1000 data points over a period of time that is 10–20-fold longer than the slowest rate constant measured (e.g., if the observed rate constant is 1 s^{-1} then the total measurement time should be >10 s and data should be collected every 10 ms). The temperature of the samples needs to be tightly controlled in the cuvette and the syringes, because fluorescence anisotropy is very sensitive to temperature changes.
4. Emission of vertically (I_{VV}) and horizontally (I_{VH}) polarized light is measured independently in two consecutive runs to obtain the required observation timescales. These values are used to calculate anisotropy values using Eq. (11.1).

5. Check data within first 10–20 ms for extra noise and discard if necessary. This can be caused by mixing artifacts that are instrument-, solution-, and temperature-dependent.
6. The resulting anisotropy data is then fitted to a first order exponential decay curve to obtain observed rate constants, which correspond here to the release rates.

5. Concluding Remarks

In this review, we have aimed to provide a detailed description of the assays used to monitor ATPase activity and the binding and release of RNA or adenosine nucleotide from the DEAD-box ATPase Dbp5. These assays have been extremely valuable in deciphering the activity and regulation of this DEAD-box ATPase functioning in mRNA export. We believe, these assays should also be useful for the study of other members of this large enzyme family.

ACKNOWLEDGMENTS

The authors want to acknowledge Christine Weirich, Jan Erzberger, and Zain Dossani for their contributions to this research. Research in the Seeliger and Weis labs is supported by NIH grants R00GM080097 (M. A. S.) and R01GM58065 (K. W.).

REFERENCES

Alcazar-Roman, A. R., Tran, E. J., Guo, S., and Wente, S. R. (2006). Inositol hexakisphosphate and Gle1 activate the DEAD-box protein Dbp5 for nuclear mRNA export. *Nat. Cell Biol.* **8,** 711–716.

Alcazar-Roman, A. R., Bolger, T. A., and Wente, S. R. (2010). Control of mRNA export and translation termination by inositol hexakisphosphate requires specific interaction with Gle1. *J. Biol. Chem.* **285,** 16683–16692.

Beisenherz, G., Bücher, T., and Garbade, K. H. (1955). α-glycerophosphate dehydrogenase from rabbit muscle: Dihydroxyacetone phosphate + DPNH + H + l-α-glycerophosphate + DPN$^+$. *Methods Enzymol.* **1,** 391.

Bücher, T., and Pfleiderer, G. (1955). Pyruvate kinase from muscle: Pyruvate phosphokinase, pyruvic phosphoferase, phosphopyruvate transphosphorylase, phosphate—Transferring enzyme II, etc. phosphoenolpyruvate + ADP ⇌ pyruvate + ATP. *Methods Enzymol.* **1,** 435.

Cantor, C. R., and Schimmel, P. R. (1980). Biophysical Chemistry. W.H. Freeman, San Francisco.

Cao, W., Coman, M. M., Ding, S., Henn, A., Middleton, E. R., Bradley, M. J., Rhoades, E., Hackney, D. D., Pyle, A. M., and De La Cruz, E. M. (2011). Mechanism of Mss116 ATPase reveals functional diversity of DEAD-box proteins. *J. Mol. Biol.* **409,** 399–414.

Collins, R., Karlberg, T., Lehtio, L., Schutz, P., van den Berg, S., Dahlgren, L. G., Hammarstrom, M., Weigelt, J., and Schuler, H. (2009). The DEXD/H-box RNA helicase DDX19 is regulated by an {alpha}-helical switch. *J. Biol. Chem.* **284,** 10296–10300.

Connors, K. A. (1987). Binding constants: The Measurement of Molecular Complex Stability. Wiley, New York, Chichester.

Cordin, O., Banroques, J., Tanner, N. K., and Linder, P. (2006). The DEAD-box protein family of RNA helicases. *Gene* **367,** 17–37.

Daneholt, B. (2001). Assembly and transport of a premessenger RNP particle. *Proc. Natl. Acad. Sci. U. S. A.* **98,** 7012–7017.

Fahrenkrog, B., and Aebi, U. (2003). The nuclear pore complex: Nucleocytoplasmic transport and beyond. *Nat. Rev. Mol. Cell Biol.* **4,** 757–766.

Fan, J. S., Cheng, Z., Zhang, J., Noble, C., Zhou, Z., Song, H., and Yang, D. (2009). Solution and crystal structures of mRNA exporter Dbp5p and its interaction with nucleotides. *J. Mol. Biol.* **388,** 1–10.

Henn, A., Cao, W., Hackney, D. D., and De La Cruz, E. M. (2008). The ATPase cycle mechanism of the DEAD-box rRNA helicase, DbpA. *J. Mol. Biol.* **377,** 193–205.

Henn, A., Cao, W., Licciardello, N., Heitkamp, S. E., Hackney, D. D., and De La Cruz, E. M. (2010). Pathway of ATP utilization and duplex rRNA unwinding by the DEAD-box helicase, DbpA. *Proc. Natl. Acad. Sci. U. S. A.* **107,** 4046–4050.

Hetzer, M. W., and Wente, S. R. (2009). Border control at the nucleus: Biogenesis and organization of the nuclear membrane and pore complexes. *Dev. Cell* **17,** 606–616.

Hodge, C. A., Colot, H. V., Stafford, P., and Cole, C. N. (1999). Rat8p/Dbp5p is a shuttling transport factor that interacts with Rat7p/Nup159p and Gle1p and suppresses the mRNA export defect of xpo1-1 cells. *EMBO J.* **18,** 5778–5788.

Iglesias, N., and Stutz, F. (2008). Regulation of mRNP dynamics along the export pathway. *FEBS Lett.* **582,** 1987–1996.

Jankowsky, E. (2011). RNA helicases at work: Binding and rearranging. *Trends Biochem. Sci.* **36,** 19–29.

Katahira, J., Strasser, K., Podtelejnikov, A., Mann, M., Jung, J. U., and Hurt, E. (1999). The Mex67p-mediated nuclear mRNA export pathway is conserved from yeast to human. *EMBO J.* **18,** 2593–2609.

Lakowicz, J. R. (2006). Principles of Fluorescence Spectroscopy. Springer, New York 954.

Lund, M. K., and Guthrie, C. (2005). The DEAD-box protein Dbp5p is required to dissociate Mex67p from exported mRNPs at the nuclear rim. *Mol. Cell* **20,** 645–651.

Montpetit, B., Thomsen, N. D., Helmke, K. J., Seeliger, M. A., Berger, J. M., and Weis, K. (2011). A conserved mechanism of DEAD-box ATPase activation by nucleoporins and InsP6 in mRNA export. *Nature* **472,** 238–242.

Moore, M. J. (2005). From birth to death: The complex lives of eukaryotic mRNAs. *Science* **309,** 1514–1518.

Noble, K. N., Tran, E. J., Alcazar-Roman, A. R., Hodge, C. A., Cole, C. N., and Wente, S. R. (2011). The Dbp5 cycle at the nuclear pore complex during mRNA export II: Nucleotide cycling and mRNP remodeling by Dbp5 are controlled by Nup159 and Gle1. *Genes Dev.* **25,** 1065–1077.

Rocak, S., and Linder, P. (2004). DEAD-box proteins: The driving forces behind RNA metabolism. *Nat. Rev. Mol. Cell Biol.* **5,** 232–241.

Schmitt, C., von Kobbe, C., Bachi, A., Pante, N., Rodrigues, J. P., Boscheron, C., Rigaut, G., Wilm, M., Seraphin, B., Carmo-Fonseca, M., and Izaurralde, E. (1999). Dbp5, A DEAD-box protein required for mRNA export, is recruited to the cytoplasmic fibrils of nuclear pore complex via a conserved interaction with CAN/Nup159p. *EMBO J.* **18,** 4332–4347.

Segref, A., Sharma, K., Doye, V., Hellwig, A., Huber, J., Luhrmann, R., and Hurt, E. (1997). Mex67p, a novel factor for nuclear mRNA export, binds to both poly(A)+ RNA and nuclear pores. *EMBO J.* **16,** 3256–3271.

Snay-Hodge, C. A., Colot, H. V., Goldstein, A. L., and Cole, C. N. (1998). Dbp5p/Rat8p is a yeast nuclear pore-associated DEAD-box protein essential for RNA export. *EMBO J.* **17,** 2663–2676.

Stewart, M. (2007a). Ratcheting mRNA out of the nucleus. *Mol. Cell* **25,** 327–330.

Stewart, M. (2007b). Molecular mechanism of the nuclear protein import cycle. *Nat. Rev. Mol. Cell Biol.* **8,** 195–208.

Strahm, Y., Fahrenkrog, B., Zenklusen, D., Rychner, E., Kantor, J., Rosbach, M., and Stutz, F. (1999). The RNA export factor Gle1p is located on the cytoplasmic fibrils of the NPC and physically interacts with the FG-nucleoporin Rip1p, the DEAD-box protein Rat8p/Dbp5p and a new protein Ymr 255p. *EMBO J.* **18,** 5761–5777.

Strambio-De-Castillia, C., Niepel, M., and Rout, M. P. (2010). The nuclear pore complex: Bridging nuclear transport and gene regulation. *Nat. Rev. Mol. Cell Biol.* **11,** 490–501.

Stutz, F., and Izaurralde, E. (2003). The interplay of nuclear mRNP assembly, mRNA surveillance and export. *Trends Cell Biol.* **13,** 319–327.

Tetenbaum-Novatt, J., and Rout, M. P. (2010). The mechanism of nucleocytoplasmic transport through the nuclear pore complex. *Cold Spring Harb. Symp. Quant. Biol.* **75,** 567–584.

Tran, E. J., Zhou, Y., Corbett, A. H., and Wente, S. R. (2007). The DEAD-box protein Dbp5 controls mRNA export by triggering specific RNA:protein remodeling events. *Mol. Cell* **28,** 850–859.

Tseng, S. S., Weaver, P. L., Liu, Y., Hitomi, M., Tartakoff, A. M., and Chang, T. H. (1998). Dbp5p, a cytosolic RNA helicase, is required for poly(A)+ RNA export. *EMBO J.* **17,** 2651–2662.

Warburg, O. H. (1948). Wasserstoffübertragende Fermente. Saenger, Berlin.

Weirich, C. S., Erzberger, J. P., Berger, J. M., and Weis, K. (2004). The N-terminal domain of Nup159 forms a beta-propeller that functions in mRNA export by tethering the helicase Dbp5 to the nuclear pore. *Mol. Cell* **16,** 749–760.

Weirich, C. S., Erzberger, J. P., Flick, J. S., Berger, J. M., Thorner, J., and Weis, K. (2006). Activation of the DExD/H-box protein Dbp5 by the nuclear-pore protein Gle1 and its coactivator InsP6 is required for mRNA export. *Nat. Cell Biol.* **8,** 668–676.

Zhao, J., Jin, S. B., Bjorkroth, B., Wieslander, L., and Daneholt, B. (2002). The mRNA export factor Dbp5 is associated with Balbiani ring mRNP from gene to cytoplasm. *EMBO J.* **21,** 1177–1187.

CHAPTER TWELVE

Biochemical Characterization of the RNA Helicase UPF1 Involved in Nonsense-Mediated mRNA Decay

Francesca Fiorini,* Fabien Bonneau,[†] *and* Hervé Le Hir*

Contents

1. Introduction	256
2. Preparation of Active UPF1 and UPF1–UPF2 Complex	257
2.1. Cloning	258
2.2. Expression	258
2.3. Purification	258
3. Complex Assembly	259
3.1. Pre-blocking affinity beads	261
3.2. Protein–protein interaction	262
3.3. RNA–protein coprecipitation	263
4. RNAse Protection Assay	264
4.1. RNA substrate preparation	264
4.2. Assay procedure	265
5. ATPase Assay	266
6. Unwinding Assay	268
6.1. Helicase substrate preparation	268
6.2. DNA oligonucleotide radiolabeling	270
6.3. RNA production	270
6.4. RNA–DNA hybrid preparation	271
6.5. Unwinding assay	271
7. Conclusions	272
Acknowledgments	273
References	273

Abstract

Degradation of eukaryotic mRNAs harboring a premature translation termination codon is ensured by the process of nonsense-mediated mRNA decay

* Institut de Biologie de l'Ecole Normale Supérieure, CNRS UMR 8197, INSERM U1024, Paris Cedex 05, France
[†] Max-Planck-Institute of Biochemistry, Department of Structural Cell Biology, Martinsried, Germany

Methods in Enzymology, Volume 511   © 2012 Elsevier Inc.
ISSN 0076-6879, DOI: 10.1016/B978-0-12-396546-2.00012-7   All rights reserved.

(NMD). The main effector of this quality-control pathway is the conserved RNA helicase UPF1 that forms a surveillance complex with the proteins UPF2 and UPF3. In all the organisms tested, the ATPase activity of UPF1 is essential for NMD. Here, we describe the expression of active recombinant UPF proteins and the reconstitution of the surveillance complex *in vitro*. To understand how UPF1 is regulated during NMD, we developed different biochemical approaches. We describe methods to monitor UPF1 binding to RNA, ATP hydrolysis and RNA unwinding in the presence of its binding partner UPF2. This functional analysis is an important complement for structural studies of protein complexes containing RNA helicases.

1. INTRODUCTION

Eukaryotic gene expression requires multiple surveillance mechanisms to ensure the integrity of the information conveyed by messenger RNA to the translation machinery. The process of nonsense-mediated mRNA decay (NMD) is a well-documented quality-control pathway that allows the elimination of mRNA carrying a premature translation termination codon (PTC) and encoding potentially aberrant truncated proteins (Chang *et al.*, 2007; Isken and Maquat, 2008). In addition to its quality-control activity, NMD controls the expression of several endogenous mRNAs at the posttranscriptional level (Amrani *et al.*, 2006; Mendell *et al.*, 2004). Although the mechanisms used to recognize PTCs may differ, the three NMD core components UPF1, UPF2, and UPF3b (UP-Frameshift proteins) are conserved from yeast to mammals (Rebbapragada and Lykke-Andersen, 2009; Wen and Brogna, 2008). The three UPF proteins form the so-called surveillance complex (He *et al.*, 1997; Serin *et al.*, 2001).

UPF1 is an RNA helicase belonging to the SF1 superfamily (Jankowsky, 2011). It possesses an RNA dependent ATPase activity essential for complete degradation of NMD substrates (Bhattacharya *et al.*, 2000; Czaplinski *et al.*, 1995; Franks *et al.*, 2010). UPF1 is composed of three domains: a conserved N-terminal region rich in cysteine and histidine residues (CH domain), a helicase core domain, and a less conserved C-terminal domain. Recent biochemical and structural studies have shed light on the mechanism by which UPF2 modulates the ATPase and RNA helicase activities of UPF1 within the surveillance complex (Chakrabarti *et al.*, 2011; Chamieh *et al.*, 2008; Clerici *et al.*, 2009; Kadlec *et al.*, 2006). In this chapter, we describe experimental approaches to reconstitute the UPF complex and to determine how UPF2 regulates UPF1 activities via its CH domain.

2. PREPARATION OF ACTIVE UPF1 AND UPF1–UPF2 COMPLEX

We expressed recombinant proteins of human origin in *Escherichia coli*. An expression vector was engineered to provide an N-terminal calmodulin binding peptide (CBP) tag as well as a C-terminal hexahistidine tag. This allows us to obtain highly homogenous proteins using a double affinity purification strategy (Fig. 12.1).

The CBP tag offers several advantages. First, CBP is a small tag that adds around 6 kDa to the N or C terminus of the protein and is thus less likely to interfere sterically during complex assembly. Second, this tag can be used as a handle to perform protein–protein interaction experiments (see Section 3). Third, CBP interaction with Calmodulin-conjugated Sepharose is relatively tight, and the elution can be performed in mild conditions to minimize the elution of proteins nonspecifically bound to the resin (Rigaut et al., 1999).

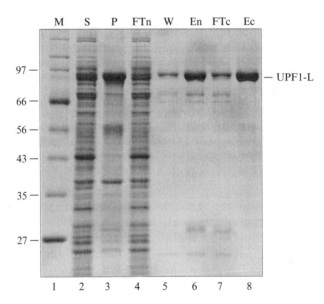

Figure 12.1 Purification of histidine and CBP-tagged UPF1-L protein from *E. coli* using gravity-flow chromatography. Fractions were collected from each of the purification steps and assayed on a 12% (w/v) SDS-PAGE and Coomassie staining (Coomassie brilliant blue R-250 Staining solution, Bio-Rad). M, broad range protein marker (2–212 kDa, New England Biolabs); S, soluble lysate fraction; P, pellet corresponding to the insoluble lysate fraction; FTn, flow through of the Ni-NTA column; W, high salt wash of the Ni-NTA column; En, elution of the Ni-NTA column; FTc, flow through of the calmodulin column; Ec, elution of the calmodulin column. (For color version of this figure, the reader is referred to the online version of this chapter.)

2.1. Cloning

The coding sequence for the CBP tag was cloned between the NcoI and NdeI sites of pET28a (Novagen), the ATG comprised within the NdeI site being in frame with the CBP tag. Coding sequences lacking their native stop codon were then cloned between the NdeI and XhoI sites.

In order to obtain a fully active UPF1/UPF2 complex the two proteins have to be coexpressed on the same plasmid and copurified using a CBP tag on one of the proteins and a hexahistidine tag on the other. The UPF1 and UPF2 coding sequences were cloned sequentially, in the same transcriptional orientation. The UPF1 coding sequence was first inserted between NdeI and EcoRI. An EcoRI site and a ribosome binding site were added by PCR to the 5′ end of the coding sequence for UPF2, which was then cloned between EcoRI and XhoI.

2.2. Expression

Each expression vector (200 ng) is transformed into chemically competent E. coli BL21(DE3) Rosetta cells (Novagen). This strain contains a supplementary plasmid encoding rare tRNA genes, designed to enhance the expression of eukaryotic proteins in E. coli. The colonies containing both plasmids are selected on LB kanamycin/chloramphenicol plates. A single colony is picked and transferred into 10 ml LB medium containing antibiotics and grown overnight at 37 °C. This starter culture is used to inoculate 1 l of LB medium in 5-l flasks. The cells are initially incubated at 37 °C under agitation (180 rpm) until the OD_{600} reaches 0.5–0.6, the temperature is then adjusted to 16 °C and the flasks are left to cool down for 1 h. Protein expression is induced by addition of IPTG to a final concentration of 250 µM and the bacteria are left shaking at 16 °C for about 16 h. The cells are subsequently harvested by centrifugation at $7500 \times g$ for 15 min. The supernatant is removed and the pellet washed in PBS and centrifuged again before freezing in an ethanol/dry ice mixture.

2.3. Purification

1. Thaw the pellet on ice and resuspend it in 10 ml lysis buffer (1.5 × PBS, pH 7.5, 225 mM NaCl, 1 mM magnesium acetate, 0.1% (w/v) NP-40, 20 mM imidazole, 10% (w/v) glycerol, 100 µg/ml egg white lysozyme (Sigma–Aldrich), 1 × protease inhibitor cocktail EDTA-Free (Sigma–Aldrich)) for each gram of pellet.
2. Transfer the suspension in a metal beaker and let it incubate 10 min on ice.
3. Lyse the bacteria by sonication at 30% amplitude and 1 s pulse cycle for 5 min (Brandson Digital Sonifier) on ice.
4. Spin down the bacterial lysate for 40 min at $27,200 \times g$ at 4 °C.

5. After centrifugation, filter the supernatant through a 0.2 μm membrane filter and mix with 1 ml Nickel-NTA Agarose slurry (Qiagen) pre-equilibrated in lysis buffer.
6. We usually perform the binding step in closed tube placed on a rotator for 1 h at 4 °C. The resin is then packed in a Poly-Prep chromatography column (Bio-Rad).
7. Wash extensively the column over 20 column volumes with buffer A (1.5 × PBS, 225 mM NaCl, 1 mM magnesium acetate, 0.1% (w/v) NP-40, 20 mM imidazole, and 10% (w/v) glycerol) and elute the bound proteins by successively adding 500 μl fractions of 500 mM imidazole in the same buffer. Pool the protein containing fractions and dialyze overnight at 4 °C against 1 l calmodulin binding buffer (1 × PBS, pH 7.5, 150 mM NaCl, 1 mM magnesium acetate, 0.1% (w/v) NP-40, 1 mM DTT, 4 mM calcium chloride, and 10% (w/v) glycerol).
8. If batch purification does not provide satisfying results, the lysate is applied to a prepacked column (HisTrap FF crude, GE Healthcare) and further processed with a chromatography machine (Äkta Purifier, GE Healthcare). In this case, wash with 20-column volumes with buffer A and elute using a gradient between 20 and 500 mM imidazole in the same buffer over 20-column volumes.
9. Incubate the dialyzed sample in batch with the resin on a rolling table for 1 h at 4 °C before packing it in a Poly-Prep column (Bio-Rad).
10. Wash out the unbound proteins with 10 column volumes of calmodulin binding buffer.
11. To elute the protein, incubate the resin with 800 μl of calmodulin elution buffer (1 × PBS, pH 7.5, 150 mM NaCl, 1 mM magnesium acetate, 0.1% (w/v) NP-40, 1 mM DTT, 20 mM EGTA, and 10% (w/v) glycerol) for 10 min before collecting the eluate. Repeat incubation/elution three times; this is usually enough to recover most of the purified protein.
12. Determine protein concentration using the Bradford assay (Bio-Rad) and assess the protein purity by SDS-PAGE (Fig. 12.1, Bio-Rad).
13. Pool the most concentrated elution fractions and dialyze against storage buffer (1 × PBS, 150 mM NaCl, 1 mM magnesium acetate, 1 mM DTT, 0.1% (w/v) NP40, 20% (w/v) glycerol, 1 μM zinc sulfate) overnight at 4 °C. After dialysis distribute the protein solution in small aliquots (50–100 μl) and snap-freeze them in liquid nitrogen.

3. Complex Assembly

Helicases rarely function as isolated units. They are usually parts of bigger protein complexes in which noncatalytic cofactors serve as regulators. The first step in understanding the role of an helicase in a particular cellular process is the identification of its direct binding partners.

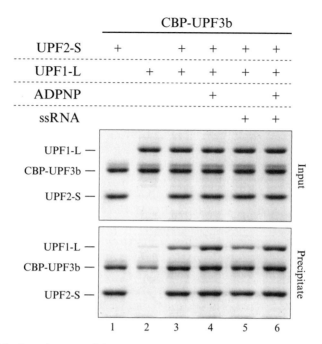

Figure 12.2 Protein coprecipitations with CBP-tagged UPF3b. Combinations of CBP-UPF1-L, CBP-UPF1-ΔCH, UPF2-S, and UPF3b were mixed with or without ADPNP and with or without single-stranded RNA (ssRNA) as indicated. Protein mixtures before (input, 20% of total) or after precipitation (precipitate) were separated by SDS-PAGE on 10% (w/v) acrylamide gels. Adapted from Chamieh et al. (2008). (For color version of this figure, the reader is referred to the online version of this chapter.)

Coimmunoprecipitation experiments from total cell lysates allow narrowing the list of putative interacting partners but do not give information about the degree of linkage between two proteins in a complex. We selected our target proteins from these short lists and confirmed or infirmed direct interactions with in vitro pull-down assays using purified proteins fused to affinity tags (Chamieh et al., 2008; Fig. 12.2).

Since RNA helicases interact with RNA, it is also possible to assemble protein complexes onto an RNA molecule (Fig. 12.3). This of course requires finding conditions where the helicase will stay tightly bound to its substrate, for example, in the presence of a nonhydrolysable nucleotide analog or in the presence of downregulating partners, or in the absence of activators. Typically, for RNA pull-down assays, we use a synthetic 3′-end biotinylated 47-mer oligoribonucleotide (5′-gcagaucagcuuggccgcguccaucuggucaucuagu-gauaucaucg-3′, Eurogentec). For both CBP and biotinylated RNA pull-downs, affinity beads are pre-blocked with unspecific proteins, RNA, and sugars prior to use in order to minimize background noise.

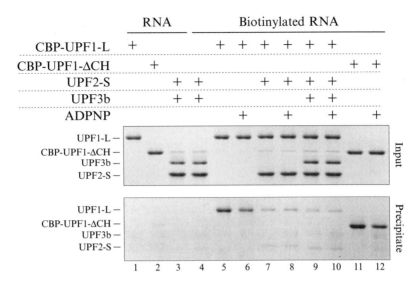

Figure 12.3 Protein coprecipitations with biotinylated RNA. Combinations of CBP-UPF1-L, CBP-UPF1-ΔCH, UPF2-S, and UPF3b were mixed with ssRNA (lanes 1–3) or with 3′-end biotinylated 47-mer ssRNA (lanes 4–12) with or without ADPNP as indicated. Before (input, 20% of total) and after coprecipitation (precipitate) protein were separated and analyzed as in Fig. 12.1 (Chamieh et al., 2008). (For color version of this figure, the reader is referred to the online version of this chapter.)

3.1. Pre-blocking affinity beads

1. A stock of pre-blocked beads can be prepared in advance and stored at 4 °C for several months as 50% (v/v) slurry. All the procedures, unless otherwise indicated, are carried out at 4 °C and all centrifugation steps are performed at 4 °C for 1 min at $800 \times g$ in a microcentrifuge.
2. Centrifuge 1 ml of Calmodulin Sepharose 4B resin (50% (v/v) slurry, GE Healthcare) or 300 μl DynabeadsMyOne Streptavidin T1 (10 mg/ml, Invitrogen) and discard the storing buffer using a thin Pasteur pipet or gel-loading tips, paying particular attention to not aspirate beads from the bottom of the tube.
3. Rinse the beads twice with 900 μl blocking buffer (20 mM HEPES, pH 7.5, 150 mM NaCl, 0.1% (w/v) NP-40).
4. Resuspend the bead pelletin 500 μl blocking buffer containing 400 mM NaCl, 50 μg of glycogen carrier (Roche), 50 μg tRNA (Sigma–Aldrich), 500 μg BSA (New England Biolabs).
5. Rotate the beads at 4 °C for more than 2 h; then wash them three times with 900 μl of blocking buffer.
6. Resuspend the Sepharose beads pellet in 500 μl BB250 buffer (20 mM HEPES, pH 7.5, 250 mM NaCl, 10% (w/v) glycerol, 2 mM magnesium acetate, 2 mM calcium chloride, 2 mM imidazole, 0.1% (w/v) NP-40,

and 1 mM DTT) and magnetic beads pellet in 300 μl of storage buffer containing 10 mM HEPES, pH 7.5, 250 mM NaCl, and 1 mM EDTA.

3.2. Protein–protein interaction

All steps hereafter up to the elution are performed on ice. For the washing steps, we use a centrifuge placed in a cold room to prevent warming of the samples during handling. Protein aliquots are thawed on ice, mixed gently by flicking the tubes, spun at 20,000 × g for 10 min at 4 °C to pellet down any precipitate that could have formed upon freezing or thawing.

Each combination of proteins is mixed in prechilled 1.5 ml microcentrifuge tubes. Approximately 2 μg of each protein are used per reaction. Only one protein per mix should carry a CBP tag. Protein concentration determination using the Bradford method only gives an estimate of what to use for a start: the actual amount of protein of interest will greatly depend on the purity of the preparation. We usually calibrate the relative amounts of proteins to be used by running a range of concentrations on a SDS-PAGE and selecting amounts that give bands of similar intensities by Coomassie staining, before performing the actual experiment. To avoid precipitation due to over dilution while mixing proteins stored in different buffers, proteins stored in high salt/glycerol buffers should be added first. Controls without CBP-tagged protein should be included to check for nonspecific binding of the putative preys to the beads. A control with the CBP-bait protein alone will sometimes be helpful to assess whether binding of interaction partners affect CBP recognition by the calmodulin beads. It is critical to ensure that all mixtures have the same final buffer. To do so, mixes can be complemented with the same volumes of the buffers used to store the proteins.

1. Mix proteins in a final volume of 30 μl. Calculate the final concentration of salt (KCl and/or NaCl) and glycerol in the sample. In our case, we adjust them to 215 mM salts and 15% (w/v) of glycerol.
2. Take a 5 μl aliquot from these 30 μl, add 2.5 μl of 3 × SDS loading dye, and keep at −20 °C until boiling and loading on the input gel. This input sample will represent one-fifth of the amount of proteins actually used in the interaction experiment.
3. Add 5 μl of RNase-free water containing 2 μM RNA and/or 20 mM NTPs if needed. In our experience, adding nucleotides before taking the input samples resulted in input gels of a much lower quality. The reaction volume is now back to 30 μl with salt and glycerol concentrations of 180 mM and 12.5% (w/v), respectively.
4. To facilitate protein–protein interactions, the salt concentration is brought down to 125 mM by diluting the samples with 30 μl of a 2 × buffer containing 40 mM HEPES, pH 7.5, 70 mM NaCl, 4 mM

magnesium acetate, 4 mM calcium chloride, 4 mM imidazole, 0.2% (w/v) NP-40, 2 mM DTT, and 12.5% (w/v) glycerol.
5. Incubate the mixes 20 min at 30 °C, or alternatively overnight at 4 °C if the proteins are thermally unstable.
6. Add 12 µl of pre-blocked Calmodulin Sepharose beads and 200 µl of BB250 buffer to the protein complexes and place the tubes on a rotator for 1 h at 4 °C.
7. Wash carefully the beads three times with 500 µl of BB250 buffer
8. After the final wash, remove as much supernatant as possible (this can be done by carefully placing the tip below the resin) before adding 20 µl elution buffer (10 mM Tris, pH 7.5, 150 mM NaCl, 14% (w/v) glycerol, 1 mM magnesium acetate, 20 mM EGTA, 2 mM imidazole, 0.1% (w/v) NP-40, and 10 mM β-mercaptoethanol).
9. The proteins are detached from the affinity beads by shaking the tubes for 5 min at 30 °C at 1400 rpm in a thermomixer.
10. Spin the tubes at $900 \times g$ to pellet the beads and collect the eluates in fresh tubes.

Concentrate the samples in a vacuum concentrator for 30 min at 45 °C. Around 4 µl glycerol should be left in the tubes. This is further diluted with 2 µl $H_2O + 3$ µl $3 \times$ SDS loading dye before boiling for 3 min and analysis by SDS-PAGE (Fig. 12.2).

Since these samples have a very high salt and glycerol content, the electrophoresis should be performed at very low intensity (6 mA per 0.75-mm-thick Bio-Rad minigel). In addition, to ensure a straight migration of all the lanes, empty wells should be loaded with concentrated elution buffer processed the same way than the reaction samples.

3.3. RNA–protein coprecipitation

The protocol is identical to the CBP pull-downs with a few modifications relative to the usage of magnetic beads and of RNA as a prey.

1. Because RNA binding to proteins often involves electrostatic interactions, high salt concentrations may interfere with the coprecipitation. The final concentration of salt after addition of the $2 \times$ buffer is adjusted to 75 mM instead of the 125 mM used for the CBP pull-down.
2. For the washing steps, BB250 is replaced by BB75 (where the NaCl concentration is lowered down to 75 mM).
3. Since the biotin–streptavidin interaction is very strong, 5 µl pre-blocked magnetic beads are enough per reaction.
4. Although beads will gather to one side of the tube when placed on the magnetic stand, we found that a brief centrifugation step ($900 \times g$ for 30 s at 4 °C) helps in speeding up the process.

5. Elution is performed by adding 7.5 μl SDS loading buffer directly onto the beads without heating step. This is sufficient to disrupt an RNA–protein interaction. Boiling the beads would result in the release of streptavidin and BSA (used for blocking) from the beads, which could be problematic if they comigrate with the proteins of interest.
6. The loading buffer is recovered from the beads using a magnetic stand and boiled separately before loading on a gel (Fig. 12.3).

4. RNAse Protection Assay

Understanding how helicases physically interact with their substrate can provide clues about their mechanism of action. The RNase protection assay allows us to define the footprint of a helicase on its substrate (Fig. 12.4). Variations of this footprint with different nucleotide analogs provide information about the conformational changes of the helicase along its reaction path. Similarly, observing the footprint change upon addition of a protein partner allows us to understand how helicase activity is regulated within the RNA surveillance complex (Chakrabarti et al., 2011; Chamieh et al., 2008).

4.1. RNA substrate preparation

For this assay, we use a 60-mer RNA composed of CU repeats. This has little secondary structure and can be degraded to single nucleotides by RNaseA treatment. The RNA is transcribed from a partially double-stranded template using the MEGAshortscript kit (Applied Biosciences) in presence of trace amounts of α-^{32}P-UTP. Because of T7 promoter requirements, 3 Gs are present at the 5′ end of the transcript. RNase T1 is used together with RNase A in the protection assay to target these nucleotides.

A 29-mer oligonucleotide encompassing the T7 promoter (cacgcatatgtaatacgactcactatagg) is annealed to the complementary sequence of T7 promoter preceded by the complement of the desired sequence (g(ga)$_{28}$ccctatagtgagtcgtattacatatgcgtg).

1. Prepare the annealing reaction mixing 200 pmol of each synthetic DNA oligonucleotides (Sigma-Aldrich) in a 10 μl final volume of TNE buffer (10 mM Tris, pH 7.5, 100 mM NaCl, and 1 mM EDTA).
2. Heat the mixture to 95 °C for 2 min and let cool down very slowly to room temperature by switching off the incubator.
3. Perform transcription in a 10 μl volume using 1 μl of the template and 4 μl α-^{32}P-UTP (3000 Ci/mmol, 10 mCi/ml, Perkin-Elmer). The final product is a 60-mer (ggg(cu)$_{28}$c) internally labeled.

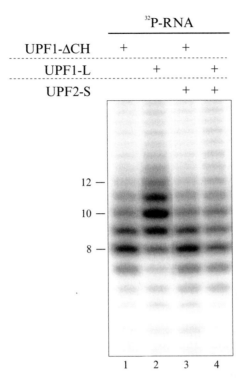

Figure 12.4 RNAse protection assay. Combination of CBP-UPF1-L, CBP-UPF1-Δ CH, and UPF2-S in the presence of ADP-BeF$_3$ were mixed with a ^{32}P body-labeled 60-mer ssRNA. After incubation, mixtures were digested with RNases A and T1 and protected RNA fragments were separated on a 22% denaturing polyacrylamide gel. Adapted from Chakrabarti et al. (2011).

4. Purify the RNA 60-mer on a 10% (w/v) polyacrylamide gel (19:1) containing 7 M Urea. The band containing the labeled product can be identified using the classical UV shadowing technique, or by exposing briefly the gel to phosphorimager plate and overlaying a full scale printout with the gel (Burton et al., 2009). A typical transcription yields around 400 pmol of purified RNA.

4.2. Assay procedure

1. Prepare the reaction in a 20 μl of final volume mixing 5 pmol of the labeled RNA with 10 pmol of protein in a buffer containing 50 mM MES, pH 6.5, 50 mM NaCl, 5 mM magnesium acetate, 10% (w/v) glycerol, 0.1% (w/v) NP-40, and 1 mM DTT.

2. When needed, 20 nmol of nucleotide or nucleotide analog is provided. If a transition state analog is desired, 0.5 μl of a metal/fluoride mix (400 nmol AlCl$_3$ (or BeCl$_2$), 2 μmol NaF in a 10 μl final volume of H$_2$O) are added to the reaction together with 20 nmol ADP.

   When using metal–fluoride complexes, keep in mind that the pH of the reaction influences their coordination state (Schlichting and Reinstein, 1999). If a transition state analog is desired, you might have to screen for the optimal pH allowing its formation.
3. Incubate for 1 h at 4 °C.

   As for the interaction assays with RNA, salt concentration is critical. The amount of salt present in the protein buffer should be taken into account when calculating the final concentrations of the reaction buffer components.
4. Perform the RNase treatment by adding 0.5 μl RNase A/T1 mix (RNase A at 2 mg/ml, RNase T1 at 5000 units/ml, Fermentas) per reaction.
5. Incubate 20 min at 20 °C.
6. Stop the reaction by diluting with 180 μl of stop buffer (100 mM Tris, pH 7.5, 150 mM NaCl, 300 mM sodium acetate, 10 mM EDTA, 1% (w/v) SDS)
7. Extract twice with 200 μl Phenol/Chloroform/Isoamyl alcohol (25:24:1, v/v).
8. After the first extraction, add 6 μg of glycogen (Roche) to help locating the RNA pellet in the next steps.
9. Add three volumes of 100% ethanol to recover the extract and let precipitate for more than 2 h at −20 °C.
10. After centrifugation at 20,000 × g for 15 min at 4 °C, wash the RNA pellets twice with 300 μl 80% (v/v) ethanol.
11. Dry in a vacuum concentrator and resuspend in 8 μl RNA loading dye (75% (v/v) formamide, 7.5 mM EDTA, 0.075% (w/v) bromophenol blue, 0.075% (w/v) xylene cyanole).
12. Heat the samples to 95 °C for 4 min before loading on a 0.5-mm thick, 20-cm long, 22% polyacrylamide gel (19:1) containing 5 M Urea.
13. Prewarm the gel 20 min at 10 W before loading the samples, and run until the slow dye reaches two-third of the length.
14. Cover the gel with Saran wrap and expose it to an image plate (Fuji) at −70 °C overnight.
15. Acquire the gel image (Fig. 12.4) with a Typhoon phosphorimager (GE Healthcare) and record it with the default software settings.

5. ATPase Assay

Most helicases hydrolyze ATP to perform their functions. An easy way to probe the effect of interaction partners on a helicase activity is to analyze variations in its ATP turnover in the presence or in the absence of these

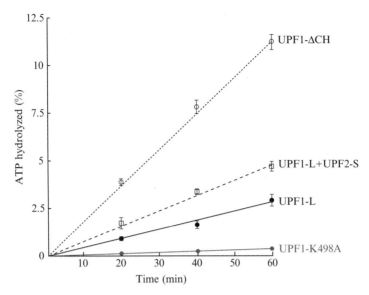

Figure 12.5 UPF1 ATPase activity. ATPase activity was determined for CBP-UPF1-L alone (filled circle) and with CBP-UPF2-S (open square) and for CBP-UPF1-ΔCH (open diamond). As a control, the fractions of hydrolyzed ATP were measured for CBP-UPF1-L K498A (filled gray circle). Data represent mean values and SD measured from three independent experiments. The best-fitting trend-line through data points were generated with the KaleidaGraph software. Adapted from Chamieh et al. (2008).

cofactors. For instance, this assay allowed us to assign a downregulatory function to the CH domain located in the N-terminal part of the Upf1. In that case, the helicase is subject to an intrinsic downregulation that can be relieved by the interaction with Upf2.

We use a charcoal-based ATPase assay (Fig. 12.5). γ^{32}P-ATP reaction with the enzyme releases radioactive inorganic phosphate that can be separated from unreacted ATP by a precipitation step with charcoal. For our purpose, the ATP and enzyme concentrations used as well as the time points for measurement have to be carefully calibrated so that less than 20% of the ATP present in solution is hydrolyzed at the end of the experiment. In these conditions, the rate of hydrolysis remains linear and allows us to derive initial reaction velocities from a four-point kinetic analysis. For UPF1, we use 1.5 pmol protein with 2 mM ATP and time points of 0, 15, 30, and 45 min.

Prepare the reaction mix in 10 μl final volume, mixing the ATPase with or without its interacting partners together with 2 μl 5× ATPase buffer (250 mM MES, pH 6.5, 250 mM potassium acetate, 25 mM magnesium acetate, 10 mM DTT, 0.5 mg/ml BSA). As for the previous assays, the protein buffer compositions have to be taken into account to ensure that all the reaction mixtures have the exact same buffer composition.

1. Start the reaction by adding 10 μl of an ATP mix, containing 2 μl 5× ATPase buffer, 4 μl ATP (10 mM), 2 μl poly(U) RNA (Sigma, 2 mg/ml in H$_2$O), 0.1 μl [γ^{32}P] ATP (Perkin-Elmer, 3000 Ci/mmol, 10 mCi/ml), and 1.9 μl H$_2$O. Both mixes should be kept on ice before the actual start of the kinetic.
2. Place the reaction vials at 30 °C sequentially every 20 s to allow sufficient time for pipetting the aliquots at desired time points.
3. Remove 4 μl aliquots and dilute them into 400 μl charcoal mix (10 mM EDTA, pH 8.0, 10% (w/v) activated charcoal (Sigma-Aldrich)). This quenches the reaction.
4. Mix samples vigorously with a vortex and keep them 1 h on ice before centrifugation at 20,000 × g for 15 min at 4 °C. The inorganic phosphate remains in the soluble fraction while AMP, ADP, and ATP precipitate with charcoal.
5. Transfer 140 μl of supernatant to a fresh vial and count in a liquid scintillation analyzer (Hidex) using the Cerenkov method.

In the experimental design, include one vial for total counts measurement (no charcoal is added in the collections tubes for this sample), as well as one negative control without protein to get a baseline for spontaneous ATP hydrolysis. From this, the percentage of ATP hydrolyzed as a function of time can be calculated and plotted to check for linearity (Fig. 12.5).

6. Unwinding Assay

UPF1 unwinds double strands of nucleic acids with a specific polarity. The assay most commonly used to measure helicase activity is an "all or none" unwinding assay (Lucius et al., 2003; Fig. 12.6A). To obtain quantitative information about the unwinding rate, the reaction is performed under presteady state kinetic conditions (single run), using an excess of enzyme over substrate. The assay consists in following the separation of a partial RNA/DNA duplex on a native polyacrylamide gel. The sequence of the substrate has to be optimized so that under the conditions required by the helicase (temperature, pH, salt, and protein concentration), the duplex is stable by itself over the course of the reaction. The length of the duplex should be short enough to ensure detection of unwinding even for an enzyme with low processivity but long enough to be stable over reaction time. The 3' or 5' position of the RNA overhang will be selected depending on the polarity of the enzyme.

6.1. Helicase substrate preparation

Human UPF1 binds 8–11 nucleotides single-stranded RNA (Chakrabarti et al., 2011; Fig. 12.4) and displays helicase activity with 5'–3' polarity (Czaplinski et al., 1995). The substrate chosen for testing UPF1 unwinding

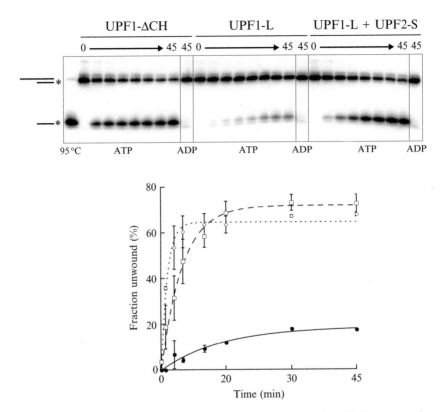

Figure 12.6 Unwinding assay-monitored UPF1 helicase activity. (A) Representative PAGE of unwinding time-course experiments over a 45 min period with CBP-UPF1-ΔCH (open diamond) and CBP-UPF1-L (filled circle) with or without CBP-UPF2-S (open square) (0, 1′, 3′, 5′, 10′, 15′, 30′, and 45′). Positions of the RNA/DNA substrate and ssDNA are indicated. The asterisk denotes the position of the ^{32}P label. Denaturation of the duplex is complete after heating at 95 °C and no unwinding is observed in the presence of ADP after 45 min incubation. The gel shows the ATP-dependent nucleic acids unwinding catalyzed by UPF1-ΔCH, UPF1-L, and UPF1-L/UPF2-S complex. Adapted from Chakrabarti et al. (2011). (B) Quantitative analysis of UPF1 helicase reaction. The graph shows the fraction of labeled DNA released from the duplex substrate by the enzyme as a function of time. The corresponding values for the reaction amplitudes (A) and rate constants (k in min^{-1}, see Section 6.5.12) are, respectively: 0.65 (±0.01) and 0.70 (±0.08) for UPF1-ΔCH, 0.19 (±0.02) and 0.07 (±0.02) for Upf1-L, 0.72 (±0.02), and 0.20 (±0.02) for UPF1-L/UPF2-S complex.

activity is a DNA/RNA duplex containing a 12 nucleotide long 5′ RNA overhang ($\Delta G°_{37} = -31.6$ kcal mol^{-1}; Sugimoto et al., 1995). We observed that UPF1 binds to even a single nucleotide overhang at the 3′-end of the substrate, compromising the unwinding test. For this reason, we use a substrate with a blunt end on the opposite side. A 21 nt-long DNA oligonucleotide (FF7) is radiolabeled and annealed to a 33 nt-long in vitro transcribed RNA (R3U33M).

6.2. DNA oligonucleotide radiolabeling

1. In a 15 μl reaction volume mix 10 pmol of DNA oligonucleotide FF7 (ggtacccgcagcttctcgagg, MWG) with 10 U of T4 Polynucleotide kinase (Fermentas), 3 μl of γ^{32}P-ATP, and 1× PNK buffer A (forward reaction).
2. Incubate the mixture at 37 °C for 1 h and add 1 μl of EDTA (0.5 M) to stop the reaction.
3. Dilute the sample with 40 μl H$_2$O and purify the probe on a G50 column (Bio-Rad) following the manufacturer's instructions.
4. Mix the eluted product with 0.1 volume of sodium acetate 3 M, pH 5.2, three volumes of ethanol 100%, 1 μg of glycogen and incubate for 1 h at −80 °C.
5. Spin the sample at 20,000 × g for 15 min and wash the pellet with 800 μl of 70% (v/v) ethanol before drying in a vacuum concentrator.
6. Resuspend the radiolabeled DNA pellet in 5 μl of DEPC treated water (Ambion).

6.3. RNA production

The R3U33M RNA (ggguaucagaucccucgagaagcugcggguacc) is transcribed from a partially double-stranded template using the MEGAshortscript kit as described in Section 4.1. This technique allows to easily vary the sequence of the substrate in order to optimize the DNA/RNA hybrid stability or to investigate specific aspect of enzyme activity. If the length of the desired oligonucleotide is not compatible with the size commercialized, the template will need to be cloned in a plasmid, downstream of a T7 promoter.

The DNA oligonucleotide containing FF7 sequence followed by the T7 complementary promoter region (ggtacccgcagcttctcgagggatctgataccctatagtgagtcgtattaatt, Eurogentec) is annealed to the T7 promoter oligonucleotide (cacgcatatgtaatacgactcactatagg).

1. Perform the transcription reaction in a final volume of 23 μl.
2. Take a 3 μl aliquot from the reaction mixture and add 0.3 μl of α^{32}P-UTP (Perkin-Elmer, 3000 Ci/mmol, 10 mCi/ml). This radiolabeled RNA will be used as a molecular weight marker for the following gel purification.
3. Purify the transcript on a 0.4-mm thick 7.5% (w/v) polyacrylamide gel (29:1) containing 7 M Urea.
4. Resuspend the purified RNA pellet in 30 μl of DEPC treated water and quantify with a nanodrop Spectrophotometer 2000C (Thermo Scientific).

6.4. RNA–DNA hybrid preparation

1. Mix 20 pmol of R3U33M RNA with 10 pmol of radiolabeled FF7 DNA oligonucleotide in 15 µl helicase buffer F-100 (20 mM MES, pH 6.0, 100 mM potassium acetate, 1 mM DTT, 0.1 mM EDTA).
2. Anneal both oligonucleotides by heating the mixture to 95 °C for 2 min and letting it cool down very slowly to room temperature by switching off the incubator.
3. Purify the hybrid duplex on a native 6% (w/v) polyacrylamide gel (29:1) in 1 × TBE buffer at 4 °C.
4. Place the band corresponding to the RNA/DNA hybrid in a tube with 800 µl elution buffer (20 mM MOPS, pH 6.0, 0.3 M sodium acetate, and 1 mM EDTA)
5. Incubate on a rolling table overnight at 4 °C. Alternatively the elution can be performed 3 h at 18 °C.

    To prevent hybrid dissociation, it is important to mix the solution gently by inverting the tubes instead of vortexing and to perform all the purification steps on ice.

6. Resuspend the nucleic acids pellet in 30 µl helicase buffer F-100 and store at −20 °C.
7. The final labeled-nucleic acid concentration is determined using a calibration curve obtained by measuring a set of known γ^{32}P-ATP dilutions with a scintillation counter.

6.5. Unwinding assay

The optimal pH for UPF1 unwinding reaction was experimentally determined at pH 6.0 in a MES-based buffer containing 50–100 mM of potassium acetate. The concentration of RNA/DNA hybrid substrate used in each reaction is fixed at 1 nM. The minimal enzyme concentration for the highest UPF1 helicase activity was determined experimentally to be around 50 nM. To prevent loss of activity during freeze-thaw cycles, we distribute the protein solution in small single-use aliquots immediately after purification. Dilution of the protein is carried out immediately before starting the reaction with the same buffer used for storage. The unwinding reaction is initiated by adding an equimolar mixture of $MgCl_2$ and ATP. To check for any contaminants, it is important to carry out a control reaction using ADP instead of ATP.

1. Prepare two unwinding reaction mixes for each protein or protein complex mixing the RNA/DNA substrate and protein in a final volume of 27 µl F-100 helicase buffer.

2. Prepare a 10× starting solution containing 20 µM of ATP (or ADP), 20 µM of MgCl$_2$, and 3 µM of cold FF7 oligonucleotide.
3. Preincubate the reaction mixes and starting solutions for 10 min at 37 °C.
4. Remove 2 µl aliquot for point 0 and mix it to 8 µl of quench buffer (150 mM NaAc, 10 mM EDTA, 0.5% (w/v) SDS, 25% (w/v) Ficoll-400, 0.05% (w/v) xylene cyanole, 0.05% (w/v) bromophenol blue). This stop buffer separates the proteins from the substrate without affecting the duplex.
5. Induce the unwinding activity of the helicase by adding 3 µl of 10× starting solution to the reaction mix. ATP induces duplex disruption by the helicase and the cold FF7 oligonucleotide excess prevents reannealing of the radiolabeled FF7 freed by the helicase.
6. Remove 2 µl aliquot at the end of each incubation time and rapidly mix them with 8 µl of stop buffer.
7. Keep the collected samples on ice until loading on an 11% (w/v) polyacrylamide gel (19:1) containing 0.5% (w/v) SDS.
8. Prerun the gel for 15 min at 130 V (20–25 mA) then load the samples and continue the run for 90 min at 130 V in a room maintained at 20 °C.
9. Wrap the gel with Saran wrap and expose it to a phosphorimaging screen overnight at −80 °C.
10. Detection (Fig. 12.6A) and quantification of the bands are carried out using a Typhoon 9400 phosphorimaging system and the ImageQuant software (GE Healthcare).
11. The unwound product fraction at time 0 is calculated using Eq. (12.1) and the product fractions at time "t" using Eq. (12.2).

$$F_0 = P_0/(R_0 + P_0) \qquad (12.1)$$

$$F_t = (P_t - F_0(P_t + R_t))/((1 - F_0)(P_t + R_t)) \qquad (12.2)$$

where P is the intensity of the band corresponding to the reaction product (lower molecular weight band) and R the intensity of the substrate (higher molecular weight band) in the helicase gel.

12. Data points were fitted with the KaleidaGraph software (Fig. 12.6B) to the equation: $y = A(1 - e^{-kt})$ in which A and k represent, respectively, the amplitude and the rate constant of the exponential (burst) phase in single-run conditions (Walmacq et al., 2004).

7. Conclusions

In this chapter, we described methods to clone, express, purify, and assemble components of the human NMD surveillance complex. Immunoprecipitations from whole cell extracts have been extremely informative to describe the dynamic aspect of the NMD process between PTC recognition to mRNA degradation (Isken and Maquat, 2008; Kashima et al., 2006). However, such an approach is less appropriate to isolate homogenous and functional protein complexes with the aim to dissect their exact function. The protocols presented here in enable the reconstitution of a minimal NMD surveillance complex composed of the purified recombinant proteins UPF1, UPF2, and UPF3 in the presence or in the absence of RNA. Hence, the enzymatic activity of different truncated forms of the essential RNA helicase UPF1 could be monitored in the presence of its direct partner UPF2. The combination of these biochemical strategies constitutes an important step for future structural and functional studies of eukaryotic mRNA surveillance.

ACKNOWLEDGMENTS

We would like to thank Marc Boudvillain for assistance with unwinding assay development. This work was supported by the Centre National de la Recherche Scientifique, l'Agence-Nationale de Recherche, the ATIP program from the Research Ministry (H. L. H.) and by the Max-Planck Institute of Biochemistry (F. B.).

REFERENCES

Amrani, N., Dong, S., He, F., Ganesan, R., Ghosh, S., Kervestin, S., Li, C., Mangus, D. A., Spatrick, P., and Jacobson, A. (2006). Aberrant termination triggers nonsense-mediated mRNA decay. *Biochem. Soc. Trans.* **34**, 39–42.

Bhattacharya, A., Czaplinski, K., Trifillis, P., He, F., Jacobson, A., and Peltz, S. W. (2000). Characterization of the biochemical properties of the human Upf1 gene product that is involved in nonsense-mediated mRNA decay. *RNA* **6**, 1226–1235.

Burton, A. S., Madix, R. A., Vaidya, N., Riley, C. A., Hayden, E. J., Chepetan, A., Arenas, C. D., Larson, B. C., and Lehman, N. (2009). Gel purification of radiolabeled nucleic acids via phosphorimaging: Dip-N-Dot. *Anal. Biochem.* **388**, 351–352.

Chakrabarti, S., Jayachandran, U., Bonneau, F., Fiorini, F., Basquin, C., Domcke, S., Le Hir, H., and Conti, E. (2011). Molecular mechanisms for the RNA-dependent ATPase activity of Upf1 and its regulation by Upf2. *Mol. Cell* **18**, 693–703.

Chamieh, H., Ballut, L., Bonneau, F., and Le Hir, H. (2008). NMD factors UPF2 and UPF3 bridge UPF1 to the exon junction complex and stimulate its RNA helicase activity. *Nat. Struct. Mol. Biol.* **15**, 85–93.

Chang, Y. F., Imam, J. S., and Wilkinson, M. F. (2007). The nonsense-mediated decay RNA surveillance pathway. *Annu. Rev. Biochem.* **76**, 51–74.

Clerici, M., Mourão, A., Gutsche, I., Gehring, N. H., Hentze, M. W., Kulozik, A., Kadlec, J., Sattler, M., and Cusack, S. (2009). Unusual bipartite mode of interaction between the nonsense-mediated decay factors, UPF1 and UPF2. *EMBO J.* **28**, 2293–2306.

Czaplinski, K., Weng, Y., Hagan, K. W., and Peltz, S. W. (1995). Purification and characterization of the Upf1 protein: A factor involved in translation and mRNA degradation. *RNA* **1**, 610–623.

Franks, T. M., Singh, G., and Lykke-Andersen, J. (2010). Upf1 ATPase-dependent mRNP disassembly is required for completion of nonsense-mediated mRNA decay. *Cell* **143**, 938–950.

He, F., Brown, A. H., and Jacobson, A. (1997). Upf1p, Nmd2p, and Upf3p are interacting components of the yeast nonsense-mediated mRNA decay pathway. *Mol. Cell. Biol.* **17**, 1580–1594.

Isken, O., and Maquat, L. E. (2008). The multiple lives of NMD factors: Balancing roles in gene and genome regulation. *Nat. Rev. Genet.* **9**, 699–712.

Jankowsky, E. (2011). RNA helicases at work: Binding and rearranging. *Trends Biochem. Sci.* **36**, 19–29.

Kadlec, J., Guilligay, D., Ravelli, R. B., and Cusack, S. (2006). Crystal structure of the UPF2-interacting domain of nonsense-mediated mRNA decay factor UPF1. *RNA* **12**, 1817–1824.

Kashima, I., Yamashita, A., Izumi, N., Kataoka, N., Morishita, R., Hoshino, S., Ohno, M., Dreyfuss, G., and Ohno, S. (2006). Binding of a novel SMG-1-Upf1-eRF1-eRF3 complex (SURF) to the exon junction complex triggers Upf1 phosphorylation and nonsense-mediated mRNA decay. *Genes Dev.* **20**, 355–367.

Lucius, A. L., Maluf, N. K., Fischer, C. J., and Lohman, T. M. (2003). General methods for analysis of sequential "n-step" kinetic mechanisms: Application to single turnover kinetics of helicase-catalyzed DNA unwinding. *Biophys. J.* **85**, 2224–2239.

Mendell, J. T., Sharifi, N. A., Meyers, J. L., Martinez-Murillo, F., and Dietz, H. C. (2004). Nonsense surveillance regulates expression of diverse classes of mammalian transcripts and mutes genomic noise. *Nat. Genet.* **36**, 1073–1078.

Rebbapragada, I., and Lykke-Andersen, J. (2009). Execution of nonsense-mediated mRNA decay: What defines a substrate? *Curr. Opin. Cell Biol.* **21**, 394–402.

Rigaut, G., Shevchenko, A., Rutz, B., Wilm, M., Mann, M., and Séraphin, B. (1999). A generic protein purification method for protein complex characterization and proteome exploration. *Nat. Biotechnol.* **17**, 1030–1032.

Schlichting, I., and Reinstein, J. (1999). pH influences fluoride coordination number of the AlFxphosphoryl transfer transition state analog. *Nat. Struct. Biol.* **6**, 721–723.

Serin, G., Gersappe, A., Black, J. D., Aronoff, R., and Maquat, L. E. (2001). Identification and characterization of human orthologues to Saccharomyces cerevisiae Upf2 protein and Upf3 protein (Caenorhabditiselegans SMG-4). *Mol. Cell. Biol.* **21**, 209–223.

Sugimoto, N., Nakano, S., Katoh, M., Matsumura, A., Nakamuta, H., Ohmichi, T., Yoneyama, M., and Sasaki, M. (1995). Thermodynamic parameters to predict stability of RNA/DNA hybrid duplexes. *Biochemistry* **34**, 11211–11216.

Walmacq, C., Rahmouni, A. R., and Boudvillain, M. (2004). Influence of substrate composition on the helicase activity of transcription termination factor Rho: Reduced processivity of Rho hexamers during unwinding of RNA-DNA hybrid regions. *J. Mol. Biol.* **342**, 403–420.

Wen, J., and Brogna, S. (2008). Nonsense-mediated mRNA decay. *Biochem. Soc. Trans.* **36**, 514–516.

CHAPTER THIRTEEN

Identification of RNA Helicase Target Sites by UV Cross-Linking and Analysis of cDNA

Markus T. Bohnsack,[*,†] David Tollervey,[‡] and Sander Granneman[§]

Contents

1. Introduction	276
2. Material	277
2.1. Oligonucleotides	278
2.2. Buffers and solutions	279
3. Methods	279
3.1. Yeast strains, cross-linking and affinity purification of RNPs	279
3.2. UV cross-linking	281
3.3. Partial RNase digestion and denaturing Nickel purification	282
3.4. Adapter ligation and radioactive labeling	283
3.5. SDS-PAGE and RNA isolation	283
3.6. Reverse transcription	284
3.7. PCR and size selection of PCR products	284
3.8. Sequencing	285
3.9. Data analysis	285
4. Protocol Adaptation and Trouble-Shooting	286
Acknowledgments	287
References	287

Abstract

Many RNA helicases have been implicated in one or more pathways of RNA metabolism, but only in a very few cases have their target sites on the RNA been identified. Here, we give a detailed description of the UV *cr*oss-linking and *a*nalysis of *c*DNA (CRAC) method, and its application to the identification of binding sites of RNA-interacting helicases. CRAC makes use of a bipartite tag on

[*] Cluster of Excellence Macromolecular Complexes, Institute for Molecular Biosciences, Goethe University Frankfurt, Frankfurt, Germany
[†] Centre for Biochemistry and Molecular Cell Biology, Göttingen University, Göttingen, Germany
[‡] Wellcome Trust Centre for Cell Biology, University of Edinburgh, Edinburgh, United Kingdom
[§] Centre for Systems Biology at Edinburgh, University of Edinburgh, Edinburgh, United Kingdom

Methods in Enzymology, Volume 511
ISSN 0076-6879, DOI: 10.1016/B978-0-12-396546-2.00013-9

© 2012 Elsevier Inc.
All rights reserved.

the protein of interest and includes a purification step under highly denaturing conditions. This is particularly important for the accurate mapping of binding sites within large RNA–protein complexes—such as spliceosomes or preribosomes. Partial RNase digestion leaves a footprint of the protein covering the interaction site, and the UV cross-linking sites are frequently highlighted by microdeletions in cDNA sequence reads. Deep sequencing of cDNA libraries generated from cross-linked RNA fragments allows a genome-wide analysis of the interactome of RNA-binding proteins. In the case of RNA helicases, this has proven to be an important step toward their functional analysis.

1. Introduction

As is the case for many other RNA-binding proteins, the precise RNA substrates and sites of action of most RNA helicases have remained elusive. Many RNA helicases play distinct roles in several different pathways; however, due to the lack of known substrates, the activities of most RNA helicases studied to date have been functionally characterized *in vitro* using model substrates. The development of tools for the precise identification of target sites is clearly important to identify the molecular functions of these proteins.

Coprecipitation techniques including RIP allow the recovery of RNAs associated with a protein of interest, but do not identify the actual RNA target sequences. These can be inferred from methods such as SELEX and yeast three-hybrid (Ellington and Szostak, 1990; SenGupta *et al.*, 1996; Tuerk and Gold, 1990). However, these approaches require interacting RNAs to bind with high affinity and stability. In case of RNA helicases, the catalytic core is thought to interact only transiently with its substrates, while specificity is often achieved by additional domains or interacting cofactors, which mediate enzyme recruitment (reviewed in Linder and Jankowsky, 2011).

Over recent years, several methods have been developed that use UV irradiation to cross-link RNAs with bound proteins, in order to capture transient and rare interactions. The Darnell lab initially published the cross-linking and *i*mmuno*p*recipitation (CLIP) technique (Ule *et al.*, 2003). Subsequent analyses reported that coverage could be increased by combination with high-throughput sequencing (HITS-CLIP; Licatalosi *et al.*, 2008), the efficiency could be enhanced by incorporation of 4-thiouridine or 6-thioguanosine (PAR-CLIP; Hafner *et al.*, 2010; Lebedeva *et al.*, 2011), and the accuracy of mapping the precise cross-linking sites is improved by making use of the tendency for reverse transcriptase to stop at this location (iCLIP; Konig *et al.*, 2010; Wang *et al.*, 2010). All of these methods employ immunoprecipitation after cross-linking to enrich the complexes containing the protein of interest and therefore depend on the availability of highly specific antibodies that bind the target protein with high affinity.

Here, we describe the application of a variant of these approaches, termed *c*ross-linking and *a*nalysis of *c*DNA (CRAC; Granneman et al., 2009), for the identification of interaction sites of RNA helicases (Bohnsack et al., 2009). CRAC does not require the generation of specific antibodies but uses a modified tandem affinity purification (TAP) protocol to isolate RNA cross-linked to the protein of interest. The use of a 6-histidine (His6) tag in combination with Ni-NTA matrix allows purification under very stringent, denaturing conditions. This is especially important for studies of complex RNA–protein particles (RNPs), such as preribosomal particles and splicing intermediates, which contain multiple RNAs and numerous RNA-binding proteins. Furthermore, the protocol utilizes adapters specifically optimized for the CRAC procedure, greatly reducing the appearance of adapter multimers in sequencing data. CRAC has been successfully used for the identification of RNA-binding sites for several NTP utilizing enzymes from yeast, including RNA helicases (Bohnsack et al., 2009), protein kinases, and putative GTPases (Granneman et al., 2010).

2. MATERIAL

Stratalinker 1800 (Agilent Technologies), "Megatron" (UVO3) or iTRIC (*in culturo* temperature regulated interaction cross-linker)
NanoDrop (Spectrophotometer; Thermo) or Qubit (Invitrogen)
Micro Bio-Spin columns (Biorad)
Snap Cap columns (Pierce)
Zirkonia beads (0.5 mm, Thistle Scientific)
IgG sepharose (Fast Flow; GE Healthcare)
Nickel-NTA agarose (Qiagen)
NuPAGE gels (4–12% polyacrylamide Bis–Tris precast gels; Invitrogen)
Hybond-C Extra (nitrocellulose membrane; GE Healthcare)
Metaphor agarose (Lonza)
MiniElute reaction cleanup kit (Qiagen)
TEV protease (tobacco etch virus protease, GST-tagged; self-made)
RNace-IT (mixture of RNase A and RNase T1; Agilent Technologies)
RNasin (RNase inhibitor; Promega)
SuperScript III (reverse transcriptase; Invitrogen)
TSAP (Thermosensitive alkaline phosphatase; Promega)
T4 RNA ligase 1 (New England Biolabs)
T4 RNA ligase 2 truncated K227Q (New England Biolabs)
T4 polynucleotide kinase (PNK; New England Biolabs)

Proteinase K (Roche)
RNase H (New England Biolabs)
LA Taq Polymerase (Takara Bio Inc.)

2.1. Oligonucleotides

Because RNA helicases are predicted to interact with many different substrates, we routinely perform Illumina high-throughput sequencing to identify target sites. The average length of CRAC cDNAs is in the range of 20–40 nucleotides and therefore 50 or 100 bp single-end sequencing is usually sufficient to identify cDNA 3' ends. Based on our experiences with yeast RNA helicases, several million reads usually provide enough coverage to unambiguously identify RNA interactions sites. For more complex genomes (i.e., mouse and human), we would expect that a higher coverage is required. At the time of writing, the Illumina HiSeq-2000 machine could output well over 100 million reads per lane in a flow cell. Hence, barcoded adapters are now routinely used, allowing multiplexing of CRAC cDNA libraries in a single lane. The 5' DNA/RNA hybrid adapter sequences listed below allow multiplexing of eight samples and can be used for both single- and paired end Illumina Solexa sequencing (Sander Granneman and David Tollervey, unpublished). To control for PCR amplification biases/artifacts, random nucleotides (indicated with "N") were included in each 5' linker. All oligonucleotides were (custom) synthesized by Integrated DNA Technologies (IDT).

"r" indicates RNA ribonucleosides.

L5Aa 5'-invddT-ACACrGrArCrGrCrUrCrUrUrCrGrArUrCrUrNrNr NrUrArArGrC-OH-3'
L5Ab 5'-invddT-ACACrGrArCrGrCrUrCrUrUrCrGrArUrCrUrNr NrNrArUrUrArGrC-OH-3'
L5Ac 5'-invddT-ACACrGrArCrGrCrUrCrUrUrCrCrGrArUrCrUrNrNr NrGrCrGrCrArGrC-OH-3'
L5Ad 5'-invddT-ACACrGrArCrGrCrUrCrUrUrCrGrArUrCrUrNrNr NrCrGrCrUrUrArGrC-OH-3'
BARCODE SET B
L5Ba 5'-invddT-ACACrGrArCrGrCrUrCrUrUrCrCrGrArUrCrUrNr NrNrArGrArGrC-OH-3'
L5Bb 5'-invddT-ACACrGrArCrGrCrUrCrUrUrCrGrArUrCrUrNr NrNrGrUrGrArGrC-OH-3'
L5Bc 5'-invddT-ACACrGrArCrGrCrUrCrUrUrCrGrArUrCrUrNrNr NrCrArCrUrArGrC-OH-3'
L5Bd 5'-invddT-ACACrGrArCrGrCrUrCrUrUrCrGrArUrCrUrNr NrNrUrCrUrCrUrArGrC-OH-3'
3' linker 5'-AppAGATCGGAAGAGCGGTTCAG/ddC/-3'

Reverse transcription (RT) primer 5′-GCTGAACCGCTCTTCCGAT-3′
Forward PCR primer 5′-AATGATACGGCGACCACCGAGATCTA
CACTCTTTCCCTACACGACGCTCTTCCGATCT
Reverse PCR primer 5′-CAAGCAGAAGACGGCATACGAGATC
GGTCTCGGCATTCCTGCTGAACCGCTCTTCCGATCT-3′

2.2. Buffers and solutions

TN150: 50 mM Tris/HCl (pH 7.6), 150 mM NaCl, 0.1% NP-40, 5 mM β-mercaptoethanol
TN1000: 50 mM Tris/HCl (pH 7.6), 1 M NaCl, 0.1% NP-40, 5 mM β-mercaptoethanol
Wash I: 50 mM Tris/HCl (pH 7.6), 300 mM NaCl, 10 mM imidazole, 6 M guanidinium-HCl, 0.1% NP-40, 5 mM β-mercaptoethanol
Wash II: 50 mM Tris/HCl (pH 7.6), 50 mM NaCl, 10 mM imidazole, 0.1% NP-40, 5 mM β-mercaptoethanol
Elution buffer: 50 mM Tris/HCl (pH 7.6), 50 mM NaCl, 200 mM imidazole, 0.1% NP-40, 5 mM β-mercaptoethanol
PNK buffer: 50 mM Tris/HCl (pH 7.6), 10 mM MgCl$_2$, 0.5% NP-40, 10 mM β-mercaptoethanol
Proteinase K buffer: 50 mM Tris/HCl (pH 7.6), 50 mM NaCl, 1% SDS, 5 mM EDTA, 0.1% NP-40, 5 mM β-mercaptoethanol.

3. METHODS

3.1. Yeast strains, cross-linking and affinity purification of RNPs

3.1.1. Choices of tags for protein fusions

The CRAC method crucially relies on the presence of a His6 tag on the protein of interest. This tag has two key functions in the protocol: (i) it allows the second round of purification to be performed under denaturing conditions and (ii) it permits the multiple steps leading to ligation of the 5′ and 3′ linkers to be performed on complexes that are immobilized on a nickel column. This greatly enhances the speed and recovery relative to performing the same steps in solution, with the attendant extraction, precipitation, and washing steps. However, the His6 tag alone allows only limited enrichment, and it was therefore combined with a cleavage site for TEV protease and with protein A (HTP tag)—as used in conventional yeast TAP protocols. In principle, these features could be exchanged for other affinity tags (e.g., Flag) and/or alternative cleavage sites (e.g., PreScission).

C-terminal tag fusions with the protein of interest would normally be generated by genomic integration (Longtine et al., 1998) under control of the endogenous promoter. This allows stable expression at levels that are expected to be close to the endogenous levels (Wlotzka et al., 2011). We have also successfully performed CRAC on proteins carrying N-terminal tags (in which the modules of the tag are reversed; His6-TEV-ProtA) in the case of proteins that do not readily accept C-terminal tags. Overall, we anticipate that proteins that can be expressed as functional TAP fusions will also function as HTP fusions. We have also successfully expressed tagged RNA helicases from plasmids under the control of regulated promoters (our unpublished results). This approach would be suitable for inducible expression of tagged RNA helicase mutants.

3.1.2. Choices of growth and cross-linking conditions

Three different basic cross-linking conditions have been used: (1) Cross-linking of isolated, partially purified complexes *in vitro* using a Stratalinker. (2) Cross-linking of intact, but nongrowing cells using a Stratalinker. (3) Cross-linking of actively growing cells, using custom-made, cross-linking devices.

Initial analyses were performed *in vitro* after purification of RNP complexes on IgG sepharose and elution by TEV protease cleavage (Bohnsack et al., 2009; Granneman et al., 2009). This yielded good cross-linking efficiencies, presumably because the protein of interest is highly enriched and can be exposed to the UV source in minimal buffer volumes. A disadvantage is that transient protein–RNA interactions are likely to be lost during the extended incubation steps required for the purification procedure. This approach will therefore bias the data toward high-affinity helicase–RNA interactions (Bohnsack et al., 2009; Granneman et al., 2009). It is, however, notable that most previous analyses of RNP complex composition—for example, by TAP purification and mass spectrometry—have similar caveats, so CRAC data generated in this way may be easier to integrate with prior knowledge in the field.

Cross-linking in living cells (Granneman et al., 2009) was initially performed by pelleting the cells and resuspending them in phosphate buffered saline, which was then spread on a Petri dish prior to cross-linking in a Stratalinker. The disadvantage of this procedure is that the cells undergo rapid cooling followed by nutrient deprivation, which will potentially change the protein–RNA interactome. In practice, this approach appears to work very well for ribosome synthesis factors but has not given good results for the small number of splicing factors tested to date. We speculate that the cold-shock leads to arrest of the ribosome synthesis machinery, capturing many of the interactions involved. In contrast, other pathways, including splicing, may not be fully arrested, potentially allowing the dissociation of preformed complexes but not the formation of new complexes.

To circumvent these problems, we developed systems for UV irradiating actively growing cell cultures in growth medium. This can increase the diversity of RNAs that are being cross-linked and allow identification of low abundant RNA substrates and transient RNA helicase–RNA interactions (Granneman et al., 2011, unpublished observations). As an example, cross-linking of growing cultures allowed the identification of RNA-binding sites for the splicing RNA helicase Brr2, which had not been detected by cross-linking in the Stratalinker (Daniela Hahn and Jean Beggs, manuscript in preparation).

Cross-linking of yeast in culture medium requires specialist equipment, due to the large volumes typically used. We have successfully used two devices; "The Megatron" (UVO3) is a modified water-sterilization unit that contains a 205 W UV-C lamp surrounded by a quartz sleeve within a tube that contains the growth medium. In the iTRIC, the UV-C lamps are arranged above the culture and the temperature of which can be closely monitored and regulated. In principle, cross-linking in culture could be performed using a standard Stratalinker, but the limited volumes that could be handled in this way would reduce the coverage and complexity of the resulting sequence data. This might not be a problem if only a limited number of binding sites are anticipated.

3.2. UV cross-linking

3.2.1. Cross-linking *in vitro*

For *in vitro* cross-linking experiments, it is important to maintain the integrity of the protein–RNA complex of interest during the IgG binding, wash, and TEV elution steps. For preribosome *in vitro* CRAC experiments, we add 1.5 mM MgCl$_2$ to the lysis and TN150 buffers to prevent dissociation of the complexes. Extracts are prepared from 0.5 to 1 l culture of OD$_{600}$ and 0.5–0.8 for each sample. Cells are pelleted and frozen in liquid nitrogen for storage. To lyse the cells, pellets are resuspended in 1.5 ml lysis buffer per gram of cells and vortexed with 3 ml Zirkonia beads for 5 min at 4 °C in the 50 ml falcon tubes. After addition of a further 3 ml of lysis buffer per gram of cells, cell debris is pelleted at 4 °C for 20 min at 4500 × g, and 20 min at 20,000 × g in Eppendorf tubes. Tagged proteins are captured from the supernatant by incubation with 200 μl of IgG sepharose beads for at least 30 min at 4 °C. Shorter incubation times can reduce nonspecific binding to the IgG beads. The beads are subsequently washed three times with 10 ml lysis buffer. Beads are resuspended in 500 μl of lysis buffer and incubated with GST-TEV protease for 2 h at 18 °C in a shaking incubator. Note that commercial preparations of TEV protease may carry His6 tags and these are not suitable for this analysis, since TEV-His6 will be retained on the

Ni-column increasing the background. Eluates are collected by spinning the bead suspension through a Micro Bio-Spin column for a few seconds at $1000 \times g$ in a micro centrifuge. The elutes are transferred to a Petri dish on a bed of ice and placed in a Stratalinker, at the same height as the UV sensor, approximately 10 cm from the light source. The samples are irradiated at 254 nm to a dose of $\sim 0.4\,\mathrm{J\,cm^{-2}}$, as determined by the internal UV monitor.

3.2.2. UV cross-linking resuspended cells
For UV cross-linking resuspended cells in a Stratalinker, cells are grown in minimal medium to an $OD_{600} \leq 0.5$ or in complete medium to $OD_{600} \leq 0.8$ and harvested by centrifugation. Cells are resuspended in ice-cold phosphate buffer saline (PBS), transferred to Petri dishes on ice, and irradiated in a Stratalinker as above at 254 nm to a dose of $\sim 1.6\,\mathrm{J\,cm^{-2}}$, with periodic shaking of the dish to help ensure even irradiation. Cells are pelleted, lysed, and TEV eluates prepared as above (Section 3.2.1).

3.2.3. UV cross-linking growing cells in culture
For UV cross-linking of growing cells, minimal medium is used (generally lacking the aromatic amino acids Tryptophan and Phenylalanine), and cells are grown to an $OD_{600} \sim 0.5$. The culture is transferred to the UV-irradiation chamber preincubated at the culture growth temperature and immediately irradiated at a dose of $1.6\,\mathrm{J\,cm^{-2}}$ (corresponding to roughly 80 s exposure in the Megatron or 5–20 min in the iTRIC), harvested, washed with ice-cold PBS, and frozen in liquid nitrogen in 50 ml falcon tubes in 1 g batches and stored at $-80\,^{\circ}\mathrm{C}$. Cell pellets are lysed and TEV elutes prepared as above (Section 3.2.1).

3.3. Partial RNase digestion and denaturing Nickel purification

Following *in vivo* or *in vitro* cross-linking, the cleared TEV eluates are incubated with RNace-IT (RNase A and T1) for 5 min at 37 °C; however, the RNace-IT concentrations need to be optimized for each protein. When performing CRAC on an RNA helicase for the first time, incubation with 1 U would be an appropriate starting point. The RNase digestion is stopped by addition of guanidinium-HCl (Sigma) powder to the reaction to a final concentration of 6 M.

After addition of NaCl to a final concentration of 300 mM and of imidazole to 10 mM, the tagged proteins are captured on Ni-NTA agarose at 4 °C for at least 3 h, but we usually incubate the beads overnight on a rotating wheel. The matrix is washed twice with 500 μl of the denaturing buffer Wash I, followed by three 500 μl washes with PNK buffer. Proteins can be eluted using 400 µl Elution buffer, or the matrix can be used in further steps for on bead reactions in Pierce Snap Cap columns.

3.4. Adapter ligation and radioactive labeling

After transferring the nickel beads to a Snap Cap column, the beads are washed twice with Wash I and three times with PNK buffer (500 µl of buffer per wash). Because RNAse A digestion releases RNAs with 3′ and 2′ phosphates, the beads are incubated in 80 µl PNK buffer containing 8 U TSAP and RNasin at 37 °C for removal of terminal phosphates before 3′ adapter ligation. To terminate the reaction, the matrix is washed with 500 µl of Wash I, followed by three 500 µl washes with PNK buffer.

For the 3′ adapter ligations, we use 5′ adenylated (App), 3′ blocked (ddC) DNA oligonucleotides that can be efficiently ligated by T4 RNA ligase in the absence of ATP. To minimize unwanted side reactions, ligation reactions are performed with a mutant T4 RNA ligase 2 (K227Q; NEB) that can only ligate adenylated adapters to RNA substrates. Ligation of the 3′ adapter (80 pmol) is performed on bead in the Snap Cap columns in 80 µl PNK buffer containing 20% PEG 4000, 80 U RNasin, 40 U T4 RNA ligase 2 (K227Q) for at least 5 h at 25 °C. The matrix is washed once with 500 µl Wash I, followed by three 500 µl washes with PNK buffer.

Cross-linked RNA is 5′ radiolabeled using 40 µCi γ^{32}P-ATP 80 U of RNasin and 20 U T4 PNK in 80 µl PNK buffer for 40 min at 37 °C. Subsequently, cold ATP is added to a final concentration of 1 mM and incubation is continued for additional 20 min to achieve almost quantitative phosphorylation prior to ligation of the 5′ adapter. The matrix is washed with 500 µl of Wash I, followed by three 500 µl washes with PNK buffer.

The 5′ adapter ligation is performed in 80 µl PNK buffer containing 80 U of RNasin, 40 U T4 RNA ligase 1, 100 pmol 5′-adapter and 1 mM ATP at 16 °C over night (at least 5 h). To terminate the ligation reaction, the matrix is washed with 500 µl of Wash I, followed by three 500 µl washes with PNK buffer.

3.5. SDS-PAGE and RNA isolation

To elute radiolabeled protein–RNA complexes, beads are incubated with 400 µl Elution buffer followed by TCA precipitation (final concentration 20%) for 20 min on ice. Cross-linked proteins are pelleted by centrifugation for 30 min at 20,000 $\times g$ at 4 °C, washed once with 800 µl 100% acetone and air dried in a hood. The pellet should not be over-dried, as it will become difficult to dissolve.

Pellets are resuspended in 1 × NuPAGE protein sample buffer, heated at 65°C for 10 min, resolved on a 4–12% NuPAGE gel and transferred onto a Hybond-C nitrocellulose membrane. Radioactive cross-linked proteins are detected by autoradiography and corresponding regions cut from the membrane. Selected regions include the bands corresponding to the protein of interest and the region directly above, which should contain cross-links to RNA fragments of different length. If the protein has a low molecular weight, to avoid enriching for very short RNAs, it is best to extract RNA only from membrane slices just above the main band.

RNA is isolated by proteinase K (100 μg) treating the nitrocellulose fragments in 400 μl Proteinase K buffer for at least 2 h at 55 °C followed by phenol/chloroform, chloroform extraction, and ethanol precipitation. We normally use 20 μg of glycogen as carrier. Higher concentrations of glycogen may inhibit the RT reaction.

3.6. Reverse transcription

The SuperScript III system is used for RT, as the enzyme tolerates incubation at 55 °C, which reduces secondary structures in the RNA, thereby increasing the RT efficiency. RT is basically performed according to the manufacturer's protocol. In short, the RNA is resuspended in 13 μl Mix1 containing 0.8 μM RT primer and 0.8 μM dNTPs and denatured by heating for 3 min at 80 °C followed by rapid cooling on ice.

Six microliters of Mix2 containing 4 μl first-strand buffer, 1 μl DTT solution (100 mM), and RNasin is added and the mixture is preincubated for 3 min at 55 °C before addition of 200 U SuperScript III and incubation for 1 h at 55 °C. After 15 min at 65 °C, 2 μl RNase H is added and the mixture incubated for further 30 min at 37 °C.

3.7. PCR and size selection of PCR products

For PCR, LA Taq polymerase (Lonza) gave the highest yield of PCR product of all polymerases tested. However, because LA Taq makes more mistakes compared to most proofreading DNA polymerases, a higher-fidelity DNA polymerase might best be tested initially (e.g., PFU (Promega) or Phusion (NEB)). A standard LA Taq PCR reaction is performed under the following conditions: 1 × LA Taq buffer, 1 μl cDNA, 0.2 μM each PCR primer, 0.25 mM dNTPs, 2.5 U LA Taq. The PCR program is: 2 min 95 °C; 25–30 cycles of 20 s each at 98, 52, and 68 °C; 5 min at 72 °C.

PCR products are separated on 2–3% Metaphor agarose gels, stained with Ethidium bromide and PCR products with insert sizes between 15 and 100 nucleotides are excised.

PCR products are gel purified using the Qiagen MiniElute reaction cleanup kit as described by the manufacturer. To minimize salt contamination in the eluted DNA, it is critical to incubate the column with wash buffer for at least 5 min at room temperature. We would recommend washing the column at least twice. DNA is eluted in 20 µl of water and DNA concentration is measured by NanoDrop or the Qubit system (Invitrogen). From our experiences, the NanoDrop has a tendency of overestimating the DNA concentration. If available, we would recommend measuring the DNA concentration using a fluorescent quantification method (like the Qubit) or analyzing the samples on a Bioanalyser (or equivalent) machine.

3.8. Sequencing

Originally, PCR products from test experiments and many CRAC samples were cloned using a TOPO TA cloning kit (Invitrogen) and sequenced by conventional Sanger sequencing in 96-well format. The requirement for statistical analysis and genome coverage combined with the increased availability of deep sequencing methodologies lead to the use of the Solexa (Illumina) sequencing technology. The use of Solexa-compatible linker and primer sequences (Section 2.1) allows direct submission of PCR products for Solexa deep sequencing, which can be performed by several companies.

3.9. Data analysis

Depending on the type of machine, Illumina Solexa deep sequencing usually results in 5–180 million reads per sample. The high number of sequences requires automated software pipelines for the analysis of the sequencing output. Most labs have written their own pipeline for sequence mapping and statistical analysis, for example, using Java, R Bioconductor or Python tools, like pyCRAC software suite (Webb, Tollervey, and Granneman, in preparation), and integrated free or commercially available programs, such as Novoalign (see, e.g., Granneman *et al.*, 2009, 2010, 2011; Wlotzka *et al.*, 2011). The pyCRAC suite of tools was designed specifically to tackle CLIP and CRAC datasets in a user-friendly way and the tools can also be used on a Galaxy server, allowing data analysis in web browser. Details on the bioinformatics will be described elsewhere (Simm, Bohnsack, and Schleiff, manuscript in preparation; Webb, Tollervey and Granneman, in preparation).

CRAC data not only reveal RNA–protein binding but the technology can also be adapted to study RNA–RNA interactions. A newly developed variant of the method called *c*ross-linking, *l*igation, *a*nd *s*equencing of *h*ybrids identified several novel *in vivo* RNA–RNA interactions (Kudla *et al.*, 2011), and this protocol could conceivably be adapted to identify interactions between RNA helicases and duplex RNA strands *in vivo*.

4. PROTOCOL ADAPTATION AND TROUBLE-SHOOTING

There can be multiple reasons why individual RNA-binding proteins might not cross-link efficiently to their cognate RNAs, or why no bands might be visible by the autoradiography of the protein eluted from the Ni-column. UV is reported to preferentially cross-link aromatic amino acids to pyrimidine nucleotides located in single-stranded regions and may therefore not work with proteins that bind double-stranded RNA. Proteins may also bind very transiently such that only a low percentage of molecules are bound to their targets at steady state. Multiple technical reasons could also impair the protein purification steps. It is therefore essential to run both negative (untagged strain to identify possible background bands) and positive controls (strains that have previously been used successfully) in each experiment. In cases where RNA-binding proteins fail to give usable cross-linking efficiencies, it may be useful to incorporate 4-thiouracil (in yeast) or 4-thiouridine (in mammalian cells) as in PAR-CLIP (Hafner et al., 2010) to enhance cross-linking.

It is critical to keep the pH of the buffers between 7 and 8. A pH below 7 could result in poor binding of His6-tagged proteins to nickel resin. A pH above 8 could lead to RNA degradation during the enzymatic reactions. Guanidine has the tendency to increase the pH of the buffer and therefore it is important to make this buffer fresh every time and confirm the pH.

Western blotting can be used to confirm that the protein is successfully tagged. In addition, we routinely analyze 5% of the TEV eluates by Western blot for successful enrichment on IgG sepharose and presence of the His6 tag.

If there are problems with genomic tagging of the protein of interest (e.g., if the C-terminus cannot be tagged), a depletion strain can be generated and the tagged protein expressed from a plasmid (Markus Bohnsack, unpublished). In this case, an appropriate promoter should be selected and expression levels controlled to avoid overexpression, which may lead to identification of false target RNAs.

If only weak cross-linking can be detected, an increase in the amount of cellular material can help, especially for less abundant proteins. We have discovered that the nickel purification works best when the protein is significantly enriched in TEV eluates. If the IgG step does not work efficiently, it might be worth considering prefractionating extracts prior to the tandem purification procedure. Most RNA helicases that we have expressed in bacteria efficiently bind to cation exchange columns. In addition, ultracentrifugation steps could be introduced after cell lysis to pellet ribosomes, greatly reducing rRNA background. This approach greatly improved the signal to noise ratio in the Brr2 CRAC data. It is also important to check the lysis efficiency. Most proteins that are closely associated with chromatin are not efficiently extracted under the standard CRAC lysis conditions described above. For these cases, we would recommend lysing cells in buffers containing 500 mM—1 M NaCl (the IgG–protein A interaction is very salt-tolerant). Different cross-linking times (Section 3.1) can be tested to optimize conditions for individual factors. We would, however, recommend using the lowest UV dose that gives good cross-linking as excessive irradiation can lead to RNA strand breaks and U–U cross-links, which will result in a large number of deletions in T-rich regions in the reads.

Many RNA helicases and other RNA-binding enzymes are thought to interact very transiently with their target RNAs. In such cases, cross-linking may be enhanced by the use of mutants that are less readily released from target RNAs, for example, in motif VI of RNA helicases.

An important parameter tht needs to be optimized for different factors is the partial RNA digest (Section 3.3). Both time and concentration of the RNase can be altered. Especially for proteins that protect only a very short stretch of sequence, the identification of target sites can otherwise be difficult. The generation of too short fragment sizes has been the most common cause of failure in CRAC analyses for proteins that otherwise give good cross-linking and recovery efficiencies.

ACKNOWLEDGMENTS

This work was funded by the Deutsche Forschungsgemeinschaft (M. T. B.) and the Wellcome Trust (S. G. and D. T.).

REFERENCES

Bohnsack, M. T., Martin, R., Granneman, S., Ruprecht, M., Schleiff, E., and Tollervey, D. (2009). Prp43 bound at different sites on the pre-rRNA performs distinct functions in ribosome synthesis. *Mol. Cell* **36,** 583–592.

Ellington, A. D., and Szostak, J. W. (1990). In vitro selection of RNA molecules that bind specific ligands. *Nature* **346,** 818–822.

Granneman, S., Kudla, G., Petfalski, E., and Tollervey, D. (2009). Identification of protein binding sites on U3 snoRNA and pre-rRNA by UV cross-linking and high-throughput analysis of cDNAs. *Proc. Natl. Acad. Sci. U. S. A.* **106,** 9613–9618.

Granneman, S., Petfalski, E., Swiatkowska, A., and Tollervey, D. (2010). Cracking pre-40S ribosomal subunit structure by systematic analyses of RNA-protein cross-linking. *EMBO J.* **29,** 2026–2036.

Granneman, S., Petfalski, E., and Tollervey, D. (2011). A cluster of ribosome synthesis factors regulate pre-rRNA folding and 5.8S rRNA maturation by the Rat1 exonuclease. *EMBO J.* **30,** 4006–4019.

Hafner, M., Landthaler, M., Burger, L., Khorshid, M., Hausser, J., Berninger, P., Rothballer, A., Ascano, M., Jr., Jungkamp, A. C., Munschauer, M., Ulrich, A., Wardle, G. S., *et al.* (2010).

Transcriptome-wide identification of RNA-binding protein and microRNA target sites by PAR-CLIP. *Cell* **141,** 129–141.

Konig, J., Zarnack, K., Rot, G., Curk, T., Kayikci, M., Zupan, B., Turner, D. J., Luscombe, N. M., and Ule, J. (2010). iCLIP reveals the function of hnRNP particles in splicing at individual nucleotide resolution. *Nat. Struct. Mol. Biol.* **17,** 909–915.

Kudla, G., Granneman, S., Hahn, D., Beggs, J. D., and Tollervey, D. (2011). Cross-linking, ligation, and sequencing of hybrids reveals RNA-RNA interactions in yeast. *Proc. Natl. Acad. Sci. U. S. A.* **108,** 10010–10015.

Lebedeva, S., Jens, M., Theil, K., Schwanhausser, B., Selbach, M., Landthaler, M., and Rajewsky, N. (2011). Transcriptome-wide analysis of regulatory interactions of the RNA-binding protein HuR. *Mol. Cell* **43,** 340–352.

Licatalosi, D. D., Mele, A., Fak, J. J., Ule, J., Kayikci, M., Chi, S. W., Clark, T. A., Schweitzer, A. C., Blume, A. C., Wang, X., Darnell, J. C., and Darnell, R. B. (2008). HITS-CLIP yields genome-wide insights into brain alternative RNA processing. *Nature* **456,** 464–469.

Linder, P., and Jankowsky, E. (2011). From unwinding to clamping—The DEAD-box RNA helicase family. *Nat. Rev. Mol. Cell Biol.* **12,** 505–516.

Longtine, M. S., McKenzie, A., Demarini, D. J., Shah, N. G., Wach, A., Brachat, A., Philippsen, P., and Pringle, J. R. (1998). Additional modules for versatile and economical PCR-based gene deletion and modification in Saccharomyces cerevisiae. *Yeast* **14,** 953–961.

SenGupta, D. J., Zhang, B., Kraemer, B., Pochart, P., Fields, S., and Wickens, M. (1996). A three-hybrid system to detect RNA-protein interactions in vivo. *Proc. Natl. Acad. Sci. U. S. A.* **93,** 8496–8501.

Tuerk, C., and Gold, L. (1990). Science, Systematic evolution of ligands by exponential enrichment: RNA ligands to bacteriophage T4 DNA polymerase. *Science* **249,** 505–510.

Ule, J., Jensen, K. B., Ruggiu, M., Mele, A., Ule, A., and Darnell, R. B. (2003). CLIP identifies Nova-regulated RNA networks in the brain. *Science* **302,** 1212–1215.

Wang, Z., Kayikci, M., Briese, M., Zarnack, K., Luscombe, N. M., Rot, G., Zupan, B., Curk, T., and Ule, J. (2010). iCLIP predicts the dual splicing effects of TIA-RNA interactions. *PLoS Biol.* **8,** e1000530.

Wlotzka, W., Kudla, G., Granneman, S., and Tollervey, D. (2011). The nuclear RNA polymerase II surveillance system targets polymerase III transcripts. *EMBO J.* **30,** 1790–1803.

CHAPTER FOURTEEN

In Vivo Approaches to Dissecting the Function of RNA Helicases in Eukaryotic Ribosome Assembly

David C. Rawling* *and* Susan J. Baserga*,†,‡

Contents

1. Introduction — 290
2. Experimental Strategies Used to Evaluate RB Helicases — 291
3. Determining Where an RNA Helicase Acts in the RB Pathway — 296
    3.1. Evaluating RB by monitoring rRNA production — 297
    3.2. Perturbation of helicase activity for RB studies — 304
4. Elucidating the Supermolecular Context of an RB Helicase: Protein–Protein Interaction Studies — 308
    4.1. Tandem affinity purification — 309
    4.2. Co-IP of tagged proteins — 313
    4.3. Yeast two-hybrid analysis — 314
References — 316

Abstract

In eukaryotes, ribosome biogenesis involves the nucleolar transcription and processing of pre-ribosomal RNA molecules (pre-rRNA) in a complex pathway requiring the participation of myriad protein and ribonucleoprotein factors. Through efforts aimed at categorizing and characterizing these factors, at least 20 RNA helicases have been shown to interact with or participate in the activities of the major ribosome biogenesis complexes. Unfortunately, little is known about the enzymatic properties of most of these helicases, and less is known about their roles in ribosome biogenesis and pre-rRNA maturation. This chapter presents approaches for characterizing RNA helicases involved in ribosome biogenesis. Included are methods for depletion of specific protein targets, with standard protocols for assaying the typical ribosome biogenesis defects that may result. Procedures and rationales for mutagenic studies of

* Department of Molecular Biophysics and Biochemistry, Yale University School of Medicine, New Haven, Connecticut, USA
† Department of Genetics, Yale University School of Medicine, New Haven, Connecticut, USA
‡ Department of Therapeutic Radiology, Yale University School of Medicine, New Haven, Connecticut, USA

target proteins are discussed, as well as several approaches for identifying protein–protein interactions in order to determine functional context and potential cofactors of RNA helicases.

1. Introduction

RNA helicases comprise a well-conserved class of NTP-dependent nucleic acid remodeling enzymes found throughout all kingdoms of life. In eukaryotes, these proteins have been shown to act in nearly every cellular processes involving RNA (Jankowsky, 2011; Tanner and Linder, 2001). RNA helicases use NTPs to bind to and act on RNA or ribonucleoprotein (RNP) substrates (Liu et al., 2008; Pyle, 2008). Although these enzymes tend to exhibit little or no substrate specificity *in vitro*, they demonstrate a high degree of specificity in the cell, and frequently play essential roles in important biological processes, including ribosome biogenesis (RB; Bernstein et al., 2006; Blum et al., 1992; Granneman et al., 2006a; Jankowsky, 2011; Wahl et al., 2009).

In eukaryotes, RB begins with the production of a large precursor transcript (pre-rRNA), which contains sequences destined for incorporation into both the small and large subunits (SSU and LSU) of the mature ribosome. This transcript is cleaved at several sites in order to separate the individual rRNA sequences, generating three fragments which are then further trimmed until only the rRNA sequences remain (Fig. 14.1; Granneman and Baserga, 2004; Henras et al., 2008; Kressler et al., 2010). The rRNA sequences also undergo a number of chemical modifications, mostly consisting of pseudouridylation and 2′-O-ribose methylation (Decatur and Fournier, 2002). Concomitant with the enzymatic modifications that take place on the nascent transcript, the individual rRNA sequences begin folding as they interact with a subset of ribosomal proteins, before both the pre-SSU and pre-LSU are exported from the nucleus for final maturation in the cytoplasm.

To date, at least 20 RNA helicases have been implicated as eukaryotic RB factors (Bleichert and Baserga, 2007; Jankowsky, 2011; Kressler et al., 2010). Given the size and complexity of rRNA and the plethora of proteins and RNPs involved in its maturation, it is easy to envision a variety of potential roles for RNA helicases in the RB process. For instance, a number of the cleavage and chemical modification events that take place in the nucleolus require hybridization of the pre-rRNA to the snoRNA factors that mediate these events, and it has been shown that the depletion of several RNA helicases—Dbp4 (Kos and Tollervey, 2005), Has1 (Liang and Fournier, 2006), and Rok1 (Bohnsack et al., 2008)—results in accumulation of snoRNPs in large pre-rRNA-containing particles. Another RNA helicase, Mtr4, aids in pre-rRNA "trimming" by preparing the transcript for degradation by the nuclear exosome (de la Cruz et al., 1998a,b). Indeed, most RB helicases

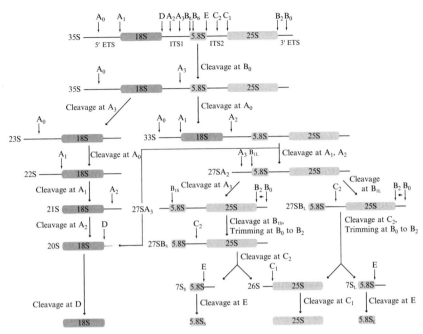

Figure 14.1 Pre-rRNA processing in *Saccharomyces cerevisiae*. Boxed regions of the transcript represent the rRNA sequences. Cleavage sites are designated by vertical arrows below their common designations. Endonucleolytic events are indicated with horizontal arrows pointing in the direction of degradation. The sedimentation designation of each pre-rRNA species is indicated adjacent to its 5′ terminus.

identified thus far have been shown to be essential (Table 14.1; Bernstein *et al.*, 2006; Granneman *et al.*, 2006a), highlighting their importance in this fundamental system. However, despite continued efforts aimed at better understanding these enzymes, a great deal remains to be learned about their specific roles in the ribosome maturation process, as well as their individual enzymatic behaviors. The following chapter presents methods for evaluating the role(s) played by RNA helicases in the pre-rRNA maturation process *in vivo*. Additionally, methods for assaying protein–protein interactions are discussed as a means of identifying potential cofactors for these enzymes and determining where a target enzyme may be acting in the context of the RB machinery.

2. Experimental Strategies Used to Evaluate RB Helicases

In determining whether an RNA helicase plays a role in RB, a first step should be to determine whether this protein localizes to the nucleolus. Nucleolar localization is a strong indicator that a target enzyme may be involved in the ribosome maturation process, and because nucleolar

Table 14.1 RNA helicases involved in ribosome biogenesis

Gene designation (S. cerevisiae)	Essential?	Helicase family	Associated pre-rRNA processing defects	References	Human homolog
Small subunit (SSU)					
Dhr1	Y	DEAH	Deficient in cleavage at A_0, A_1, and A_2	Colley et al. (2000), Granneman et al. (2006a)	DHX37
Dhr2	Y	DEAH	Deficient in cleavage at A_0, A_1, and A_2	Colley et al. (2000), Granneman et al. (2006a)	DHX32
Dbp4	Y	DEAD	Deficient in cleavage at A2, likely due to non-productive accumulation of U14 snoRNP on pre-rRNA transcript	Kos and Tollervey (2005)	DDX10
Dbp8	Y	DEAD	Deficient in cleavage at A_0, A_1, and A_2	Daugeron and Linder (2001), Granneman et al. (2006a)	DDX49
Fal1	Y	DEAD	Deficient in cleavage at A_0, A_1, and A_2	Kressler et al. (1997), Granneman et al. (2006a)	DDX48
Rok1	Y	DEAD	Deficient in cleavage at A_0, A_1, and A_2, possibly due to non-productive accumulation of snR30 on pre-rRNA transcript	Venema et al. (1997), Bohnsack et al. (2008)	DDX52
Rrp3	Y	DEAD	Deficient in cleavage at A_1 and A_2	O'Day et al. (1996), Granneman et al. (2006a)	DDX47
Large subunit (LSU)					
Dbp3	N	DEAD	Deficient in cleavage at A_3	Weaver et al. (1997)	N/A

Table 14.1 (Continued)

Gene designation (S. cerevisiae)	Essential?	Helicase family	Associated pre-rRNA processing defects	References	Human homolog
Dbp6	Y	DEAD	Reduction in 27S and 7S pre-rRNA levels	Kressler et al. (1998)	DDX51
Dbp7	N	DEAD	Reduction in 27S and 7S pre-rRNA levels	Daugeron and Linder (1998)	DDX31
Dbp9	Y	DEAD	Reduction in 27S and 7S pre-rRNA levels	Daugeron et al. (2001)	DDX56
Dbp10	Y	DEAD	Deficient in processing 27SB intermediates	Burger et al. (2000)	DDX54
Drs1	Y	DEAD	Reduction in mature 25S rRNA levels	Ripmaster et al. (1992), Bernstein et al. (2006)	DDX27
Mak5	Y	DEAD	Deficient in cleavage at A_0, A_1, and A_2	Bernstein et al. (2006)	DDX24
Spb4	Y	DEAD	Deficient in cleavage at A_0, A_1, A_2, C_1, and C_2	de la Cruz et al. (1998a,b), Bernstein et al. (2006)	DDX55
Mtr4	Y	Ski2-like	7S pre-rRNA accumulation; deficient in cleavage at A_0, A_1, A_2, and C_2	de la Cruz et al. (1998), Bernstein et al. (2006)	SKIL2
Both subunits					
Has1	Y	DEAH	Deficient in cleavage at A_0, A_1, and A_2; Deficient in processing 27SA$_3$ and 27SB intermediates; Accumulation of snoRNAs in pre-rRNA particles	Emery et al. (2004), Liang and Fournier (2006)	DDX18
Prp43	Y	DEAH	Deficient in processing 35S, 27S, and 20S intermediates	Combs et al. (2006), Leeds et al. (2006)	DHX15

localization has already been demonstrated for a number of putative RNA helicases in both yeast and mammalian systems, it may be possible to identify potential targets by querying a public database. In yeast, protein localization has been investigated for much of the proteome (Huh *et al.*, 2003; Kumar *et al.*, 2002; Ross-Macdonald *et al.*, 1999), and localization information for helicases identified in *Saccharomyces cerevisiae* can be obtained through the *Saccharomyces* Genome Database at (http://www.yeastgenome.org). Several studies have also identified a number of proteins present in the human nucleolus (Anderson *et al.*, 2002, 2005; Scherl *et al.*, 2002), including many RNA helicases that exhibit homology to those implicated in yeast RB. The results of these studies have been compiled by Leung *et al.* (2006) and are available at the Nucleolar Protein Database (http://www.lamondlab.com/NOPdb).

Although the presence of a protein in the nucleolus is suggestive of a potential role in RB, it is by no means a definitive indicator. In order to firmly establish an enzyme as acting in the RB pathway, two complementary strategies are frequently employed. The first involves perturbing the target protein through either depletion or mutagenesis and measuring changes in levels of mature rRNA molecules as well as in levels of intermediate precursors that may accumulate due to defects in pre-rRNA processing (Colley *et al.*, 2000; Combs *et al.*, 2006; Daugeron and Linder, 2001; Leeds *et al.*, 2006). In this way, it is possible to determine whether the enzyme in question is required for efficient production of mature rRNA, and whether its activity is required for transcription, early cleavage events, or later events in the maturation of the SSU, the LSU, or both.

A second, orthogonal approach used to definitively place a helicase in the RB pathway involves identifying the protein or RNP factors with which the helicase associates. A number of the proteins involved in the production and processing of pre-rRNA have already been identified in yeast and mammals. Some of these proteins can serve as markers for a RB pathway (LSU or SSU processing), or, in the case of several LSU proteins, for a particular stage of processing the pathway. Association with these factors can be assayed in either an ensemble or binary manner, yielding information about the super molecular context of the protein and the specific binding partner, respectively (Colley *et al.*, 2000). Examining the interaction network of a target helicase will likely provide enough information to determine whether it acts predominantly in LSU or SSU maturation and may even serve to identify potential cofactors for the enzyme. Unlike the first approach, however, studying binding partners will provide relatively little information about the particular pre-rRNA maturation events in which the target helicase participates.

In conjunction, these strategies make it possible to broadly define the role of a target helicase by elucidating the steps of pre-rRNA processing that are perturbed by disruption of the enzyme's activity and by providing a context for the protein among the large, dynamic cellular machinery involved in RB. In addition, binding experiments may identify cofactors that can modulate the catalytic activities of the helicase and/or aid in substrate binding. This information can then be used to formulate testable hypotheses regarding the biochemical properties and mechanistic behavior of the target enzyme.

Work done on the yeast DEAH-box protein Prp43 represents an excellent example of experimental development which can serve as a guide for future studies of putative RB helicases. Prp43 was initially identified as a spliceosome disassembly factor (Arenas and Abelson, 1997), but was later shown to be present in preribosome particles as well (Lebaron et al., 2005). Further work showed that Prp43 was required for efficient processing of both SSU and LSU components of pre-rRNA by demonstrating that several rRNA precursors accumulated when a mutant form of the enzyme was introduced into a cell depleted of the wild-type enzyme (Combs et al., 2006; Leeds et al., 2006). Based on the enrichment of the Prp43 binding partner, Pfa1, observed by Lebaron et al. and published in their 2005 study, this group went on to more thoroughly characterize the biochemical consequences of this interaction, eventually demonstrating that Pfa1 stimulates the ATPase and helicase activities of the Prp43 enzyme (Lebaron et al., 2009).

Most recently, a modified version of the high-throughput sequencing with cross linking assisted immunoprecipitation (HITS-CLIP) (Licatalosi et al., 2008; Ule et al., 2005) protocol was developed and optimized for use in yeast (Granneman et al., 2009). This procedure, termed CRAC for cross-linking and cDNA analysis, was used to identify putative target sites for Prp43 on the pre-rRNA in an effort to determine more precisely how its activity contributes to proper pre-rRNA maturation (Bohnsack et al., 2009; see also Chapter 13).

In this work, we will outline a typical experimental workflow designed to elucidate the biological and biochemical properties of a putative RNA helicase involved in RB (Fig 14.2). A number of the methods employed in these experiments have been thoroughly described elsewhere; in these cases, citations for appropriate protocols will be included; however, the discussion presented will focus on considerations that are particular to working with nucleolar helicases and the RB system. RNA helicases involved in eukaryotic RB have been most thoroughly characterized in budding yeast and humans, therefore the majority of the logical and practical considerations presented in this work will refer to and be derived from methods employed in these systems, with an emphasis on those described for *S. cerevisiae*. Nevertheless, the general reasoning and experimental strategy discussed here will be broadly applicable to eukaryotic systems, though actual protocols may differ significantly depending upon the organism in question.

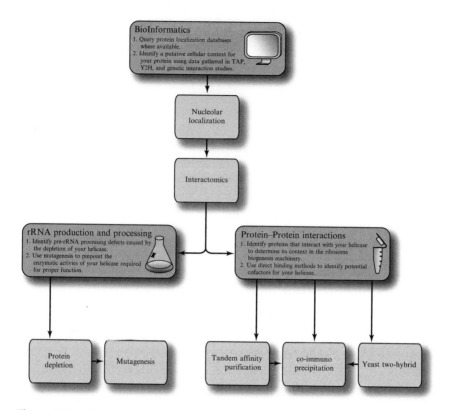

Figure 14.2 Flowchart of an experimental strategy designed to efficiently elucidate the role of a putative helicase protein in eukaryotic ribosome biogenesis. Tandem affinity purification is denoted as TAP. Yeast two-hybrid is denoted as Y2H.

3. Determining Where an RNA Helicase Acts in the RB Pathway

In yeast, RB begins with RNA polymerase I-mediated transcription of a 35S precursor that contains the 18S, 5.8S, and 25S rRNA sequences, in addition to two external transcribed spacer (ETS) and two internal transcribed spacer (ITS) sequences. This transcript undergoes a number of site specific nucleolytic cleavage events that are mediated by RB machinery and result in the stepwise production of several well-defined pre-rRNA processing intermediates (Fig. 14.1; Granneman and Baserga, 2004; Henras et al., 2008; Kressler et al., 2010; Venema and Tollervey, 1999). Normally, these intermediates are relatively short lived and are present at low levels during productive ribosome synthesis. Interfering with the normal activity of RB helicases often leads to diminished activity at one or more of the nucleolytic stages in the ribosome maturation process, which therefore results in the accumulation of one or more pre-rRNA precursors concomitant with a decrease in levels of mature rRNA (Combs et al., 2006; de la Cruz et al., 1998a,b; Emery et al., 2004; Kressler et al., 1997; Lee et al., 1992; Leeds et al., 2006; Ripmaster et al., 1992, and others).

Monitoring cellular levels of mature rRNA and pre-rRNA intermediates has become a standard tool used to evaluate RB defects. Because the process of pre-rRNA trimming has been exceptionally well characterized, it is possible to associate a target helicase with a specific stage in ribosome maturation by interfering with its enzymatic activity and observing any changes in rRNA and pre-rRNA levels (Bernstein et al., 2006; Granneman et al., 2006a). Methods for monitoring cellular RNA levels will be discussed, followed by methods for depleting a target helicase and introducing a mutagenized version into the cell, including important considerations for targeting mutagenesis.

3.1. Evaluating RB by monitoring rRNA production

In yeast, collecting and evaluating rRNA from a strain of interest is a fairly straightforward procedure that involves four basic steps: (i) the selected strain is grown to an optimal optical density, (ii) cells are harvested and total RNA is extracted, (iii) RNA species are separated by gel electrophoresis, and (iv) rRNA levels are quantified using either ethidium bromide staining or northern analysis. Important considerations will be discussed and protocols presented for each of these steps.

3.1.1. Selecting and culturing a yeast strain

In selecting a yeast strain, it is important to consider the availability of various metabolic markers for use in subsequent experiments in which the successful transformation of a plasmid or proper chromosomal insertion will have to be selected. A typical example of such a strain is YPH499 (Sikorski and Hieter, 1989), a mating type a strain auxotrophic for uridine, adenine, tryptophan, histidine, lysine, and leucine (denoted *MATa ura3-52 lys2-80 ade2-101 trp1-Δ63 his3-Δ200 leu2-Δ1*), although a wide variety of backgrounds are acceptable as long as they are compatible with plasmids chosen for use in future experiments. General guidelines for working with yeast are presented below:

1. Prior to culturing, a new strain should be struck out from a frozen stock onto rich media, as well as media lacking one of each of the purported selectable markers and incubated at 30 °C. Cells should grow only on the rich media. This procedure is used to verify the genotype of a new or freshly thawed strain. If cells grow on media lacking a purported marker, the strain may be contaminated or the genotype may be incorrect.

2. From a frozen stock or freshly struck plate, prepare an overnight culture by collecting a small amount of yeast on the tip of a sterile pipette and inoculating 10 ml liquid YPD media under sterile conditions. Incubate the culture on a shaking platform overnight at 30 °C.

   Prepare YPD media by dissolving 20 g/l peptone and 10 g/l Yeast extract in water. Adjust the pH of the solution to 6.5 if necessary and autoclave. After autoclaving, allow the solution to cool to approximately 55 °C. Add glucose to a final concentration of 2% (w/v) from a sterile 20% (w/v) stock solution. Note that in this context, glucose is usually referred to by the antiquated designation dextrose (hence, YPD). This convention has been retained so that media containing galactose, another carbon source used in yeast culture, can be denoted YPG.
3. The following morning, assay the optical density at 600 nm (OD_{600}) of the culture. Note that the culture may need to be diluted in order to achieve a reading in the linear range of the spectrophotometer.
4. Dilute an aliquot of the overnight culture into 25 ml of sterile YPD media in a new flask such that the final $OD_{600} = 0.1$ AU. Note that the doubling time for most strains of *S. cerevisiae* in liquid YPD is approximately 2 h at 30 °C; however it is recommended that doubling time be determined experimentally under the precise conditions in which cultures will be grown.
5. Continue incubating the cells at 30 °C. Allow at least two doubling events to take place, monitoring the optical density occasionally during incubation.
6. Once the cells have reached an OD_{600} between 0.4 and 0.5 AU, cells should be harvested by centrifugation. The amount of culture harvested depends on the number of cells needed for the particular application. In order to maintain consistency, an equal number of optical density units (ODU) are harvested for each sample. ODU can be calculated by multiplying the volume harvested in ml by the optical density in AU, for example, 10 ml of cells at $OD_{600} = 0.5$ AU yields 5 ODU.

   For RB studies, cells must not be allowed to grow beyond an $OD_{600} = 0.5$, as ribosome synthesis becomes attenuated at this phase of growth, which can significantly alter experimental results (Ju and Warner, 1994).
7. Resuspend pelleted cells in 1 ml sterile water and transfer to 1.5 ml microcentrifuge tube. Pellet cells again and remove supernatant. Pellet can be flash frozen and stored at -80 °C, although it is recommended that the RNA extraction process be initiated prior to storage; see below.

3.1.2. Total RNA extraction from yeast cells

Due to the inherent instability of RNA and the high possibility of RNase contamination, it is recommended that equipment to be used for RNA extraction is kept free of total cellular material and/or wiped clean using 70% ethanol prior to use. All solutions should be prepared using RNase-free water that has been autoclaved or passed through a 0.22-μm filter. Where possible, work should be performed in a fume hood or designated RNase-free area.

1. Resuspend 10 ODU of cells in 400 μl TES solution (10 mM Tris (pH 7.5), 10 mM EDTA, 0.5% SDS).
2. Add 400 μl acid phenol and vortex suspension for 10 s at maximum intensity.
3. Incubate suspension at 65 °C for 45 min, vortexing briefly every 10 min to homogenize sample.
4. Centrifuge sample at maximum speed for 5 min at 4 °C.
5. After centrifugation, two layers will be discernable in the microcentrifuge tube. The top layer is aqueous, while the bottom will be composed of the organic phenol. Remove the aqueous phase and transfer to a clean 1.5 ml tube.
6. Add 400 μl acid phenol, briefly vortex the tube and centrifuge as in step 5; transfer the aqueous phase to a clean 1.5 ml microcentrifuge tube.
7. Add 400 μl chloroform, briefly vortex the tube and centrifuge as in step 5; transfer the aqueous phase to a 1.5-ml microcentrifuge tube.
8. The RNA is precipitated to concentrate the sample and remove residual phenol/chloroform. Add 1 ml of ice cold 100% ethanol and 40 μl of 3 M NaOAc.
    At this point the sample can be stored at −80 °C with less potential for sample loss or degradation than if frozen in the total cell pellet after harvesting.
9. If the sample is not stored at −80 °C, incubate in a dry ice/ethanol bath for 30 min.
10. Pellet the RNA by centrifugation at max speed for 20 min at 4 °C.
11. Remove the supernatant, being careful to disturb the pellet as little as possible.
12. Add 1 ml ice cold 70% ethanol to the pellet, microfuge for an additional 2 min at max speed.
13. Remove the supernatant as thoroughly as possible without disturbing the pellet.
14. Allow the remaining ethanol to evaporate off the pellet by leaving the tube open on the bench for 10 min.
15. Resuspend the pellet in 50–100 μl of RNase-free water.
16. RNA can be quantified by measuring the absorbance of the sample at 260 nm, where an A_{260} of 1 represents 40 μg or RNA per ml of solution.

3.1.3. Evaluating ribosomal RNA levels by gel electrophoresis and northern blotting

Depending on the application, rRNA can be readily visualized by either ethidium bromide staining with UV detection, or by northern blotting and densitometry. The mature 18S and 25S rRNA species can easily be resolved and visualized using a 1.25% agarose TAE gel with ethidium bromide staining (Freed and Baserga, 2010); however, this method is not sufficiently sensitive for accurate determination of the levels of rRNA processing intermediates. To monitor accumulation of various pre-rRNA species, samples should be subjected to electrophoresis in a 1.25% agarose/formaldehyde gel then transferred to a nylon membrane for northern analysis with appropriate radiolabeled oligonucleotides.

3.1.3.1. Gel electrophoresis and transfer

1. To prepare a 100 ml, 1.25% agarose/formaldehyde gel, begin by dissolving 1.25 g high grade agarose into 72 ml water by heating in a microwave.
2. Working in a fume hood, allow the agarose solution to cool to the touch. Add 18 ml of 37% formaldehyde solution and 10 ml $10 \times$ MOPS buffer (0.4 M MOPS (pH 7.0), 0.1 M NaOAc, 10 mM EDTA); mix well and pour gel.
3. Prepare sample loading buffer by combining 1 μl $10 \times$ MOPS buffer, 10 μl formamide, and 3.5 μl 37% formaldehyde per sample. Add 15–20 μg of total RNA in no more than 3 μl total volume to the sample buffer.
4. Denature RNA by incubating the sample at 65 °C for 15 min.
5. Add 2 μl loading dye (50% glycerol, 10 mM EDTA, 0.25% (w/v), bromophenol blue) to each sample.
6. Load samples on 1.25% agarose/formaldehyde gel and run for 22–24 h at 59 V in $1 \times$ MOPS buffer.
7. Once running is complete, transfer the gel to a clean tray and incubate in 75 mM NaOH for 20 min on an orbital shaker, ensuring that the gel is immersed in the solution.
8. Wash the gel $2 \times$ with water, then incubate in $10 \times$ SSC buffer (0.3 M NaCl, 30 mM sodium citrate, pH 7.0) for 45 min.
9. Transfer overnight to a Hybond N+nylon membrane (www.gelifesciences.com) equilibrated in $10 \times$ SSC using a standard capillary blotting apparatus.
10. Cross-link RNA to the membrane by exposure to UV light at 0.125 J for 30 s on each side of the membrane.
11. The blot can be stored at -20 °C in aluminum foil until use in northern analysis.

3.1.3.2. Northern analysis
Oligonucleotide sequences to be used as probes in northern blotting depend greatly on the particular rRNA species under consideration. In general, levels of 18S and 25S rRNA will be considerably greater than levels of any pre-rRNA; therefore, while a probe complementary to a sequence within the 18S rRNA could

theoretically be used to visualize a number of processing intermediates, exposure times necessary to observe pre-rRNA transcripts on a film will result in excessive exposure in the region corresponding to the mature rRNAs. Additionally, certain intermediates—such as 27SA and 27SB—may not be well resolved on the gel, and it might therefore be necessary to select a probe unique to one of these transcripts to definitively locate it on the gel. Probes have already been designed, which can distinguish between many of the important pre-rRNA species and may serve as a useful starting point for further analysis (Table 14.2).

1. Prepare a $5\times$ SSPE solution (0.9 M NaCl, 50 mM NaH$_2$PO$_4$ (pH 7.7), 1 mM EDTA).
2. Place cross-linked membrane containing RNA samples into a hybridization tube with the RNA facing the center. Immediately prior to use, add 20% (w/v) SDS solution to an aliquot of $5\times$ SSPE to achieve a final concentration of 0.1% SDS. Wet the membrane with this solution and remove any air bubbles using a sterile pipette.
3. Add between 10 and 25 ml of $5\times$ SSPE $-$ 0.1% SDS solution depending on the size of the membrane; ensure that the portion of the membrane at the nadir of the tube will be sufficiently immersed.
4. Place the tube in a hybridization oven at 60 °C and allow it to rotate for 4 h.
5. Denature 10^6 cpm of 5′-end-labeled oligonucleotide(s) per ml of $5\times$ SSPE solution by incubating at 95 °C for 3 min.
6. Dilute labeled oligonucleotides(s) in 500 μl of $5\times$ SSPE to ensure better dispersion the hybridization tube.
7. Add probe mixture to the hybridization tube and incubate in the hybridization oven at 60 °C overnight.
8. The following day, remove the hybridization tubes from the oven and discard the hybridization buffer, taking care to observe all pertinent radioactive safety protocols.
9. Wash the blot by adding an equal volume of $5\times$ SSPE $-$ 0.1% SDS back to the tube and incubating at 60 °C for 15 min.
10. Remove the wash buffer and repeat the wash.
11. Remove the wash buffer and add an equal volume of $1\times$ SSPE $-$ 0.1% SDS to the tube. Incubate at 60 °C for 5 min.
12. Carefully remove the blot from the hybridization tube and wrap it in clear plastic wrap.
13. Expose the blot using an appropriate film or intensifying screen for densitometric analysis.

By running samples from the same conditions on multiple gels and blotting with different probes, alone or in combinations, it is possible to monitor the levels of rRNA processing intermediates and to make direct comparisons between the levels of different pre-rRNA species. Additionally,

Table 14.2 Oligonucleotide probes used to detect pre-rRNA processing intermediates

Designation as published	Sequence (shown 5′ to 3′)	Region of the pre-rRNA detected	References
a	CATGGCTTAATCTTTGAGAC	18S rRNA	Berges et al. (1994)
b	GCTCTTTGCTCTTGCC	ITS1 between the 18S and A_2	
c	ATGAAAACTCCACAGTG	ITS1 between A_2 and A_3	Venema and Tollervey (1996)
d	CCAGTTACGAAAATTCTTG	ITS1 at A_2	
e	GGCCAG CAATTT CAAGT	ITS2 between E and C_2	
f	AGATTAGCCGCAGTTGG	ITS2 between C_2 and C_1	
g	CTCCGCTTATTGATATGC	25S rRNA	
1	TCGGGTCTCTCTGCTGC	5′ ETS before A_0	
3	CATGGCTTAATCTTTGAGAC	5′ region of the 18S rRNA	
4	CTCCGCTTATTGATATGC	5′ region of the 25S rRNA	
5	GCTCTTTGCTCTTGCC	ITS1 between 18S and A_2	
6	TGTTACCTCTGGGCCC	ITS1 between A_2 and A_3	
7	TTTCGCTGCGTTCTTCATC	5.8S 3′ of B_{1S}	
8	AACAGAATGTTTGAGAAGG	ITS2 3′ of E	
9	GGCCAGCAATTTCAAGTTA	ITS2 5′ of C_2	
1	GGTCTCTCTGCCGG	5′ ETS before A_0	Kressler et al. (1997)
2	CATGGCTTAATCTTTGAGAC	18S rRNA	
3	CGGTTTAATTGTCCTA	ITS1 between D and A_2	
4	TGTTACCTCTGGGCC	ITS1 between A_2 and A_3	
5	AATTTCCAGTTACGAAAATTCTTG	ITS1 between A_3 and B_1	
6	TTTCGCTGCGTTCTTCAT	5.8S rRNA	
7	GGCCAGCAATTTCAAGTT	ITS2 between E and C_2	
8	GAACATTGTTCGCCTAGA	ITS2 between C_1 and C_2	

Table 14.2 (Continued)

Designation as published	Sequence (shown 5′ to 3′)	Region of the pre-rRNA detected	References
9	CTCCGCTTATTGATATGC	25S rRNA	Benard et al. (1998)
p18S	CGTCCTATTCTATTATTCCATG	18S rRNA	
p5.8S	TTTCGCTGGGTTCTTCATC	5.8S rRNA	
p25S	GCCCGTTCCCTTGGCTGTG	25S rRNA nucleotides 2359 to 2377	
p25S5′	GCGGGTACTCCTACCTGATTTGAGGTC	25S rRNA nucleotides 5 to 31	
p25S3′	CAGCAGATCGTAACAACAAGGCTACTCTAC	25S rRNA nucleotides 3336 to 3365	
y	GCCCGTTCCCTTGGCTGTG	Mid-25S rRNA	Wehner et al. (2002)

samples can be taken at intervals subsequent to helicase depletion (see below) to monitor the change in levels of each species of rRNA over time.

3.2. Perturbation of helicase activity for RB studies

Of the 19 RNA helicases implicated as RB factors in yeast, 18 belong to the DExD/H-box family of enzymes (with one Ski2-like helicase, Mtr4) (Bleichert and Baserga, 2007; Jankowsky, 2011; Kressler et al., 2010). The following discussion will therefore be focused on studies, methods, and applications that pertain to this group of proteins.

As mentioned previously, the basic principle governing the standard methodology for investigating RB helicases is simply to somehow abrogate the function of a target enzyme then to determine how this perturbation has affected the quantity and maturation state of rRNA within the cell. Because the majority of RB helicases are essential (Table 14.1), it is necessary to deplete the target enzyme conditionally using genetic manipulation, as simply disrupting the gene would be lethal. Once a strain is developed in which the chromosomal copy of the enzyme can be conditionally depleted successfully, the gene can be reintroduced into the cell on a plasmid, making it possible to perform mutagenic studies *in vivo*.

3.2.1. Chromosomal insertion of a GAL1 promoter

In order to conditionally reduce the levels of protein in a yeast cell without excessively perturbing the system at large, a gene of interest can be placed under the control of an alternative endogenous promoter that can be activated or deactivated without drastically altering the cellular environment. In yeast, several promoters have been successfully adapted to this task (Belli et al., 1998; Etcheverry, 1990; Longtine et al., 1998; Mumberg et al., 1995), but as the methods employed to put a target gene under the control of any of these promoters are essentially identical, only the *GAL1* promoter system will be discussed here. The *GAL1* promoter normally controls transcription of genes necessary to metabolize galactose when glucose is no longer available as a carbon source. Taking advantage of this promoter is particularly convenient because activation/deactivation of genes under its control simply requires changing the carbon source with which a culture has been supplemented, a relatively simple and inexpensive process.

Placing a gene under the control of the *GAL1* promoter is accomplished by transforming yeast cells with linear, double stranded DNA encoding the promoter, and a selectable marker flanked by sequences that are homologous to a target site in the yeast chromosome (Fig. 14.3). Once in the cell, the linear DNA undergoes a homologous recombination event, resulting in a chromosomal insertion at a location determined by the flanking sequences (Baudin et al., 1993; Schneider et al., 1995). DNA encoding the promoter

Oligonucleotide design: 50 bp sequences designed to target recombination are appended to 20–25 bp sequences homologous to the tagging cassette.

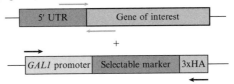

+

Amplification: Primers containing chromosomal target sequences are used to amplify the tagging cassette by PCR.

Incorporation: The amplified cassette DNA is transformed into yeast cells and incorporated into the chromosome by recombination.

Figure 14.3 Chromosomal insertion of a *GAL1* promoter. Sequences to be selected for incorporation into final oligonucleotide PCR primers are represented by arrows pointing in a 5′ to 3′ direction in relation to the primer sequence. Regions of the primer where the sequence is homologous to the chromosomal target site are shaded gray, while regions of the primer homologous to the tagging cassette are black. The final primer containing sequences from both regions is denoted as a chimera of both gray and black arrows. The final panel represents a recombination event mediated by the homologous sequences in the amplified product and the chromosomal target site.

and selectable marker can be amplified from plasmids produced by Longtine *et al.* which have been designed to allow amplification of a "cassette" containing a selectable marker and optional epitope tag in addition to the *GAL1* promoter sequence (Longtine *et al.*, 1998).

1. PCR primers should be designed with the forward primer containing a 50-bp segment homologous to the sense strand of chromosomal DNA immediately 5′ to the gene of interest, followed by 20–30 bp of sequence designed to anneal to the most 5′ portion of the template cassette.
2. The reverse primer should be the reverse complement of a 50-bp sequence homologous to the 5′ portion of the gene followed by 20–30 bp homologous to the 3′ portion of the template cassette (see Fig. 14.3 and Longtine *et al.*, 1998).
3. Double stranded DNA for transformation should be generated using standard PCR techniques with a high fidelity polymerase in a 200-μl total reaction volume.

4. PCR product should be pooled and ethanol precipitated as described in Section 3.1.2. Precipitated DNA should be resuspended in 5–10 μl of sterile water depending on the amount of PCR product produced.
5. To transform the desired yeast strain, begin by growing a 25-ml culture in YPD media to an OD_{600} of 0.3–0.6 AU.
6. Harvest the cells by centrifugation, resuspend in 1 ml sterile water, and transfer to a 1.5-ml microcentrifuge tube.
7. Pellet the cells again and remove the supernatant.
8. Prepare a fresh LiAc/TE solution (150 μl 1 M LiAc, 150 μl TE, 1.2 ml sterile water) and LiAc/TE/PEG solution (150 μl 1 M LiAc, 150 μl TE, 1.2 ml 80% PEG).
9. Resuspend cell pellet in 250 μl LiAc/TE solution, and separate into 50 μl aliquots.
10. Denature a stock sample of sonicated salmon sperm DNA by incubating at 95 °C for 3 min.
11. To a single aliquot, add 2.5 μg of DNA (approximately 3–5 μl of the precipitated PCR product), 5 μl sonicated salmon sperm DNA, and 300 μl of the LiAc/TE/PEG solution.
12. Incubate the mixture at 30 °C for 30 min.
13. Move the sample to 42 °C and incubate for a further 15 min.
14. Pellet the cells by centrifugation and remove the supernatant.
15. Resuspend the pellet in 150 μl of sterile water and plate on an appropriate restrictive media.
16. Incubate the plate at 30 °C until colonies appear.

3.2.2. Depleting a protein under control of the GAL1 promoter

After successful insertion of the *GAL1* promoter is accomplished, a gene product can be readily depleted from the cell by growing the cell in a media lacking galactose. Normally, yeast are cultured in a 2% dextrose solution for optimal growth; however, a strain containing an essential gene that has been placed under the control of the *GAL1* promoter must be grown in galactose (Gal) prior to protein depletion. Because galactose is such a poor carbon source, Gal media is usually supplemented with 2% raffinose (Raf). Upon commencement of a depletion experiment, cells can be moved from a Gal/Raf medium to a dextrose medium, and protein levels can be monitored by western analysis. Protein depletion is accomplished as follows:

1. Recall that for RB studies, cells should not be allowed to reach an $OD_{600} > 0.5$ AU less than two doubling events prior to analysis. To begin a depletion experiment, start a 10-ml overnight culture in YPG/R media from yeast grown on a medium selecting for proper chromosomal insertion of the *GAL1* promoter. Incubate the culture on a shaking platform overnight at 30 °C.

2. The following morning, measure the OD_{600} of the culture. If the optical density is <0.5 AU, proceed to step 4, if not, dilute the cells into a new 25 ml culture of YPG/R media to a final $OD_{600} = 0.1$ AU. Continue incubating the culture at 30 °C.
3. Monitor the OD_{600} of the culture over the course of the day. Once the culture is approaching an optical density of 0.5, harvest the cells by centrifugation. Dilute the cells into a new 25 ml overnight culture of YPG/R such that the final optical density is 0.01 AU. The cells should not exceed an optical density of 0.5 AU for at least 12–14 h, although this will need to be optimized experimentally. If carried out properly, this will ensure that the cells can be used the following morning without requiring additional dilution and doubling events.
4. Harvest 5 ODU from the YPG/R culture by centrifugation.
5. Wash the cells by resuspending the pellet in 1 ml of sterile water. Pellet the cells again.
6. Resuspend the cells in 1 ml of sterile water. The OD_{600} of these cells is now 5 AU.
7. Dilute 250 μl of the cells into a fresh 25 ml culture of YPD, resulting in a final $OD_{600} = 0.01$ AU. Incubate the cells on a shaking platform at 30 °C.
8. Protein depletion can be monitored by removing equal ODU from the culture at several time points over a 24-h period to be analyzed by western blot. If no antibody is available for the protein of interest, an epitope tag should be appended when transforming the target with the *GAL1* chromosomal insertion.
9. Once the protocol has been optimized for a target helicase, protein-depleted cells can be analyzed for rRNA production and processing as described in Section 3.1.

3.2.3. Protein mutagenesis and expression in a native-protein-depleted background

Depletion of a target helicase is designed to show that this protein plays some role in some RB event or events, but it gives little insight into the mechanistic behavior of the enzyme. Because DExD/H-box helicases have been so thoroughly characterized and because they share such striking homology among their conserved motifs, a number of potential residues can be readily identified as targets for site directed mutagenesis studies designed to elucidate the particular enzymatic activities required for productive action. The roles and consensus sequences for each of the motifs present in any DExD/H-box protein have been reviewed extensively elsewhere (Cordin *et al.*, 2006; Fairman-Williams *et al.*, 2010; Linder and Jankowsky, 2011, and others). Homologous residues on any protein of interest can be identified by a combination of homology mapping (www.ncbi.nlm.nih.gov/BLAST) and sequence inspection.

Mutagenized proteins can be introduced into a cell depleted of its native protein in order to assay for complementation. If a particular activity of the protein can be perturbed without loss of complementation, it is reasonable to conclude that this activity is not essential for RB. Several studies have already been undertaken to identify mechanistically important motifs in a number of yeast RB helicases, and these can serve as starting points for further studies as well as examples of how these studies are accomplished (e.g., Bernstein et al., 2006, Daugeron and Linder, 2001; Granneman et al., 2006a; Kos and Tollervey, 2005; Rocak et al., 2005). The majority of techniques required for the introduction of a mutagenized protein have been discussed elsewhere in this manuscript, therefore only a few essential considerations will be detailed here.

1. Standard cloning techniques should be used to insert the gene of interest into a yeast expression vector compatible with the parental strain as discussed in Section 3.1.1.
2. For convenience, the vector chosen should express the gene under the control of a constitutive promoter, allowing concomitant depletion of the endogenous copy of a target gene and expression of the exogenous copy. Note that a mutagenized protein may confer a dominant negative phenotype when constitutively expressed in the cell. In these cases, protein expression must be placed under the control of an inducible promoter that can be activated subsequent to native protein depletion.
3. Using the material cited above as a guide, one or more residues should be selected for mutagenesis based on the enzymatic activities of the protein for which they are required.
4. Mutagenesis can be carried out using commercially available kits (e.g., GeneArt® Site Directed Mutagenesis Kit, www.invitrogen.com). Mutagenized enzymes can then be screened for the requirement of each activity individually (e.g., RNA binding, ATP hydrolysis, etc.).
5. Plasmids containing mutagenized proteins can be transformed into yeast cells and successful transformants selected as described for the linear recombination cassette in Section 3.2.1.
6. Protein depletion and rRNA processing studies can be carried out as previously described.

4. ELUCIDATING THE SUPERMOLECULAR CONTEXT OF AN RB HELICASE: PROTEIN–PROTEIN INTERACTION STUDIES

Processing of the pre-rRNA transcript is accomplished primarily in the context of large protein or RNP machines in which numerous factors act together to properly facilitate the many cleavage, modification, and folding events that comprise RB. The macromolecular complexes that mediate pre-rRNA maturation are broadly defined by their participation in small ribosomal subunit processing (i.e., the SSU processome), large ribosomal subunit processing, or chemical modification of the rRNA transcripts (e.g., snoRNPs). Many of the protein and snoRNA components involved in these tasks have already been identified and grouped based on their copurification with larger complex(es) known to mediate one of the above activities (Bassler *et al.*, 2001; Dragon *et al.*, 2002; Gallagher *et al.*, 2004; Harnpicharchai *et al.*, 2001; Krogan *et al.*, 2004; Nissan *et al.*, 2002; Pérez-Fernández *et al.*, 2007). It is therefore possible to determine the general pathway in which a target enzyme participates by identifying the factors that are associated with that enzyme and comparing them to those that have previously been categorized.

Association with a given protein can occur directly—through a binary protein–protein interaction—or indirectly—via an intermediary protein or nucleic acid. It is important to recognize what kind of interactions will be observed using a chosen technique as the implications of a protein–protein association can vary drastically depending on whether the association is direct or indirect. This is particularly true when working with RB factors, as these factors are almost always found associated with extremely large complexes and can therefore coprecipitate many proteins that they do not interact with directly. This is inconsequential when attempting to determine whether a protein acts in SSU or LSU biogenesis but becomes important when attempting to identify potential cofactors for a target enzyme.

A variety of strategies exist for assaying both direct and indirect protein–protein interactions. Some of the most powerful and frequently applied methods will be discussed here, including Tandem Affinity Purification (TAP), co-immunoprecipitation (co-IP), and yeast two-hybrid (Y2H) analysis. TAP analysis is frequently coupled to protein identification by mass spectrometry and is most often used to select and identify members of multiprotein complexes, yielding results that include both direct and indirect binding partners (Hoang *et al.*, 2005; Lebaron *et al.*, 2005). Co-IP experiments can provide information on direct or indirect interactions depending on experimental conditions (Champion *et al.*, 2008; Charette and Baserga, 2010; Yoshikawa *et al.*, 2011). Interpretation of co-IP results involving RB proteins requires careful analysis and a thorough understanding of the macromolecular complexes involved in this system. Finally, the Y2H approach has been employed with great success as a high-throughput system for screening and identifying protein pairs involved in direct binding interactions (Champion *et al.*, 2008; Gallagher and Baserga, 2004; Granneman *et al.*, 2006b; Lebaron *et al.*, 2005).

4.1. Tandem affinity purification

The TAP method was designed for the selection and identification of large protein complexes. TAP protocols involve relatively gentle conditions designed to minimize perturbation of native protein complexes while maintaining a high level of selectivity by employing a multistep approach to purification.

4.1.1. Epitope tagging target proteins

In order to purify a protein complex containing a helicase or cofactor of interest using the TAP method, an appropriate fusion construct of the target protein must be introduced into the host cell. In yeast, this type of tagging is frequently accomplished by incorporating a sequence into the chromosome at a position flanking either terminus of the target gene (Baudin et al., 1993; Knop et al., 1999; Schneider et al., 1995). This procedure involves transforming an appropriate yeast strain with double stranded DNA containing sequences that encode the desired tag and a selectable marker flanked by 50 bp sequences homologous to regions of the chromosome where the insertion event is to take place, and the method is discussed in detail in Section 3.2.1 (Fig. 14.3). Cells which have successfully incorporated the tagging cassette are selected via the marker contained therein. The presence of the fusion protein in the cell can be assayed by western blot.

Amplification of the TAP sequence in conjunction with a selectable marker can be achieved using one of two plasmids designed by Puig et al. (2001). These vectors—designated pBS1479 and pBS1539—are designed to accommodate either C- or N-terminal tagging, respectively. In designing primers, it is important to ensure that the cassette will be incorporated in frame with the target gene, and that in the case of C-terminal fusions, the native stop codon is not present in the primer sequence so that translation can proceed through the tag. It is also important to note that while the TAP epitope can be fused to either terminus of the target protein, the choice of terminus may be important when examining higher order complex formation. It is conceivable that tagging a particular terminus may perturb or abrogate important protein–protein interactions, resulting in low yields of native binding partners after purification.

4.1.2. TAP protocols

Protocols for TAP have been published in extensive detail elsewhere. Several large scale, genome-wide studies of protein complex formation have been conducted in yeast using the TAP methodology (Gavin et al., 2002, 2006; Krogan, et al., 2006), and the refined protocol developed through these studies has been exhaustively described in Babu et al. (2009).

The TAP tag, as originally described in Puig et al. (2001) and Rigaut et al. (1999), consists of two IgG binding domains of protein A (ProtA) from *Staphylococcus aureus* and a calmodulin binding peptide (CBP) with a TEV protease cleavage site separating these two motifs. The TAP method involves selective immunoprecipitation of tagged proteins using IgG-Sepharose beads, followed by washing in experimentally optimized conditions. Bound complexes are eluted from the bead by TEV protease cleavage,

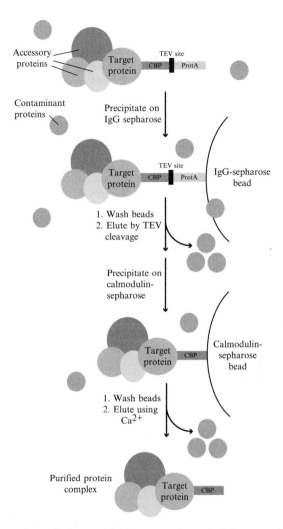

Figure 14.4 Tandem affinity purification procedure. A fusion protein construct is expressed *in vivo* and purified from contaminating cell lyste. The fusion protein in complex with associate factors is precipitated on IgG-Sepharose beads and washed prior to specific elution by TEV cleavage. In a second purification step, the fusion protein and associate factors are precipitated on calmodulin-Sepharose beads, washed, and eluted with by increasing the concentration of Ca^{2+}.

which separates the ProtA portion of the tag (bound to the IgG beads) from the remaining CBP portion and the target protein. This eluate is then precipitated on calmodulin sepharose beads, washed, and eluted using increasing concentrations of Ca^{2+} (Fig. 14.4).

Despite the existence of a somewhat standardized approach to TAP, truly optimal experimental conditions can vary greatly depending on the nature of the target protein and the system in which it acts. RNA helicases can be particularly problematic in TAP studies due to the fact that interactions between these proteins and their binding partners can often be weak or transient. Indeed, a number of TAP studies using tagged constructs of other SSU processome proteins have failed to identify more than a small subset of the RNA helicases that have been shown to act in this system by alternative techniques (Dragon et al., 2002; Gavin et al., 2006; Krogan, et al., 2006).

The most successful examples of identifying protein–protein interactions involving these enzymes have relied on tagging the helicase (Bernstein et al., 2006; Granneman et al., 2006a; Lebaron et al., 2005) or its cofactor (Hoang et al., 2005) directly. It is therefore recommended that when assaying protein complex formation among RNA helicases in this system, the helicase should always be tagged, and particularly permissive conditions should be used in binding and washing buffers (e.g., salt concentrations at or below 100 mM, less than 0.1% detergent). It is also important to attempt each experiment in a variety of conditions, as it is probable that optimal conditions for any protein tested will differ significantly. Finally, as mentioned previously, it is crucial that cultures grown for these studies are never allowed to exceed an OD_{600} of greater than 0.5 AU, as RB in *S. cerevisiae* is significantly attenuated when cells reach this phase of growth (Ju and Warner, 1994). Allowing cells to grow beyond this point induces dissipation of the RB machinery and might result in false negatives, as the helicase of interest is no longer partaking in the synthesis of ribosomes and may therefore not be in physical contact with other RB factors.

It should be noted that in addition to canonical TAP protocols, alternative procedures have been developed to provide different tagging options and more robust purification and recovery of target complexes. Several modified systems have been designed to improve yields and specificity when working in mammalian systems (Burckstummer et al., 2006; Drakas et al., 2005; Gloeckner et al., 2009). Most recently, a new three tag system consisting of a CBP, a streptavidin-binding peptide, and a 6× histidine tag optimized for use in mammalian cells was described in Li et al. (2011). In addition to alternative tagging strategies, new protocols have been developed to improve the sensitivity and selectivity of current TAP systems. A notable example comes from Oeffinger et al. in which antibody-conjugated magnetic beads are used rather than a conjugated resin such as Sepharose, resulting in recovery of larger complexes with fewer nonspecifically bound contaminants (Oeffinger et al., 2007). In cases where typical procedures fail to produce quality results turning to a more specific or highly optimized method may provide a successful alternative.

4.2. Co-IP of tagged proteins

The general strategy employed in performing a co-IP experiment in yeast consists of growing cells in liquid culture to an optimal density, harvesting and lysing the cells, and incubating the soluble lysate with an IP media conjugated to an appropriate antibody to select for proteins containing the conjugate epitope tag (Steitz, 1989). The precipitated proteins can be eluted from the media and separated by SDS-PAGE, then detected by western blotting.

1. From a glycerol stock or freshly struck plate, prepare an overnight culture by collecting a small amount of yeast on the tip of a sterile pipette and inoculating 10 ml liquid YPD media under sterile conditions. Incubate the culture on a shaking platform overnight at 30 °C.
2. Also prepare IP beads by incubating 3 mg of ProtA-sepharose beads per sample in 500 µl NET-2 buffer with a saturating amount of IP antibody overnight at 4 °C. Before use, pellet and wash the beads 3 × with 1 ml NET-2 per 3 mg beads.
3. The following morning, measure the OD_{600} of the culture. Dilute an appropriate amount of the overnight culture into 25 ml of fresh YPD media such that the final optical density will be 0.1 AU. Continue incubating the culture at 30 °C.
4. Monitor the optical density, allowing the cells to grow to an OD_{600} of 0.4–0.5 AU.
5. Harvest at least 10 ODU of culture (e.g., 20 ml of culture at $OD_{600} = 0.5$ AU) by centrifugation.
6. Resuspended pelleted cells in 1 ml sterile water and transfer the sample to a 1.5-ml microcentrifuge tube. Pellet the cells again and remove the supernatant by pippetting.
7. Resuspend the cells in 600 µl NET-2 buffer (20 mM Tris–HCl (pH 7.5), 150 mM NaCl, 0.05% Nonidet P-40) containing the appropriate dilution of a protease inhibitor cocktail [Roche]
8. Break the cells by adding 100 µl 0.5 mm glass beads and vortexing at 4 °C for five times at 45 s intervals. Place the cells on ice for 45 s between each round of vortexing.
9. Pellet the cell debris by centrifugation at maximum speed for 20 min in a microcentrifuge at 4 °C. Retain a 20-µl sample of the resulting supernatant for subsequent analysis.
10. Add 500 µl of the remaining lysate to 3 mg ProtA-sepharose conjugated to the IP antibody.
11. Incubate the sample for 1–2 h at 4 °C, preferably on a rocking or nutating platform.
12. Pellet the media by spinning at max speed for 1 min in a microcentrifuge. Remove the supernatant and discard.

13. Wash the beads by adding 1 ml NET-2 to the sample. Repeat steps 12 and 13 four more times. At this point, the protein selected by the antibody conjugated to the media will be precipitated on the resin. If the target protein acts in RB, a portion of the RB marker protein will also be precipitated via its interaction with the target.
14. To elute bound proteins, resuspend the media in 10 µl 5× SDS Loading Buffer (200 mM Tris–HCl (pH 6.8), 10% (w/v) SDS, 20% (v/v) glycerol, 10 mM β-mercaptoethanol, 0.05% (w/v) bromophenol blue) and incubate the sample 3 min at 95 °C.
15. Perform SDS-PAGE with the total input and eluate samples in adjacent positions on the gel.
16. Bound proteins can be detected by western analysis.

4.3. Yeast two-hybrid analysis

Y2H analysis can be used as a high-throughput assay for detecting direct protein–protein interactions among large numbers of potential binding partners. Y2H analysis is made possible by the modular nature of many eukaryotic transcription factors. In these proteins, the domains responsible for binding DNA and recruiting transcription machinery can be physically separated and their activity reconstituted *in trans*. Y2H analysis consists of fusing one protein to the DNA binding domain of a transcription factor, and another to the activation domain, then coexpressing the fusion constructs in a yeast strain containing a reporter gene under the control of the parsed transcription factor. If the proteins interact with each other, that interaction will reconstitute the transcription factor, resulting in transcription of the reporter gene (Fields and Song, 1989).

As in the case of the TAP protocol, Y2H procedures vary widely and are largely dependent on the type of equipment available for use in the study, therefore a specific protocol will not be detailed here. Instead, a general framework for constructing a Y2H system will be discussed, with emphasis on important considerations for screening in systems with known protein components. Descriptions of high-throughput protocols employing various mechanical and biological techniques have been published and can be found in Cagney *et al.* (2000) and Rajagopala and Uetz (2009, 2011).

1. Because many of the factors involved in RB have already been identified, it is possible to perform experiments using a directed approach, in which a tractable list of potential targets can be compiled prior to analysis. Proteins that should be included in the study of a RB helicase will depend on results of other analyses. For RNA helicases involved in yeast ribosome maturation, the pathway in which each protein acts has been elucidated; hence, a number of ribosome synthesis factors can be immediately eliminated from consideration for a number of these

proteins. Information gained in TAP or other binding partner studies can be used to further focus a study to a small group of potential binding partners.

2. Genes corresponding to all proteins of interest must be cloned into expression plasmids designed to produce a fusion construct containing either the DNA binding or activating domain of an appropriate transcription factor. The *GAL4*-responsive promoters are frequently used as the control element for various reporters within common Y2H strains, including Y2H Gold and Y187 (www.clontech.com), and the conjugate plasmids will therefore contain either the *GAL4* DNA binding domain (*GAL4* DNA-BD) or the *GAL4* activating domain (*GAL4* AD). Fusion proteins containing the DNA-BD are referred to as baits, while those containing the AD are considered preys. The choice of plasmid will depend on the markers available in the selected strain, as well as the desired cloning strategy; multiple vectors are available.

3. To facilitate screening of large numbers of proteins, it is preferable to use a yeast mating system, in which plasmids expressing one set of fusion proteins—for example, all *GAL4* AD fusions—are transformed individually into a mating type a (*MATa*) version of the Y2H strain, while the other set of fusion plasmids is transformed into mating type alpha (*MATα*)cells. In this way, cultures of each transformed strain can be separately grown and mated in a one-to-one fashion until all possible pairings have been accomplished.

Previously, pooling strategies have been employed to screen large libraries of potential interactors (Ito et al., 2001; Uetz et al., 2000; Yu et al., 2008); however, this strategy should be avoided when assaying smaller cohorts, as it introduces bias into the system, selecting the most robust or least cytotoxic interacting partners over other potential positives (Koegl and Uetz, 2007; Lim et al., 2011; Rajagopala and Uetz, 2009).

4. Successful mating can be determined by growth on media selective for both plasmid markers. Once mating has been accomplished, cells containing both plasmids can be deposited on media selective for one or more of the reporter genes under control of the *GAL4* promoter. Growth on this media implies successful reconstitution of the transcription factor and, by extension, a physical interaction between the two fusion proteins.

5. A number of selectable markers are available for use in most Y2H strains. Common markers used include but are not limited to *his3* and *ade2*, which select for the ability to grow on media lacking histidine and adenine, respectively, as well as *mel1*, which metabolizes X-α-Gal resulting in blue colonies. In order to improve confidence in positive results, screens can be replicated on different markers or a combination of markers can be used in a single screen.

Histidine selection is somewhat leaky and a minimal amount of histidine biosynthesis can still occur in the cell without activation of the reporter gene. To minimize spurious positives related to this effect, an experimentally optimized amount (usually to a final concentration between 3 and 6 mM) of 3-aminotriazole is added to the media to dampen background histidine production.

REFERENCES

Anderson, J. S., Lyon, C. E., Fox, A. H., Leung, A. K., Lam, Y. W., Steen, H., Mann, M., and Lamond, A. I. (2002). Directed proteomic analysis of the human nucleolus. *Curr. Biol.* **12**, 1–11.

Anderson, J. S., Lam, Y. W., Leung, A. K., Ong, S. E., Lyon, C. E., Lamond, A. I., and Mann, M. (2005). Nucleolar proteome dynamics. *Nature* **433**, 77–83.

Arenas, J. E., and Abelson, J. N. (1997). Prp43: An RNA helicase-like factor involved in spliceosome disassembly. *Proc. Natl. Acad. Sci. USA* **94**, 11798–11802.

Babu, M., Krogan, N. J., Awrey, D. E., Andre, E., and Greenblatt, J. F. (2009). Systematic characterization of the protein interaction network and protein complexes in *Saccharomyces cerevisiae* using tandem affinity purification and mass spectrometry. *Methods Mol. Biol.* **548**, 187–207.

Bassler, J., Grandi, P., Gadal, O., Lessmann, T., Petfalski, E., Tollervey, D., Lechner, J., and Hurt, E. (2001). Identification of a 60S preribosomal particle that is closely linked to nuclear export. *Mol. Cell* **8**, 517–529.

Baudin, A., Ozier-Kalogeropoulos, O., Denouel, A., Lacroute, F., and Culin, C. (1993). A simple and efficient method for direct gene deletion in *Saccaromyces cerevisiae*. *Nucleic Acids Res.* **21**, 3329–3330.

Belli, G., Gari, E., Piedrafita, L., Aldea, M., and Herrero, E. (1998). An activator/repressor dual system allows tight tetracycline-regulated gene expression in budding yeast. *Nucleic Acids Res.* **26**, 942–947.

Benard, L., Carroll, K., Valle, R. C., and Wickner, R. B. (1998). Ski6p is a homolog of RNA-processing enzymes that affects translation of non-poly(A) mRNAs and 60S ribosomal subunit biogenesis. *Mol. Cell. Biol.* **18**, 2688–2696.

Berges, T., Petfalski, E., Tollervey, D., and Hurt, E. C. (1994). Synthetic lethality with fibrillarin identifies NOP77p, a nucleolar protein required for pre-rRNA processing and modification. *EMBO J.* **13**, 3136–3148.

Bernstein, K. A., Granneman, S., Lee, A. V., Manickman, S., and Baserga, S. J. (2006). Comprehensive mutational analysis of yeast DEXD/H box RNA helicases involved in large ribosomal subunit biogenesis. *Mol. Cell. Biol.* **26**, 1195–1208.

Bleichert, F., and Baserga, S. J. (2007). The long unwinding road of RNA helicases. *Mol. Cell* **27**, 339–352.

Blum, S., Schmid, S. R., Pause, A., Buser, P., Linder, P., Sonenberg, N., and Trachsel, H. (1992). ATP hydrolysis by initiation factor 4A is required for translation initiation in Saccharomyces cerevisiae. *Proc. Natl. Acad. Sci. USA* **89**, 7664–7668.

Bohnsack, M. T., Kos, M., and Tollervey, D. (2008). Quantitative analysis of snoRNA association with pre-ribosomes and release of snR30 by Rok1 helicase. *EMBO Rep.* **9**, 1230–1236.

Bohnsack, M. T., Martin, R., Granneman, S., Ruprecht, M., Schleiff, E., and Tollervey, D. (2009). Prp43 bound at different sites on the pre-rRNA performs distinct functions in ribosome synthesis. *Mol. Cell* **36**, 583–592.

Burckstummer, T., Bennett, K. L., Preradovic, A., Schutze, G., Hantschel, O., Superti-Furga, G., and Bauch, A. (2006). An efficient tandem affinity purification procedure for interaction proteomics in mammalian cells. *Nat. Methods* **3,** 1013–1019.

Burger, F., Daugeron, M. C., and Linder, P. (2000). Dbp10, a putative RNA helicase from *Saccharomyces cerevisiae*, is required for ribosome biogenesis. *Nucleic Acids Res.* **28,** 2315–2323.

Cagney, G., Uetz, P., and Fields, S. (2000). High-throughput screening for protein-protein interactions using two-hybrid assay. *Methods Enzymol.* **328,** 3–14.

Champion, E. A., Lane, B. H., Jackrel, M. E., Regan, L., and Baserga, S. J. (2008). A direct interaction between the Utp6 half-a-tetratricopeptide repeat domain and a specific peptide in Utp21 is essential for efficient pre-rRNA processing. *Mol. Cell. Biol.* **28,** 6547–6556.

Charette, J. M., and Baserga, S. J. (2010). The DEAD-box RNA helicase-like Utp25 is an SSU processome component. *RNA* **16,** 2156–2169.

Colley, A., Beggs, J. D., Tollervey, D., and Lafontaine, D. L. (2000). Dhr1p, a putative DEAH-box RNA helicase, is associated with the box C+D snoRNP U3. *Mol. Cell. Biol.* **20,** 7238–7246.

Combs, D. J., Nagel, R. J., Ares, M., Jr., and Stevens, S. W. (2006). Prp43p is a DEAH-box spliceosome disassembly factor essential for ribosome biogenesis. *Mol. Cell. Biol.* **26,** 523–534.

Cordin, O., Banroques, J., Tanenr, N. K., and Linder, P. (2006). The DEAD-box protein family of RNA helicases. *Gene* **15,** 17–37.

Daugeron, M. C., and Linder, P. (1998). Dbp7p, a putative ATP-dependent RNA helicase from *Saccharomyces cerevisiae*, is required for 60S ribosomal subunit assembly. *RNA* **4,** 566–581.

Daugeron, M. C., and Linder, P. (2001). Characterization and mutational analysis of yeast Dbp8p, a putative RNA helicase involved in ribosome biogenesis. *Nucleic Acids Res.* **29,** 1144–1155.

Daugeron, M. C., Kressler, D., and Linder, P. (2001). Dbp9p, a putative ATP-dependent RNA helicase involved in 60S-ribosomal-subunit biogenesis, functionally interacts with Dbp6p. *RNA* **7,** 1317–1334.

de la Cruz, J., Kressler, D., Rojo, M., Tollervey, D., and Linder, P. (1998a). Spb4, an essential putative RNA helicase, is required for a late step in the assembly of 60S ribosomal subunits in *Saccharomyces cerevisiae*. *RNA* **4,** 1268–1281.

de la Cruz, J., Kressler, D., Tollervey, D., and Linder, P. (1998b). Dob1p (Mtr4p) is a putative ATP-dependent RNA helicase required for the 3′ end formation of 5.8S rRNA in *Saccharomyces cerevisiae*. *EMBO J.* **17,** 1128–1140.

Decatur, W. A., and Fournier, M. J. (2002). rRNA modifications and ribosome function. *Trends Biochem. Sci.* **27,** 344–351.

Dragon, F., Gallagher, J. E., Compagnone-Post, P. A., Mitchell, B. M., Porwancher, K. A., Wehner, K. A., Wormsley, S., Settlage, R. E., Shabanowitz, J., Osheim, Y., Beyer, A. L., Hunt, D. F., et al. (2002). A large nucleolar U3 ribonucleoprotein required for 18S ribosomal RNA biogenesis. *Nature* **417,** 967–970.

Drakas, R., Prisco, M., and Baserga, R. (2005). A modified tandem affinity purification tag technique for the purification of protein complexes in mammalian cells. *Proteomics* **5,** 132–137.

Emery, B., de la Cruz, J., Rocak, S., Deloche, O., and Linder, P. (2004). Has1p, a member of the DEAD-box family, is required for 40S ribosomal subunit biogenesis in *Saccharomyces cerevisiae*. *Mol. Microbiol.* **52,** 141–158.

Etcheverry, T. (1990). Induced expression using yeast copper metallothionein promoter. *Methods Enzymol.* **185,** 319–329.

Fairman-Williams, M. E., Guenther, U. P., and Jankowsky, E. (2010). SF1 and SF2 helicases: Family matters. *Curr. Opin. Struct. Biol.* **20,** 313–324.

Fields, S., and Song, O. (1989). A Novel genetic system to detect protein–protein interactions. *Nature* **340**, 245–246.

Freed, E. F., and Baserga, S. J. (2010). The C-terminus of Utp4, mutated in childhood cirrhosis, is essential for ribosome biogenesis. *Nucleic Acids Res.* **38**, 4798–4806.

Gallagher, J. E., and Baserga, S. J. (2004). Two-hybrid Mpp 10p interaction-defective Imp4 proteins are not interaction defective *in vivo* but do confer specific pre-rRNA processing defects in Saccharomyces cerevisiae. *Nucleic Acids Res.* **32**, 1404–1413.

Gallagher, J., Dunbar, D., Granneman, S., Mitchess, B., Osheim, Y., Beyer, A., and Baserga, S. J. (2004). RNA polymerase I transcritption and pre-rRNA processing are linked by specific SSU processome components. *Genes Dev.* **18**, 2506–2517.

Gavin, A. C., Bosche, M., Krause, R., Grandi, P., Marzioch, M., Bauer, A., Schultz, J., Rick, J. M., Michon, A. M., Cruciat, C. M., Remor, M., Hofert, C., et al. (2002). Functional organization of the yeast proteome by systematic analysis of protein complexes. *Nature* **415**, 141–147.

Gavin, A. C., Aloy, P., Grandi, P., Krause, R., Boesche, M., Marzioch, M., Rau, C., Jensen, L. J., Bastuck, S., Dumpelfeld, B., Edelmann, A., Heurtier, M. A., et al. (2006). Proteome survey reveals modularity of the yeast cell machinery. *Nature* **440**, 631–636.

Gloeckner, C. J., Boldt, K., Schumacher, A., and Ueffing, M. (2009). Tandem affinity purification of protein complexes from mammalian cells by the Strep/FLAG (SF)-TAP tag. *Methods Mol. Biol.* **564**, 359–372.

Granneman, S., and Baserga, S. J. (2004). Ribosome biogenesis: Of knobs and RNA processing. *Exp. Cell Res.* **15**, 43–50.

Granneman, S., Bernstein, K. A., Bleichert, F., and Baserga, S. J. (2006a). Comprehensive mutational analysis of yeast DExD/H box RNA helicases required for small ribosomal subunit synthesis. *Mol. Cell. Biol.* **26**, 1183–1194.

Granneman, S., Lin, C., Champion, E. A., Nandineni, M. R., Zorca, C., and amd Baserga, S. J. (2006b). The nucleolar protein Esf2 interacts directly with the DExD/H box RNA helicase, Dbp8, to stimulate RNA hydrolysis. *Nuc. Acids Res* **34**, 3189–3199.

Granneman, S., Kudla, G., Petfalski, E., and Tollervery, D. (2009). Identification of protein binding sites on U3 snoRNA and pre-rRNA by UV cross-linking and high-throughput analysis of cDNAs. *Proc. Natl. Acad. Sci. USA* **106**, 9613–9618.

Harnpicharchai, P., Jakovljevic, J., Horsey, E., Miles, T., Roman, J., Rout, M., Meacher, D., Imai, B., Guo, Y., Brame, C. J., Shabanowitz, J., Hunt, D. F., et al. (2001). Composition and functional characterization of yeast 66S ribosome assembly intermediates. *Mol. Cell* **8**, 505–515.

Henras, A. K., Soudet, J., Gerus, M., Lebaron, S., Caizergues-Ferrer, M., Mougin, A., and Henry, Y. (2008). The post-transcriptional steps of eukaryotic ribosome biogenesis. *Cell. Mol. Life Sci.* **65**, 2334–2359.

Hoang, T., Peng, W. T., Vanrobays, E., Krogan, N., Hiley, S., Beyer, A. L., Osheim, Y. M., Breenblatt, J., Hughes, T. R., and Lafontaine, D. L. (2005). Esf2p, a U3-associated factor required for small-subunit processome assembly and compaction. *Mol. Cell. Biol.* **25**, 5523–5534.

Huh, W. K., Falvo, J. V., Gerke, L. C., Carroll, A. S., Howson, R. W., Weissman, J. S., and O'Shea, E. K. (2003). Global analysis of protein localization in budding yeast. *Nature* **425**, 686–691.

Ito, T., Chiba, T., Ozawa, R., Yoshida, M., Hattori, M., and Sakaki, Y. (2001). A comprehensive two-hybrid analysis to explore the yeast protein interactome. *Proc. Natl. Acad. Sci. USA* **98**, 4569–4574.

Jankowsky, E. (2011). RNA helicases at work: Binding and rearranging. *Trends Biochem. Sci.* **36**, 19–29.

Ju, Q., and Warner, J. R. (1994). Ribosome synthesis during the growth cycle of *Saccharomyces cerevisiae*. *Yeast* **10**, 151–157.

Knop, M., Siegers, K., Pereira, G., Zachariae, W., Winsor, B., Nasmyth, K., and Schiebel, E. (1999). Epitope tagging of yeast genes using a PCR-based strategy: More tags and improved practical routines. *Yeast* **15**, 963–972.

Koegl, M., and Uetz, P. (2007). Improving yeast two-hybrid screening systems. *Brief. Funct. Genomic. Proteomic.* **6**, 302–312.

Kos, M., and Tollervey, D. (2005). The Putative RNA Helicase Dbp4p is Required for Release of the U14 snoRNA from Preribosomes in Saccharomyces cerevisiae. *Mol. Cell* **20**, 53–64.

Kressler, D., de la Cruz, J., Rojo, M., and Linder, P. (1997). Fal1p is an essential DEAD-box protein involved in 40S-ribosomal-subunit biogenesis in *Saccharomyces cerevisiae*. *Mol. Cell. Biol.* **17**, 7283–7294.

Kressler, D., de la Cruz, J., Rojo, M., and Linder, P. (1998). Dbp6p is an essential putative ATP-dependent RNA helicase required for 60S-ribosomal-subunit assembly in *Saccharomyces cerevisiae*. *Mol. Cell. Biol.* **18**, 1855–1865.

Kressler, D., Hurt, E., and Bassler, J. (2010). Driving ribosome assembly. *Biochim. Biophys. Acta* **1803**, 673–683.

Krogan, N. J., Peng, W. T., Cagney, G., Robinson, M. D., Haw, R., Zhong, G., Guo, X., Zhang, X., Canadien, V., Richards, D. P., Beattie, B. K., Lalev, A., et al. (2004). High-definition macromolecular composition of yeast RNA-processing complexes. *Mol. Cell* **13**, 225–239.

Krogan, N. J., Cagney, G., Yu, H., Zhong, G., Guo, X., Ignatchenko, A., Li, J., Pu, S., Datta, N., Tikuisis, A. P., Punna, T., Peregrin-Alvarez, J. M., et al. (2006). Global landscape of protein complexes in the yeast *Saccharomyces cerevisiae*. *Nature* **440**, 637–643.

Kumar, A., Agarwal, S., Heyman, J. A., Matson, S., Heidtman, M., Piccirillo, S., Umansky, L., Drawid, A., Jansen, R., Liu, Y., Cheung, K. H., Miller, P., et al. (2002). Subcellular localization of the yeast proteome. *Genes Dev.* **16**, 707–719.

Lebaron, S., Froment, C., Fromont-Racine, M., Rain, J., Monsarrat, B., and Caizergues-Ferrer, M. (2005). The splicing ATPase Prp43p is a component of multiple preribosomal particles. *Mol. Cell. Biol.* **25**, 9269–9282.

Lebaron, S., Papin, C., Capeyrou, R., Chen, Y. L., Froment, C., Monsarrat, B., Caizergues-Ferrer, M., Grigoriev, M., and Henry, Y. (2009). The ATPase and helicase activities of Prp43p are stimulated by the G-patch protein Pfa1p during yeast ribosome biogenesis. *EMBO J.* **28**, 3808–3819.

Lee, W. C., Zabetakis, D., and Melese, T. (1992). NSR1 is required for pre-rRNA processing and for the proper maintenance of the steady-state levels of ribosomal subunits. *Mol. Cell. Biol.* **12**, 3865–3871.

Leeds, N. B., Small, E. C., Hiley, S. L., Hughes, T. R., and Staely, J. P. (2006). The splicing factor Prp43p, a DEAH box ATPase, functions in ribosome biogenesis. *Mol. Cell. Biol.* **26**, 513–522.

Leung, A. K., Trinkle-Mulcahy, L., Lam, Y. W., Andersen, J. S., Mann, M., and Lamond, A. I. (2006). NOPdb: Nucleolar proteome database. *Nucleic Acids Res.* **34**, D218–D220.

Li, Y., Franklin, S., Zhang, M. J., and Vondriska, T. M. (2011). Highly efficient purification of protein complexes from mammalian cells using a novel streptavidin-binding peptide and hexahistidine tandem tag system: Application to Bruton's tyrosine kinase. *Protein Sci.* **20**, 140–149.

Liang, X. H., and Fournier, M. J. (2006). The helicase Has1p is required for snoRNA release from pre-rRNA. *Mol. Cell. Biol.* **26**, 7437–7450.

Licatalosi, D. D., Mele, A., Fak, J. J., Ule, J., Kayikci, M., Chi, S. W., Clark, T. A., Schweitzer, A. C., Blume, J. E., Wang, X., Darnell, J. C., and Darnell, R. B. (2008).

HITS-CLIP yields genome-wide insights into brain alternative RNA processing. *Nature* **456,** 464–469.

Lim, Y. H., Charette, J. M., and Baserga, S. J. (2011). Assembling a protein–protein interaction map of the SSU processome from existing datasets. *PLoS One* **6,** e17701.

Linder, P., and Jankowsky, E. (2011). From unwinding to clamping—The DEAD box RNA helicase family. *Nat. Rev. Mol. Cell Biol.* **12,** 505–516.

Liu, F., Putnam, A., and Jankowsky, E. (2008). ATP hydrolysis is required for DEAD-box protein recycling but not for duplex unwinding. *Proc. Natl. Acad. Sci. USA* **105,** 20209–20214.

Longtine, M. S., McKenzie, A., 3rd, Demarini, D. J., Shah, N. G., Wach, A., Brachat, A., Philippsen, P., and Pringle, J. R. (1998). Additional modules for versatile and economical PCR-based gene deletion and modification in *Saccharomyces cerevisiae*. *Yeast* **14,** 953–961.

Mumberg, D., Mailer, R., and Funk, M. (1995). Yeast vectors for the controlled expression of heterologous proteins in different genetic backgrounds. *Gene* **156,** 119–122.

Nissan, T. A., Bassler, J., Petfalski, E., Tollervey, D., and Hurt, E. (2002). 60S pre-ribosome formation viewed from assembly in the nucleolus until export to the cytoplasm. *EMBO J.* **21,** 5539–5547.

O'Day, C. L., Chavanikamannil, F., and Abelson, J. (1996). 18S rRNA processing requires the RNA helicase-like protein Rrp3. *Nucleic Acids Res.* **24,** 3201–3207.

Oeffinger, M., Wei, K. E., Rogers, R., DeGrasse, J. A., Chait, B. T., Aitchison, J. D., and Rout, M. P. (2007). Comprehensive analysis of diverse ribonucleoprotein complexes. *Nat. Methods* **4,** 951–956.

Pérez-Fernández, J., Román, A., De Las Rivas, J., Bustelo, X., and Dosil, M. (2007). The 90S preribosome is a multimodular structure that is assembled through a hierarchical mechanism. *Mol. Cell. Biol* **27,** 5415–5429.

Puig, O., Caspary, F., Rigaut, G., Rutz, B., Bouveret, E., Bragado-Nilsson, E., Wilm, M., and Seraphin, B. (2001). The tandem affinity purification (TAP) method: A general procedure of protein complex purification. *Methods* **24,** 218–229.

Pyle, A. M. (2008). Translocation and unwinding mechanisms of RNA and DNA helicases. *Annu. Rev. Biophys.* **37,** 317–336.

Rajagopala, S. V., and Uetz, P. (2009). Analysis of protein-protein interactions using array-based yeast two-hybrid screens. *Methods Mol. Biol.* **548,** 223–245.

Rajagopala, S. V., and Uetz, P. (2011). Analysis of protein-protein interactions using high-throughput yeast two-hybrid screens. *Methods Mol. Biol.* **781,** 1–29.

Rigaut, G., Shevchenko, A., Rutz, B., Wilm, M., Mann, M., and Seaphin, B. (1999). A generic protein purification method for protein complex characterization and proteome exploration. *Nat. Biotechnol.* **17,** 1030–1032.

Ripmaster, T. L., Vaughn, G. P., and Woolford, J. L., Jr. (1992). A putative ATP-dependent RNA helicase involved in *Saccharomyces cerevisiae* ribosome assembly. *Proc. Natl. Acad. Sci. USA* **89,** 11131–11135.

Rocak, S., Emery, B., Tanner, N. K., and Linder, P. (2005). Characterization of the ATPase and unwinding activities of the yeast DEAD-box protein Has1p and the analysis of the roles of the conserved motifs. *Nucleic Acids Res.* **33,** 999–1009.

Ross-Macdonald, P., Coelho, P. S., Roemer, T., Agarwal, S., Kumar, A., Jansen, R., Cheung, K. H., Sheehan, A., Symoniatis, D., Umansky, L., Heidtman, M., Nelson, F. K., *et al.* (1999). Large-scale analysis of the yeast genome by transposon tagging and gene disruption. *Nature* **402,** 413–418.

Scherl, A., Coute, Y., Deon, C., Calle, A., Kindbeiter, K., Sanchez, J. C., Greco, A., Hochstrasser, D., and Diaz, J. J. (2002). Functional proteomic analysis of human nucleolus. *Mol. Biol. Cell* **13,** 4100–4109.

Schneider, B. L., Seufert, W., Steiner, B., Yang, Q. H., and Futcher, A. B. (1995). Use of polymerase chain reaction epitope tagging for protein tagging in *Saccharomyces cerevisiae*. *Yeast* **11,** 1265–1274.

Sikorski, R. S., and Hieter, P. (1989). A system of shuttle vectors and yeast host strains designed for efficient manipulation of DNA in *Saccharomyces cerevisiae*. *Genetics* **122,** 19–27.

Steitz, J. A. (1989). Immunoprecipitation of ribonucleoproteins using autoantibodies. *Methods Enzymol.* **180,** 468–481.

Tanner, N. K., and Linder, P. (2001). DExD/H box RNA helicases: From generic motors to specific dissociation functions. *Mol. Cell* **8,** 251–262.

Uetz, P., Giot, L., Cagney, G., Mansfield, T. A., Judson, R. S., Knight, J. R., Lockshon, D., Narayan, V., Srinivasan, M., Pochart, P., Qureshi-Emili, A., Li, Y., *et al.* (2000). A comprehensive analysis of protein-protein interactions in *Saccharomyces cerevisiae*. *Nature* **403,** 623–627.

Ule, J., Jensen, K., Mele, A., and Darnell, R. B. (2005). CLIP: A method for identifying protein-RNA interaction sites in living cells. *Methods* **37,** 376–386.

Venema, J., and Tollervery, D. (1996). RRP5 is required for formation of both 18S and 5.8S rRNA in yeast. *EMBO J.* **15,** 5701–5714.

Venema, J., and Tollervey, D. (1999). Ribosome synthesis in *Saccharomyces cerevisiae*. *Annu. Rev. Genet.* **33,** 261–311.

Venema, J., Bousquet-Antonelli, C., Gelugne, J. P., Caizerques-Ferrer, M., and Tollervey, D. (1997). Rok1p is a putative RNA helicase required for rRNA processing. *Mol. Cell. Biol.* **17,** 3398–3407.

Wahl, M. C., Will, C. L., and Lührmann, R. (2009). The spliceosome: Design principles of a dynamic RNP machine. *Cell* **136,** 701–718.

Weaver, P. L., Sun, C., and Chang, T. H. (1997). Dbp3p, a putative RNA helicase in *Saccharomyces cerevisiae*, is required for efficient pre-rRNA processing predominantly at site A3. *Mol. Cell. Biol.* **17,** 1354–1365.

Wehner, K. A., Gallagher, J. E., and Baserga, S. J. (2002). Components of an interdependent unit within the SSU processome regulate and mediate its activity. *Mol. Cell. Biol.* **22,** 7258–7267.

Yoshikawa, H., Komatsu, W., Hayano, T., Miura, Y., Homma, K., Izumikawa, K., Ishikawa, H., Miyazawa, N., Tachikawa, H., Yamauchi, Y., Isobe, T., and Takahashi, N. (2011). Splicing factor 2-associated protein p32 participates in ribosome biogenesis by regulating the binding of Nop52 and Fibrillarin to preribosome particles. *Mol. Cell. Proteomics* **10,** M110.006148.

Yu, H., Braun, P., Yildirim, M. A., Lemmens, I., Venkatesan, K., Sahalie, J., Hirozane-Kishikawa, T., Gebreab, F., Li, N., Simonis, N., Hao, T., Rual, J. F., *et al.* (2008). High-quality binary protein interaction map of the yeast interactome network. *Science* **322,** 104–110.

CHAPTER FIFTEEN

Analysis of RNA Helicases in P-Bodies and Stress Granules

Angela Hilliker

Contents

1. Introduction	324
2. RNA Helicases in Cytoplasmic mRNP Granules	325
2.1. RNA helicases that localize to cytoplasmic mRNP granules	325
2.2. Mechanism of RNA helicase localization to cytoplasmic mRNP granules	328
3. RNA Helicases That Affect P-Bodies or SGs	329
4. Determining Whether an RNA Helicase Can Localize to Cytoplasmic mRNP Granules	331
4.1. Markers of cytoplasmic mRNP granules in *Saccharomyces cerevisiae*	331
4.2. Induction of cytoplasmic mRNP granules in *S. cerevisiae*	332
4.3. Markers of cytoplasmic mRNP granules in mammalian cell culture	334
4.4. Induction of cytoplasmic mRNP granules in mammalian cell culture	334
4.5. Quantification of cytoplasmic granules	335
5. Determining Whether an RNA Helicase Affects Cytoplasmic mRNP Granules	337
5.1. Genetic depletion or deletion of RNA helicases	337
5.2. Overexpression of RNA helicases	338
5.3. ATPase domain mutations in RNA helicases	339
5.4. N and C terminal mutations in RNA helicases	340
5.5. Small molecule inhibitors of RNA helicases	340
6. Discussion and Perspective	340
Acknowledgment	341
References	341

Abstract

Cytoplasmic mRNA protein complexes (mRNPs) can assemble in granules, such as processing bodies (P-bodies) and stress granules (SGs). Both P-bodies and

Department of Biology, The University of Richmond, Richmond, Virginia, USA

Methods in Enzymology, Volume 511
ISSN 0076-6879, DOI: 10.1016/B978-0-12-396546-2.00015-2

© 2012 Elsevier Inc.
All rights reserved.

SGs contain repressed messenger RNAs (mRNAs) and proteins that regulate the fate of the mRNA. P-bodies contain factors involved in translation repression and mRNA decay; SGs contain a subset of translation initiation factors and mRNA-binding proteins. mRNAs cycle in and out of granules and can return to translation. RNA helicases are found in both P-bodies and SGs. These enzymes are prime candidates for facilitating the changes in mRNP structure and composition that may determine whether an mRNA is translated, stored, or degraded. This chapter focuses on the RNA helicases that localize to cytoplasmic granules. I outline approaches to define how the helicases affect the granules and the mRNAs within them, and I explain how analysis of cytoplasmic granules provides insight into physiological function and targets of RNA helicases.

1. Introduction

The regulation of messenger RNAs (mRNAs) in the cytoplasm is an important aspect in the control of gene expression. Cytoplasmic mRNAs associate with different sets of proteins to form mRNA protein complexes (mRNPs) that influence the localization, translatability, and stability of mRNA. The mRNP composition is dynamic and changes in response to cellular signals, allowing an mRNA to move in and out of the translating pool. RNA helicases are strong candidates for altering the protein composition of an mRNP and thus affecting the cytoplasmic fate of an mRNA.

A dramatic example of mRNP rearrangement occurs during cellular stress, when mRNAs accumulate in cytoplasmic foci as nontranslating mRNAs. The two most prevalent of these cytoplasmic granules are processing bodies (P-bodies) and stress granules (SGs). P-bodies are present constitutively in cells but increase upon certain stresses. P-bodies contain nontranslating mRNAs and factors involved in translation repression, mRNA decay, and nonsense-mediated decay (NMD). SGs form upon a variety of stresses that inhibit translation initiation. SGs contain nontranslating mRNAs, a subset of translation initiation factors, and modification enzymes (reviewed in Kedersha and Anderson, 2009). Based on their composition, P-bodies may be sites of mRNA decay and storage, while SGs may store mRNAs that are stalled early in translation initiation (reviewed in Buchan and Parker, 2009).

P-bodies and SGs are dynamic. For example, proteins shuttle in and out of these granules (reviewed in Buchan and Parker, 2009). P-bodies and SGs can contain the same mRNA and can dock, suggesting mRNAs may also shuttle between them (Kedersha et al., 2005). Upon stress, P-bodies and SGs can change in size and in their association with one another. Depending on the stress, composition of the granule or its assembly pathway may change (Buchan and Parker, 2009; Buchan et al., 2008; Grousl et al., 2009; Hoyle *et al.*, 2007). Upon relief from stress, the granules disassemble and the mRNAs return to the translating pool (Brengues *et al.*, 2005).

Given the dynamic nature of cytoplasmic granules and the extensive mRNP remodeling that is likely to occur to move an mRNA in and out of translation, it is intriguing that multiple RNA helicases accumulate in these cytoplasmic granules. They could affect mRNA entry to, accumulation in, or exit from granules by altering the protein composition of an mRNP. Additionally, they could be targets for regulation that affects whether an mRNA is translated, stored, or decayed.

2. RNA HELICASES IN CYTOPLASMIC mRNP GRANULES

2.1. RNA helicases that localize to cytoplasmic mRNP granules

While there has been no systematic screen to test for the localization of RNA helicases in cytoplasmic mRNP granules, a few family members localize to one or more type of granule (Table 15.1). Some RNA helicases appear to be constitutive members of cytoplasmic mRNP granules, while others likely associate only transiently.

A handful of RNA helicases have been shown to localize primarily to P-bodies; these include Dhh1/RCK, Upf1, and Mov10. Like other components of P-bodies, these helicases have been implicated in translation repression or mRNA decay pathways. For example, the DEAD-box helicase Dhh1/RCK (yeast/mammalian nomenclature) has been shown to localize to P-bodies and promote mRNA translation repression in several different organisms (Cougot *et al.*, 2004; Hillebrand *et al.*, 2010; Navarro *et al.*, 2001; Sheth and Parker, 2003). Upf1, an SF1 helicase involved in NMD, also colocalizes to P-bodies in both yeast and humans, but only when NMD factors are overexpressed or mRNA decay factors are deleted (Luke *et al.*, 2007; Sheth and Parker, 2006; Unterholzner and Izaurralde, 2004).These results suggest that, in wild-type cells, Upf1 may localize to P-bodies transiently or in small numbers. Finally, Mov10, a Upf1-like helicase (Fairman-Williams *et al.*, 2010) required for miRNA-mediated silencing *in vivo*, colocalizes to P-bodies (Meister *et al.*, 2005).

RNA helicases that colocalize to SGs either have a role in translation initiation and/or translation repression. This includes the DEAD-box helicase eIF4A/DDX2 (Buchan *et al.*, 2010; Low *et al.*, 2005; Mazroui *et al.*, 2006), which is a component of eIF4F and thought to unwind secondary structure in the 5′ UTR to aid translation initiation (reviewed in Parsyan *et al.*, 2011). The DEAD-box helicase Ded1/DDX3, which has been implicated in both translation initiation and translation repression, has also been shown to accumulate in SGs in yeast (Fig. 15.1A; Hilliker *et al.*, 2011)

Table 15.1 RNA helicases that localize to P-bodies and stress granules

RNA helicase (yeast/*mammalian* orthologs, unless otherwise indicated)	Present in	Note
Dhh1/*RCK*[a]	PB, SG, RTG[b]	Presence in SGs after prolonged stress
Upf1/*hUpf1*[c]	PB	Presence in PB only in mutant conditions
Mov10[d]	PB	
eIF4A/*DDX2*[e]	SG	In yeast, intensity varies with type of stress
Ded1/*DDX3*[f]	SG, RTG, germ granules	Overlaps with P-bodies in yeast only[g]
RHAU[h]	SG	
DDX1[i]	SG, RTG	
ISE2 (Arabidopsis)[j]	PB and SG?[k]	
Dbp5/Rat8[l]	PB and SG?[k]	
DDX5[m]	RTG	
Vasa (Drosophila)[n]	Germ granules	
Dicer (Drosophila)[o]	Germ granules	

[a] Budding yeast/mammalian ortholog name indicated here; Sheth and Parker (2003), Wilczynska et al. (2005), Buchan et al. (2008), Mollet et al. (2008), Swisher and Parker (2010).
[b] PB, P-bodies; SG, stress granules; RTG, RNA transport granules (in neurons).
[c] Only localizes to P-bodies in mutant conditions; Sheth and Parker (2006).
[d] Meister et al. (2005).
[e] Mazroui et al. (2006), Buchan et al. (2011).
[f] Goulet et al. (2008), Lai et al. (2008), Hilliker et al. (2011).
[g] Beckham et al. (2008).
[h] Chalupnikova et al. (2008).
[i] Onishi et al. (2008), Elvira et al. (2006), Kanai et al. (2004).
[j] Kobayashi et al. (2007).
[k] In granules that associate with P-bodies, as stress granules do, but were characterized before the availability of stress granule markers.
[l] Scarcelli et al. (2008).
[m] Kanai et al. (2004), Miller et al. (2009).
[n] Johnstone et al. (2005), Kotaja et al. (2006).
[o] Kotaja et al. (2006).

and mammalian cells (Goulet et al., 2008; Lai et al., 2008). The DEAH/RHA helicase RHAU/DHX36 colocalizes to SGs in mammalian cells (Chalupnikova et al., 2008) and binds guanine–quadruplex RNA motifs (Creacy et al., 2008; Vaughn et al., 2005) that promote translation repression (Kumari et al., 2007; Morris and Basu, 2009). The DEAD-box ATPase DDX1 colocalized to SGs in mammalian cells (Onishi et al., 2008). While DDX1 has no known role in translation repression, it accumulates in RNA

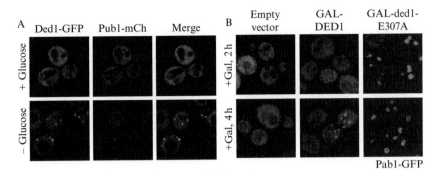

Figure 15.1 Ded1, an RNA helicase, accumulates in and promotes formation of SGs. (A) Yeast cells containing plasmids expressing Ded1-GFP and Pub1-mCh, an SG marker, were grown to mid-log condition. SG accumulation was assayed without stress (top) or after a 15-min glucose deprivation stress. Ded1 colocalizes with Pub1-mCh and other SG markers (data not shown), indicating that it accumulates in SGs. (B) Yeast cells containing a plasmid expressing Pab1-GFP, an SG marker, and a second plasmid containing a galactose-inducible promoter controlling no gene, wild-type *DED1*, or an ATPase-deficient allele (*ded1-E307A*). A time course of galactose-driven overexpression (0, 2, and 4 h) shows that overexpression of wild-type Ded1 induces SGs. Overexpression of the ATPase-deficient allele induces SGs earlier. Reproduced from Hilliker *et al.* (2011), with permission. (See the Color Insert.)

transport granules in neurites (Kanai *et al.*, 2004), which contain repressed mRNAs.

Recently, it was shown that at least one RNA helicase, Dhh1/RCK, colocalizes to both P-bodies and SGs but only after long periods of stress (Buchan *et al.*, 2008; Mollet *et al.*, 2008; Swisher and Parker, 2010; Wilczynska *et al.*, 2005). Its presence in both P-bodies and SGs in multiple organisms makes it a strong candidate for facilitating the exchange of mRNAs between granules and/or altering the protein composition of the mRNPs in each granule.

Some RNA helicases are seen both in P-bodies and in some other, as yet undefined, foci. These foci may be SGs, which were not well described in the organism at the time. For example, ISE2, an SF2 helicase in *Arabidopsis thaliana* that is important for posttranslational gene silencing (Kobayashi *et al.*, 2007), shows limited colocalization (20%) with P-bodies. Most ISE2-containing foci are distinct from, but dock near P-bodies, much like a SG (Kobayashi *et al.*, 2007). Subsequent work defining SGs in *Arabidopsis* provided the markers needed to evaluate whether ISE2 localizes to SGs (Weber *et al.*, 2008). Likewise, a mutant of Dbp5/Rat8, a DEAD-box ATPase involved in mRNA nuclear export in budding yeast, localizes to cytoplasmic foci along with at least one other nuclear export factor, Mex67 (Scarcelli *et al.*, 2008), prompting the proposal that these granules contain mRNPs that have not been remodeled after export. These granules are often proximal to P-bodies and show strong overlap with Pab1 (Scarcelli et al., 2008), similar to yeast SGs. Additionally, export factors have been found in SGs in yeast and mammalian cells (Buchan et al., 2008; Lai et al., 2008).

RNA transport granules in neurons and germ cell granules share overlapping protein composition to P-bodies and SGs (reviewed in Buchan and Parker, 2009) and reveal the presence of RNA helicases that colocalize with repressed mRNPs, namely, DDX1, Ded1/DDX3, DDX5, and Dhh1/RCK (Elvira et al., 2006; Kanai et al., 2004; Miller et al., 2009). Like neurons, germ cells also contain dense localizations of repressed mRNA and proteins that also accumulate in P-bodies and SGs. These foci are called nuage or germplasm (*Drosophila melanogaster*), P granules (*Caenorhabditis elegans*), or chromatoid body (mouse). These foci contain orthologs of Ded1/DDX3 (Drosophila; Johnstone et al., 2005), the DEAD-box RNA helicase Vasa (Drosophila and mouse; Johnstone et al., 2005; Kotaja et al., 2006), and the RIG-I-like ATPase Dicer, which processes miRNAs (mouse; Kotaja et al., 2006).

Several RNA helicases localize to more than one type of cytoplasmic mRNP granule, suggesting that a given RNA helicase can be part of more than one complex. Consistent with this idea, colocalization studies among the RNA helicases in neuronal transport granules show that there are several distinct helicase-containing RNP granules in neurites (Miller et al., 2009). These different RNP granules may reflect that each RNA helicase has unique target mRNAs or that the same mRNAs can aggregate in different mRNPs with unique helicase compositions.

2.2. Mechanism of RNA helicase localization to cytoplasmic mRNP granules

There are two major principles emerging about how RNA helicases localize to different granules. First, ATPase activity is not important for RNA helicase entry into cytoplasmic granules but perhaps for remodeling of mRNPs within granules. Second, N and C termini, which differ among helicase family members, are important for formation of cytoplasmic granules, suggesting that there are diverse paths of entry into granules.

ATPase activity is not critical for accumulation in granules, for most helicases tested (tested for Upf1, Ded1, Dhh1, RHAU; Carroll et al., 2011; Chalupnikova et al., 2008; Dutta et al., 2011; Franks et al., 2010; Hilliker et al., 2011; Stalder and Muhlemann, 2009). However, one study of Dhh1 in *Trypanosoma brucei* contradicts this trend (Kramer et al., 2010). Generally, ATPase activity seems to be important for rearrangements of the mRNP within granules, as ATPase-deficient mutants cause the accumulation of one or more cytoplasmic granule, suggesting a defect in disassembly (Dutta et al., 2011; Hilliker et al., 2011; Sheth and Parker, 2006). Additionally, FRAP (fluorescence recovery after photobleaching) experiments show slow recovery of ATPase-deficient RHAU in bleached SGs (Chalupnikova et al., 2008) and slow recovery of ATPase-defective Dhh1 in bleached P-bodies (Carroll et al., 2011), consistent with slow turnover of ATPase-defective mutants from mRNPs within cytoplasmic granules. These data are consistent with in vitro work that shows ATP turnover is important for helicase release from RNA (Henn et al., 2010; Liu et al., 2008). These results suggest generally that the ATPase activity of RNA helicases is not required to recruit the helicases to cytoplasmic granules, but instead to promote disassembly of the granules or exit of mRNPs from the granules.

While ATP turnover may not be important for an RNA helicase to accumulate in granules, ATP binding, which is tightly coupled to RNA binding, might be. A Dhh1 mutant defective in ATP binding shows a slight defect in Dhh1's accumulation in P-bodies; in contrast, mutants that have no defect in ATP binding, but a large defect in ATP hydrolysis, accumulate in P-bodies better than wild-type Dhh1 (Dutta et al., 2011). Since ATP binding is coupled to RNA binding (reviewed in Pan and Russell, 2010), RNA-binding mutants in Dhh1 also show a slight defect in Dhh1's accumulation in P-bodies (Dutta et al., 2011). Additionally, Upf1 mutants that abolish RNA binding also diminish the accumulation of P-bodies (Cheng et al., 2007).

As ATP hydrolysis does not appear to promote RNA helicase recruitment to granules, the N and C termini are likely important for recruitment. As the termini vary among RNA helicases, there are likely many different modes of entry into cytoplasmic granules, which may contribute to the specificity of helicase function. Three regions of Ded1 (two N and one C terminal) are important for inducing SGs (Hilliker et al., 2011). Additionally, the N terminus of RHAU is necessary and sufficient for accumulation in SGs (Chalupnikova et al., 2008). The N terminus of RHAU also contains an RGG-box domain, which has been implicated in SG localization of other factors, such as Caprin-1 (Mazroui et al., 2002; Solomon et al., 2007) and Scd6 (Rajyaguru et al., 2011).

3. RNA HELICASES THAT AFFECT P-BODIES OR SGS

As RNA helicases have been implicated in RNP and RNA remodeling, they may alter mRNP composition to affect formation, stability, or disassembly of cytoplasmic mRNP granules. Consistent with this idea, RNA helicases promote P-body and SG assembly and disassembly (Table 15.2).

Both Ded1/DDX3 and Dhh1/RCK affect P-bodies. Deletions in N and C terminal portions of Ded1 confer cold sensitivity, which correlates with an increase in P-bodies, but not SGs, suggesting that Ded1 may play a role in disassembling P-bodies (Hilliker et al., 2011). Dhh1/RCK appears to

Table 15.2 RNA helicases that effect P-bodies and stress granules

RNA helicase (yeast/*mammalian* orthologs, unless otherwise indicated)	Proposed effect on granules[a]
Dhh1/*RCK*[b]	PB assembly and disassembly, formation of PB-dependent SGs, separation of PBs and SGs
Ded1/*DDX3*[c]	PB disassembly, SG assembly, and disassembly
Upf1/*hUpf1*[d]	Promotes mRNA recruitment to PBs

[a] Data to support these proposed effects are discussed in the text.
[b] Andrei et al. (2005), Serman et al. (2007), Teixeira and Parker (2007), Carroll et al. (2011), Buchan et al. (2008).
[c] Lai et al. (2008), Hilliker et al. (2011).
[d] Sheth and Parker (2006).

have dual roles in promoting the formation and disassembly of P-bodies. First, Dhh1/RCK appears to promote P-body formation, as knock down or deletion of RCK/Dhh1 reduces P-bodies (Andrei et al., 2005; Serman et al., 2007; Teixeira and Parker, 2007), while overexpression of Dhh1 induces P-bodies in the absence of stress (Coller and Parker, 2005). However, Dhh1/RCK1 may also promote the disassembly of P-bodies, as the *Dhh1* null causes an increase in P-bodies (Teixeira and Parker, 2007) and an ATPase-defective Dhh1 mutant induces the formation of P-bodies (Carroll et al., 2011) in the absence of stress. Perhaps, Dhh1 acts as an ATP-dependent switch to move mRNAs in and out of P-bodies.

SG dynamics are influenced by Ded1/DDX3, which may promote both the assembly and the disassembly of SGs by distinct domains (Hilliker et al., 2011). Overexpression of *DED1* causes the accumulation of SGs in the absence of stress (Fig. 15.1B; Hilliker et al., 2011; Lai et al., 2008), suggesting that Ded1 promotes SG assembly. This accumulation is inhibited by mutations in the N and C terminus of Ded1 (Hilliker et al., 2011), suggesting that these areas of Ded1 promote SG formation. Overexpression of an ATPase-deficient mutant of *ded1* causes a faster accumulation of SGs, compared to overexpression of wild-type *DED1* (Fig. 15.1B), suggesting that the ATPase domain either antagonizes SG assembly or promotes SG disassembly (Hilliker et al., 2011). The latter interpretation suggests that Ded1 acts as an ATP-dependent switch to allow mRNPs to exit SGs.

As P-bodies and SGs can dock with one another or overlap (Buchan et al., 2008; Hoyle et al., 2007; Kedersha et al., 2005), some RNA helicases could affect the interaction between P-bodies and SGs. During glucose deprivation in budding yeast, the formation of SGs requires the presence of

P-bodies; under these conditions, Dhh1 is required for formation of P-body-dependent SGs (Buchan et al., 2008). Dhh1/RCK also appears to be important for the separation of P-bodies and SGs in mammalian cells, as knockdown of RCK leads to the overlap of P-body and SG markers (Serman et al., 2007).

RNA helicases can also promote the accumulation of individual factors in granules without affecting overall granule dynamics. For example, Upf1 is required for recruiting normal mRNA and PTC (premature termination codon)-containing mRNAs into P-bodies. This dependence is unique to Upf1 and not its partners in NMD (Sheth and Parker, 2006).

4. Determining Whether an RNA Helicase Can Localize to Cytoplasmic mRNP Granules

Most work on cytoplasmic granules has been done in budding yeast and mammalian cell culture. This review will focus on the methodology of visualizing cytoplasmic granules in these systems. The issues discussed here are applicable to other organisms.

4.1. Markers of cytoplasmic mRNP granules in *Saccharomyces cerevisiae*

To determine if an RNA helicase localizes to or affects cytoplasmic mRNP granules, one must have a marker unique for each granule type. In budding yeast, the mRNA decay factors Edc3 or Dcp2 are often used as P-body markers, as they colocalize exclusively to P-bodies. For SGs, Pab1 and Pub1 are reliable marker choices. Their orthologs have been found in mammalian SGs (Kedersha et al., 1999), and they are found in yeast SGs induced by multiple stresses (Buchan et al., 2011; Grousl et al., 2009; Hoyle et al., 2007). These factors are fused to fluorescent proteins and available on plasmids (see Buchan et al., 2010 for complete list) that can be transformed into the strain of interest. Additionally, there are plasmids available that contain both a P-body factor conjugated to mCherry and an SG factor conjugated to GFP (Buchan et al., 2008, 2010). These plasmids are useful to assess the effect of an RNA helicase on both granule types simultaneously. Note that, unless the target strain is deleted for the marker gene, introducing these fluorescently tagged versions result in slight overexpression of the granule factor, which could lead to a greater induction of P-bodies and/or SGs.

To test for accumulation of an RNA helicase in P-bodies or SGs, one needs to introduce the granule markers in a strain containing the RNA helicase fused to a different fluorescent protein. Most yeast ORFs are readily available as GFP fusions from Invitrogen (Huh et al., 2003); these fusions are chromosomal and under the control of the endogenous promoter, and should avoid overexpression artifacts. P-body and SG markers fused to mCherry can be used with these GFP fusions (Buchan et al., 2010). In any localization study, one should verify that the helicase fusion is functional, as nonfunctional proteins can be mislocalized. In the case of essential RNA helicases, this test is most simply done by showing that the GFP fusion can complement a null strain. Otherwise, one can purify recombinant helicase-GFP fusions and perform *in vitro* ATPase or helicase assays to determine the functionality of the fusion protein. Finally, nonstressed yeast will show few P-bodies and no SGs, so to assess localization to granules, the granules need to be induced by stress, as discussed below.

It is important to note that P-bodies and SGs overlap significantly in yeast, making the determination of a true SG factor difficult. In fact, all SG factors assessed in yeast to date show overlap with P-bodies (Buchan et al., 2008, 2011; Hilliker et al., 2011; Swisher and Parker, 2010). Therefore, some of these factors are classified as P-body factors. Now that there are clear SG markers in yeast, it is important to assess the colocalization of any new granule component with both a P-body and SG marker and quantitate the localization with P-body and SG markers. As P-bodies are present in low numbers in the absence of stress, any helicase that accumulates in P-bodies in nonstress conditions is clearly a P-body factor.

4.2. Induction of cytoplasmic mRNP granules in *S. cerevisiae*

To determine if an RNA helicase colocalizes with granule markers, one needs to induce the granules through stress. In yeast, both cytoplasmic granules can be induced through a variety of stresses, including glucose deprivation, growth to saturation, osmotic stress, heat shock, and sodium azide treatment (Buchan et al., 2008, 2011; Grousl et al., 2009; Sheth and Parker, 2003; Teixeira et al., 2005). Each stress increases P-bodies but has been characterized to differing degrees and has varying effects on the composition of SGs. Depending on the application, one or more stresses may be preferred.

The best characterized stress for inducing P-bodies and SGs is glucose deprivation (Buchan et al., 2008; Teixeira et al., 2005), followed by sodium azide treatment (Buchan et al., 2011) and high heat stress (Grousl et al., 2009). Each stress leads to subtle differences in the factors that colocalize to SGs. For example, SGs formed under glucose deprivation lack eIF3 components (Buchan et al., 2008) and show faint colocalization of eIF4A, eIF4B, and eIF5B (Buchan et al., 2011). However, SGs formed under sodium azide treatment contain robust levels of eIF4A, eIF4B, and eIF5B as well as eIF1A and some eIF3 components (Buchan et al., 2011). High heat shock granules contain an eIF3 component and a small ribosomal protein, suggesting the presence of the 40S ribosome (Grousl et al., 2009). The composition of SGs under the different stress types suggests that the mRNPs accumulating in each are stalled at different rate limiting steps. One should assess colocalization under several stresses, as an RNA helicase may colocalize to granules in one type of stress, but not all, as is the case for eIF4A.

Comprehensive methods for P-body and SG formation in glucose-deprived budding yeast have been published (Buchan et al., 2010; Nissan and Parker, 2008). While the methods for stressing the yeast cells are described here briefly, refer to these reviews for in depth descriptions. To apply a glucose deprivation stress, yeast are grown at 30 °C with shaking in minimal media containing 2% glucose to mid-log phase (an OD, or optical density, of 0.3–0.5), washed quickly in minimal media lacking glucose, and then incubated at 30 °C with shaking in minimal media lacking glucose for 10 min. This sample should be compared to cells from the same parent culture that were mock stressed, that is, washed and incubated in fresh minimal media containing 2% glucose.

To apply a sodium azide stress, cells are grown to mid-log phase as described above. The cultures are split, with half being treated with a final concentration of 0.5% sodium azide (w/v; from a 10% stock) and half being treated by the addition of the equivalent volume of water. The cells are incubated at 30 °C with shaking for 30 min (Buchan et al., 2011).

Finally, high heat stress is applied by resuspending mid-log phase cells in rich media (YPD, which contains Yeast Extract Peptone and Dextrose) preheated to 46 °C and incubating with shaking for an additional 10 min at 46 °C. Cells were washed with minimal media before being placed on the slide (Grousl et al., 2009), as YPD autofluorescences in the GFP channel (Buchan et al., 2010). After any stress, the cells are concentrated and placed on a slide for immediate observation by fluorescence microscopy.

P-body and SG formation is stochastic, and only some factors that lead to variability in induction are understood. The amount of induction can vary between different "wild-type" laboratory strains (Buchan et al., 2010). The number, intensity, and composition of granules vary both with the type of stress and the duration of stress (Buchan et al., 2008, 2011; Swisher and Parker, 2010). These variables must be held constant between samples. Additionally, the appearance of granules increases over time on a microscope slide (more so in stressed cells than nonstressed cells), so the time spent analyzing cells should be kept to a minimum and held constant between samples. Additionally, one should hold constant the temperature of growth, the media, flask size, shaker speed, growth history, centrifugation speed and time, etc. Even by keeping all these details constant, there is usually variability in the stress response from experiment to experiment. Therefore, all controls should be processed in parallel within each experiment, and strains from different experiments should not be compared without normalization to a control. Because of the stochastic nature of granules (in both yeast and mammalian cells) and the variability from cell to cell in the same population, it is important to (1) take images blindly, that is, without viewing the fluorescent channels before taking the pictures so as not to bias your sample with more granule-containing cells compared to the population, (2) sample a large number of cells (at least 100), and (3) repeat experiments on different days to rule out any effects from variabilities in media batch, temperature fluctuation, etc.

4.3. Markers of cytoplasmic mRNP granules in mammalian cell culture

In mammalian cell culture, the most common markers are DCP1a and RCK for P-bodies and TIA and PABP for SGs. However, given the recent evidence that RCK is present in SGs upon prolonged stress (Buchan et al., 2008; Mollet et al., 2008; Swisher and Parker, 2010; Wilczynska et al., 2005), other P-body markers, such as GE-1/hedls may be preferred (Kedersha and Anderson, 2007). In mammalian cells, unlike yeast, cytoplasmic granules are often fixed and stained using antibodies to endogenous proteins; detailed protocols as well as a list of endogenous markers and antibodies have already been described (Kedersha and Anderson, 2007). Alternatively, one can transiently express factors that are fused to fluorescent proteins. However, the overexpression of P-body or SG components will often cause the appearance of granules in the absence of stress (reviewed in Kedersha and Anderson, 2007). This artifact is not a detriment if one is looking for colocalization of an RNA helicase in granule, but is problematic if one wants to assess how an RNA helicase influences granule dynamics. The markers and their uses in transient expression in mammalian cells have been described (Kedersha and Anderson, 2007).

Recently, cell lines that stably express fluorescently tagged P-body or SG markers have been produced, including GFP-Ago2, which accumulates in both SGs and P-bodies, in HeLa cells (Leung et al., 2006) and a broader series of P-body and SG markers in U2OS cells (Kedersha et al., 2008). Particularly useful are lines that stably express mRFP-Dcp1a (red fluorescent protein; P-body marker), YFP-eIF3b (yellow fluorescent protein; SG marker), and a double marker strain with stably expressed mRFP-Dcp1a and YFP-eIF3b (Kedersha et al., 2008). These cell lines can be used to assess colocalization of target RNA helicases or study the effect of RNA helicases on cytoplasmic granule dynamics.

4.4. Induction of cytoplasmic mRNP granules in mammalian cell culture

Detailed protocols and advice for imaging cytoplasmic granules in either fixed mammalian cells (Kedersha and Anderson, 2007) or live mammalian cells (Kedersha et al., 2008) have already been reviewed. There are several stresses that can induce P-bodies and SGs, including pateamine A and hippuristanol treatment (Dang et al., 2006; Mazroui et al., 2006), heat shock, and overexpression of granule components (reviewed in Kedersha and Anderson, 2007). However, the most common stress used in mammalian cells is treatment with arsenite, which has been well-characterized, causes robust induction of both SGs and P-bodies, and is reversible by washing out the drug. To stress the cells, incubate them in 0.5–1.0 mM arsenite (Sigma S-7400) for 30 min to 1 h in the same media the cells were cultured in to prevent artifacts from flooding the cells with fresh nutrients (Kedersha and Anderson, 2007).

As is the case with budding yeast, the best stress response stems from cells that were growing and translating well. Individual strains will have different granule induction responses, so a time course of treatment may be necessary to find the optimal induction. Finally, as discussed above for yeast, because of the stochastic nature of granule formation and the connections between granules and cell metabolism, it is important to hold constant the growth conditions, growth rate/confluency of cells, environmental conditions, etc. (Kedersha and Anderson, 2007).

4.5. Quantification of cytoplasmic granules

P-bodies and SGs can change in many different ways. The percentage of cells with granules may change. The granules themselves can change in number per cell, size, or intensity of fluorescent signal. Finally, the granules may change in the timing of their appearance upon stress or the timing of their disappearance upon recovery from stress. Therefore, when assessing whether a mutant condition alters cytoplasmic granules, there are many aspects of granule appearance that can be quantitated. Because of the stochastic nature of granules and the variability from cell to cell in the same population, it is important to quantitate a reasonable sample size (100 cells) to avoid unintentional bias.

The quantitation of P-bodies and SGs in yeast has been reviewed (Buchan et al., 2010) and presents challenges that are relevant to cytoplasmic granules in any organism. This section focus on using ImageJ, a free downloadable image analysis package from the NIH, to quantitate cytoplasmic mRNP granules (Abramoff et al., 2004). An alternative version of ImageJ, which is preloaded with useful plug-ins, is available from the Wright Cell Imaging Facility (http://www.uhnres.utoronto.ca/facilities/wcif/download.php).

A step-by-step method for analyzing P-bodies and SGs using ImageJ has already been published (Buchan et al., 2010); however, the major steps and issues will be discussed here. Quantitation of the entire cell is preferable; when each focal plane of the cell has been imaged, collapse the Z-stack image in ImageJ. The contrast setting should be normalized for each individual marker in all images in the experiment; the explicit contrast values will vary depending on the marker and the microscope. Then, use the smooth and subtract functions to make the foci more distinct from background fluorescence. Select the P-bodies in the image using the threshold mask, which is set by the user. The program uses the mask to calculate a number of characteristics about the foci, the most useful being the number of foci, area, and integrated density (intensity) of each focus.

The process to set the threshold is subjective, however, and will vary from image-to-image depending on the ratio of signal in foci to the background cytoplasmic levels. This variability is substantial in yeast as the fluorescently tagged markers are usually expressed from low copy plasmids. As each yeast cell can vary in copy number of the plasmid, the intensity of the fluorescent protein signal can vary, too. Additionally, variability in P-body size complicates the selection of an appropriate threshold. If the threshold filter is set low enough to capture the faintest P-bodies, the mask will inflate the area of the larger P-bodies, causing an inaccurate P-body area value. If one selects the mask to stay within the boundaries of the largest P-bodies, the mask may not select the smallest P-bodies, resulting in a conservative count of P-body number. Therefore, one can count all P-bodies with a generous mask but ignore the area and brightness settings under those conditions. To fairly quantitate the area, the Otsu Threshold filter is suitable. This filter will automatically choose the mask, preventing bias in user selection (Otsu, 1979). This automatically generated threshold can be used to calculate the size (area) or the integrated density (intensity) of the P-bodies. If one is interested in calculating the number of P-bodies per cell or the percentage of cells with P-bodies, the number of cells in the image (and the number of cells containing P-bodies) must be counted and recorded by hand. ImageJ does not aggregate the data from several images, so these numbers should be exported to a spreadsheet program for processing.

Quantitating SGs in yeast presents a different challenge because many SG markers also show a strong cytoplasmic signal. An added difficulty arises from use of plasmid-based markers, which cause variability in cytoplasmic signal from cell to cell. Although the use of chromosomally integrated markers reduces this problem, it does not entirely abrogate it. SGs can be quantitated like P-bodies, above, but it may not be possible to choose a single threshold that works for every cell in the image. One can select small portions of the image with similar cytoplasmic background levels and quantitate each portion of the image separately, as described for P-bodies, defining the threshold independently in each section of the image (see also Buchan *et al.*, 2010).

Because of the subjective aspects of this data analysis, that is, threshold setting, it is best to score these images blindly. Additionally, it is recommended to train by quantitating the same small data set a few times to make sure that the quantitation is reproducible.

5. Determining Whether an RNA Helicase Affects Cytoplasmic mRNP Granules

Different types of mutations are useful for learning more about RNA helicase function *in vivo*. By introducing an RNA helicase mutant into a strain that has both a P-body and an SG marker, one can determine whether the RNA helicase mutant affects the formation, dissociation, or composition of both granules as well as the interaction between the granules. Through examination of effects of RNA helicases on cytoplasmic mRNPs granules, we can obtain insight into the role of RNA helicases in modulating mRNPs. Additionally, we may gain insight into physiological targets of these RNA helicases. The following section will explain the different types of mutations and their benefits.

5.1. Genetic depletion or deletion of RNA helicases

Creating a null in a gene is a clean and powerful way to assess its importance in granule biology. Upon deleting or downregulating an RNA helicase gene, one can assess whether that RNA helicase (1) promotes granule assembly generally, (2) promotes granule disassembly, or (3) affects the accumulation of other proteins or of mRNA into granules. The way to achieve a null phenotype varies based on the type of gene and the host organism. In most model organisms, it is easiest to downregulate the RNA helicase of interest by RNA interference. In budding yeast, nonessential genes can be readily deleted in a haploid strain. In fact, a collection containing deletions of each nonessential gene in *S. cerevisiae* is available (Invitrogen; Winzeler *et al.*, 1999). For essential genes, one can put the gene under the control of a regulatable promoter. These strains are commercially available as well (Open Biosystems; Mnaimneh *et al.*, 2004).

Genetic nulls are useful for many reasons. First, one can assess whether the RNA helicase promotes granule assembly. For example, deletion of *Dhh1* (or knock down of RCK) decreases the number of stress-induced P-bodies, suggesting that Dhh1/RCK promotes P-body formation (Andrei *et al.*, 2005; Serman *et al.*, 2007; Teixeira and Parker, 2007). Second, one can assess whether the RNA helicase is required to recruit other factors to granules. For example, a *upf1* deletion illustrates that Upf1 is required for PTC-containing mRNA to accumulate in P-bodies (Sheth and Parker, 2006). Finally, deletion mutants are useful for ordering the function of the helicase compared to other factors through an epistasis analysis. In an epistasis experiment, one can order the functions of two genes that have distinct mutant phenotypes by combining the two mutants and assessing whether one gene's phenotype masks the other. If this is the case, the gene whose mutant phenotype masked the other is considered epistatic, or working upstream of, the other gene. For example, $upf1\Delta$ cells have a few P-bodies in nonstressed conditions, while $upf2\Delta$ or $upf3\Delta$ causes accumulation of P-bodies in nonstressed conditions, likely because NMD substrates are accumulating (Sheth and Parker, 2006). The double mutants $upf1\Delta upf2\Delta$ and $upf1\Delta upf3\Delta$ look like $upf1\Delta$ alone; that is, there is no accumulation of P-bodies, suggesting that *UPF1* must function upstream of *UPF2* and *UPF3*.

5.2. Overexpression of RNA helicases

If a factor is thought to promote cytoplasmic granules, it is traditionally transfected and overexpressed, especially in mammalian cells. If this overexpression induces cytoplasmic granules, it is usually concluded that the factor promotes granule formation. This reasoning has been applied to Dhh1/RCK and Upf1, whose overexpression induces P-bodies (Coller and Parker, 2005; Luke *et al.*, 2007; Sheth and Parker, 2006), and Ded1/DDX3, whose overexpression induces SGs (Hilliker *et al.*, 2011; Lai *et al.*, 2008). Since the RNA helicase is wild type, it will still be able to bind its target. For RNA helicases involved in cytoplasmic granules, we can visualize and characterize the targets by fluorescence microscopy.

Overexpression of *DED1* serves as a good example of this principle. Overexpression of *DED1* causes the accumulation of SGs that contain eIF4E, eIF4G, eIF4A, and eIF4B, but not eIF3, suggesting that mRNPs in these SGs have assembled all of eIF4F, but have not associated with the 43S ribosome, which includes eIF3 and the small subunit of the ribosome; these SGs appear to be stalled between a "glucose deprivation" SG and a "sodium azide" SG (discussed above). Consistent with this idea, *in vitro* studies show that excess Ded1 blocks the addition of the 43S ribosome on the mRNA (Hilliker *et al.*, 2011). Additionally, Ded1 has been shown to interact directly with eIF4G (Hilliker *et al.*, 2011), while its mammalian counterpart, DDX3, has been shown to interact with eIF4E (Shih *et al.*, 2008). This example illustrates that microscopy of cytoplasmic granules can reveal potential targets of RNA helicases. Factors that colocalize with the RNA helicase could be part of the helicase's mRNP. Of course, this result is not definitive, as cytoplasmic granules may contain a diverse set of mRNPs.

As the physiological targets of most RNA helicases are unknown, simple fluorescence microscopy experiments can narrow the field of possible targets. While granules allow coarse mapping of mRNP composition, they do not reveal the details of those interactions. Other techniques, such as immunoprecipitation and CLIP (cross-linking and immunoprecipitation), can be used to identify proteins or RNAs bound to the helicase (Hafner *et al.*, 2010; Zhang and Darnell, 2011).

5.3. ATPase domain mutations in RNA helicases

An alternative way to trap an RNA helicase in a physiologically relevant intermediate and to identify cellular helicase targets are mutations in the conserved ATPase domain of an RNA helicase. For example, *in vitro* studies with Ded1 show that ATP binding is important for Ded1's association with RNA and that ATP hydrolysis is important for Ded1's turnover (Liu *et al.*, 2008). This work suggests that an ATPase-defective helicase mutant may bind its target RNA (and proteins) and allow this target RNP to accumulate. This accumulation should be augmented by overexpressing the ATPase-deficient mutant. Consistent with this notion, overexpression of wild-type *DED1* causes the accumulation of SGs, while overexpression of an ATPase-defective allele causes faster accumulation of SGs (Fig. 15.1B).

ATPase-defective helicase mutants have several important uses. First, they may trap the RNA helicase on the physiological target. As discussed above, overexpression of wild-type or mutant *DED1* allowed staging of Ded1 in granules that contain only a subset of SG factors (Hilliker *et al.*, 2011). Second, ATPase-defective mutants may trap the RNA helicase in granules when it normally localizes to granules transiently. For example, Upf1 does not colocalize to P-bodies in normal cells. While overexpression of *UPF1* causes a low level of Upf1 localization to P-bodies, overexpression of ATPase-deficient mutants of *upf1* induces much larger P-bodies with more Upf1 (Cheng *et al.*, 2007; Sheth and Parker, 2006). Third, ATPase-defective alleles can illuminate the physiological role of the RNA helicase. For example, overexpression of ATPase-deficient *upf1* alleles leads to the accumulation of P-bodies that contain mRNA (normal and PTC-containing) and normal mRNA decay factors, but not NMD factors, such as Upf2 and Upf3 (Sheth and Parker, 2006). In mammalian cells, similar results are seen with strong accumulation of the mRNA decay factor Xrn1 and NMD factors Smg5, 6, and 7, but only weak accumulation of Smg1 and hUpf2 (Franks *et al.*, 2010). These results suggest that ATPase-defective upf1 can form an mRNP that contains PTC-containing mRNA, but that this mRNP is not yet ready for NMD. Additionally, these microscopy results suggest that Upf1's ATP turnover is necessary to recruit Upf2 and Upf3 to the mRNP.

Overexpression of ATPase-defective RNA helicase mutants may also help to enrich an RNA helicase bound to their target RNPs and may be useful prior to, affinity chromatography and mass spectrometry to identify protein partners, or CLIP to define RNA targets. Studies of exogenous expression of hUpf1 ATPase-defective mutants (but not wild-type hUpf1) showed an accumulation of NMD factors Smg5, Smg6, Smg7, and hUpf2 in P-bodies, and affinity chromatography of hupf1 ATPase-deficient mutants copurified with more of these factors, compared to wild-type hUpf1 (Franks *et al.*, 2010).

The studies on *UPF1* and *DED1* examined mutations in motif II of the conserved ATPase domain. Mutations within the other conserved motifs are likely to be informative for this type of analysis as well. However, mutations in the ATP-binding site that affect ATP hydrolysis are often accompanied by defects in ATP binding. Therefore, it may be difficult to deconvolute effects of ATP binding, hydrolysis, or product release.

5.4. N and C terminal mutations in RNA helicases

As the catalytic core of SF1 and SF2 helicases is highly conserved, the variable N and C termini likely confer specificity. Mutations in these C and N termini may thus provide insight into targets and functions of RNA helicases. For example, analysis of RHAU revealed an N terminal RNA-binding site outside the ATPase domain that is important both for RNA binding and accumulation in SGs (Chalupnikova et al., 2008). Mutations in both the N and C terminus of Ded1 promote the induction of SGs upon *DED1* overexpression and are required either for Ded1's interaction with eIF4G (Hilliker et al., 2011) or for the interaction of the Ded1 ortholog DDX3 with eIF4E (Shih et al., 2008).

5.5. Small molecule inhibitors of RNA helicases

As RNA helicases are often essential and mutations in conserved motifs are lethal, another approach to understanding their function is to use small molecule inhibitors. For example, NMDI1, a small molecule inhibitor of hUpf1, causes the accumulation of hUpf1 in P-bodies that contain other NMD factors. In contrast, NMDI1 prevents P-body accumulation of Smg5, a protein involved in Upf1 dephosphorylation, which is usually found in P-bodies. Consistent with this granule composition, NMDI1 precludes copurification of hUpf1 and Smg5 from cells and promotes the accumulation of hyperphosphorylated hUpf1 (Durand et al., 2007). NMDI1 is thus a useful tool to study the effect of hyperphosphorylated hUpf1, without the caveats of knock down and transfection experiments. Other RNA helicases are targets of small molecule inhibitors, including DDX3 (Maga et al., 2008), eIF4A (Bordeleau et al., 2006), and RNA helicase A (RHA; Erkizan et al., 2009). For a detailed discussion of inhibitors of RNA helicases, see Chapter 20.

6. Discussion and Perspective

P-bodies and SGs represent pools of translationally repressed mRNAs stalled in different types of protein complexes. P-body mRNPs include translation repression factors, while SG mRNPs include a subset of translation initiation factors and may be primed for reentry into translation. Interestingly, the composition of P-bodies, SGs, and related RNA granules can vary depending on stress or genetic condition, suggesting that granules can be used to broadly stage the protein complex with which mRNAs associate. Cytoplasmic mRNP granules are highly dynamic and likely require energy to form and disassemble these granules (Grousl et al., 2009). Given the role of RNA helicases in remodeling RNA/RNA or RNA/protein complexes, the RNA helicases within these granules may recruit or release mRNAs from cytoplasmic granules.

These RNA helicases accumulate in foci that are tractable by microscopy. Thus, a simple combination of genetics and microscopy can illuminate the physiological function and targets of these helicases. For example, loss-of-function mutations in RNA helicases will illuminate how the RNA helicase is recruited to granules or how the RNA helicase affects granule dynamics. Alternatively, overexpression of ATPase-deficient mutants may allow binding of a given RNA helicase to its targets without turnover, potentially causing the accumulation of target-bound helicase in the cytoplasmic granules. These approaches allow us to quickly devise models of RNA helicase targets and functions that can be further tested with more direct methods including CLIP or co-immunoprecipitation. Conceptually identical approaches could also be used with RNA helicases that do not accumulate in P-bodies or SGs. Overexpression of their ATPase-deficient mutants may cause aggregation of their RNP targets, which could then be characterized by numerous methods, including microscopy, affinity chromatography, or CLIP.

Last, it is important to understand how RNA helicases are modulated *in vivo*. Previous work suggests that Dhh1 (discussed above) and Ded1 (Hilliker et al., 2011) may act as ATP-dependent switches to move mRNPs through P-bodies and SGs, respectively. Thus, by controlling when an RNA helicase hydrolyzes and turns over ATP, the storage and release of an mRNA from a translationally repressed state can be tightly controlled. Since RNA helicases that accumulate in P-bodies and SGs are also present in RNP granules in germ cells and in neurons, modulation of the ATPase domain of these RNA helicases may be an important part of translational control during stress, early development, and memory formation.

ACKNOWLEDGMENT

I would like to thank Roy Parker for helpful comments on this chapter.

REFERENCES

Abramoff, M. D., Magalhães, P. J., and Ram, S. J. (2004). Image processing with imageJ. *Biophotonics International* **11**, 36–42.
Andrei, M. A., Ingelfinger, D., Heintzmann, R., Achsel, T., Rivera-Pomar, R., and Luhrmann, R. (2005). A role for eIF4E and eIF4E-transporter in targeting mRNPs to mammalian processing bodies. *RNA* **11**, 717–727.

Beckham, C., Hilliker, A., Cziko, A. M., Noueiry, A., Ramaswami, M., and Parker, R. (2008). The DEAD-Box RNA Helicase Ded1p Affects and Accumulates in Saccharomyces cerevisiae P-Bodies. *Mol. Biol. Cell* **19**, 984–993.

Bordeleau, M. E., Mori, A., Oberer, M., Lindqvist, L., Chard, L. S., Higa, T., Belsham, G. J., Wagner, G., Tanaka, J., and Pelletier, J. (2006). Functional characterization of IRESes by an inhibitor of the RNA helicase eIF4A. *Nat. Chem. Biol.* **2**, 213–220.

Brengues, M., Teixeira, D., and Parker, R. (2005). Movement of eukaryotic mRNAs between polysomes and cytoplasmic processing bodies. *Science* **310**, 486–489.

Buchan, J. R., and Parker, R. (2009). Eukaryotic stress granules: The ins and outs of translation. *Mol. Cell* **36**, 932–941.

Buchan, J. R., Muhlrad, D., and Parker, R. (2008). P bodies promote stress granule assembly in Saccharomyces cerevisiae. *J. Cell Biol.* **183**, 441–455.

Buchan, J. R., Nissan, T., and Parker, R. (2010). Analyzing P-bodies and stress granules in Saccharomyces cerevisiae. *Methods Enzymol.* **470**, 619–640.

Buchan, J. R., Yoon, J. H., and Parker, R. (2011). Stress-specific composition, assembly and kinetics of stress granules in Saccharomyces cerevisiae. *J. Cell Sci.* **124**, 228–239.

Carroll, J. S., Munchel, S. E., and Weis, K. (2011). The DExD/H box ATPase Dhh1 functions in translational repression, mRNA decay, and processing body dynamics. *J. Cell Biol.* **194**, 527–537.

Chalupnikova, K., Lattmann, S., Selak, N., Iwamoto, F., Fujiki, Y., and Nagamine, Y. (2008). Recruitment of the RNA helicase RHAU to stress granules via a unique RNA-binding domain. *J. Biol. Chem.* **283**, 35186–35198.

Cheng, Z., Muhlrad, D., Lim, M. K., Parker, R., and Song, H. (2007). Structural and functional insights into the human Upf1 helicase core. *EMBO J.* **26**, 253–264.

Coller, J., and Parker, R. (2005). General translational repression by activators of mRNA decapping. *Cell* **122**, 875–886.

Cougot, N., Babajko, S., and Seraphin, B. (2004). Cytoplasmic foci are sites of mRNA decay in human cells. *J. Cell Biol.* **165**, 31–40.

Creacy, S. D., Routh, E. D., Iwamoto, F., Nagamine, Y., Akman, S. A., and Vaughn, J. P. (2008). G4 resolvase 1 binds both DNA and RNA tetramolecular quadruplex with high affinity and is the major source of tetramolecular quadruplex G4-DNA and G4-RNA resolving activity in HeLa cell lysates. *J. Biol. Chem.* **283**, 34626–34634.

Dang, Y., Kedersha, N., Low, W. K., Romo, D., Gorospe, M., Kaufman, R., Anderson, P., and Liu, J. O. (2006). Eukaryotic initiation factor 2alpha-independent pathway of stress granule induction by the natural product pateamine A. *J. Biol. Chem.* **281**, 32870–32878.

Durand, S., Cougot, N., Mahuteau-Betzer, F., Nguyen, C. H., Grierson, D. S., Bertrand, E., Tazi, J., and Lejeune, F. (2007). Inhibition of nonsense-mediated mRNA decay (NMD) by a new chemical molecule reveals the dynamic of NMD factors in P-bodies. *J. Cell Biol.* **178**, 1145–1160.

Dutta, A., Zheng, S., Jain, D., Cameron, C. E., and Reese, J. C. (2011). Intermolecular interactions within the abundant DEAD-box protein Dhh1 regulate its activity in vivo. *J. Biol. Chem.* **286**, 27454–27470.

Elvira, G., Wasiak, S., Blandford, V., Tong, X. K., Serrano, A., Fan, X., del Rayo Sanchez-Carbente, M., Servant, F., Bell, A. W., Boismenu, D., Lacaille, J. C., McPherson, P. S., et al. (2006). Characterization of an RNA granule from developing brain. *Mol. Cell. Proteomics* **5**, 635–651.

Erkizan, H. V., Kong, Y., Merchant, M., Schlottmann, S., Barber-Rotenberg, J. S., Yuan, L., Abaan, O. D., Chou, T. H., Dakshanamurthy, S., Brown, M. L., Uren, A., and Toretsky, J. A. (2009). A small molecule blocking oncogenic protein EWS-FLI1 interaction with RNA helicase A inhibits growth of Ewing's sarcoma. *Nat. Med.* **15**, 750–756.

Fairman-Williams, M. E., Guenther, U. P., and Jankowsky, E. (2010). SF1 and SF2 helicases: Family matters. *Curr. Opin. Struct. Biol.* **20**, 313–324.

Franks, T. M., Singh, G., and Lykke-Andersen, J. (2010). Upf1 ATPase-dependent mRNP disassembly is required for completion of nonsense-mediated mRNA decay. *Cell* **143**, 938–950.

Goulet, I., Boisvenue, S., Mokas, S., Mazroui, R., and Cote, J. (2008). TDRD3, a novel tudor domain-containing protein, localizes to cytoplasmic stress granules. *Hum. Mol. Genet.* **17**, 3055–3074.

Grousl, T., Ivanov, P., Frydlova, I., Vasicova, P., Janda, F., Vojtova, J., Malinska, K., Malcova, I., Novakova, L., Janoskova, D., Valasek, L., and Hasek, J. (2009). Robust heat shock induces eIF2alpha-phosphorylation-independent assembly of stress granules containing eIF3 and 40S ribosomal subunits in budding yeast Saccharomyces cerevisiae. *J. Cell Sci.* **122**, 2078–2088.

Hafner, M., Landthaler, M., Burger, L., Khorshid, M., Hausser, J., Berninger, P., Rothballer, A., Ascano, M., Jungkamp, A. C., Munschauer, M., Ulrich, A., Wardle, G. S., et al. (2010). PAR-CliP—A method to identify transcriptome-wide the binding sites of RNA binding proteins. *J. Vis. Exp.* 2034.

Henn, A., Cao, W., Licciardello, N., Heitkamp, S. E., Hackney, D. D., and De La Cruz, E. M. (2010). Pathway of ATP utilization and duplex rRNA unwinding by the DEAD-box helicase, DbpA. *Proc. Natl. Acad. Sci. USA* **107**, 4046–4050.

Hillebrand, J., Pan, K., Kokaram, A., Barbee, S., Parker, R., and Ramaswami, M. (2010). The Me31B DEAD-Box helicase localizes to postsynaptic foci and regulates expression of a CaMKII reporter mRNA in dendrites of drosophila olfactory projection neurons. *Front. Neural Circuits* **4**, 121.

Hilliker, A., Gao, Z., Jankowsky, E., and Parker, R. (2011). The DEAD-Box protein Ded1 modulates translation by the formation and resolution of an eIF4F-mRNA complex. *Mol. Cell* **43**, 962–972.

Hoyle, N. P., Castelli, L. M., Campbell, S. G., Holmes, L. E., and Ashe, M. P. (2007). Stress-dependent relocalization of translationally primed mRNPs to cytoplasmic granules that are kinetically and spatially distinct from P-bodies. *J. Cell Biol.* **179**, 65–74.

Huh, W. K., Falvo, J. V., Gerke, L. C., Carroll, A. S., Howson, R. W., Weissman, J. S., and O'Shea, E. K. (2003). Global analysis of protein localization in budding yeast. *Nature* **425**, 686–691.

Johnstone, O., Deuring, R., Bock, R., Linder, P., Fuller, M. T., and Lasko, P. (2005). Belle is a Drosophila DEAD-box protein required for viability and in the germ line. *Dev. Biol.* **277**, 92–101.

Kanai, Y., Dohmae, N., and Hirokawa, N. (2004). Kinesin transports RNA: Isolation and characterization of an RNA-transporting granule. *Neuron* **43**, 513–525.

Kedersha, N., and Anderson, P. (2007). Mammalian stress granules and processing bodies. *Methods Enzymol.* **431**, 61–81.

Kedersha, N., and Anderson, P. (2009). Regulation of translation by stress granules and processing bodies. *Prog. Mol. Biol. Transl. Sci.* **90**, 155–185.

Kedersha, N. L., Gupta, M., Li, W., Miller, I., and Anderson, P. (1999). RNA-binding proteins TIA-1 and TIAR link the phosphorylation of eIF-2 alpha to the assembly of mammalian stress granules. *J. Cell Biol.* **147**, 1431–1442.

Kedersha, N., Stoecklin, G., Ayodele, M., Yacono, P., Lykke-Andersen, J., Fritzler, M. J., Scheuner, D., Kaufman, R. J., Golan, D. E., and Anderson, P. (2005). Stress granules and processing bodies are dynamically linked sites of mRNP remodeling. *J. Cell Biol.* **169**, 871–884.

Kedersha, N., Tisdale, S., Hickman, T., and Anderson, P. (2008). Real-time and quantitative imaging of mammalian stress granules and processing bodies. *Methods Enzymol.* **448**, 521–552.

Kobayashi, K., Otegui, M. S., Krishnakumar, S., Mindrinos, M., and Zambryski, P. (2007). INCREASED SIZE EXCLUSION LIMIT 2 encodes a putative DEVH box RNA helicase involved in plasmodesmata function during Arabidopsis embryogenesis. *Plant Cell* **19,** 1885–1897.

Kotaja, N., Bhattacharyya, S. N., Jaskiewicz, L., Kimmins, S., Parvinen, M., Filipowicz, W., and Sassone-Corsi, P. (2006). The chromatoid body of male germ cells: Similarity with processing bodies and presence of Dicer and microRNA pathway components. *Proc. Natl. Acad. Sci. USA* **103,** 2647–2652.

Kramer, S., Queiroz, R., Ellis, L., Hoheisel, J. D., Clayton, C., and Carrington, M. (2010). The RNA helicase DHH1 is central to the correct expression of many developmentally regulated mRNAs in trypanosomes. *J. Cell Sci.* **123,** 699–711.

Kumari, S., Bugaut, A., Huppert, J. L., and Balasubramanian, S. (2007). An RNA G-quadruplex in the 5′ UTR of the NRAS proto-oncogene modulates translation. *Nat. Chem. Biol.* **3,** 218–221.

Lai, M. C., Lee, Y. H., and Tarn, W. Y. (2008). The DEAD-box RNA helicase DDX3 associates with export messenger ribonucleoproteins as well as tip-associated protein and participates in translational control. *Mol. Biol. Cell* **19,** 3847–3858.

Leung, A. K., Calabrese, J. M., and Sharp, P. A. (2006). Quantitative analysis of Argonaute protein reveals microRNA-dependent localization to stress granules. *Proc. Natl. Acad. Sci. USA* **103,** 18125–18130.

Liu, F., Putnam, A., and Jankowsky, E. (2008). ATP hydrolysis is required for DEAD-box protein recycling but not for duplex unwinding. *Proc. Natl. Acad. Sci. USA* **105,** 20209–20214.

Low, W. K., Dang, Y., Schneider-Poetsch, T., Shi, Z., Choi, N. S., Merrick, W. C., Romo, D., and Liu, J. O. (2005). Inhibition of eukaryotic translation initiation by the marine natural product pateamine A. *Mol. Cell* **20,** 709–722.

Luke, B., Azzalin, C. M., Hug, N., Deplazes, A., Peter, M., and Lingner, J. (2007). Saccharomyces cerevisiae Ebs1p is a putative ortholog of human Smg7 and promotes nonsense-mediated mRNA decay. *Nucleic Acids Res.* **35,** 7688–7697.

Maga, G., Falchi, F., Garbelli, A., Belfiore, A., Witvrouw, M., Manetti, F., and Botta, M. (2008). Pharmacophore modeling and molecular docking led to the discovery of inhibitors of human immunodeficiency virus-1 replication targeting the human cellular aspartic acid-glutamic acid-alanine-aspartic acid box polypeptide 3. *J. Med. Chem.* **51,** 6635–6638.

Mazroui, R., Huot, M. E., Tremblay, S., Filion, C., Labelle, Y., and Khandjian, E. W. (2002). Trapping of messenger RNA by Fragile X Mental Retardation protein into cytoplasmic granules induces translation repression. *Hum. Mol. Genet.* **11,** 3007–3017.

Mazroui, R., Sukarieh, R., Bordeleau, M. E., Kaufman, R. J., Northcote, P., Tanaka, J., Gallouzi, I., and Pelletier, J. (2006). Inhibition of ribosome recruitment induces stress granule formation independently of eukaryotic initiation factor 2alpha phosphorylation. *Mol. Biol. Cell* **17,** 4212–4219.

Meister, G., Landthaler, M., Peters, L., Chen, P. Y., Urlaub, H., Luhrmann, R., and Tuschl, T. (2005). Identification of novel argonaute-associated proteins. *Curr. Biol.* **15,** 2149–2155.

Miller, L. C., Blandford, V., McAdam, R., Sanchez-Carbente, M. R., Badeaux, F., DesGroseillers, L., and Sossin, W. S. (2009). Combinations of DEAD box proteins distinguish distinct types of RNA: Protein complexes in neurons. *Mol. Cell. Neurosci.* **40,** 485–495.

Mnaimneh, S., Davierwala, A. P., Haynes, J., Moffat, J., Peng, W. T., Zhang, W., Yang, X., Pootoolal, J., Chua, G., Lopez, A., Trochesset, M., Morse, D., *et al.* (2004). Exploration of essential gene functions via titratable promoter alleles. *Cell* **118,** 31–44.

Mollet, S., Cougot, N., Wilczynska, A., Dautry, F., Kress, M., Bertrand, E., and Weil, D. (2008). Translationally repressed mRNA transiently cycles through stress granules during stress. *Mol. Biol. Cell* **19,** 4469–4479.

Morris, M. J., and Basu, S. (2009). An unusually stable G-quadruplex within the 5′-UTR of the MT3 matrix metalloproteinase mRNA represses translation in eukaryotic cells. *Biochemistry* **48,** 5313–5319.

Navarro, R. E., Shim, E. Y., Kohara, Y., Singson, A., and Blackwell, T. K. (2001). cgh-1, a conserved predicted RNA helicase required for gametogenesis and protection from physiological germline apoptosis in C. elegans. *Development* **128,** 3221–3232.

Nissan, T., and Parker, R. (2008). Analyzing P-bodies in Saccharomyces cerevisiae. *Methods Enzymol.* **448,** 507–520.

Onishi, H., Kino, Y., Morita, T., Futai, E., Sasagawa, N., and Ishiura, S. (2008). MBNL1 associates with YB-1 in cytoplasmic stress granules. *J. Neurosci. Res.* **86,** 1994–2002.

Otsu, N. (1979). Threshold selection methods from gray-level histograms. *IEEE Trans. Syst. Man Cybern.* **9,** 62–66.

Pan, C., and Russell, R. (2010). Roles of DEAD-box proteins in RNA and RNP Folding. *RNA Biol.* **7,** 667–676.

Parsyan, A., Svitkin, Y., Shahbazian, D., Gkogkas, C., Lasko, P., Merrick, W. C., and Sonenberg, N. (2011). mRNA helicases: The tacticians of translational control. *Nat. Rev. Mol. Cell Biol.* **12,** 235–245.

Rajyaguru, P., She, M., and Parker, R. (2011). Scd6 targets eIF4G to repress translation: RGG-motif proteins as a class of eIF4G-binding proteins. *Mol. Cell* **45,** 244–254.

Scarcelli, J. J., Viggiano, S., Hodge, C. A., Heath, C. V., Amberg, D. C., and Cole, C. N. (2008). Synthetic genetic array analysis in Saccharomyces cerevisiae provides evidence for an interaction between RAT8/DBP5 and genes encoding P-body components. *Genetics* **179,** 1945–1955.

Serman, A., Le Roy, F., Aigueperse, C., Kress, M., Dautry, F., and Weil, D. (2007). GW body disassembly triggered by siRNAs independently of their silencing activity. *Nucleic Acids Res.* **35,** 4715–4727.

Sheth, U., and Parker, R. (2003). Decapping and decay of messenger RNA occur in cytoplasmic processing bodies. *Science* **300,** 805–808.

Sheth, U., and Parker, R. (2006). Targeting of aberrant mRNAs to cytoplasmic processing bodies. *Cell* **125,** 1095–1109.

Shih, J. W., Tsai, T. Y., Chao, C. H., and Wu Lee, Y. H. (2008). Candidate tumor suppressor DDX3 RNA helicase specifically represses cap-dependent translation by acting as an eIF4E inhibitory protein. *Oncogene* **27,** 700–714.

Solomon, S., Xu, Y., Wang, B., David, M. D., Schubert, P., Kennedy, D., and Schrader, J. W. (2007). Distinct structural features of caprin-1 mediate its interaction with G3BP-1 and its induction of phosphorylation of eukaryotic translation initiation factor 2alpha, entry to cytoplasmic stress granules, and selective interaction with a subset of mRNAs. *Mol. Cell. Biol.* **27,** 2324–2342.

Stalder, L., and Muhlemann, O. (2009). Processing bodies are not required for mammalian nonsense-mediated mRNA decay. *RNA* **15,** 1265–1273.

Swisher, K. D., and Parker, R. (2010). Localization to, and effects of Pbp1, Pbp4, Lsm12, Dhh1, and Pab1 on stress granules in Saccharomyces cerevisiae. *PLoS One* **5,** e10006.

Teixeira, D., and Parker, R. (2007). Analysis of P-body assembly in Saccharomyces cerevisiae. *Mol. Biol. Cell* **18,** 2274–2287.

Teixeira, D., Sheth, U., Valencia-Sanchez, M. A., Brengues, M., and Parker, R. (2005). Processing bodies require RNA for assembly and contain nontranslating mRNAs. *RNA* **11,** 371–382.

Unterholzner, L., and Izaurralde, E. (2004). SMG7 acts as a molecular link between mRNA surveillance and mRNA decay. *Mol. Cell* **16,** 587–596.

Vaughn, J. P., Creacy, S. D., Routh, E. D., Joyner-Butt, C., Jenkins, G. S., Pauli, S., Nagamine, Y., and Akman, S. A. (2005). The DEXH protein product of the DHX36 gene is the major source of tetramolecular quadruplex G4-DNA resolving activity in HeLa cell lysates. *J. Biol. Chem.* **280,** 38117–38120.

Weber, C., Nover, L., and Fauth, M. (2008). Plant stress granules and mRNA processing bodies are distinct from heat stress granules. *Plant J.* **56,** 517–530.

Wilczynska, A., Aigueperse, C., Kress, M., Dautry, F., and Weil, D. (2005). The translational regulator CPEB1 provides a link between dcp1 bodies and stress granules. *J. Cell Sci.* **118,** 981–992.

Winzeler, E. A., Shoemaker, D. D., Astromoff, A., Liang, H., Anderson, K., Andre, B., Bangham, R., Benito, R., Boeke, J. D., Bussey, H., Chu, A. M., Connelly, C., et al. (1999). Functional characterization of the S. cerevisiae genome by gene deletion and parallel analysis. *Science* **285,** 901–906.

Zhang, C., and Darnell, R. B. (2011). Mapping in vivo protein-RNA interactions at single-nucleotide resolution from HITS-CLIP data. *Nat. Biotechnol.* **29,** 607–614.

CHAPTER SIXTEEN

DEAD-Box RNA Helicases as Transcription Cofactors

Frances V. Fuller-Pace *and* Samantha M. Nicol

Contents

1. Introduction	348
2. Analysis of Interactions Between RNA Helicases and Transcription Factors *In Vitro* and in Cell Lines	349
2.1. Interactions *in vitro*: GST- and Nickel-pull down	349
2.2. Interactions in mammalian cell lines: Co-immunoprecipitation of proteins from nuclear extracts	355
3. Analysis of p68/p72 Sumoylation in Cell Lines	358
3.1. Expression of myc-tagged p68/p72 and His-tagged SUMO in mammalian cells	358
3.2. Ni^{2+} isolation of His-tagged proteins expressed in mammalian cells	359
4. Analysis of Transcriptional Coactivator/Corepressor Activity of p68 and p72	361
4.1. Transfection of cells with RNA helicase, transcription factor, and promoter/reporter plasmids	362
4.2. Luciferase assays	362
5. Summary	363
Acknowledgments	365
References	365

Abstract

It is established that several DEAD box RNA helicases perform multiple functions in the cell, often through interactions with different partner proteins in a context-dependent manner. Several studies have shown that some DEAD box proteins play important roles as regulators of transcription, particularly as coactivators or cosuppressors of transcription factors that are themselves highly regulated. Two such RNA helicases are DDX5 (p68) and DDX17 (p72). These proteins are known to function in RNA processing/alternative splicing, but they have also been shown to interact with, and act as coregulators of,

Division of Cancer Research, Medical Research Institute, University of Dundee, Ninewells Hospital and Medical School, Dundee, United Kingdom

Methods in Enzymology, Volume 511 © 2012 Elsevier Inc.
ISSN 0076-6879, DOI: 10.1016/B978-0-12-396546-2.00016-4 All rights reserved.

transcription factors that are themselves highly regulated. In this chapter, we shall describe protocols we have used to investigate the factors that influence the function of p68 and p72 in transcriptional regulation. These include the interactions of p68 and p72 with transcription factors and/or components of the transcription machinery and posttranslational modification by the small *u*biquitin-related *mo*difier, SUMO.

1. INTRODUCTION

DEAD box RNA helicases have well-documented roles in many cellular processes that require manipulation of RNA structures including RNA processing, nuclear export of RNA, ribosome assembly, and translation. It has been established that the characteristic ATPase and RNA helicase activities are essential for the proteins' function in these processes. However, in the past several years, a large body of evidence has accumulated demonstrating that several members of the DEAD box family are multifunctional and they also play key roles in transcription regulation, often acting as coactivators or corepressors of transcription factors in a manner that, in many cases, does not appear to require helicase activity.

DEAD box proteins known to exhibit additional functions as transcriptional coregulators include DDX3, DDX20 (DP103), DDX5 (p68), and DDX17 (p72). These proteins have been shown to regulate transcription, at least in part, by acting at the promoters of responsive genes. To date, there is no conclusive evidence that the RNA helicases themselves bind directly to the promoter; instead, it is thought that they are recruited to specific sequences at responsive promoters through interaction with transcription factors. Interestingly, in most cases so far studied, RNA helicase activity does not generally appear to be required.

DDX3, which is involved in several processes regulating gene expression and plays a role in innate immune signaling pathways (reviewed in Schroder, 2010), has been shown to coactivate transcription from the $p21^{waf1}$ promoter (Chao *et al.*, 2006) and the interferon promoter (Schroder *et al.*, 2008; Soulat *et al.*, 2008), while it represses expression from the E-cadherin promoter (Botlagunta *et al.*, 2008). DDX20 (DP103), which was shown to be important for transcriptional activation by the Epstein Barr virus proteins EBNA2 and EBNA3C (Grundhoff *et al.*, 1999), has been found to repress the activity of both the Egr2/Krox20 and the steroidogenic factor 1 (SF-1) transcription factors (Gillian and Svaren, 2004; Ou *et al.*, 2001). Interestingly, DP103 was shown to repress SF-1 activity by promoting its sumoylation (Lee *et al.*, 2005). The highly related DDX5 (p68) and DDX17 (p72) proteins are multifunctional proteins with reported roles in development, pre-mRNA processing/alternative splicing, microRNA processing, and transcription (reviewed in Fuller-Pace and Moore, 2011). Both p68 and p72 have been shown to be aberrantly expressed in a range of cancers, suggesting that changes in their function(s) may contribute to tumor development. In terms of their role in transcription, several studies have demonstrated that p68 and p72 interact with, and function as coactivators of, several transcription factors that are themselves highly regulated, including estrogen receptor alpha (ERα) (Endoh et al., 1999; Watanabe et al., 2001; Wortham et al., 2009), androgen receptor (Clark et al., 2008), tumor suppressor p53 (Bates et al., 2005), MyoD (Caretti et al., 2006), and Runx2 (Jensen et al., 2008). However, in some contexts, p68 and p72 can also clearly function as transcriptional corepressors (Guo et al., 2010; Wilson et al., 2004). Both p68 and p72 are modified by SUMO at a single equivalent site (K53 in p68 and K50 in p72); sumoylation at this site modulates their coactivator/corepressor activity in a context-dependent manner and promoted their interaction with HDAC1, suggesting a possible mechanism for the modulation of transcriptional activity (Jacobs et al., 2007; Mooney et al., 2010b).

In this chapter, we shall describe protocols we have used for the analysis of (i) interactions of p68 with transcription factors, (ii) sumoylation of p68 and p72, and (iii) coactivation and corepression activities of p68 and p72. Methods for the analysis of p68 siRNA knockdown and chromatin immunoprecipitation (ChIP) have been described previously (Nicol and Fuller-Pace, 2010).

2. Analysis of Interactions Between RNA Helicases and Transcription Factors *In Vitro* and in Cell Lines

2.1. Interactions *in vitro*: GST- and Nickel-pull down

The following protocol has been used successfully for the analysis of direct interactions between the RNA helicase p68 and the tumor suppressor p53 (Moore et al., 2010).

2.1.1. Bacterial GST-tagged protein expression and purification

Expression

There are several vectors that allow isopropyl β-D-1-thiogalactopyranoside (IPTG)—inducible expression of GST-tagged fusion proteins. We generally express GST-tagged proteins (cloned in pGEX-4T: GE Healthcare) in BL21(DE3) bacteria; there are several strains that can be tested to optimize expression and protein solubility.

1. Inoculate 50 ml of L-broth with a single colony of the desired expression plasmid along with the relevant selection antibiotic and shake overnight at 225 rpm, 37 °C.
2. The following day, inoculate 500 ml of L-broth, plus antibiotic, with 15 ml of overnight culture and incubate with agitation at 225 rpm, 37 °C, until the culture has reached an OD (600 nm) of 0.3–0.4. (At this stage, remove 1 ml to serve as uninduced control, centrifuge at $1500 \times g$ for 5 min, and store pellet at -20 °C.)
3. Induce the 500 ml culture with 0.5 mM IPTG and leave shaking at 225 rpm, at 22 °C, for 3 h. (The lower temperature and resulting slower growth have been shown to yield protein of higher solubility.)
4. Following the 3-h incubation, remove 1 ml of culture (to serve as the induced sample) and centrifuge at $1500 \times g$ for 5 min. The uninduced and induced samples can be analyzed by SDS-PAGE gel electrophoresis/Coomassie staining to assess the efficiency of expression and IPTG induction.
5. Centrifuge the remaining 500 ml culture at $4000 \times g$ for 20 min, 4 °C. Discard the supernatant and resuspend the pellet in 30 ml of 50 mM Tris–HCl (pH 7.5), transfer the sample to a 50-ml Falcon tube and centrifuge at $1700 \times g$ for 20 min, 4 °C, discard the supernatant and store the pellet at -80 °C.

Purification

1. Resuspend the bacterial pellet in 10 ml of lysis buffer, add 150 µl of 10 mg/ml lysozyme, and place on ice/salt water for 30 min until the bacteria have completely lysed.
2. Warm the lysate slowly in a 37 °C water bath with gentle rolling for 1 min and sonicate twice for 30 s (on ice to prevent protein denaturation), centrifuge at $10,000 \times g$ for 20 min, 4 °C.
3. Aliquot the supernatant in 1 ml amounts and snap-freeze on dry ice. The aliquots can be stored at -80 °C until required.

Lysis buffer

50 mM Tris–HCl (pH 7.5).
50 mM NaCl.
10% Sucrose.
1 mM Benzamidine.

2.1.2. Preparation of GST-beads and binding of GST-tagged proteins

1. Place 1.5 ml of GST-beads (Gluthatione Sepharose™ 4B Beads, GE Healthcare—available as a suspension in ethanol) in a Falcon tube with 8.5 ml of ice-cold phosphate buffered saline (PBS—tablets available

from Sigma–Aldrich), mix by gentle inversion, and centrifuge at $400 \times g$ for 3 min, 4 °C. Repeat this step to ensure complete removal of ethanol.
2. Equilibrate the beads by gentle inversion and centrifugation as above, three times in 1 ml of ice-cold GST-Buffer A. Following the final wash, resuspend the beads in 800 μl of ice-cold GST-Buffer A.
3. To check binding efficiency of the lysate, add 200 μl of equilibrated GST-beads to 800 μl of purified lysate (from one of the stored aliquots) and place on a rotating wheel for 1 h, 4 °C.
4. Centrifuge the sample at $400 \times g$ for 3 min, 4 °C. At this stage, retain the supernatant on ice as it contains the "unbound protein". Wash the bead/lysate pellet by gentle inversion and centrifugation as above, three times with 1 ml of ice-cold GST-Buffer A and a further three times with 1 ml of ice-cold GST-Buffer B. Following the final wash, resuspend the bead/lysate pellet in an appropriate sample buffer for analysis by SDS-PAGE gel electrophoresis and Coomassie staining or western blotting to assess efficiency of binding and purity of bound fraction. Once conditions of expression/purification have been optimized, the bead/lysate pellet can be used directly to study the interactions with other protein partners, as described below in Section 2.1.6.

GST-Buffer A	GST-Buffer B
50 mM Tris–HCl (pH 8.0)	50 mM Tris–HCl (pH 8.0)
500 mM NaCl	100 mM NaCl
10 mM β-Mercaptoethanol	5 mM MgCl$_2$
1% Triton-X-100	10 mM β-Mercaptoethanol
1 × Protease inhibitor (PI) cocktail[a]	1 × PI cocktail

[a]Available as 25 × (standard or EDTA-free) from Roche.

2.1.3. Bacterial His-tagged protein expression and purification

The same protocol is used to express His-tagged proteins as described above for GST-tagged proteins with the following adaptations for purification. (His-tagged proteins are generated by cloning the gene of interest with N- or C-terminal His tags: there are several suitable vectors available; e.g., pET302/NT-His and pET303/CT-His (Invitrogen).)

Purification

1. Remove a pellet containing the lysate from bacteria expressing the relevant His-tagged protein from -80 °C and resuspend in 10 ml of ice-cold His-Buffer A, sonicate twice (on ice to prevent protein denaturation) for 30 s each time, and centrifuge at $4000 \times g$ for 10 min, 4 °C.

2. Decant the supernatant into a 50-ml Falcon tube and keep on ice. Resuspend the pellet in 5 ml ice-cold His-Buffer A, sonicate, and centrifuge as above. Combine the supernatants and aliquot in 1 ml amounts, snap-freeze on dry ice, and store at $-80\,^{\circ}\text{C}$.

2.1.4. Preparation of Ni^{2+} agarose beads and binding of His-tagged proteins

1. Place 1.5 ml of Ni^{2+} beads (Qiagen—available as a suspension in ethanol) in a Falcon tube with 8.5 ml of ice-cold PBS, mix by gentle inversion, and centrifuge at $400 \times g$ for 3 min, $4\,^{\circ}\text{C}$. Repeat this step to ensure complete removal of ethanol.
2. Equilibrate the beads by gentle inversion and centrifuge as above, three times in 1 ml of ice-cold His-Buffer A. Following the final wash, resuspend the beads in 800 µl of ice-cold His-Buffer A.
3. To check the binding efficiency, centrifuge the His-lysate at $10,000 \times g$ for 20 min, $4\,^{\circ}\text{C}$ and filter the supernatant through a $0.22\,\mu M$ filter (Millipore). Add 200 µl of equilibrated Ni^{2+} beads to 800 µl of lysate and rotate for 1 h, $4\,^{\circ}\text{C}$. Centrifuge the sample at $400 \times g$ for 3 min, $4\,^{\circ}\text{C}$. At this stage, retain the supernatant on ice as it contains the "unbound protein". Wash the bead/lysate pellet, by gentle inversion and centrifugation as above, with 1 ml of ice-cold His-Buffer A and a further three times with 1 ml of ice-cold His-Buffer B. Following the final wash, resuspend the bead/lysate pellet in an appropriate sample buffer for analysis by SDS-PAGE gel electrophoresis and Coomassie staining or western blotting to assess the efficiency of binding and purity of bound fraction—see Fig. 16.1. As for GST-tagged proteins, once conditions of expression/purification have been optimized, the bead/lysate pellet can be used directly to study the interactions with other protein partners as described below in Section 2.1.6.

His-Buffer A	His-Buffer B
50 mM Tris–HCl (pH 8.0)	50 mM Tris–HCl (pH 8.0)
300 mM NaCl	100 mM NaCl
10 mM Imidazol	10 mM β-Mercaptoethanol
1 mM DTT	5 mM MgCl$_2$
1 × PI Cocktail	1 × PI Cocktail

Note: For both GST- and His-tagged protein-binding experiments, appropriate vector constructs are used as a negative control.

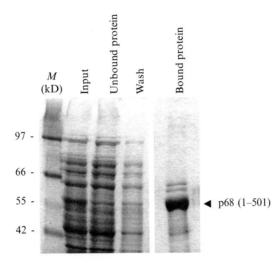

Figure 16.1 Purification of His-tagged proteins: Coomassie-stained SDS-PAGE gel showing purification of a His-tagged p68 deletion derivative (residues 1–501) by binding to Ni^{2+} agarose beads. Shown also are the input bacterial lysate and the unbound proteins.

2.1.5. Preparation of ^{35}[S]-labeled, *in vitro*-translated proteins

This is a standard procedure for generating ^{35}[S]-labeled proteins from genes cloned in plasmid vectors containing the T7 promoter, for example pcDNA3.

Generally we have used 0.5 μg of plasmid DNA encoding the gene of interest together with ^{35}[S]-Methionine (1000 Ci/mmol at 10 mCi/ml) and the TNT® T7 Quick Coupled Transcription/Translation System from Promega. The reaction is incubated at 30 °C for 60–90 min, according to the manufacturer's instructions.

2.1.6. Binding of GST-tagged/His-tagged proteins to *in vitro*-translated proteins

1. To each pellet of GST/His-tagged protein bound beads, add 200 μl of binding buffer and 10–20 μl of "*in vitro*-translated" ^{35}[S]-Methionine-labeled protein and incubate on a rotating wheel for 2 h, 4 °C.
2. Following the 2-h incubation, centrifuge the samples at $400 \times g$ for 3 min and wash the pellets once with 1 ml of wash buffer 1 and twice with 1 ml of wash buffer 2.
3. Resuspend the pellets in 15 μl of SDS-PAGE sample buffer and analyze by SDS-PAGE.

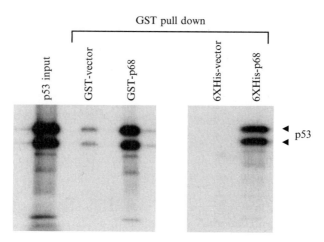

Figure 16.2 Interactions of GST- and His-tagged p68 with ^{35}[S]-labeled p53: Also shown is the p53 input signal. *Note*: *In vitro*-translated p53 gives two products, the lower one resulting from an internal translation initiation.

4. Fix the gel in 10% methanol/10% acetic acid solution for 20 min, after which place the gel in Amplify solution to enhance the ^{35}[S] signal (GE Healthcare) for 30 min.
5. Dry the gel and expose to film for the appropriate length of time to visualize signal—see Fig. 16.2.

    Note: If the binding of different GST- or His-tagged proteins to a labeled partner protein is being compared, it is important that equivalent amounts of bound proteins are used. This can be achieved by comparing expression and binding efficiency as described above and adjusting the amount of lysate used for binding accordingly—see Fig. 16.3.

Binding buffer

20 mM Tris–HCl (pH 7.5).
100 mM NaCl.
2 mM EDTA (pH 8.0).
2 mM DTT.
5% (v/v) Glycerol.
0.1% Igepal CA630.
0.05% BSA.
1 × PI Cocktail (EDTA-free).

Wash buffer 1

50 mM Tris–HCl (pH 7.5).
150 mM NaCl.
5 mM EDTA (pH 8.0).

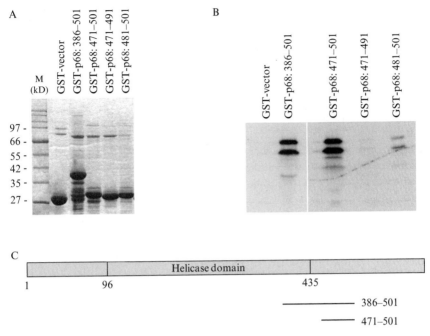

Figure 16.3 Identification of interacting regions/domains: interaction of GST-tagged deletion derivatives of p68 with ^{35}[S]-labeled p53. (A) Equalization of input GST-tagged p68. (B) GST pull downs showing deletion derivatives that interact with p53. (C) Cartoon showing the interaction regions relative to the full-length p68 protein (not to scale).

1× PI Cocktail (EDTA-free).

Wash buffer 2

50 mM Tris–HCl (pH 7.5).
150 mM NaCl.
2 mM EDTA (pH 8.0).
0.1% Igepal CA630.
1× PI Cocktail (EDTA-free).

2.2. Interactions in mammalian cell lines: Co-immunoprecipitation of proteins from nuclear extracts

While the *in vitro* protocols described above can provide very useful data for identifying regions of interactions between RNA helicases and protein partners and can also demonstrate whether interactions are direct or via other factors, it is important to show that such interactions can also occur *physiologically* in the cell. Co-immunoprecipitation (co-IP) from cell extracts can provide such information, although co-IP of two proteins does not indicate *direct* interaction. Ideally, one should demonstrate the interaction by using both *in vitro* interaction studies and co-IP.

We have found that co-IP of p68/p72 and transcription factors such as p53 or ERα was considerably more efficient when performed from nuclear extracts rather than from whole cell lysates. The protocol for the preparation of nuclear extracts given below has been adapted from that described by Dignam *et al.* (1983). We have found that changing the NaCl concentration in Buffer B from 400 to 330 mM (see below) efficiently disrupts the nuclei but maintains the more "transient" interactions between nuclear proteins (Bates *et al.*, 2005).

2.2.1. Preparation of nuclear extracts

Note: When preparing nuclear extracts, it is essential to keep all extracts/lysates and buffers cold at all times. This protocol is for tissue culture cells grown as a monolayer. We usually prepare extracts from 15 cm plates in batches of 5–10 plates at a time. However, it can be easily adapted for cells grown in suspension.

1. Wash cell monolayers three times in PBS and trypsinize [in trypsin EDTA (Invitrogen)] to detach, using the same protocol used for standard propagation of the cell line being used.
2. Once the cells have detached, harvest in PBS (10 ml per 15 cm plate), transfer to a 50-ml Falcon tube, and centrifuge at $1000 \times g$ for 5 min at room temperature.
3. Resuspend the cell pellet in ice-cold PBS (2× pellet volume) and transfer to microfuge tubes. Resuspend and wash the cells three times in PBS (2× pellet volume), each time centrifuging at $400 \times g$ for 5 min, 4 °C.
4. Following the final wash, remove the supernatant and resuspend pellet in Buffer A (2× pellet volume); leave on ice for 5 min to allow the cells to swell.
5. Lyse cells by manual homogenization, using a Dounce homogenizer until a homogeneous suspension is achieved (usually approximately 10×), transfer to new microfuge tubes, and centrifuge at $10,000 \times g$ for 10 min, 4 °C.
6. Resuspend the nuclear pellet in Buffer B with gentle pipetting to prevent the extract from foaming and place on a rotating wheel for 15 min, 4 °C. Centrifuge at $10,000 \times g$ for 10 min, 4 °C. Carefully take supernatant: this is the nuclear extract. Dilute to a final concentration of 150 mM NaCl with Buffer C. Aliquot into appropriate amounts, depending on the sort of experiments envisaged (we usually store in aliquots of 500 μl), snap-freeze on dry ice, and store at −80 °C.

2.2.2. Co-immunoprecipitation

The following protocol is based on a 500-μl aliquot derived from a nuclear extract (NE) preparation.

1. DNase/RNase-treat 500 μl of NE by adding 10 μl of RQ-1 DNase (Promega) and 5 μl of RNase A (Sigma), rotating for 10 min at 4 °C.

    Note: We found that the RQ-1 DNase is more efficient in this protocol than other DNases. DNase/RNase treatment, although not strictly necessary, gave much cleaner co-IP results.

2. Preclear the NE by adding 50 μl Protein A or Protein G Sepharose beads (Sigma) (equilibrated by washing three times in co-IP buffer and resuspended in an appropriate volume to yield a 50% bead slurry) and mix by rotation for 30 min at 4 °C. The use of Protein A or Protein G will depend on the antibody to be used in the IP; the affinity of the antibody for Protein A or G should be checked in test IPs.

3. Immunoprecipitate the precleared NE (the amount of NE to be used needs to be optimized depending on expression of the protein of interest and whether an overexpressed or endogenous protein is being studied) in a 500-μl volume containing 50 μl Protein A/G Sepharose beads, the appropriate antibody, and co-IP buffer. An irrelevant IgG α-rabbit or α-mouse antibody can be used as a negative control. Rotate the IP reaction for 1 h at 4 °C. Alternatively, this can be performed overnight; the optimal time should be determined empirically.

4. Centrifuge the IP reaction at $800 \times g$ for 3 min, 4 °C, and remove the supernatant.

5. Wash the beads in 500 μl co-IP buffer, containing the appropriate amount of Igepal (in the range of 0.5–1%; again this should be tested empirically and will vary according to the antibody used), vortex briefly to resuspend the beads, and rotate for 5 min at 4 °C. Repeat this step three times before resuspending the bead pellet in 50 μl of SDS sample buffer for subsequent analysis by SDS-PAGE and western blotting with reciprocal antibodies. For example, for studying p68/p53 interactions, a p68-specific antibody can be used for the IP and a p53-specific antibody can be used for western blotting, and *vice versa*—see Fig. 16.4.

    Note: It is advisable to use antibodies from different species (e.g., rabbit and mouse) for the IP and western blotting to avoid cross-reaction with the antibody heavy and light chains in the western blotting.

Co-IP buffer

20 mM Hepes (Na Salt) (pH 7.9).
150 mM NaCl.

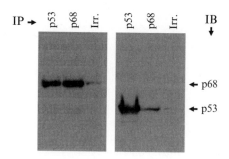

Figure 16.4 Co-immunoprecipitation of endogenous p68 and p53: Western blots showing reciprocal co-immunoprecipitation of p68 and p53 from nuclear extracts. Left panel shows a p68 western blot; right panel shows a p53 western blot.

0.5 mM DTT.
20% (v/v) Glycerol.
10 mM NaF.
1 × PI Cocktail (EDTA-free).

3. Analysis of p68/p72 Sumoylation in Cell Lines

Others and we have shown that sumoylation influences the coactivator/corepressor activity of p68 and p72 (Jacobs et al., 2007; Mooney et al., 2010b). This protocol describes the procedures used for the analysis of sumoylation of exogenously expressed p68/p72 in mammalian cell lines. As SUMO is covalently attached to target lysines on sumoylation sites [ψKX(D/E)], it is possible to analyze sumoylation of specific proteins by expression of plasmids encoding myc-tagged p68/p72 (to distinguish between expressed and endogenous protein) together with His-tagged SUMO; sumoylated proteins will contain attached His-tagged SUMO and can be isolated by binding to Ni^{2+} agarose beads. These procedures also allow the identification of physiological sumoylation sites by testing the sumoylation efficiency of proteins in which consensus potential sumoylation sites have been mutated. Finally, the relative efficiency of sumoylation by the different isoforms SUMO-1, SUMO-2, or SUMO-3 can be estimated (Gareau and Lima, 2010).

3.1. Expression of myc-tagged p68/p72 and His-tagged SUMO in mammalian cells

In our studies, we have generally used COS-7 cells for expression of plasmids encoding myc-tagged p68 or p72 and His-tagged SUMO. However, this protocol is suitable for many cell lines and the choice will depend on the proteins to be tested and the conditions to be used, for example, stress conditions. Cells can be transfected using a variety of transfection reagents; we find FuGENE®6 (Promega) transfection works well and allows high levels of expression when harvesting cells 48 h after transfection. The cell line, transfection reagent, and time of harvesting should be optimized for the proteins/cell lines under study.

3.2. Ni^{2+} isolation of His-tagged proteins expressed in mammalian cells

This method describes the isolation of His-tagged proteins through the ability of these proteins to bind Ni^{2+} agarose beads. As the expressed SUMO construct is His-tagged, all sumoylated cellular proteins will bind to the Ni^{2+} agarose beads. However, it is possible to distinguish the sumoylated protein of interest (in this case, p68 or p72) by western blotting using specific antibodies. The use of plasmids expressing myc-tagged p68/p72 will additionally distinguish between plasmid encoded p68/p72 and endogenous p68/p72. It is also possible to use alternative tags for the expressed proteins, for example, HA or FLAG tags. Additionally, western blotting with a His-specific antibody will give an indication of the level of expression of the SUMO plasmid and of overall sumoylation efficiency.

All steps and solutions are at room temperature, but note extraction is under denaturing conditions to minimize removal of SUMO from substrates by SUMO proteases.

1. Wash cells 3× with PBS.
2. Add 5 ml guanidine extraction buffer, scrape cells, and transfer into 15 ml Falcon tubes.
3. Sonicate briefly to ensure complete cell lysis.
4. Add 75 µl of Qiagen Ni^{2+} agarose beads and rotate for 4 h.
5. Centrifuge bead/lysate pellet at $800 \times g$ and remove supernatant.
6. Resuspend the bead/lysate pellet in 750 µl of guanidine wash buffer and transfer into a microfuge tube.
7. Wash the beads by rotation for 5 min, pellet the beads by pulse centrifugation for 10 s at $10,000 \times g$, and remove supernatant.
8. Repeat steps 5 and 6 for urea wash buffers 1–3.
9. Elute lysate from beads in 75 µl of elution buffer by rotation for 20 min.
10. Centrifuge sample at $800 \times g$; remove supernatant into a new microfuge tube. The sample can be analyzed/visualized by SDS-PAGE/western blotting or stored at $-20\ °C$ for analysis at a later date. As the expressed p68/p72 are myc-tagged, it is possible to specifically identify sumoylated p68/p72 by western blotting using an antibody specific for the myc epitope—see Fig. 16.5. Western blotting using a His-specific antibody will provide a good control to assess efficiency of expression of the His-tagged SUMO.

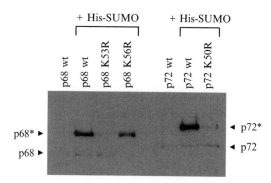

Figure 16.5 Sumoylation of p68 and p72: Western blot (using a myc-specific antibody) to detect exogenously expressed sumoylated p68/p72 that have been purified by binding to Ni^{2+} beads. *Note*: the expressed SUMO construct is His-tagged and therefore only sumoylated proteins will bind the Ni^{2+} beads. This western blot identifies K53 and K50 as the sumoylation targets on p68 and p72 respectively, as the p68 K53R and p72 K50R mutants are not sumoylated. (The p68 K56R mutant is efficiently sumoylated indicating that K56 is not a major sumoylation site.). * Denotes sumoylated proteins. The faint p68/p72 bands show background of nonspecific binding to the Ni^{2+} beads.

Guanidinium stock buffer (pH 8.0)

6 M Guanidine.
0.1 M Na_2HPO_4/NaH_2PO_4.
0.01 M Tris–HCl.

Urea buffer (pH 8.0)

8 M Urea.
0.1 M Na_2HPO_4/NaH_2PO_4.
0.01 M Tris–HCl.

Urea buffer (pH 6.3)

8 M Urea.
0.1 M Na_2HPO_4/NaH_2PO_4.
0.01 M Tris–HCl.

Note: Dissolve initially in a minimal volume of water; on dissolution the temperature of these solutions drop. Allow to return to room temperature before adjusting pH.

Guanidine extraction buffer

Add fresh per ml of guanidine stock buffer.
10 mM Iodoacetamide.
5 mM Imidazole.

Guanidine wash buffer

Add fresh per ml of guanidine stock buffer.
10 mM Iodoacetamide.

Urea wash buffer 1

Add fresh per ml of urea buffer (pH 8.0).
10 mM Iodoacetamide.

Urea wash buffer 2

Add fresh per ml of urea buffer (pH 6.3).
10 mM Iodoacetamide.
0.2% Triton-X.

Urea wash buffer 3

Add fresh per ml of urea buffer (pH 6.3).
10 mM Iodoacetamide.
0.1% Triton-X.

Elution buffer

200 mM Imidazole.
5% SDS.
150 mM Tris–HCl (pH 6.7).
30% Glycerol.
0.72 M β-Mercaptoethanol.

Elution buffer can be made up in large quantities, aliquoted, and stored at $-20\,°C$.

4. Analysis of Transcriptional Coactivator/ Corepressor Activity of p68 and p72

As discussed above, several studies have shown that p68 and p72 will act as coactivators of several transcription factors and, in some contexts, to act as repressor/corepressors. Interestingly, while sumoylation, acetylation and phosphorylation have been shown to modulate p68/p72 transcriptional activity in a context-dependent manner (Clark *et al.*, 2008; Jacobs *et al.*,

2007; Mooney et al., 2010a,b), RNA helicase activity does not appear to be important for transcriptional activity in the majority of cases (reviewed in Fuller-Pace and Moore, 2011). The protocol below describes a standard method for assessing p68/p72 transcriptional coactivator or repressor/corepressor activity using standard promoter/luciferase reporter assays. As the method utilizes cotransfection of p68 (or p72), together with the transcription factor of interest (e.g., p53) and the appropriate promoter/reporter construct, it allows the determination of the effects of mutations in p68/p72 on their transcriptional activity.

4.1. Transfection of cells with RNA helicase, transcription factor, and promoter/reporter plasmids

It is important to optimize the conditions for achieving the correct background activity with the promoter/reporter construct so that activation, as well as repression, can be measured; this can be achieved by titration of the promoter/reporter construct DNA to obtain a suitable "basal" reading. It is also important to titer the amounts of DNA for both the transcription factor to be tested and the RNA helicase (e.g., p68/p72) to ensure that measurements are within the linear range.

1. Seed cells in 24-well plates, one day prior to transfecting, so that the cells are approximately 50–80% confluent on the day of transfection. This will need to be determined empirically for the cells to be used. (We have used the p53-null H1299 cells for studying the effect of p68/p72 on p53 transcriptional activity and U2OS (p53 wt) for measuring p68/p72 repressor activity.) Cells are seeded in a total of 500 μl appropriate antibiotic-free tissue culture medium per well and incubated at 37 °C, 5% CO_2 overnight.
2. The following day, transfect cells with an appropriate amount of DNA for the promoter/reporter construct, the RNA helicase, and the transcription factor to be studied, using a preferred transfection reagent according to manufacturer's instructions. (We have found FuGENE®6 transfection reagent to work well with our cell lines of choice.)
3. Incubate for 24–48 h. Again this needs to be checked to achieve the optimal expression levels of the transfected plasmid constructs and will depend on the cell line and proteins under investigation.

4.2. Luciferase assays

There are several kits available for the detection of luciferase activity. We have used the Dual-Luciferase® Reporter (DLR™) Assay System from Promega, but any of the commercially available systems can be used. This system allows the measurement of Firefly luciferase and Renilla luciferase.

In many studies, a plasmid expressing Renilla luciferase is used as a transfection control. However, in several of our studies, we found that p68 can affect the expression of the Renilla plasmid. Therefore, in the protocol below, we did not incorporate the Renilla as a transfection control; instead, we performed transfections in several independent experiments and also checked transfection efficiency of the various p68/p72/p53 constructs by western blotting (Bates et al., 2005; Jacobs et al., 2007).

All steps are performed at room temperature, and kit components, which are stored frozen ($-20\ °C$), should be warmed to room temperature prior to use. All components are provided with the kit.

1. Following the required incubation, wash cells twice with PBS at room temperature.
2. To each well of the 24-well plate, add 100 µl of room temperature 1× Passive Lysis Buffer and place plate on a rocking platform for 15 min in order to lyse the cells (Note: At this point, the plate can be stored at $-20\ °C$ for analysis at a later date.)
3. Following incubation in lysis buffer, transfer 20 µl of the sample to individual wells of a 96-well dish.
4. Measure luminescence using an automated luminometer and the luciferase substrate supplied in the kit, according to the manufacturer's instructions—see Fig. 16.6.

5. Summary

It is clear that p68 and p72 have multiple functions in the cell and that their roles are often context dependent. This is particularly evident when one investigates their function as transcriptional regulators where they can act as coactivators or corepressors and indeed appear to themselves have some "intrinsic" transcriptional repression activity (Wilson et al., 2004; reviewed in Fuller-Pace and Moore, 2011). Although p68 and p72 appear to exhibit some functional redundancy in some contexts (Jalal et al., 2007), each protein also has discrete essential functions as demonstrated by the differences in the phenotypes of individual knockout mice (Fukuda et al., 2007). Moreover, p68 and p72 not only appear to exhibit promoter-specific effects but also have clearly different functions in their coactivation of transcription factors (Bates et al., 2005; Wortham et al., 2009). This specificity is likely to be imparted, at least to some extent, by interacting protein partners. Although p68 and p72 share remarkable homology (90% identity at the amino acid level across the central conserved helicase domain), they have significantly different N- and C-terminal extensions, and it is clear that, as in several other RNA helicases, these are involved in interacting

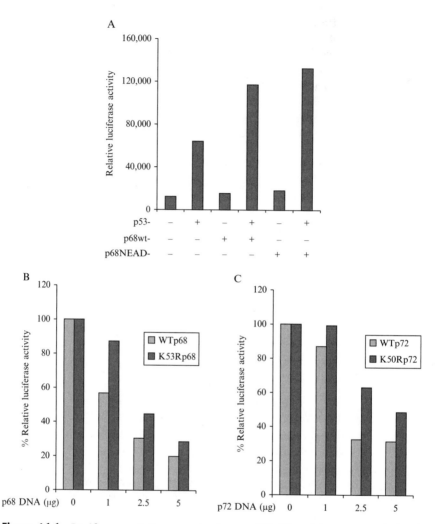

Figure 16.6 Luciferase activity assays to detect p68/p72 activity as transcriptional regulators: (A) Stimulation of p53 transcriptional activity of wt and helicase inactive p68 (NEAD, mutation in the DEAD motif) from the $p21^{waf1}$ promoter. *Note*: p21 is a cellular target of the p53 tumor suppressor. This experiment shows that p68 RNA helicase activity is not required to stimulate p53 transcriptional activity in this system. (B) and (C) Transcriptional repressor activity of p68 and p72 and their corresponding sumoylation-deficient mutants (p68 K53R and p72 K50R) using the Tymidine kinase promoter linked to luciferase (TK-Luc) as a model system. In this experiment, the p68 and p72 constructs were GAL4-tagged, and the pTK-Luc construct has GAL4-binding sites to allow the p68 and p72 proteins to bind. *Note*: as discussed in the text, p68 and p72 are not thought to bind directly to the promoter sequences. These experiments show that sumoylation stimulates p68/p72 transcriptional repressor activity.

with partners (reviewed in Fuller-Pace, 2006). Therefore, the analysis of such interactions and the importance of posttranslational modifications are likely to provide a better understanding of the role of p68 and p72 in transcription regulation.

ACKNOWLEDGMENTS

This work was supported by grants from the Association for International Cancer Research (06-613) and Cancer Research UK (C8745/A11216).

REFERENCES

Bates, G. J., Nicol, S. M., Wilson, B. J., Jacobs, A. M., Bourdon, J. C., Wardrop, J., Gregory, D. J., Lane, D. P., Perkins, N. D., and Fuller-Pace, F. V. (2005). The DEAD box protein p68: A novel transcriptional coactivator of the p53 tumour suppressor. *EMBO J.* **24,** 543–553.

Botlagunta, M., Vesuna, F., Mironchik, Y., Raman, A., Lisok, A., Winnard, P., Jr., Mukadam, S., Van Diest, P., Chen, J. H., Farabaugh, P., Patel, A. H., and Raman, V. (2008). Oncogenic role of DDX3 in breast cancer biogenesis. *Oncogene* **27,** 3912–3922.

Caretti, G., Schiltz, R. L., Dilworth, F. J., Di Padova, M., Zhao, P., Ogryzko, V., Fuller-Pace, F. V., Hoffman, E. P., Tapscott, S. J., and Sartorelli, V. (2006). The RNA helicases p68/p72 and the noncoding RNA SRA are coregulators of MyoD and skeletal muscle differentiation. *Dev. Cell* **11,** 547–560.

Chao, C. H., Chen, C. M., Cheng, P. L., Shih, J. W., Tsou, A. P., and Lee, Y. H. (2006). DDX3, a DEAD box RNA helicase with tumor growth-suppressive property and transcriptional regulation activity of the p21waf1/cip1 promoter, is a candidate tumor suppressor. *Cancer Res.* **66,** 6579–6588.

Clark, E. L., Coulson, A., Dalgliesh, C., Rajan, P., Nicol, S. M., Fleming, S., Heer, R., Gaughan, L., Leung, H. Y., Elliott, D. J., Fuller-Pace, F. V., and Robson, C. N. (2008). The RNA helicase p68 is a novel androgen receptor coactivator involved in splicing and is overexpressed in prostate cancer. *Cancer Res.* **68,** 7938–7946.

Dignam, J. D., Lebovitz, R. M., and Roeder, R. G. (1983). Accurate transcription initiation by RNA polymerase II in a soluble extract from isolated mammalian nuclei. *Nucleic Acids Res.* **11,** 1475–1489.

Endoh, H., Maruyama, K., Masuhiro, Y., Kobayashi, Y., Goto, M., Tai, H., Yanagisawa, J., Metzger, D., Hashimoto, S., and Kato, S. (1999). Purification and identification of p68 RNA helicase acting as a transcriptional coactivator specific for the activation function 1 of human estrogen receptor alpha. *Mol. Cell. Biol.* **19,** 5363–5372.

Fukuda, T., Yamagata, K., Fujiyama, S., Matsumoto, T., Koshida, I., Yoshimura, K., Mihara, M., Naitou, M., Endoh, H., Nakamura, T., Akimoto, C., Yamamoto, Y., et al. (2007). DEAD-box RNA helicase subunits of the Drosha complex are required for processing of rRNA and a subset of microRNAs. *Nat. Cell Biol.* **9,** 604–611.

Fuller-Pace, F. V. (2006). DExD/H box RNA helicases: Multifunctional proteins with important roles in transcriptional regulation. *Nucleic Acids Res.* **34,** 4206–4215.

Fuller-Pace, F. V., and Moore, H. C. (2011). RNA helicases p68 and p72: Multifunctional proteins with important implications for cancer development. *Future Oncol.* **7,** 239–251.

Gareau, J. R., and Lima, C. D. (2010). The SUMO pathway: Emerging mechanisms that shape specificity, conjugation and recognition. *Nat. Rev. Mol. Cell Biol.* **11,** 861–871.

Gillian, A. L., and Svaren, J. (2004). The Ddx20/DP103 dead box protein represses transcriptional activation by Egr2/Krox-20. *J. Biol. Chem.* **279,** 9056–9063.

Grundhoff, A. T., Kremmer, E., Tureci, O., Glieden, A., Gindorf, C., Atz, J., Mueller-Lantzsch, N., Schubach, W. H., and Grasser, F. A. (1999). Characterization of DP103, a novel DEAD box protein that binds to the Epstein-Barr virus nuclear proteins EBNA2 and EBNA3C. *J. Biol. Chem.* **274,** 19136–19144.

Guo, J., Hong, F., Loke, J., Yea, S., Lim, C. L., Lee, U., Mann, D. A., Walsh, M. J., Sninsky, J. J., and Friedman, S. L. (2010). A DDX5 S480A polymorphism is associated with increased transcription of fibrogenic genes in hepatic stellate cells. *J. Biol. Chem.* **285,** 5428–5437.

Jacobs, A. M., Nicol, S. M., Hislop, R. G., Jaffray, E. G., Hay, R. T., and Fuller-Pace, F. V. (2007). SUMO modification of the DEAD box protein p68 modulates its transcriptional activity and promotes its interaction with HDAC1. *Oncogene* **26,** 5866–5876.

Jalal, C., Uhlmann-Schiffler, H., and Stahl, H. (2007). Redundant role of DEAD box proteins p68 (Ddx5) and p72/p82 (Ddx17) in ribosome biogenesis and cell proliferation. *Nucleic Acids Res.* **35,** 3590–35601.

Jensen, E. D., Niu, L., Caretti, G., Nicol, S. M., Teplyuk, N., Stein, G. S., Sartorelli, V., van Wijnen, A. J., Fuller-Pace, F. V., and Westendorf, J. J. (2008). p68 (Ddx5) interacts with Runx2 and regulates osteoblast differentiation. *J. Cell. Biochem.* **103,** 1438–1451.

Lee, M. B., Lebedeva, L. A., Suzawa, M., Wadekar, S. A., Desclozeaux, M., and Ingraham, H. A. (2005). The DEAD-box protein DP103 (Ddx20 or Gemin-3) represses orphan nuclear receptor activity via SUMO modification. *Mol. Cell. Biol.* **25,** 1879–1890.

Mooney, S. M., Goel, A., D'Assoro, A. B., Salisbury, J. L., and Janknecht, R. (2010a). Pleiotropic effects of p300-mediated acetylation on p68 and p72 RNA helicase. *J. Biol. Chem.* **285,** 30443–30452.

Mooney, S. M., Grande, J. P., Salisbury, J. L., and Janknecht, R. (2010b). Sumoylation of p68 and p72 RNA helicases affects protein stability and transactivation potential. *Biochemistry* **49,** 1–10.

Moore, H. C., Jordan, L. B., Bray, S. E., Baker, L., Quinlan, P. R., Purdie, C. A., Thompson, A. M., Bourdon, J. C., and Fuller-Pace, F. V. (2010). The RNA helicase p68 modulates expression and function of the Delta133 isoform(s) of p53, and is inversely associated with Delta133p53 expression in breast cancer. *Oncogene* **29,** 6475–6484.

Nicol, S. M., and Fuller-Pace, F. V. (2010). Analysis of the RNA helicase p68 (Ddx5) as a transcriptional regulator. *Methods Mol. Biol.* **587,** 265–279.

Ou, Q., Mouillet, J. F., Yan, X., Dorn, C., Crawford, P. A., and Sadovsky, Y. (2001). The DEAD box protein DP103 is a regulator of steroidogenic factor-1. *Mol. Endocrinol.* **15,** 69–79.

Schroder, M. (2010). Human DEAD-box protein 3 has multiple functions in gene regulation and cell cycle control and is a prime target for viral manipulation. *Biochem. Pharmacol.* **79,** 297–306.

Schroder, M., Baran, M., and Bowie, A. G. (2008). Viral targeting of DEAD box protein 3 reveals its role in TBK1/IKKepsilon-mediated IRF activation. *EMBO J.* **27,** 2147–2157.

Soulat, D., Burckstummer, T., Westermayer, S., Goncalves, A., Bauch, A., Stefanovic, A., Hantschel, O., Bennett, K. L., Decker, T., and Superti-Furga, G. (2008). The DEAD-box helicase DDX3X is a critical component of the TANK-binding kinase 1-dependent innate immune response. *EMBO J.* **27,** 2135–2146.

Watanabe, M., Yanagisawa, J., Kitagawa, H., Takeyama, K., Ogawa, S., Arao, Y., Suzawa, M., Kobayashi, Y., Yano, T., Yoshikawa, H., Masuhiro, Y., and Kato, S. (2001). A subfamily of RNA-binding DEAD-box proteins acts as an estrogen receptor alpha coactivator through the N-terminal activation domain (AF-1) with an RNA coactivator, SRA. *EMBO J.* **20,** 1341–1352.

Wilson, B. J., Bates, G. J., Nicol, S. M., Gregory, D. J., Perkins, N. D., and Fuller-Pace, F. V. (2004). The p68 and p72 DEAD box RNA helicases interact with HDAC1 and repress transcription in a promoter-specific manner. *BMC Mol. Biol.* **5,** 11.

Wortham, N. C., Ahamed, E., Nicol, S. M., Thomas, R. S., Periyasamy, M., Jiang, J., Ochocka, A. M., Shousha, S., Huson, L., Bray, S. E., Coombes, R. C., Ali, S., *et al.* (2009). The DEAD-box protein p72 regulates ERalpha-/oestrogen-dependent transcription and cell growth, and is associated with improved survival in ERalpha-positive breast cancer. *Oncogene* **28,** 4053–4064.

CHAPTER SEVENTEEN

DEAD-Box RNA Helicases in Gram-Positive RNA Decay

Peter Redder *and* Patrick Linder

Contents

1. Introduction	369
2. Measuring mRNA Decay	371
2.1. RNA preparation	371
2.2. qRT-PCR determination of RNA decay	372
2.3. Using northern blotting to determine RNA decay	374
2.4. Analyses of RNA-decay data	374
3. Phenotypic Readouts	377
3.1. Biofilm assay	378
3.2. Hemolysis assay	380
4. Concluding Remarks	381
Acknowledgments	381
References	381

Abstract

DEAD-box RNA helicases are important players in eukaryotic and bacterial RNA metabolism. A helicase from *Staphylococcus aureus* was recently shown to affect RNA decay, most likely via its interaction with the proposed Gram-positive degradosome. Some, but not all, RNAs are stabilized when the helicase CshA is mutated, and among the affected RNAs is the *agrBDCA* mRNA, which is responsible for quorum sensing in *S. aureus*. We describe how the stabilization of *agr* mRNA (and others) can be measured and how to conduct assays to measure the effects of quorum-sensing defects, such as biofilm formation and hemolysin production.

1. Introduction

Proteins from the DEAD-box RNA helicase family represent the largest family of helicase proteins throughout eukaryotic and prokaryotic systems (Linder and Jankowsky, 2011; Tanner *et al.*, 2003). DEAD-box

Department of Microbiology and Molecular Medicine, University of Geneva, Genève, Switzerland proteins can be defined by the presence of 12 conserved motifs (Jankowsky, 2011). Although individual motifs may vary as, for example, the DECD motif instead of DEAD in Sub2, the proteins are easily identified by the overall presence of these motifs. Several of these motifs are required for binding and hydrolysis of the ATP, whereas others are involved in RNA binding or intraprotein interactions.

Originally, the DEAD-box protein family was proposed to be a collection of RNA helicases (Linder *et al.*, 1989). This was based on the ability of DEAD-box proteins to alter RNA secondary structures and their presence in dynamic processes such as translation initiation and ribosome biogenesis. Consistent with this idea, a multitude of DEAD-box proteins are required in processes that use guide RNAs to perform RNA modification, RNA processing, or RNA editing. Moreover, genetic evidences suggest that DEAD-box proteins are indeed required for dissociation of guide RNA containing complexes *in vivo* (Chen *et al.*, 2001; Kistler and Guthrie, 2001). Nevertheless, a bona fide dsRNA dissociation activity *in vivo* is extremely difficult to demonstrate.

Whereas eukaryotic genomes encode a multitude of DEAD-box proteins (25 in yeast, 36 in humans), bacteria encode only a few of them (5 in *Escherichia coli*, 4 in *Bacillus subtilis*, 2 in *Staphylococcus aureus*, 6 in *Pseudomonas aeruginosa*, 9 in *Pseudoalteromonas haloplanktis*, 10 in *Vibrio fischeri*, 11 in *Colwellia psychrerythraea*; Iost and Dreyfus, 2006; Lopez-Ramirez *et al.*, 2011). Moreover, many DEAD-box proteins are essential in eukaryotes, but mutations in bacterial DEAD-box protein genes at most give a cold-sensitive phenotype. Interestingly, even a quintuple mutant in *E. coli* is viable, suggesting that these proteins do not perform essential functions under normal laboratory conditions (Jagessar and Jain, 2010).

So far, in bacteria only DEAD-box proteins from *E. coli* were studied in detail: CsdA (originally called DeaD) and SrmB were identified in genetic screens using mutations in ribosomal proteins and their role in ribosome biogenesis has since been confirmed. Extensive and beautiful work has pinpointed a CsdA function to the biogenesis of the 50S subunit (Iost and Dreyfus, 2006). The DbpA protein is famous for its requirement for a specific RNA substrate for efficient ATP hydrolysis (Fuller-Pace *et al.*, 1993). Indeed, this DEAD-box protein is exclusively and heavily stimulated in its activity by the hairpin h92 from the 23S rRNA (Diges and Uhlenbeck, 2001; Nicol and Fuller-Pace, 1995). Recently, it was shown that a dominant-negative mutation in motif VI of *dbpA* results in a ribosome biogenesis defect (Elles and Uhlenbeck, 2008). While the above proteins are involved in ribosome biogenesis, the RhlB protein is part of the degradosome, a multicomponent complex involved in the targeted degradation of a large subset of mRNAs (Carpousis, 2007; Py *et al.*, 1996). The RNA helicase in this complex is associated on the scaffold protein RNase E and is generally assumed to assist the degradosome on structured RNAs. Nevertheless, the precise role of the helicase in RNA degradation is not known. Interestingly, CsdA can also associate with the degradosome under cold-shock conditions, but this interaction does not use the same region of RNase E involved in the interaction with RhlB (Prud'homme-Genereux *et al.*, 2004). Finally, RhlE is the least-characterized RNA helicase in *E. coli*. Genetic data indicate that RhlE might be involved in ribosome biogenesis together with CsdA and SrmB (Jain, 2008). Other data show that RhlE is also able to associate with the degradosome (Khemici *et al.*, 2004). Clearly further work will be required to elucidate the function(s) of RhlE.

RNA helicases are important players in RNA metabolism and participate in the dynamic rearrangement of RNP (ribonucleoprotein) complexes. It is therefore not surprising that several bacterial DEAD-box proteins popped up in a variety of screens. In cyanobacteria, the RNA helicases CrhC and CrhR are induced by cold and redox stress, respectively (Owttrim, 2006). In *Bacillus cereus*, RNA helicases from the DEAD-box family were identified by mutagenesis to be required for survival at low temperatures, pH, and oxidative stress (Pandiani *et al.*, 2010). Similarly, Tn916 mutagenesis identified an RNA helicase from the Gram-positive *Clostridium perfringens* to be involved in the adaptive response to oxidative stress (Briolat and Reysset, 2002), and random transposon mutagenesis identified a RNA helicase involved in the regulation of phenolic acid metabolism in *Lactobacillus plantarum* (Gury *et al.*, 2004) and a RNA helicase (CshA) in *S. aureus* involved in biofilm formation (Tu Quoc *et al.*, 2007). So far, however, their molecular function remains mostly elusive.

Very recently, bacterial two-hybrid analysis in *B. subtilis* and *S. aureus* identified the CshA RNA helicase as a potential member of a degradosome in Gram-positive bacteria (Lehnik-Habrink *et al.*, 2010; Roux *et al.*, 2011). These results are nicely consistent with data from our laboratory that show that a mutation in *cshA* results in a stabilization of the *agr* mRNA (Oun *et al.*, in preparation). This transcript is part of a quorum-sensing system that is induced at high-cell density to repress the expression of surface proteins and to induce synthesis of secreted virulence factors. Here, we give examples of how to measure RNA decay and phenotypic readouts caused by differences in the mRNA of the *agr* quorum-sensing system.

2. Measuring mRNA Decay

2.1. RNA preparation

In order to measure an effect on RNA decay in bacterial systems, new RNA synthesis is blocked by rifampicin treatment (400 μg/ml), whereupon the level of the RNA in question is measured at different time points (e.g., 0,

2.5, 5, 15, and 30 min; Oun et al., in preparation). Obviously, the strain used needs to be checked for sensitivity to rifampicin, since Rif^R mutations occur readily, and various S. aureus strains may differ in the rifampicin concentration needed.

Many RNAs are expressed differently depending on growth phase (Dunman et al., 2001), and while the initial amount of a given RNA might not influence its decay rate significantly, the potential differences in intracellular enzyme composition between a stationary and an exponentially growing cell certainly might. Therefore, it is important to ensure that the wild-type and the helicase mutant are at similar growth phases when rifampicin is added. In praxis, we usually dilute overnight cultures 200 times in fresh medium and then follow the growth until it reaches approximately $OD_{600} = 0.4$, at which point we add the rifampicin. The culture will still be in exponential phase, but we have enough cells to get a sufficient amount of RNA.

For sample preparation, an aliquot (e.g., 1 ml) is taken from the bacterial culture at the indicated time points, immediately centrifuged briefly to remove excess liquid (30 s at 10,000 g), and then half a volume of 1:1 ethanol:acetone is added to stop further degradation of the RNA. The samples can then be stored at $-80\ ^\circ C$ until they are ready for RNA extraction (Fig. 17.1A). We find that thorough and immediate mixing with rifampicin is very important for obtaining consistent results, especially if short time intervals are chosen. It should also be noted that rifampicin only blocks initiation of new RNA synthesis, thus allowing ongoing RNA synthesis to finish, which can lead to a small increase in RNA levels immediately after the rifampicin treatment.

RNA extraction (Fig. 17.1B), which can be done using a commercially available kit, such as RNeasy Mini Kit (Qiagen, Cat N°74104), should also include a DNase treatment, to remove genomic DNA that will otherwise interfere with subsequent measurements. The RNA should then be checked for quality (lack of degradation) by agarose gel electrophoresis or using a bioanalyzer.

2.2. qRT-PCR determination of RNA decay

Quantitative reverse transcription PCR (qRT-PCR) can be used to determine the amount of the RNA of interest, by using a specific primer-set (Fig. 17.1C). However, care must be taken when interpreting the results, and a second set of primers must be used as an internal reference (choice of reference(s) for RNA decay is discussed below). Total RNA from the various time points and from mutant and wild-type cultures is used as template for the qRT-PCR reaction, and the individual relative quantities for the time points can then be plotted. For comparison of the wild-type and mutant data, we find it convenient to normalize the steady-state levels (e.g., time = 0) to 100% (Fig. 17.2). An exponential regression curve can be

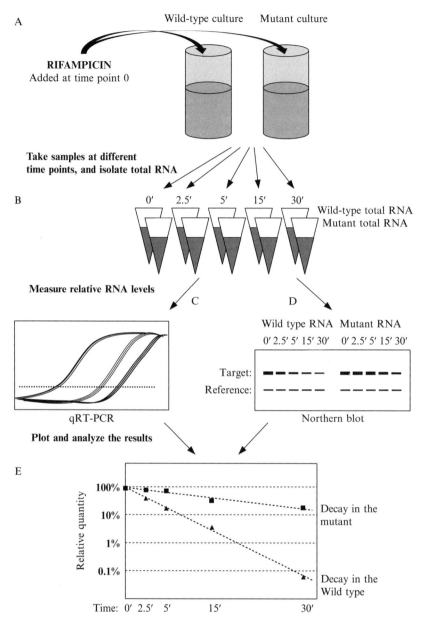

Figure 17.1 Flowchart for RNA-decay assay. (A) Rifampicin is added to exponentially growing cultures of the mutant and the wild type. (B) Samples are taken at specific time points, for example, 0, 2.5, 5, 15, and 30 min after addition of the rifampicin, and RNA is isolated from each sample. (C) qRT-PCR can be used to determine the quantity of the target RNA, using a nonaffected RNA as internal reference. (D) Alternatively, northern blotting can be used to quantify the target RNA. (E) The RNA quantities, relative to the reference RNA, can then be plotted in a semilogarithmic

calculated from this plot, or alternatively, the mutant and the wild type can be compared for each individual time point (Figs. 17.1E and 17 2).

In spite of the strength and accuracy of qRT-PCR method, or perhaps due to it, there are several caveats to observe (Bustin et al., 2009). Not only does the chosen reference RNA have to be "unaffected" by the RNA-decay mechanism being studied (more on choice of reference below), but also the primers used to amplify the reference RNA must have a similar efficiency to the primers used for the RNA of interest. Further, the abundance of reference RNA must be such that the two RNAs can be observed within the same range of detection in the qRT-PCR amplification plot (see below).

2.3. Using northern blotting to determine RNA decay

The use of qRT-PCR allows for more accurate measurements and is also generally less labor intensive than northern blotting. It is therefore the method of choice for the majority of experimental setups. However, northern blotting additionally provides information about the length and particularly on the integrity of the RNA, whereas qRT-PCR does not. Therefore, northern blotting may be the superior technique for situations where it is suspected that one RNA segment is decaying more rapidly than another. An RNA-decay measurement using northern blotting is carried out in essentially the same way as qRT-PCR, using both a probe for the RNA of interest and a reference probe (Fig. 17.1D). Experimental details for northern blotting can be found in Sambrook and Russell (2001).

2.4. Analyses of RNA-decay data

Once RNA levels have been measured for several time points after rifampicin treatment, either by northern blot or by qRT-PCR, it is possible to calculate relative half-lives of the RNAs. Relatively, because each RNA measurement is measured in relation to the reference RNA, and it is therefore not possible to calculate an absolute half-life, but only possible to calculate how much faster (or slower) the RNA of interest decays in comparison to the decay of the reference RNA (Fig. 17.3).

When the relative quantity values are plotted in a semilogarithmic fashion, the decay curves should appear linear (Fig. 17.1E), and it should be possible to generate a linear regression (using Microsoft Excel,

---

plot, after normalization by defining the time point 0 as having 100% RNA. Black squares indicate the data for the mutant, and black triangles the data for the wild type. Dotted lines indicate exponential regression curves fitted to the data points. A relative RNA half-life can be determined based on the regression curves.

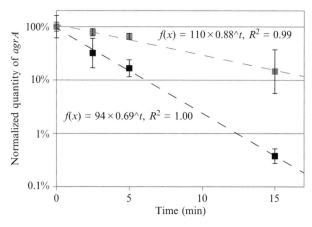

Figure 17.2 Analyzing RNA-decay data. The graph represents an example of an output from OpenOffice.org Calc, where the data from a single qRT-PCR experiment were analyzed (taken from Oun et al., in preparation). Each measurement of *agrA* and HU (the reference gene) was carried out in triplicate on the same qRT-PCR plate, and the error-bars indicate the 95% confidence level. Black squares: SA564 wild type. Gray squares: *cshA* mutant. Dotted lines show the exponential regression curves with the formulas and R^2 value indicated. Mutant and wild type do not have the same steady-state levels of *agrA* mRNA, and the data was therefore normalized to the measurements at time point 0.

OpenOffice.org Calc, or similar) with an R^2 value close to 1. For example, using OpenOffice.org Calc the formula for the curve will be in the format

$$Y = A \times B^t, \qquad (17.1)$$

where A is the offset at $t=0$, t is the time, and B is a constant that defines the slope of the curve. To calculate the relative half-life, one uses the formula

$$1/2 = B^{t_{1/2}} \qquad (17.2)$$

resulting in

$$\ln(1/2)/\ln(B) = t_{1/2}. \qquad (17.3)$$

2.4.1. Choosing a reference RNA for RNA-decay assays

A DEAD-box helicase is one of the core components of Gram-positive degradosome, and although very little is as yet known about the function of the degradosome in these organisms, it must be presumed that this protein complex affects the decay rate of many RNAs in the cell. Therefore, it is a

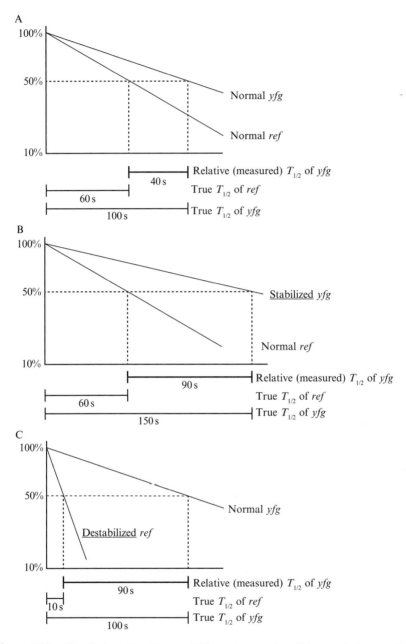

Figure 17.3 The choice of a reference RNA and examples of dangers in interpreting decay data. *yfg*, the gene of interest; *ref*, the chosen reference RNA. Thin lines indicate the decay curves obtained for *yfg* RNA and *ref* RNA. Measured RNA levels of both *yfg* and *ref* are decreasing over time, as expected. However, Rifampicin is an antibiotic and is killing the bacteria during our time course, and part of the observed decrease might be due to fewer cells harvested. Moreover, even with when care is taken to have the same

non-trivial task to choose an "unaffected" RNA to use as a reference RNA for either northern blotting or qRT-PCR (Eleaume and Jabbouri, 2004; Vaudaux et al., 2002). In most publications, rRNA is used as an internal standard, but in this case, it has to be considered that this reference is very abundant and will not appear in a similar abundance window and that ribosome biogenesis and the highly structured nature of rRNA may be influenced by the RNA helicase. Nevertheless, if several mRNAs are shown to be affected to the same degree and to similar extents in wild type and mutant (i.e., the decay rate of one stays the same in relation to the others), then one of these RNAs can be used for normalizing affected RNAs. As an example, in *S. aureus*, a mutation in the DEAD-box helicase *cshA* gene appears to stabilize the *agrBDCA* mRNA, when compared to the *HU* mRNA (compare Fig. 17.3A and B). However, if one only compares the two mRNAs, then the effect might equally be that *HU* mRNA is destabilized and *agrBDCA* mRNA stays unaffected (Fig. 17.3C). *HU* mRNA decay was therefore compared to a number of other RNAs to show that it is indeed unaffected by the *cshA* mutation (or at least is affected to the same degree as the other RNAs). Further, in a transcriptome-wide RNA-decay assay, the *HU* mRNA belongs to the (mostly) unaffected group of RNAs, whereas *agrBDCA* mRNA was stabilized in comparison to the global mRNA decay (Oun et al., in preparation).

3. PHENOTYPIC READOUTS

In *S. aureus*, disruption of the *cshA* gene causes the cells to exhibit several phenotypic changes, especially regarding expression of virulence factors. This was first noticed by Tu Quoc et al. (2007) who observed a reduced capacity for producing biofilm in a *cshA* mutant of the *S. aureus* S30 strain, which is normally a strong biofilm producer. Moreover, *cshA* mutants exhibit stabilization of their *agrBDCA* mRNA, which is responsible for quorum sensing in *S. aureus* (Oun et al., in preparation). The *agr*-system induces a regulatory RNA, RNAIII, which is responsible for

---

amount of total RNA in each well, it is almost impossible to avoid small fluctuations. Therefore, it is important to use an internal reference RNA (*ref* RNA) to standardize the amount of *yfg* RNA. One consequence of this is that it is not possible with this method to determine an absolute half-life ($T_{1/2}$) for *yfg*, but only a $T_{1/2}$ relative to *ref* RNA. The figure displays three different scenarios: (A) For a wild-type situation, the true $T_{1/2}$ of *yfg* RNA is 100 s and that of *ref* RNA is 60 s, resulting in a relative $T_{1/2}$ of 40 s for *yfg* RNA. (B) In a mutant, where *yfg* RNA is stabilized, the true $T_{1/2}$ of *yfg* RNA has increased to 120 s while the true $T_{1/2}$ of *ref* RNA stays at 60 s, leading to a relative $T_{1/2}$ for *yfg* of 60 s. (C) In another mutant, *yfg* RNA stability stay the same as in wild type ($T_{1/2}=100$ s), whereas *ref* RNA is destabilized and now only has a $T_{1/2}$ of 40 s. The relative $T_{1/2}$ of *yfg* RNA in this mutant is therefore 60 s, indistinguishable from the mutant in example B, unless additional experiments are carried out.

up- and downregulation of a number of virulence factors, such as Rot, protein A, and hemolysin alpha (Boisset et al., 2007; Kong et al., 2006). Further, RNAIII includes a small open reading frame, which encodes the hemolysin delta protein. As a consequence, hemolysin alpha and delta are useful markers for disruptions of the cshA helicase gene (Fig. 17.4, agr pathway).

Hemolysins are small proteins that lyse erythrocytes, and it is possible to distinguish between the alpha and delta hemolysin by the difference in efficiency against rabbit and horse erythrocytes. Hemolysin alpha is highly active against rabbit blood, whereas horse erythrocytes are highly vulnerable to hemolysin delta (note: hemolysin beta can be assayed using sheep blood; Garvis et al., 2002).

Mutations in genes to be analyzed in clinical S. aureus strains can be obtained by a variety of methods. In the laboratory, we have used the targetron system (Yao et al., 2006) and for making clean deletions, a highly powerful selection/antiselection system that takes advantage of the sensitivity of bacteria to 5-FOA (Redder and Linder, 2012).

3.1. Biofilm assay

To determine the level of biofilm production for a mutant of S. aureus, the classical method is via crystal violet staining of biofilm after a specific period of nonagitated growth. The method is described in Mack et al. (2001), but an adaptation used in our laboratory can be briefly summarized as follows: Mutant and wild-type cultures are diluted to OD = 0.1 and 1 ml is incubated for 6 h in a polystyrene tube without agitation at 37 °C. It should be noted that biofilm formation depends on the number of cells and moreover may be influenced by the quorum-sensing system, and it is therefore important to take into account potential differences in growth rate and adapt initial dilutions accordingly. For staining, 100 µl of a solution of 1% w/v crystal violet and 2% ethanol are then added and allowed to stain for 15 min, whereupon the liquid is removed and the tube is washed with water. The crystal violet that is fixed by the biofilm is then dissolved in 400 µl ethanol and transferred to a cuvette with 600 µl water. The amount of biofilm can be measured by the absorption at 570 nm (we find that 600-nm work for most purposes as well). Alternatively, biofilm can be grown on glass slides, which can then be studied under the microscope, after staining with crystal violet.

The best possible but also most laborious way of studying biofilm is probably when it is monitored by 3D imaging in a flow cell. If the bacteria are fluorescent, then this method even allows the forming of the biofilm matrix to be followed live. Detailed protocols for growing biofilm in flow cells and examination by 3D imaging can be found in Christensen et al. (1999).

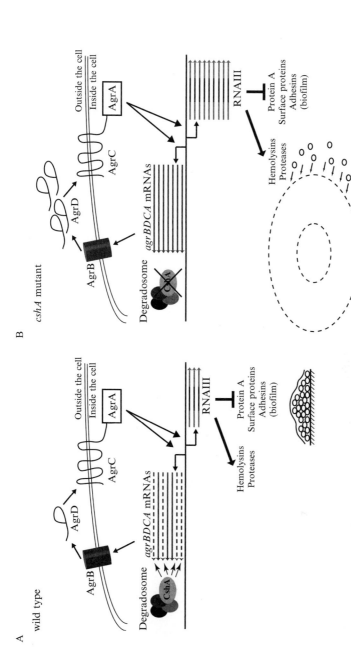

Figure 17.4 The role of the degradosome in quorum sensing. (A) In the wild type, with normal levels of CshA, *agrBDCA* mRNAs are produced (gray arrows) but the degradosome is able to degrade a significant portion of them (dotted gray arrows). The remaining *agrBDCA* mRNAs are translated and the quorum-sensing system is able to respond correctly to cell density to regulate the amount of RNAIII (light gray arrows). The hemolysin delta open reading frame, encoded by RNAIII, is shown in black. (B) In the *cshA* mutant, the degradosome is unable to degrade the *agrBDCA* mRNAs correctly, leading to a much higher level of Agr proteins and results in an overactive quorum sensing that leads to elevated RNAIII levels. The RNAIII in turn overly stimulates production of hemolysins and extracellular proteases and inhibits the production of biofilm components.

3.2. Hemolysis assay

The easiest way of testing hemolysis activity is via blood-agar plates, which will give a semiquantitative estimate of activity (Garvis et al., 2002). Rich medium (e.g., Mueller-Hinton, Becton Dickinson, NJ, USA), autoclaved with 10 g/l agar, is cooled to 43 °C and mixed with 7% defibrinated blood, preheated to 43 °C (TCS Biosciences Ltd., Buckingham, UK), whereupon the mix is immediately poured into petri dishes (antibiotics can be added if desired). The plates should be dried well at 37 °C before use.

For the assay, 10 µl of wild type and mutant overnight cultures are then spotted on the same plate, and after the spot has dried, the plate is incubated at 37 °C for about 20 h and then transferred to 5 °C for 2–3 days to allow hemolysis on the plate to occur. The efficiency of the hemolysis can then be monitored by the size of the zones of transparency around the spots (Fig. 17.5).

When interpreting data from the plate assay, it is important to take into account that one strain might grow significantly slower than another, which means that there might be significantly fewer cells in one spot than in another, until stationary phase is reached. This means that one strain might have much more time to produce hemolysins than another, especially taking into

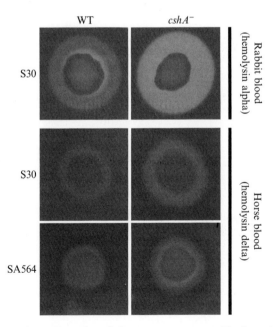

Figure 17.5 Hemolysis. Examples of clearance zones caused by hemolysins on rabbit blood plates and horse blood plates. The clearance zones are consistently larger in the cshA mutants, on both rabbit and horse blood. However, there are strain-specific differences in the appearance of the clearance zones, exemplified here by two clinical strains S30 and SA564.

consideration that hemolysins in *S. aureus* are produced in a growth-phase- and quorum-sensing-dependent manner (Dunman *et al.*, 2001; Kong *et al.*, 2006).

An alternative to the blood-agar assay is to measure hemolytic activity in a liquid solution, with the advantage that it is possible to quantify the activity by spectrophotometry. The assay is described in detail in Bernheimer (1988). To prepare the hemolysin for the liquid assay, 2 ml of bacterial culture is centrifuged for 5 min at 8000 rpm, whereupon the supernatant is sterilized with a 0.22-μm filter. The removal of the hemolysin-producing bacteria, by centrifugation and filtration, ensures that only hemolysins that were present at the time of harvesting will be measured in the assay, thus eliminating the experimental error that arises from differences in doubling times, as is the case in the spots of the blood-agar-plate assay. However, the difference in growth rate should still be taken into account when deciding at what point the cells are removed, since fewer cells and/or cells in exponential phase will produce less hemolysins.

4. Concluding Remarks

The Gram-positive degradosome has only recently been discovered by bacterial two-hybrid analyses. Its function and the exact role of the individual components, including the DEAD-box helicase CshA, are still very much on the speculative stage. Here, we have presented various methods and their advantages and caveats necessary for measuring the effects of a *cshA* mutation, via the stabilization of the *agrBDCA* mRNA, although many other RNAs are likely to be affected as well. The field of mRNA turnover is very dynamic and many variations or even alternative methods from those described here can be found in the literature.

ACKNOWLEDGMENTS

We are grateful to Patrice François, Jacques Schrenzel, and the members of the Linder laboratory for helpful discussions. Work in the author's laboratory is supported by the Swiss National Science Foundation (P. L.) and the canton of Geneva.

REFERENCES

Bernheimer, A. W. (1988). Assay of hemolytic toxins. *Methods Enzymol.* **165,** 213–217.
Boisset, S., Geissmann, T., Huntzinger, E., Fechter, P., Bendridi, N., Possedko, M., Chevalier, C., Helfer, A. C., Benito, Y., Jacquier, A., Gaspin, C., and Vandenesch, F. (2007). Staphylococcus aureus RNAIII coordinately represses the synthesis of virulence factors and the transcription regulator Rot by an antisense mechanism. *Genes Dev.* **21,** 1353–1366.

Briolat, V., and Reysset, G. (2002). Identification of the Clostridium perfringens genes involved in the adaptive response to oxidative stress. *J. Bacteriol.* **184,** 2333–2343.

Bustin, S. A., Benes, V., Garson, J. A., Hellemans, J., Huggett, J., Kubista, M., Mueller, R., Nolan, T., Pfaffl, M. W., Shipley, G. L., Vandesompele, J., and Wittwer, C. T. (2009). The MIQE guidelines: Minimum information for publication of quantitative real-time PCR experiments. *Clin. Chem.* **55,** 611–622.

Carpousis, A. J. (2007). The RNA degradosome of Escherichia coli: An mRNA-degrading machine assembled on RNase E. *Annu. Rev. Microbiol.* **61,** 71–87.

Chen, J. Y.-F., Stands, L., Staley, J. P., Jackups, R. R., Jr., Latus, L. J., and Chang, T.-H. (2001). Specific alterations of U1-C protein or U1 small nuclear RNA can eliminate the requirement of Prp28p, an essential DEAD box splicing factor. *Mol. Cell* **7,** 227–232.

Christensen, B. B., Sternberg, C., Andersen, J. B., Palmer, R. J., Jr., Nielsen, A. T., Givskov, M., and Molin, S. (1999). Molecular tools for study of biofilm physiology. *Methods Enzymol.* **310,** 20–42.

Diges, C. M., and Uhlenbeck, O. C. (2001). Escherichia coli DbpA is an RNA helicase that requires hairpin 92 of 23S rRNA. *EMBO J.* **20,** 5503–5512.

Dunman, P. M., Murphy, E., Haney, S., Palacios, D., Tucker-Kellogg, G., Wu, S., Brown, E. L., Zagursky, R. J., Shlaes, D., and Projan, S. J. (2001). Transcription profiling-based identification of Staphylococcus aureus genes regulated by the agr and/or sarA loci. *J. Bacteriol.* **183,** 7341–7353.

Eleaume, H., and Jabbouri, S. (2004). Comparison of two standardisation methods in real-time quantitative RT-PCR to follow Staphylococcus aureus genes expression during in vitro growth. *J. Microbiol. Methods* **59,** 363–370.

Elles, L. M., and Uhlenbeck, O. C. (2008). Mutation of the arginine finger in the active site of Escherichia coli DbpA abolishes ATPase and helicase activity and confers a dominant slow growth phenotype. *Nucleic Acids Res.* **36,** 41–50.

Fuller-Pace, F. V., Nicol, S. M., Reid, A. D., and Lane, D. P. (1993). DbpA: A DEAD box protein specifically activated by 23S rRNA. *EMBO J.* **12,** 3619–3626.

Garvis, S., Mei, J. M., Ruiz-Albert, J., and Holden, D. W. (2002). Staphylococcus aureus svrA: A gene required for virulence and expression of the agr locus. *Microbiology* **148,** 3235–3243.

Gury, J., Barthelmebs, L., and Cavin, J. F. (2004). Random transposon mutagenesis of Lactobacillus plantarum by using the pGh9:IS S1 vector to clone genes involved in the regulation of phenolic acid metabolism. *Arch. Microbiol.* **182,** 337–345.

Iost, I., and Dreyfus, M. (2006). DEAD-box RNA helicases in Escherichia coli. *Nucleic Acids Res.* **34,** 4189–4197.

Jagessar, K. L., and Jain, C. (2010). Functional and molecular analysis of Escherichia coli strains lacking multiple DEAD-box helicases. *RNA* **16,** 1386–1392.

Jain, C. (2008). The E. coli RhlE RNA helicase regulates the function of related RNA helicases during ribosome assembly. *RNA* **14,** 381–389.

Jankowsky, E. (2011). RNA helicases at work: Binding and rearranging. *Trends Biochem. Sci.* **36,** 19–29.

Khemici, V., Toesca, I., Poljak, L., Vanzo, N. F., and Carpousis, A. J. (2004). The RNase E of Escherichia coli has at least two binding sites for DEAD-box RNA helicases: Functional replacement of RhlB by RhlE. *Mol. Microbiol.* **54,** 1422–1430.

Kistler, A. L., and Guthrie, C. (2001). Deletion of MUD2, the yeast homolog of U2AF65, can bypass the requirement for Sub2, an essential spliceosomal ATPase. *Genes Dev.* **15,** 42–49.

Kong, K. F., Vuong, C., and Otto, M. (2006). Staphylococcus quorum sensing in biofilm formation and infection. *Int. J. Med. Microbiol.* **296,** 133–139.

Lehnik-Habrink, M., Pfortner, H., Rempeters, L., Pietack, N., Herzberg, C., and Stulke, J. (2010). The RNA degradosome in Bacillus subtilis: Identification of CshA as the major RNA helicase in the multiprotein complex. *Mol. Microbiol.* **77,** 958–971.

Linder, P., and Jankowsky, E. (2011). From unwinding to clamping—The DEAD box RNA helicase family. *Nat. Rev. Mol. Cell Biol.* **12,** 505–516.
Linder, P., Lasko, P. F., Ashburner, M., Leroy, P., Nielsen, P. J., Nishi, K., Schnier, J., and Slonimski, P. P. (1989). Birth of the D-E-A-D box. *Nature* **337,** 121–122.
Lopez-Ramirez, V., Alcaraz, L. D., Moreno-Hagelsieb, G., and Olmedo-Alvarez, G. (2011). Phylogenetic distribution and evolutionary history of bacterial DEAD-Box proteins. *J. Mol. Evol.* **72,** 413–431.
Mack, D., Bartscht, K., Fischer, C., Rohde, H., de Grahl, C., Dobinsky, S., Horstkotte, M. A., Kiel, K., and Knobloch, J. K. (2001). Genetic and biochemical analysis of Staphylococcus epidermidis biofilm accumulation. *Methods Enzymol.* **336,** 215–239.
Nicol, S. M., and Fuller-Pace, F. V. (1995). The "DEAD box" protein DbpA interacts specifically with the peptidyltransferase center in 23S rRNA. *Proc. Natl. Acad. Sci. USA* **92,** 11681–11685.
Oun, S., Redder, P., François, P., Corvaglia, A. R., Didier, J. P., Buttazzoni, E., Giraud, C., Girard, M., Schrenzel, J., and Linder, P. (in preparation). A DEAD-box RNA helicase is a key element in persistence and virulence regulation in *Staphylococcus aureus*.
Owttrim, G. W. (2006). RNA helicases and abiotic stress. *Nucleic Acids Res.* **34,** 3220–3230.
Pandiani, F., Brillard, J., Bornard, I., Michaud, C., Chamot, S., Nguyen-the, C., and Broussolle, V. (2010). Differential involvement of the five RNA helicases in adaptation of Bacillus cereus ATCC 14579 to low growth temperatures. *Appl. Environ. Microbiol.* **76,** 6692–6697.
Prud'homme-Genereux, A., Beran, R. K., Iost, I., Ramey, C. S., Mackie, G. A., and Simons, R. W. (2004). Physical and functional interactions among RNase E, polynucleotide phosphorylase and the cold-shock protein, CsdA: Evidence for a 'cold shock degradosome'. *Mol. Microbiol.* **54,** 1409–1421.
Py, B., Higgins, C. F., Krisch, H. M., and Carpousis, A. J. (1996). A DEAD-box RNA helicase in the *Escherichia coli* RNA degradosome. *Nature* **381,** 169–172.
Redder, P., and Linder, P. (2012). A new range of vectors with a stringent 5-fluoro orotic acid-based antiselection system, for generating mutants by allelic replacement in Staphylococcus aureus. *Appl. Environ. Microbiol.* Mar 23.
Roux, C. M., Demuth, J. P., and Dunman, P. M. (2011). Characterization of components of the Staphylococcus aureus messenger RNA degradosome holoenzyme-like complex. *J. Bacteriol.* **193,** 5520–5526.
Sambrook, J., and Russell, D. R. (2001). Molecular Cloning: A Laboratory Manual. Cold Spring Harbor Laboratory Press, Cold Spring Harbor, N.Y.
Tanner, N. K., Cordin, O., Banroques, J., Doere, M., and Linder, P. (2003). The Q motif: A newly identified motif in DEAD box helicases may regulate ATP binding and hydrolysis. *Mol. Cell* **11,** 127–138.
Tu Quoc, P. H., Genevaux, P., Pajunen, M., Savilahti, H., Georgopoulos, C., Schrenzel, J., and Kelley, W. L. (2007). Isolation and characterization of biofilm formation-defective mutants of Staphylococcus aureus. *Infect. Immun.* **75,** 1079–1088.
Vaudaux, P., Francois, P., Bisognano, C., Kelley, W. L., Lew, D. P., Schrenzel, J., Proctor, R. A., McNamara, P. J., Peters, G., and Von Eiff, C. (2002). Increased expression of clumping factor and fibronectin-binding proteins by hemB mutants of Staphylococcus aureus expressing small colony variant phenotypes. *Infect. Immun.* **70,** 5428–5437.
Yao, J., Zhong, J., Fang, Y., Geisinger, E., Novick, R. P., and Lambowitz, A. M. (2006). Use of targetrons to disrupt essential and nonessential genes in Staphylococcus aureus reveals temperature sensitivity of Ll.LtrB group II intron splicing. *RNA* **12,** 1271–1281.

CHAPTER EIGHTEEN

RNA Helicases in Cyanobacteria: Biochemical and Molecular Approaches

George W. Owttrim

Contents

1. Introduction	386
2. Preparation of Cyanobacterial Extracts	387
2.1. Cyanobacterial growth	387
2.2. Cautionary note: Cyanobacterial growth	388
2.3. RNA isolation	388
2.4. Protein isolation	389
3. Alteration of RNA Helicase Expression in Cyanobacteria	390
3.1. *Anabaena* sp. strain PCC 7120 transformation: Triparental mating	390
3.2. *Synechocystis* sp. strain PCC 6803 transformation	391
3.3. RNA helicase gene inactivation	392
3.4. Inducible RNA helicase expression	392
3.5. Confirmation of genetic alteration	393
4. Northern Analysis of Transcript Levels	394
4.1. Electrophoresis and blotting	394
4.2. Transcript detection	395
4.3. RNA loading control	395
5. Western Analysis of Protein Levels	396
5.1. Western analysis	396
5.2. Protein loading controls	397
6. Cellular Ultrastructure and *In Situ* RNA Helicase Localization	397
6.1. Cellular ultrastructure	397
6.2. *In situ* RNA helicase localization	398
Acknowledgments	400
References	400

Department of Biological Sciences, University of Alberta, Edmonton, Alberta, Canada

Abstract

RNA helicases are associated with every aspect of RNA metabolism and function. A diverse range of RNA helicases are encoded by essentially every organism. While RNA helicases alter gene expression, RNA helicase expression is itself regulated, frequently in response to abiotic stress. Photosynthetic cyanobacteria present a unique model system to investigate RNA helicase expression and function. This chapter describes methodology to study the expression and cellular localization of RNA helicases, providing insights into the metabolic pathway(s) in which these enzymes function in cyanobacteria. The approaches are applicable to other systems as well.

1. INTRODUCTION

Cyanobacteria are Gram-negative eubacteria (basically green *Escherichia coli*) which perform oxygenic photosynthesis essentially identical to that observed in chloroplasts of higher plants. In fact, chloroplasts evolved from cyanobacteria by endosymbiosis (Criscuolo and Gribaldo, 2011). An increasing number of cyanobacterial genomes have been sequenced (see http://genome.kazusa.or.jp/cyanobase/). Cyanobacteria can be genetically manipulated, with routine procedures existing for transformation, gene inactivation, etc. Cyanobacteria also exhibit a diverse range of interesting genetic characteristics including the ability to fix atmospheric nitrogen, terminal cellular differentiation, robust circadian rhythms and are amenable to production of secondary products such as biodiesel and therapeutic drugs, all while scavenging atmospheric carbon dioxide and releasing oxygen as a byproduct (Dong and Golden, 2008; Kumar et al., 2010; Ruffing, 2011). For these and other reasons, cyanobacteria are attractive model systems to study photosynthesis, cell development, and differentiation and the "green," environmentally friendly production of commercial products.

Cyanobacteria encode members of multiple RNA helicase families. However, to date, investigation of RNA helicases in cyanobacteria has been limited to members of the DEAD-box family. One DEAD-box helicase is encoded in the *Synechocystis* genome (*crhR*, slr0083; Kujat and Owttrim, 2000) and two in *Anabaena* (*crhB*, alr1223 and *crhC*, alr4718; Chamot et al., 1999). Expression of these genes is enhanced in response to abiotic stress (Owttrim, 2006). *crhC* is upregulated solely in response to low temperature (20 °C; Chamot and Owttrim, 2000; Chamot et al., 1999), while *crhR* is regulated in response to conditions which elicit reduction of the electron transport chain including light, cold, and salt stress (Kujat and Owttrim, 2000; Suzuki et al., 2001; Vinnemeier and Hagemann, 1999). *crhB* appears to be regulated in a similar manner (Chamot et al., 1999).

The CrhR and CrhC proteins exhibit interesting biochemical characteristics. CrhC unwinds short dsRNA templates with $5'$ to $3'$ polarity (Yu and Owttrim, 2000), while CrhR catalyzes both dsRNA unwinding and annealing through an RNA strange-exchange mechanism (Chamot et al., 2005). In addition, CrhC is a polar-localized, membrane-associated protein in *Anabaena*, potentially involved in transport of proteins into or across the plasma membrane (El-Fahmawi and Owttrim, 2003). This chapter concentrates on techniques required to investigate RNA helicases and their localization in cyanobacteria. The outlined techniques are also applicable to other genes and to other bacterial systems.

2. Preparation of Cyanobacterial Extracts

2.1. Cyanobacterial growth

This chapter focuses on experiments with the freshwater cyanobacterial strains *Synechocystis* sp. strain Pasteur Culture Collection (PCC) 6803 and *Anabaena* sp. strain PCC 7120, both commonly used as model systems. The unicellular, nonnitrogen-fixing strain, *Synechocystis* sp. strain PCC 6803, was originally isolated from a fresh water lake in California by Riyo Kunisawa and deposited in the PCC as ATCC 27184 PCC 6803 in 1968 (Rippka et al., 1979). The strain was first described and used by Stanier et al. (1971). The origins of the nitrogen fixing strain, *Anabaena* sp. strain PCC 7120 (ATCC 27893) are unclear but the strain was first described by Adolph and Haselkorn (1971).

Freshwater cyanobacteria are generally grown in BG-11 liquid or on agar plates (1.0–1.5%), as originally described by Mary Mennes Allen (Allen, 1968; Allen and Stanier, 1968) and modified slightly by Castenholz (1988). *Anabaena* is capable of nitrogen fixation and BG-11 components containing fixed N are generally replaced with the chloride counterparts to give nitrogen-free BG-11$_0$ (Rippka et al., 1979). For cyanobacteria transformed with plasmids or mutants created by insertional inactivation, BG-11 is frequently modified by addition of a buffer (the most common example is TES (N-[tris(hydroxymethyl) methyl]-2-aminoethanesulfonic acid) pH 8.0 at 10–25 mM although MES and Tricine are also used) and sodium thiosulfate (1 mM), the latter of which is added only to agar media. Media should also be buffered if liquid cultures are continuously sparged with air enriched with CO_2 (1–5% v/v), a variation that decreases the CO_2 stress and thereby increases the growth rate. Cell growth can be estimated by measuring light scattering at OD 730 or 750 nm. The quality and quantity of light is also an important consideration. Frequently, cyanobacteria are grown under continuous illumination provided by cool-white fluorescent lamps (400–700 nm) at an irradiance level of 50 μmol photons/m^2/s. Cells can also be grown under low light, LL = 10–20 μmol photons/m^2/s or high light, HL = 150–200 μmol photons/m^2/s, although levels up to 2500 μmol photons/m^2/s have been used for short periods (Vavilin et al., 2007). Light stress, induced in response to self-shading at densities above OD 750 nm ~0.8–1.0, should be avoided. Normal or warm grown cells are incubated at 30 °C and cold stress is induced by incubation at 20 °C. A more detailed description of cyanobacterial media composition and considerations is provided in articles by Rippka (1988) and Castenholz (1988).

Cyanobacteria can be preserved for extended periods by cryopreservation. Liquid culture (2 ml) is pelleted in a cryovial (Nalgene) and gently suspended in an equal volume of half-strength BG-11 containing either 5% methanol or 8% DMSO. The cells are light sensitive at this point, so they must be kept in subdued light during this procedure. The solution is slowly cooled by initially placing the sample in a Mr. Frosty (Nalgene) freezing container prechilled to 4 °C. The container is placed at −80 °C for 2 h, the cryovial is removed, immersed, and continuously maintained in liquid nitrogen for long-term storage. To recover viable cells, thaw tubes rapidly, gently pellet cells, and suspend in 1 ml BG-11. Vials are incubated at 30 °C in the dark for 1 day before plating on BG-11 plates under reduced light for the first 2 days, thereafter use normal conditions.

2.2. Cautionary note: Cyanobacterial growth

It is quite evident that cyanobacteria are under "stress" no matter what the growth condition. They also sense and respond to any change in the environment, rapidly altering their genetic profile to optimize growth under the new conditions. Therefore, all attempts should be made to maintain the cultures at conditions identical to the growth conditions for biological replicates and especially during cell harvesting. This is particularly important to obtain reproducibility when doing, for example, microarray or proteomic experiments. Cells in each lab will be exquisitely fine tuned to growth under the exact conditions present in each individual lab, to the point that each lab is most likely propagating a lab-specific strain.

These genetic changes should also be kept in mind when initiating growth under different conditions, for example, at different temperature, light, or fixed nitrogen regimes. Cells should be allowed to acclimate to these changes by passing them through a number of growth cycles (~10 generations) under the new conditions before initiating experiments.

2.3. RNA isolation

All solutions, plasticware, glass, etc., must be RNase free. Water can be made RNase free by stirring in the presence of diethyl pyrocarbonate (DEPC; 0.5–0.1%) for 4 h to overnight in a fume hood and then extensive autoclaving (15–45 min) to remove residual DEPC. DEPC inactivates proteins by reacting with primary amine, hydroxy, and thiol groups, therefore solutions containing these compounds (e.g., Tris, HEPES, and mercaptans) cannot be effectively treated with DEPC. Note that DEPC is carcinogenic and should be handled appropriately. A less hazardous alternative is dimethyl-propyl carbonate, which is used as described for DEPC. It has also been reported that DEPC-treated water affects some downstream processes for example, *in vivo* and *in vitro* translation applications. An alternative is autoclaved Milli-Q water which has been reported to be RNase free (Huang *et al.*, 1995).

Before performing the RNA isolation, an aliquot (5 ml) of each culture is harvested for protein extraction in order to allow comparison of transcript and protein levels. Cells for RNA isolation are rapidly killed and RNases inactivated by mixing a cyanobacterial culture (50 ml) with an equal volume of an ice-cold mixture of buffer-saturated phenol (5%) in ethanol (100%). Mixing is performed directly in the growth chamber in order to not alter the growth conditions. Cells are harvested by centrifugation (7 min, $7000 \times g$) at 4 °C and washed once with TE (50 mM Tris–HCl, 100 mM EDTA, pH 8.0) buffer. Pellets are suspended in breakage buffer (500 µl; TE (50/100), containing Triton X-100, Sarkosyl, and SDS at 0.5% each). Glass beads (300 µl, 0.2 mm, Impandex Inc.) and TE-buffered phenol (400 µl) are added and cells lysed by vortexing at full speed for 10 cycles of 30 s followed by 30 s in an ice-water bath. The mixture is clarified by centrifugation at 13,000 rpm for 5 min at 4 °C (supernatant will be light pink/purplish in color) and the supernatant sequentially extracted with equal volumes of phenol, twice with a 1:1 mixture of phenol–chloroform (chloroform–isoamyl alcohol, 25:24:1) and finally chloroform. RNA is precipitated from the final aqueous layer by addition of an equal volume of 4 M LiCl and incubation at -20 °C for 16 h. RNA is pelleted by centrifugation at 13,000 rpm for 30 min at 4 °C, suspended in RNase-free water and precipitated by addition of 1/10th volume of sodium acetate (3 M, pH 5.2) and ethanol (2.5 volumes of 100%) at -20 °C. Store RNA in ethanol at -80 °C. Wash RNA pellets with 80% ethanol and dry before use. Yield is ~ 2 µg of total RNA per milliliter culture.

2.4. Protein isolation

Cyanobacterial cultures (5 ml) are harvested by centrifugation (7 min, $7000 \times g$) and washed once with BG-11 or TE (10 mM Tris–HCl, 1 mM EDTA, pH 8.0), all at the growth temperature used in the experiment. All subsequent procedures are performed at 4 °C. Cell pellets are suspended in five times their volume in ice-cold extraction buffer (20 mM Tris–HCl (pH 8.0), 100 mM KCl, 1 mM MgCl$_2$, 2 mM DTT) containing protease inhibitors (Complete^{TM} Mini Protease Inhibitor Cocktail Tablet (Roche)). Cells are mechanically lysed by either sonication (70% power, 10 cycles of

30 s followed by 30 s in an ice-water bath) or vortexing in the presence of ~3/4 volume glass beads (0.2 mm) at full speed for 10 cycles of 30 s followed by 30 s in an ice-water bath. For glass bead lysis, a half volume of extraction buffer is added and the lysate clarified by centrifugation at 13,000 rpm for 5 min at 4 °C. The resulting dark blue supernatant will contain soluble protein extract. A total protein extract can be obtained by supplementing the extraction buffer with Triton X-100, Sarkosyl, and SDS at 0.5% each. Protein concentration is measured using either the Lowry or Bradford procedure using BSA as the standard with typical yields of 500 μg/ml of culture.

Photosynthetic or enzymatic activity or cell number is frequently standardized to the chlorophyll concentration in a cyanobacterial culture. Chl *a* is extracted by suspension of the cell pellet obtained from 1 ml of culture in an equal volume of 100% methanol and incubation for 16 h at $-20\ °C$. Following clarification, the OD_{665} and OD_{652} are measured and Chl *a* concentration determined using the equation of Porra *et al.* (1989) (Chl *a* (μg/ml) = $(16.29 \times OD_{665}) - (8.54 \times OD_{652})$).

3. Alteration of RNA Helicase Expression in Cyanobacteria

3.1. *Anabaena* sp. strain PCC 7120 transformation: Triparental mating

Conjugation provides a number of advantages over transformation for DNA transfer into cyanobacteria, even naturally competent strains (Elhai and Wolk, 1988a). *Anabaena* is not naturally competent and must be transformed using a triparental mating procedure that transfers plasmids between three parent bacteria via conjugation. The plasmids required for conjugal transfer into cyanobacteria are termed (i) conjugal, for example, RP4 or derivatives such as pRL443, encoding the proteins required for conjugal transfer, (ii) a helper plasmid carrying restriction–modification system methytransferase genes, and (iii) a mobilizable cargo plasmid carrying the DNA to be transferred. *Anabaena* possesses a number of restriction–modification systems (*Ava*I, II, III), which significantly reduce transformation efficiency. Cargo plasmid transformation into *E. coli* containing the helper plasmid, pRL623 encoding the three *Ava* methyltransferases methylates the *Ava* endonuclease sites in the transferred DNA, resulting in significantly enhanced transformation efficiency (Elhai and Wolk, 1988b; Elhai *et al.*, 1997). Cargo plasmids can either be targeted for insertion into the genome (suicide plasmids) or autonomous replication, the two plasmids differing by the absence or presence, respectively, of an origin of replication (*ori*) which functions in the target cyanobacterium.

To prepare the *E. coli* parental cells, HB101 carrying the conjugative plasmid and containing both a cargo and a helper plasmid were grown overnight. Aliquots (1 ml of each strain for each mating) were washed twice with LB without antibiotics and suspended in LB (100 μl). The cyanobacterial parent is prepared by harvesting cells from a mid- to late-log phase culture (1 ml per conjugation; age does not affect the results significantly) and resuspension in 1/10th volume of BG-11. Triparental mating is performed by mixing 100 μl of each of the three parent bacteria, *E. coli* carrying the conjugal plasmid, *E. coli* carrying the helper plus cargo plasmids and the cyanobacteria, and incubating under nonselective conditions in the light with gentle shaking at 30 °C for 24 h. Transformants are selected on BG-11 plates supplemented with the appropriate antibiotic with colonies appearing after ~10 days. BG-11, especially liquid culture, does not support the growth of *E. coli* and thus selection against the *E. coli* parents is not required. More extensive discussions on conjugal transfer to cyanobacteria have been provided by Thiel and Wolk (1987) and Elhai and Wolk (1988a).

3.2. *Synechocystis* sp. strain PCC 6803 transformation

Synechocystis sp. strain PCC 6803 is naturally competent for DNA uptake and can be transformed simply by mixing plasmid DNA with cells (Barten and Lill, 1995; Grigorieva and Shestakov, 1982). A detailed method has been described by Eaton-Rye (2011). Note, however, that the efficiency of *Synechocystis* transformation appears to differ between different lab strains. This is apparently related to the evolution of a nonmotile variant of the original *Synechocystis* sp. strain PCC 6803 that exhibits reduced competence, as phototactic-induced mobility is linked to competence (Nakasugi et al., 2006). Possibly for this reason, optimized conditions for *Synechocystis* sp. strain PCC 6803 transformation have recently been described (Zang et al., 2007) which significantly enhance efficiency in our hands.

Natural transformation is based on the procedure described by Zang et al. (2007). Mid-log phase cells (5 ml, $OD_{750} \sim 0.8$) are harvested at 6000 rpm for 15 min, washed with BG-11 and suspended in 200 μl BG-11. Plasmid DNA is added to a final concentration of 10 μg/ml and the mixture incubated in the light at 30 °C for 5 h at which time aliquots are plated on BG-11 plates and incubated in the light at 30 °C. After 24 h, the appropriate antibiotic (in 1 ml BG-11) is sterially provided under the agar and incubation continued under normal growth conditions. Colonies will appear in ~10 days.

If natural transformation is not successful, triparental mating-based mechanisms, based on the procedures described above for *Anabaena* (Elhai and Wolk, 1988a), have also been described for *Synechocystis* (Ng et al., 2000).

3.3. RNA helicase gene inactivation

Gene disruption in cyanobacteria is accomplished by homologous recombination between plasmid and genomic sequences. To construct the inactivation plasmid, two fragments of genomic DNA representing ~1500 bp sequences both upstream and downstream of the translation start and stop codons of the gene to be inactivated are amplified by PCR. The two fragments are cloned into a mobilizable suicide plasmid, typically pBR322, so that they are joined by a unique restriction endonuclease site. An antibiotic resistance cassette is cloned into this unique restriction site; typically we utilize the Ω cassettes coding for resistance to spectinomycin/streptomycin or kanamycin obtained from the corresponding plasmids generated by Alexeyev et al. (1995). The resulting inactivation plasmid can be used to transform *Synechocystis* directly or as the cargo plasmid for conjugal transfer into either cyanobacterial strain. Once transformed, cyanobacteria possess an efficient homologous recombination system that generates gene replacement via double recombination. Cyanobacteria also contain multiple genome copies per cell and thus recombinants must be restreaked numerous times on selective plates containing increasing concentrations of the selective antibiotic before a completely inactivated segregant is obtained. Typically, we aim to grow transformants on the highest antibiotic concentrations possible: for example, a mixture of spectinomycin and streptomycin at 50 µg/ml each and kanamycin at 25 µg/ml. Complete inactivation of all genomic copies is confirmed using southern and/or PCR analysis. Failure to generate a complete inactivant provides evidence that the targeted gene is essential. If your gene of interest is in an operon, ensure that polar effects are not introduced by the concomitant inactivation of gene expression downstream of the target when designing the inactivation construct.

3.4. Inducible RNA helicase expression

Autonomously replicating plasmids possessing inducible promoters are available for protein overexpression or complementation of mutants. In *Anabaena*, inducible gene expression is regulated metabolically by alteration of either the fixed nitrogen source (Desplancq et al., 2005) or copper (Buikema and Haselkorn, 2001). Plasmids based on the pNIR promoter are tightly regulated by the fixed nitrogen source, repression occurring in the presence of ammonia and derepression in the presence of nitrate, as first shown by Maeda et al. (1998). Expression from the NIR promoter is repressed by growing transformants in BG-11 in which the nitrate has been replaced by ammonia (1–8 mM ammonium chloride or sulfate). Variations of these plasmids encoding affinity tags (His, GST, and MBP) are also available (Desplancq et al., 2005). An alternative is plasmids containing the Cu^{2+} inducible promoter,

PpetE, which is repressed in the absence and derepressed in the presence of Cu^{2+} (0.3–1 μM $CuSO_4$; Buikema and Haselkorn, 2001; Callahan and Buikema, 2001).

The following issues should be considered when using inducible plasmids. Alteration of media composition, ammonia addition, or Cu^{2+} depletion, will affect gene expression. In addition, production of copper-deplete media is complicated as it is difficult to remove Cu^{2+} from water and glassware and it is also a minor contaminant in other BG-11 components. This is crucial as the *petE* promoter is induced by trace levels of Cu^{2+} (Callahan and Buikema, 2001). Copper can also potentially produce unexpected side-effects, as cyanobacteria are very sensitive to Cu^{2+}, even at the levels present in BG-11 (Rippka, 1988; Fig. 18.1).

3.5. Confirmation of genetic alteration

A combination of PCR and Southern analysis using genomic DNA is used to verify transformation or gene inactivation. For medium- to large-scale isolation of genomic DNA, cells are harvested by centrifugation at $7000 \times g$ for 10 min at room temperature. *Note*: *Anabaena* cell pellets are unstable. Wash cell pellet with 1/20th volume SE (120 mM NaCl, 50 mM EDTA, pH 8.0), pellet as above and suspend in 1/40th volume SE. Add powdered lysozyme (3 mg/ml) and incubate at 37 °C for 1 h. SDS (1% final concentration) is added and incubation continued for 1 h. The mixture is heated to 60 °C for

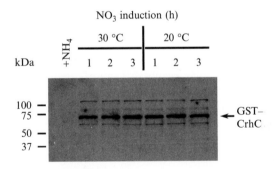

Figure 18.1 *CrhC overexpression in* Anabaena *sp. strain PCC 7120*. CrhC, cloned into the pNIR-GST plasmid (Desplancq *et al.*, 2005) as a translational fusion, was transferred into *Anabaena* by triparental mating. Growth of transformed cells in the presence of ammonia as the fixed nitrogen source represses the pNIR promoter and the GST-CrhC fusion protein is not detected (+NH$_4$). Growth in the presence of nitrate (NO$_3$–) deregulates the pNIR promoter and the 71-kDa GST-CrhC fusion protein is detected at a constant level. Note that expression of the GST-CrhC fusion protein occurs in response to the fixed nitrogen composition of the media and not the growth temperature, as observed from the native CrhC promoter (Chamot and Owttrim, 2000).

5 min to aid release of DNA from membrane components. SDS and proteins are removed by incubation at 4 °C for 10 min with gentle stirring in the presence of sodium perchlorate (0.5 M; Note: sodium perchlorate is toxic). The mixture is extracted sequentially with equal volumes of phenol, phenol–chloroform, and chloroform. Nucleic acid is precipitated with 2 volumes of ethanol (100%), collected by centrifugation (13,000 rpm for 10 min at 4 °C) and washed with 70% ethanol. The dry pellet is suspended in TE (200 μl) and contaminating RNA removed by adding RNase A (50 μg/ml) and incubation at 37 °C for 1 h. The phenol–chloroform extractions and ethanol precipitation is repeated, the DNA suspended in ultrapure water (e.g., Milli-Q water (Millipore Corporation)) or TE and quantified spectrophotometrically. Typical yield is 0.5–1 μg genomic DNA per milliliter culture.

Verification of autonomous plasmid transformation can be performed using colony PCR. A loop full of cells is removed from an agar plate and suspended in 50 μl Milli-Q water so that it is visibly green. Boil for 15 min, vortex 30 s and clarify by centrifugation at 13,000 rpm for 5 min. Perform PCR with 2–8 μl of the supernatant.

4. Northern Analysis of Transcript Levels

4.1. Electrophoresis and blotting

Precipitated total RNA (5 μg) is suspended in RNA sample buffer (25 mM EDTA, pH 8.0, 0.1% SDS) and denatured by addition of formaldehyde loading buffer (1× MOPS [20 mM MOPS, 5 mM sodium acetate, 1 mM EDTA, pH 7.0], 50% deionized formamide, 15% formaldehyde, 7% glycerol, 0.25% bromophenol blue (w/v)) and heating to 65 °C for 15 min. RNA is separated on a formaldehyde (2%)-agarose gel (1%) prepared and separated in 1× MOPS buffer essentially as described by Fourney et al. (1988). RNA is transferred vertically to a nylon membrane using 20× SSC (3 M NaCl, 300 mM sodium citrate; Sambrook and Russell, 2001), UV cross-linked (120,000 μJ/cm^2; Stratalinker®, Stratagene) and stored dry at room temperature. Alternatively, for small RNA analysis (60–250 nt using this protocol), precipitated total RNA (5 μg) is suspended in 2 volumes of 1.5× urea loading buffer (10 M urea, 1.5× TBE, 0.015% (w/v) bromophenol blue, 0.015% (w/v) xylene cyanol) and denatured by incubation at 65 °C for 15 min. RNA is separated on a 1× TBE – 6% polyacrylamide gel containing 8.3 M urea. RNA is transferred to a positively charged nylon membrane in 0.5× TBE using a semidry transfer apparatus at 100 mA for 0.5 h followed by UV cross-linking. An appropriate size marker should be included on each gel, originating from either RNA or dsDNA denatured as described above for RNA. Note that the cyanobacterial 23S rRNA undergoes a specific endonucleolytic cleavage and thus four ethidium-stained

RNA bands (23S, 20S, 16S, 5S) are visible on agarose gels stained with ethidium bromide (Doolittle, 1973).

4.2. Transcript detection

Transcript levels on northern blots can be detected using either DNA or RNA probes. The enhanced stability of RNA:RNA over DNA:RNA duplexes necessitates the use of increased hybridization and washing stringencies when using RNA probes. DNA probes, produced by restriction digestion or PCR, are most efficiently labeled with ^{32}P-dCTP using a random primer kit. ssRNA probes code for the antisense sequence corresponding to the gene of interest and are produced by transcription from a phage RNA pol promoter in the presence of ^{32}P-UTP. The RNA pol promoter is either present on the plasmid carrying the gene of interest and RNA pol sites, for example, T7 or T3 in pBluescript or has been incorporated into one of the primers used for PCR amplification of a portion of the gene of interest. If a plasmid is used as the source for RNA probe production, it must be linear, created by cleavage with a restriction endonuclease at the 5′ end of the RNA probe sequence.

For DNA probes, northern blots are prehybridized for a minimum of 4 h in aqueous RNA hyb buffer (5 × SSPE [3 M NaCl, 0.2 M sodium dihydrogen phosphate, 0.02 M EDTA, pH 7.4], 5 × Denhardts solution [Ficoll, polyvinylpyrrolidone, and BSA at 0.1% each], 0.1% SDS, 25 µg/ml sheared single-stranded salmon sperm DNA or yeast tRNA) at 65 °C. The DNA probe is boiled for 5 min immediately before addition to fresh RNA hyb buffer and hybridization allowed to proceed overnight at 65 °C. The membrane is washed at room temperature for 15 min each successively with 1 × SSC, 0.1% SDS followed by 0.1 × SSC, 0.1% SDS. For RNA probes, membranes are prehybridized and hybridized in RNA hyb buffer containing 50% formamide at 65 °C as described above. Membranes are washed at 65 °C for 15 min each successively with 1 × SSPE, 0.5% SDS followed by 0.1 × SSPE, 0.5% SDS. Hybridized transcripts are detected using a phosphorimager or by exposure to X-ray film at −80 °C using intensifying screens.

4.3. RNA loading control

In order to compare transcript levels between samples, it is crucial that differences in RNA loading can be accurately determined. In cyanobacteria, transcript levels of the *mpB* gene, coding for the functional RNA present in the RNase P endoribonuclease, are frequently used to quantify RNA loads (Chamot and Owttrim, 2000; Kujat and Owttrim, 2000; Vioque, 1992). Phosphorimager detection of hybridizations performed with both the gene of interest and *mpB* allows calibration of transcript levels (Kujat and Owttrim, 2000; Fig. 18.2).

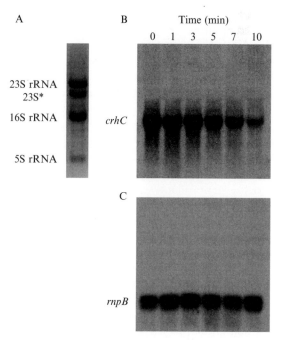

Figure 18.2 (A) Cyanobacterial total RNA. Total RNA isolated from *Anabaena* sp. strain PCC 7120 cells was denatured, separated on a 1% agarose gel, stained with ethidium bromide and a digital image obtained using an image documentation system. Visible are the three rRNAs, 23S, 16S, and 5S plus the 23S rRNA breakdown product (23S*) typically observed in total RNA preparations from cyanobacteria. (B) Transcript detection. Northern analysis was used to determine *crhC* transcript half-life. *Anabaena* sp. strain PCC 7120 cells were subjected to cold stress at 20 °C for 1 h at which time rifampin (400 μg/ml) was added and samples taken at the indicated times. Total RNA (15 μg) was separated on a 1% agarose gel, transferred to nylon membrane and probed with ^{32}P-labeled *crhC* ORF. (C) *rnpB* RNA loading control. The blot from (B) above was stripped and probed with *rnpB*, coding for the constitutively expressed RNA component of RNaseP, to control for RNA loading. The relative intensities of the *rnpB* and *crhC* signals, determined using a phosphorimager, can then be used to correct for differences in RNA loading. For additional details see Chamot and Owttrim (2000). Figures B and C are adapted from Figure 6 in Chamot and Owttrim (2000).

5. Western Analysis of Protein Levels

5.1. Western analysis

For the 47–55 kDa DEAD-box RNA helicase proteins we study, 10% SDS-PAGE gels are optimal for good separation. Separated proteins (5–30 μg per lane) are transferred to reinforced nitrocellulose using a semidry transfer apparatus using protein transfer buffer (25 mM Tris-base, 150 mM glycine, 20% methanol) at 52 mA for 1 h. Membranes are blocked in 1× BLOTTO (1× TBS [50 mM Tris–HCl, pH 7.4, 150 mM NaCl], 5% (w/v) skimmed

milk powder) for 1 h at room temperature. Blots are incubated in 1×
BLOTTO containing primary polyclonal antibodies (1:5000) for 2–16 h at
room temperature, washed and incubated with HRP-conjugated secondary
antibodies (1:20,000) for 30 min at room temperature. Membranes are washed
between each step in 1×TBS, 1×TBST (1×TBS, 0.1% Tween®-20),
1×TBS for 15 min each at room temperature. Complexes are detected using
an ECL kit and detected with X-ray film or an ECL-plus kit whose chemiluminescent signal can be detected with X-ray film or CCD or whose fluorescent
signal can be captured with fluorescence imagers to allow quantification. This is
a very sensitive method; if background is a problem, membranes can be blocked
with purified protein, for example, bovine serum albumin (BSA, 0.5–5%), or
casein (0.5%) and/or a nonionic detergent (typically Tween®-20, 0.05–0.1%)
can be added. *Note*: do not include sodium azide in the blocking buffer as azide
irreversibly inhibits HRP.

5.2. Protein loading controls

It is also advantageous to control for protein loading on western blots however,
until recently, antibodies to prokaryotic proteins whose levels remain constant
in response to a range of conditions were not available. Recently, a few
polypeptides have been used as loading controls under specific growth conditions. These include the beta subunit of the DNA-dependent RNA polymerase (RpoB) in *Synechocystis* (Imamura et al., 2006), Thioredoxin A (TrxA) in
Anabaena (Galmozzi et al., 2010), and the cell division protein FtsZ in *Microcystis
aeruginosa* PCC 7806 (Zilliges et al., 2011). Note that these controls have only
been tested under the growth conditions specified in each manuscript and have
not been widely used in the literature to date. For example, constitutive
expression of TrxA (Ehira and Ohmori, 2006) and RpoB (Imamura et al.,
2006) is reported to occur independently of the nitrogen source and in
response to nitrogen deprivation, respectively, while FtsZ is underexpressed
in heterocysts in *Anabaena* (Kuhn et al., 2000). We have also used antibodies to
E. coli Rps1 coding for ribosomal protein S1 as a control in electron microscopy (El-Fahmawi and Owttrim, 2003) and western blot analysis where Rps1
levels remain relatively constant in response to growth temperature in *Synechocystis* (A. R. R. Rosana and G. W. Owttrim unpublished; Fig. 18.3).

6. Cellular Ultrastructure and *In Situ* RNA Helicase Localization

6.1. Cellular ultrastructure

Ultrastructural analysis of cyanobacterial RNA helicase mutants is performed by cryo-EM. Cryofixation was performed using a high-pressure,
low-temperature protocol. Briefly, cells were rapidly frozen in liquid nitrogen

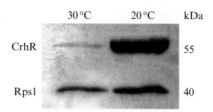

Figure 18.3 Protein detection. *Synechocystis* sp. strain PCC 6803 cells were grown to mid-log phase at 30 °C at which time one-half of the culture was subjected to cold stress at 20 °C for 1 h. Extracted proteins (25 μg) were separated on 10% SDS-PAGE gels, electroblotted to nylon-reinforced nitrocellulose membrane and detected simultaneously with polyclonal antibodies against CrhR and Rps1. CrhR is induced in response to low temperature while Rps1 levels remain relatively constant (Rosana and Owttrim, unpublished).

at high pressure (2100 bar) in a Bal-Tec HPM100 (Leica Mirosystems, Wetzlar Germany), freeze substituted at −85 °C for 72 h in 1% glutaraldehyde and 1% tannic acid in HPLC grade acetone and washed with anhydrous acetone at −85 °C for 1 h. Staining with osmium (1% OsO_4 in acetone) was performed consecutively at −85 °C for 1 h, samples were slowly warmed to −20 °C over 6 h, incubation continued at −20 °C for 2 h, the temperature increased slowly to 4 °C and incubation continued for 2 h and a final 1-h incubation at room temperature. Samples were washed in anhydrous acetone twice for 15 min each, incubated in propylene oxide/anhydrous acetone (1:1) for 15 min, propylene oxide for 15 min. Samples were infiltrated with resin by incubation at 60 °C in propylene oxide:Spurr resin (1:1) followed by propylene oxide:Spurr resin (1:3) for 1 day each and polymerized in 100% Spurr resin for 2 days, all under vacuum. Ultrathin sections (40–70 nm, Ultracut E, Reichert-Jung) are mounted on copper grids (G300-CP, Electron Microscopy Sciences). Sections were stained with uranyl acetate (4%) for 20 min and lead citrate (2%) for 4 min at room temperature. Sections are viewed with a Philips Morgagni 268 transmission electron microscope at 80 kV. Digital images were obtained with a Gatan Orius CCD camera and DigitalMicrograph software.

6.2. *In situ* RNA helicase localization

Intracellular localization of RNA helicases can be achieved using immunoelectron microscopy (IEM) essentially as described by El-Fahmawi and Owttrim (2003). Harvest cyanobacterial cells by centrifugation at $8500 \times g$ for 10 min and fix in a mixture of freshly prepared 4% formaldehyde with 0.8% glutaraldehyde in 0.1 M sodium phosphate buffer (pH 7.2) for 1 h. Samples were washed in phosphate buffer and dehydrated by incubation in an ascending ethanol series (20–90%) for 30 min each. Infiltration and

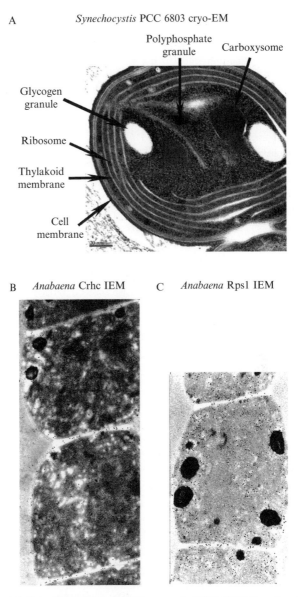

Figure 18.4 (A) Cryo-EM of *Synechocystis* sp. strain PCC 6803. *Synechocystis* cells were grown at 30 °C and subjected to cryofixation and staining at high pressure (2100 bar) and low temperature (−85 °C). After embedding in Spur resin, microtome sectioning and staining with uranyl acetate and lead citrate, sections are viewed with a Philips Morgagni 268 transmission electron microscope at 80 kV. Digital images were obtained with a Gatan Orius CCD camera and DigitalMicrograph software. G. W. Owttrim (unpublished). (B) Immunoelectron microscopy (IEM) cellular localization. IEM was used to determine the location of CrhC within *Anabaena* sp. strain PCC 7120 cells

subjected to cold stress at 20 °C for 6 h. IEM indicates that CrhC is predominately associated with the cell membrane at the septa between adjacent cells in *Anabaena*, that is, a polar-localized protein. Note that CrhC is not detected on the intracellular thylakoid membranes indicating that antibodies in the rabbit serum do not have a nonspecific affinity for *Anabaena* membranes. (C) IEM control. As a control for (B) above, *Anabaena* cells were subjected to IEM using *E. coli* Rps1 polyclonal antisera. Rps1 exhibits intracellular localization although not in a uniform manner, being concentrated outside the nucleolar region, as expected for a cytoplasmic protein. Note that the gold articles are not specifically concentrated in the membrane region indicating that rabbit polyclonal antibodies do not associate with *Anabaena* cell membranes in a nonspecific manner. Figures B and C are adapted from El-Fahmawi and Owttrim (2003).

embedding in LR white resin (London Resin Company Ltd., Reading Berkshire, England) were performed by overnight incubation in LR White–ethanol (1:1) at room temperature, two changes of pure LR white over 2 h and twice in pure LR white for 12 h at 50 °C. Serial ultrathin sections (40–70 nm) were mounted on copper grids and incubated in PBG blocking solution (10 mM sodium phosphate (pH 7.2), 600 mM NaCl, and 0.5% fish gelatin) for 10 min. Grids were incubated with primary antibody (rabbit; 1:100 dilution) in PGB buffer at room temperature for 1 h. Grids were extensively washed with 1 × PBS and incubated at room temperature for 1 h with goat anti-rabbit IgG coupled to 5-nm colloidal gold particles (Sigma, USA) diluted 1:20 in PBG as the secondary antibody for 1 h. Sections were washed four times for 5 min each in PBST (0.1% Tween 20 in PBS) and stained with uranyl acetate (4%) for 20 min at room temperature. Sections were examined in an electron microscope at 80 kV as described above (Fig. 18.4).

ACKNOWLEDGMENTS

This publication originated from research supported by funding from the Natural Sciences and Engineering Research Council of Canada (NSERC), grant number 171319.

REFERENCES

Adolph, K. W., and Haselkorn, R. (1971). Isolation and characterization of a virus infecting the blue-green alga nostoc muscorum. *Virology* **46,** 200–208.
Alexeyev, M. F., Shokolenko, I. N., and Croughan, T. P. (1995). Improved antibiotic-resistance gene cassettes and omega elements for escherichia coli vector construction and in vitro deletion/insertion mutagenesis. *Gene* **160,** 63–67.
Allen, M. M. (1968). Simple conditions for growth of unicellular blue-green algae on plates. *J. Phycol.* **4,** 1–4.
Allen, M. M., and Stanier, R. Y. (1968). Selective isolation of blue-green algae from water and soil. *J. Gen. Microbiol.* **51,** 203–209.

Barten, R., and Lill, H. (1995). DNA-uptake in the naturally competent cyanobacterium, *Synechocystis* sp. PCC 6803. *FEMS Microbiol. Lett.* **129**, 83–88.

Buikema, W. J., and Haselkorn, R. (2001). Expression of the *Anabaena hetR* gene from a copper-regulated promoter leads to heterocyst differentiation under repressing conditions. *Proc. Natl. Acad. Sci. U.S.A.* **98**, 2729–2734.

Callahan, S. M., and Buikema, W. J. (2001). The role of HetN in maintenance of the heterocyst pattern in *Anabaena* sp. PCC 7120. *Mol. Microbiol.* **40**, 941–950.

Castenholz, R. W. (1988). Culturing methods for cyanobacteria. *Methods Enzymol.* **167**, 68–93.

Chamot, D., and Owttrim, G. W. (2000). Regulation of cold shock-induced RNA helicase gene expression in the cyanobacterium *Anabaena* sp. strain PCC 7120. *J. Bacteriol.* **182**, 1251–1256.

Chamot, D., Magee, W. C., Yu, E., and Owttrim, G. W. (1999). A cold shock-induced cyanobacterial RNA helicase. *J. Bacteriol.* **181**, 1728–1732.

Chamot, D., Colvin, K. R., Kujat-Choy, S. L., and Owttrim, G. W. (2005). RNA structural rearrangement via unwinding and annealing by the cyanobacterial RNA helicase, CrhR. *J. Biol. Chem.* **280**, 2036–2044.

Criscuolo, A., and Gribaldo, S. (2011). Large-scale phylogenomic analyses indicate a deep origin of primary plastids within cyanobacteria. *Mol. Biol. Evol.* **28**, 3019–3032.

Desplancq, D., Bernard, C., Sibler, A. P., Kieffer, B., Miguet, L., Potier, N., Van Dorsselaer, A., and Weiss, E. (2005). Combining inducible protein overexpression with NMR-grade triple isotope labeling in the cyanobacterium *Anabaena* sp. PCC 7120. *Biotechniques* **39**, 405–411.

Dong, G., and Golden, S. S. (2008). How a cyanobacterium tells time. *Curr. Opin. Microbiol.* **11**, 541–546.

Doolittle, W. F. (1973). Postmaturational cleavage of 23s ribosomal ribonucleic acid and its metabolic control in the blue-green alga *Anacystis nidulans*. *J. Bacteriol.* **113**, 1256–1263.

Eaton-Rye, J. J. (2011). Construction of gene interruptions and gene deletions in the cyanobacterium *Synechocystis* sp. strain PCC 6803. *Methods Mol. Biol.* **684**, 295–312.

Ehira, S., and Ohmori, M. (2006). NrrA, a nitrogen-responsive response regulator facilitates heterocyst development in the cyanobacterium *Anabaena* sp. strain PCC 7120. *Mol. Microbiol.* **59**, 1692–1703.

El-Fahmawi, B., and Owttrim, G. W. (2003). Polar-biased localization of the cold stress-induced RNA helicase, CrhC, in the cyanobacterium *Anabaena* sp. strain PCC 7120. *Mol. Microbiol.* **50**, 1439–1448.

Elhai, J., and Wolk, C. P. (1988a). Conjugal transfer of DNA to cyanobacteria. *Methods Enzymol.* **167**, 747–754.

Elhai, J., and Wolk, C. P. (1988b). A versatile class of positive-selection vectors based on the nonviability of palindrome-containing plasmids that allows cloning into long polylinkers. *Gene* **68**, 119–138.

Elhai, J., Vepritskiy, A., Muro-Pastor, A. M., Flores, E., and Wolk, C. P. (1997). Reduction of conjugal transfer efficiency by three restriction activities of *Anabaena* sp. strain PCC 7120. *J. Bacteriol.* **179**, 1998–2005.

Fourney, R. M., Miyakoshi, J., Day, R. S., III, and Paterson, M. C. (1988). Northern blotting: Efficient RNA staining and transfer. *Focus* **10**, 5–7.

Galmozzi, C. V., Saelices, L., Florencio, F. J., and Muro-Pastor, M. I. (2010). Posttranscriptional regulation of glutamine synthetase in the filamentous cyanobacterium *Anabaena* sp. PCC 7120: Differential expression between vegetative cells and heterocysts. *J. Bacteriol.* **192**, 4701–4711.

Grigorieva, G., and Shestakov, S. (1982). Transformation in the cyanobacterium *Synechocystis* sp. 6803. *FEMS Microbiol. Lett.* **13**, 367–370.

Huang, Y. H., Leblanc, P., Apostolou, V., Stewart, B., and Moreland, R. B. (1995). Comparison of milli-Q PF plus water with DEPC-treated water in the preparation and analysis of RNA. *Nucleic Acids Symp. Ser.* **33,** 129–133.

Imamura, S., Tanaka, K., Shirai, M., and Asayama, M. (2006). Growth phase-dependent activation of nitrogen-related genes by a control network of group 1 and group 2 sigma factors in a cyanobacterium. *J. Biol. Chem.* **281,** 2668–2675.

Kuhn, I., Peng, L., Bedu, S., and Zhang, C. C. (2000). Developmental regulation of the cell division protein FtsZ in *Anabaena* sp. strain PCC 7120, a cyanobacterium capable of terminal differentiation. *J. Bacteriol.* **182,** 4640–4643.

Kujat, S. L., and Owttrim, G. W. (2000). Redox-regulated RNA helicase expression. *Plant Physiol.* **124,** 703–714.

Kumar, K., Mella-Herrera, R. A., and Golden, J. W. (2010). Cyanobacterial heterocysts. *Cold Spring Harb. Perspect. Biol.* **2,** a000315.

Maeda, S., Kawaguchi, Y., Ohe, T. A., and Omata, T. (1998). Cis-acting sequences required for NtcB-dependent, nitrite-responsive positive regulation of the nitrate assimilation operon in the cyanobacterium *Synechococcus* sp. strain PCC 7942. *J. Bacteriol.* **180,** 4080–4088.

Nakasugi, K., Svenson, C. J., and Neilan, B. A. (2006). The competence gene, *comF*, from *Synechocystis* sp. strain PCC 6803 is involved in natural transformation, phototactic motility and piliation. *Microbiology* **152,** 3623–3631.

Ng, W. O., Zentella, R., Wang, Y., Taylor, J. S., and Pakrasi, H. B. (2000). PhrA, the major photoreactivating factor in the cyanobacterium *Synechocystis* sp. strain PCC 6803 codes for a cyclobutane-pyrimidine-dimer-specific DNA photolyase. *Arch. Microbiol.* **173,** 412–417.

Owttrim, G. W. (2006). RNA helicases and abiotic stress. *Nucleic Acids Res.* **34,** 3220–3230.

Porra, R. J., Thompson, W. A., and Kriedemann, P. E. (1989). Determination of accurate extinction coefficients and simultaneous equations for assaying chlorophylls a and b extracted with four different solvents: Verification of the concentration of chlorophyll standards by atomic absorption spectroscopy. *BBA Bioen.* **975,** 384–394.

Rippka, R. (1988). Isolation and purification of cyanobacteria. *Methods Enzymol.* **167,** 3–27.

Rippka, R., Deruelles, J., Waterbury, J. B., Herdman, M., and Stanier, R. Y. (1979). Generic assignments, strain histories and properties of pure cultures of cyanobacteria. *J. Gen. Microbiol.* **111,** 1–61.

Ruffing, A. M. (2011). Engineered cyanobacteria: Teaching an old bug new tricks. *Bioeng. Bugs* **2,** 136–149.

Sambrook, J., and Russell, D. W. (eds.), (2000). Molecular Cloning: A Laboratory Manual, Cold Spring Harbor laboratory, Cold Spring Harbor, NY.

Stanier, R. Y., Kunisawa, R., Mandel, M., and Cohen-Bazire, G. (1971). Purification and properties of unicellular blue-green algae (order chroococcales). *Bacteriol. Rev.* **35,** 171–205.

Suzuki, I., Kanesaki, Y., Mikami, K., Kanehisa, M., and Murata, N. (2001). Cold-regulated genes under control of the cold sensor Hik33 in *Synechocystis*. *Mol. Microbiol.* **40,** 235–244.

Thiel, T., and Wolk, C. P. (1987). Conjugal transfer of plasmids to cyanobacteria. *Methods Enzymol.* **153,** 232–243.

Vavilin, D., Yao, D., and Vermaas, W. (2007). Small cab-like proteins retard degradation of photosystem II-associated chlorophyll in *Synechocystis* sp. PCC 6803: Kinetic analysis of pigment labeling with 15N and 13C. *J. Biol. Chem.* **282,** 37660–37668.

Vinnemeier, J., and Hagemann, M. (1999). Identification of salt-regulated genes in the genome of the cyanobacterium *Synechocystis* sp. strain PCC 6803 by subtractive RNA hybridization. *Arch. Microbiol.* **172,** 377–386.

Vioque, A. (1992). Analysis of the gene encoding the RNA subunit of ribonuclease P from cyanobacteria. *Nucleic Acids Res.* **20**, 6331–6337.

Yu, E., and Owttrim, G. W. (2000). Characterization of the cold stress-induced cyanobacterial DEAD-box protein CrhC as an RNA helicase. *Nucleic Acids Res.* **28**, 3926–3934.

Zang, X., Liu, B., Liu, S., Arunakumara, K. K., and Zhang, X. (2007). Optimum conditions for transformation of *Synechocystis* sp. PCC 6803. *J. Microbiol.* **45**, 241–245.

Zilliges, Y., Kehr, J. C., Meissner, S., Ishida, K., Mikkat, S., Hagemann, M., Kaplan, A., Borner, T., and Dittmann, E. (2011). The cyanobacterial hepatotoxin microcystin binds to proteins and increases the fitness of *Microcystis* under oxidative stress conditions. *PLoS One* **6**, e17615.

CHAPTER NINETEEN

Determination of Host RNA Helicases Activity in Viral Replication

Amit Sharma[*,†,‡,§] and Kathleen Boris-Lawrie[*,†,‡,§]

Contents

1. Introduction	406
1.1. RNA helicases are ubiquitously active at the virus–host interface	406
1.2. RNA helicases exhibit enzymatic and nonenzymatic functions	408
1.3. The experimental design to characterize the role of an RNA helicase in viral replication is focused on four issues	409
2. Methods Used To Study Cell-Associated RNA Helicase in Cultured Mammalian Cells	409
2.1. Downregulation with siRNA and rescue by exogenous expression of the RNA helicase	409
2.2. Coprecipitation of RNA helicase with target RNA or protein cofactors	411
2.3. Polysome association of RNA helicase with target RNA or protein cofactors	415
3. Biochemical and Biophysical Methods to Study RNA Helicase	417
3.1. RNA-affinity chromatography for cofactor identification	417
3.2. Determination of RNA-binding activity using EMSA	419
3.3. Determination of RNA-binding activity using FA	421
4. Methods Used to Study Virion-Associated RNA Helicase in Cultured Mammalian Cells	423
4.1. Detection of RNA helicase in virion preparations	423
4.2. Detection of virion-associated protein and RNA: Host and viral factors	426
4.3. Measurement of infectivity of RNA helicase-deficient virions	428
5. Concluding Remarks	430
Acknowledgments	431
References	431

[*] Department of Veterinary Biosciences, Ohio State University, Columbus, Ohio, USA
[†] Center for Retrovirus Research, Ohio State University, Columbus, Ohio, USA
[‡] Center for RNA Biology, Ohio State University, Columbus, Ohio, USA
[§] Comprehensive Cancer Center, Ohio State University, Columbus, Ohio, USA

Methods in Enzymology, Volume 511
ISSN 0076-6879, DOI: 10.1016/B978-0-12-396546-2.00019-X

© 2012 Elsevier Inc.
All rights reserved.

Abstract

RNA helicases are encoded by all eukaryotic and prokaryotic cells and a minority of viruses. Activity of RNA helicases is necessary for all steps in the expression of cells and viruses and the host innate response to virus infection. Their vast functional repertoire is attributable to the core ATP-dependent helicase domain in conjunction with flanking domains that are interchangeable and engage viral and cellular cofactors. Here, we address the important issue of host RNA helicases that are necessary for replication of a virus. This chapter covers approaches to identification and characterization of candidate helicases and methods to define the biochemical and biophysical parameters of specificity and functional activity of the enzymes. We discuss the context of cellular RNA helicase activity and virion-associated RNA helicases. The methodology and choice of controls fosters the assessment of the virologic scope of RNA helicases across divergent cell lineages and viral replication cycles.

1. INTRODUCTION

1.1. RNA helicases are ubiquitously active at the virus–host interface

Viruses are intracellular parasites that require many of the same essential processes as their host. All eukaryotic cells require the function of RNA helicases in all processes involving RNA (Jankowsky and Jankowsky, 2000; Linder, 2006; Linder and Jankowsky, 2011). To date, the activity of 13 host-encoded helicases and three virus-encoded RNA helicases in virus biology has been documented (Table 19.1). The scope of activities in all steps of viral replication and the interface with the antiviral host response demonstrates their remarkable versatility and potential utility as antiviral targets (Jeang and Yedavalli, 2006; Linder and Jankowsky, 2011; Ranji and Boris-Lawrie, 2010).

RNA helicases provide malleable connections between a viral infection and the host innate response (Ranji and Boris-Lawrie, 2010). In some viral infections, RNA helicase activity benefits the virus by promoting viral gene expression and squelching the antiviral response. In others, RNA helicase activity benefits the host by sensing viral nucleic acid and triggering antiviral response. The mechanisms that explain the versatility of RNA helicases are important to define but poorly understood. Experiments to determine the role of RNA helicases in viral replication are the focus of this chapter and stand from the perspective of the virus. We address RNA helicase activity in the steps of viral replication: gene expression, morphogenesis, and replication of viral nucleic acid. The complete elucidation of the role of RNA helicases in viral replication and the pathogenesis of disease is a priority in biomedical research.

Table 19.1 Overview of the role of cellular RNA helicases in virus replication

Cellular RNA helicase superfamily member	Virus	References
DDX1	Human immunodeficiency virus type 1 (HIV-1)	Edgcomb et al. (2011), Fang et al. (2004, 2005), and Robertson-Anderson et al. (2011)
	John Cunningham virus (JCV)	Sunden et al. (2007)
	Infectious bronchitis virus	Xu et al. (2010)
DDX3	HIV-1	Garbelli et al. (2011), Ishaq et al. (2008), Liu et al. (2011), and Yedavalli et al. (2004)
	Vaccinia virus	Kalverda et al. (2009) and Schroder et al. (2008)
	Hepatitis B virus (HBV)	Wang and Ryu (2010), Wang et al. (2009), and Yu et al. (2010)
	Hepatitis C virus (HCV)	Angus et al. (2010), Ariumi et al. (2007), Chang et al. (2006), Oshiumi et al. (2010), and Owsianka and Patel (1999)
DDX5/p68	HCV	Goh et al. (2004)
	SARS coronavirus (SARS-CoV)	Chen et al. (2009)
DDX6/Rck/p54	HCV	Ariumi et al. (2011), Jangra et al. (2010), Miyaji et al. (2003), and Scheller et al. (2009)
	Dengue virus (DENV)	Ward et al. (2011)
	Adenovirus	Greer et al. (2011)
	Retroviruses: HIV-1, prototype foamy virus (PFV)	Chable-Bessia et al. (2009) and Yu et al. (2011)
DHX9/RHA	Retroviruses: HIV-1, human T-cell leukemia virus type 1, bovine leukemia virus, spleen necrosis virus, feline leukemia virus, Mason-Pfizer monkey virus (MPMV)	Bolinger et al. (2007, 2010), Fujii et al. (2001), Hartman et al. (2006), Li et al. (1999), Ranji et al. (2011), Roy et al. (2006), Sadler et al. (2009), Tang and Wong-Staal (2000), and Westberg et al. (2000)

(Continued)

Table 19.1 (Continued)

Cellular RNA helicase superfamily member	Virus	References
	Herpes simplex virus 1 (HSV-1)	Kim et al. (2010)
	Bovine viral diarrhea virus	Isken et al. (2003)
	HCV	He et al. (2008)
	Foot and mouth disease virus	Lawrence and Rieder (2009)
	Kaposi sarcoma-associated herpesvirus (KSHV)	Jong et al. (2010)
DDX24	HIV-1	Ma et al. (2008) and Roy et al. (2006)
DHX30	HIV-1	Zhou et al. (2008)
DDX41	HSV-1	Zhang et al. (2011)
DDX56	West Nile virus	Xu et al. (2011)
DDX60	Vesicular stomatitis virus (VSV), poliovirus, Sendai virus (SeV), HSV-1	Miyashita et al. (2011)
Mov10	Hepatitis delta virus (HDV)	Haussecker et al. (2008)
	Retroviruses: HIV-1, simian immunodeficiency virus, murine leukemia virus, feline immunodeficiency virus, equine infectious anemia virus	Abudu et al. (2011), Burdick et al. (2010), Furtak et al. (2010), and Wang et al. (2010)
RH116	HIV-1	Cocude et al. (2003)
UAP56	Influenza A virus	Kawaguchi et al. (2011), Momose et al. (2001), and Wisskirchen et al. (2011)
	KSHV	Majerciak et al. (2010)

1.2. RNA helicases exhibit enzymatic and nonenzymatic functions

Viral and cellular RNA helicases uniformly display modular, genetically separable, catalytic, and scaffold domains (Ranji and Boris-Lawrie, 2010). Some RNA helicases exhibit processive, ATP-dependent unwinding activity on nucleic acid, while others unwind RNA duplexes in a nonprocessive but also ATP-dependent fashion (Linder and Jankowsky, 2011). RNA helicases have also been shown to remodel ribonucleoprotein complexes (RNPs; Fuller-Pace, 2006; Jankowsky and Bowers, 2006; Linder, 2006; Linder et al., 2001). RNP remodeling can be accomplished independent of duplex unwinding (Jankowsky and Bowers, 2006). In addition, RNA helicases can function as binding partners of other proteins, and these interactions are not always dependent on catalytic helicase or ATPase activities (see Chapter 16).

Therefore, it is important to devise experiments that aim to distinguish functions of RNA helicases that require enzymatic capacity (e.g., unwinding, RNP remodeling, ATPase) from nonenzymatic functions (e.g., protein-binding partners).

1.3. The experimental design to characterize the role of an RNA helicase in viral replication is focused on four issues

Elucidation of RNA helicase activity in replication of viruses has the potential to produce important fundamental information broadly significant to cell biology and unveil newly appreciated targets for therapeutic drugs (Ranji and Boris-Lawrie, 2010). Four critical issues are:

(1) To determine whether the candidate RNA helicase is necessary for viral replication,
(2) To define features of helicase that are necessary for selective and specific recognition of viral RNA and viral protein,
(3) To measure the biochemical and biophysical requirements for productive interaction,
(4) To document whether the candidate RNA helicase is a component of viral particle and potential to antagonize induction of the antiviral state in target cells.

These issues enumerate the scope of this chapter. The experimental methods discussed have been utilized to demonstrate RNA helicase A (RHA) is an important host factor in viruses that infect humans and animals. RHA activity has been demonstrated in retroviruses, hepatitis C virus, foot and mouth disease virus, and bovine viral diarrhea virus (Bolinger *et al.*, 2007, 2010; Hartman *et al.*, 2006; He *et al.*, 2008; Isken *et al.*, 2003; Lawrence and Rieder, 2009).

2. METHODS USED TO STUDY CELL-ASSOCIATED RNA HELICASE IN CULTURED MAMMALIAN CELLS

2.1. Downregulation with siRNA and rescue by exogenous expression of the RNA helicase

This method was employed to evaluate whether RHA is necessary for efficient translation of retroviruses (Bolinger *et al.*, 2007, 2010; Hartman *et al.*, 2006). The first component of the experiment is transfection with siRNA complementary to the RHA mRNA or a nonsilencing control siRNA that is a scrambled (Sc) sequence lacking any match in Genbank (sequences for siRNAs are delineated below). Transfections were performed in HEK293, COS7, or HeLa cells. The second component of the experiment is evaluation of rescue by exogenously expressed siRNA-resistant epitope-tagged RNA helicase (e.g., FLAG-RHA). Taken together, these components test whether a cellular factor is necessary for viral activity. Furthermore, the evaluation of mutant siRNA-resistant epitope-tagged RHA reveals alleles that are sufficient for activity. RHA downregulation studies in HEK293 cells demonstrated that RHA is necessary for efficient translation of HIV-1 mRNA. The rescue observed by siRNA-resistant FLAG-RHA but not a mutant allele (K417R) deficient in the ATP binding determined the necessary role of ATP-dependent helicase activity in viral mRNA translation (Bolinger et al., 2010).

The advantage of siRNA/shRNA downregulation is the ability to investigate the necessary and sufficient role of an RNA helicase. The major caveat to siRNA downregulation is the potential for off-target effects. Two sets of controls are requisite to address this issue. First, evaluation of two or more distinct siRNA that target different sequences is necessary. Likewise, at least two different shRNA vectors are standard in experiments that select cells with long-term downregulated helicase. In either case, similar results of downregulation and rescue are evidence of targeted downregulation of the gene of interest. In either case, the loss-of-function phenotype is rescued by exogenously expressed RNA helicase.

Second, it is important to determine whether the siRNA effect is selective and not secondary to collateral damage to cell viability, steady state mRNA, global protein synthesis, or another biological process. An approach to evaluate the potential effect of helicase downregulation within the time frame of siRNA treatment on cellular proliferation is a colorimetric MTT assay. Metabolic labeling is useful to assess global mRNA synthesis and stability or global protein synthesis. Briefly, cells are incubated with [^{3}H]-uridine or [^{35}S]-cysteine/methionine to label newly synthesized RNA or proteins, respectively. Levels are determined by precipitation with trichloroacetic acid and scintillation. These assays determined that cellular proliferation, mRNA levels, or translation are not significantly increased or decreased upon downregulation of RHA for 48 h (Hartman et al., 2006).

2.1.1. Method for RHA silencing with siRNA and rescue

1. Incubate 1×10^6 HEK293 or COS7 cells in a 100-mm dish.
2. Incubate overnight and then perform siRNA treatment:

siRNA preparation: For 1 plate,
- Combine 120 μl OptiMEM media (Gibco) + 20 μM siRNA in a sterile microfuge tube.
  i. Tube 1: RHA target #1 and #2; 20 μM combined

ii. Tube 2: Sc siRNA
siRNA sequences:
RHA target #1: UAGAAUGGGUGGAGAAGAAUU
RHA target #2: GGCUAUAUCCAUCGAAAUUUU
Sc siRNA: UAGACUAGCUGACGAGAAAUU
- Combine 1.8 ml OptiMEM media and 50 μl oligofectamine (Invitrogen) in 15-ml sterile tube.
- Incubate at room temperature for 15 min.
- Add the siRNA–OptiMEM mix to the 15-ml tube and incubate at room temperature for 25 min. During this incubation, remove media from the cultured cells and wash twice with DMEM that is not supplemented with FBS or antibiotics. Add 5 ml DMEM.
- After the 25-min period, add the siRNA mix to cultured cells; Incubate for 3–5 h at 37 °C.
3. Aspirate the siRNA-containing medium and add 10 ml of DMEM with 10% FBS and 1% antibiotic. Incubate at 37 °C for 48 h.

   Note: adequate downregulation of helicases may require a sequential siRNA treatment. In this case, incubate the treated culture for 48 h and then replate cells at the density of 1×10^6 cells/100-mm dish. Incubate overnight and repeat the siRNA treatment as described in step 2. Cotransfection of rescue plasmid is advised at 6–18 h after the final siRNA treatment.
4. After overnight incubation, transfect with 10 μg of HIV-1 provirus and 30 μl FuGene6 (Roche). Perform replicate transfections with and without 5 μg of siRNA-resistant pcDNA-FLAG-RHA or pcDNA-FLAG.
5. Incubate at 37 °C for 48 h. Harvest cells from each 100-mm plate in 1 ml of 1 × PBS. Resuspend cells in 250–500 μl of RIPA cell lysis buffer (50 mM Tris, pH 8.0, 0.1% SDS, 1% Triton-X, 150 mM NaCl, 1% deoxycholic acid, 2 mM PMSF).

An approach complementary to downregulation of the cellular RNA helicase is exogenous expression of mutated alleles of the RNA helicase. In the case of RHA, expression of the amino-terminal RNA-binding domain dominantly interferes with translation of retrovirus mRNA (Ranji et al., 2011).

2.2. Coprecipitation of RNA helicase with target RNA or protein cofactors

RHA is a ubiquitous RNA-binding protein that modulates posttranscriptional expression of retroviruses (Ranji and Boris-Lawrie, 2010). RHA recognizes structural features of a 5′ terminal posttranscriptional control element (PCE) within the complex 5′ UTR to facilitate polyribosome loading and efficient virion protein synthesis. Evidence for convergent cellular adaptation of the PCE/RHA RNA switch has been presented for the *junD* proto-oncogene.

junD is representative of cellular transcripts that contain a complex 5′ UTR, yet are reliant on cap-dependent translation initiation.

To assess the scope of RHA translational control in cells, microarray screens evaluated RNAs that coprecipitate with FLAG epitope-tagged RHA (Marcela Hernandez and Kathleen Boris-Lawrie, submitted). In parallel, transcripts that were depleted from polyribosomes upon RHA down-regulation were also identified. These microarrays are applicable to any helicase of interest and the outcome candidate genes are validated by complementary RT-PCR and/or quantitative real-time PCR. These screens are useful to identify distinct RNA helicase–mRNP complexes, including those in viral particles.

2.2.1. Methods for RNA and protein immunoprecipitation
2.2.1.1. Epitope immunoprecipitation

1. Typically four 150-mm plates of HEK293 or COS7 cells are transfected per IP. Four 150-mm plates of 2×10^6 COS7 cells are incubated overnight. As above, transfect with 15 μg of pcDNA-FLAG-RHA or pcDNA-FLAG and 45 μl FuGene6 (Roche) for 48 h.
2. Harvest cells from each plate in 2 ml of $1 \times$ PBS.
3. Pool the cells from the four plates into a 15-ml sterile tube. *Remove 1 ml aliquot for western blot analysis to verify expression of FLAG-RHA.*
4. Pellet by centrifugation at $500 \times g$ for 5 min at 4 °C, washing three times with 4 ml of ice-cold $1 \times$ PBS.
5. Resuspend washed cell pellet in 1 ml of $1 \times$ PBS and move to a 1.5-ml microcentrifuge tube. Pellet and add an approximate equal volume of polysome lysis buffer (100 mM KCl, 5 mM MgCl$_2$, 10 mM HEPES (pH 7.0), 0.5% NP40, 1 mM DTT). Supplement with RNase inhibitors and protease inhibitors (10 μl of 100 U/ml RNaseOUT, 10 μl of protease inhibitor cocktail and 2 μl of 200 mM vanadyl ribonucleoside complexes per milliliter of polysome lysis buffer). Pipette the mixture gently to resolve clumps of cells.
6. Incubate mRNP preparation on ice for 5 min and freeze promptly at -180 °C (liquid nitrogen). *Keep frozen until FLAG-RHA expression is verified by immunoblotting.*
7. Distribute 1 ml of FLAG beads (2 ml of 50% slurry) in a 15-ml sterile tube. Centrifuge at $350 \times g$ for 2 min at 4 °C. Remove supernatant.
8. Wash beads three times with 4 ml ice-cold NT2 buffer (50 mM Tris–HCl, pH 7.4, 150 mM NaCl, 1 mM MgCl$_2$, 0.05% NP40). To wash, centrifuge $350 \times g$ for 2 min at 4 °C, remove liquid with aspirator and resuspend in ice-cold NT2 buffer and invert the tube several times.
9. Thaw mRNP lysate on ice and centrifuge at 13,000 rpm for 15 min. If the preparation is cloudy, then the volume should be increased.

For instance, add 200 μl of polysome lysis buffer and freeze at −80 °C for 15 min. Thaw in 37 °C water bath and centrifuge at 13,000 rpm for 15 min.
10. Reserve aliquots for control treatments: one-tenth of the preparation for RNA analysis and one-tenth for protein analysis at −80 °C. Dilute remainder in 4 ml of ice-cold NT2 buffer. Supplement with 1000 units of an RNase inhibitor (25 μl RNaseOUT), 400 μM vanadyl ribonucleoside complexes (8 μl), and 20 mM EDTA.
11. Incubate the bead/lysate slurry overnight at 4 °C tumbling end over end.
12. Centrifuge 350 × g for 2 min at 4 °C and reserve supernatant at −20 °C for subsequent analysis.
13. Wash beads three times with 3 ml of ice-cold NT2 buffer (supplemented with 1000 units RNAseOUT, 10 μl 200 mM VRC, and EDTA to 20 mM) by tumbling end over end at 4 °C for 2–5 min. Centrifuge at 350 × g for 2 min at 4 °C. Aspirate the supernatant.
14. Prepare 3× FLAG elution buffer (Sigma FLAG Tagged Protein Immunoprecipitation Kit, Product Code FLAGIPT-1) by combining 60 μl of 5 μg/μl 3 × FLAG peptide solution to 2 ml of 1 × wash buffer. Add 200 units of RNAseOUT.
15. Add 2 ml of 3× FLAG elution buffer to each IP reaction. Incubate with gentle shaking for 30 min at 4 °C.
16. Centrifuge 350 × g for 2 min at 4 °C. Decant the supernatant to a fresh tube.

2.2.1.2. RNA extraction for genome-wide or candidate target mRNA identification

TRIzol (Invitrogen) may be used to extract the RNA from the beads independent of elution from the FLAG beads for a small-scale RNA IP or in cases where target RNA is being validated simply by RT-PCR. However, elution is useful for RNA preparation for selected applications, including microarray. In this case, we recommend following the RNA isolation procedure from Qiagen's RNeasy MinElute Cleanup Kit (Cat. #74204) after following the steps:

1. Precipitate the RNA by adding 5 ml 100% EtOH and 200 μl of 3 M NaOAc to the supernatants from step 16 (above). Incubate on ice for 10 min followed by centrifugation at 10,000 rpm for 15 min.
2. Resuspend each pellet in 200 μl water.
3. Add 150 μl of water to 50 μl of lysate saved for total RNA.

We recommend final elution volumes to be 20 μl for IP samples and 50 μl for total RNA samples. Eluates may be assessed for concentration by nanospectrometer. Freeze samples at −80 °C.

All of the above solutions and buffers are prepared in RNase–DNase-free H_2O.

2.2.1.3. Protein extraction for proteomic analysis or candidate target protein identification The approach of RNA-epitope coprecipitation described above (Section 2.2.1.1) is useful to prepare samples for mass spectrometry. The outcomes have identified cofactors that are coprecipitated with epitope-tagged helicase (Wei Jing, Mamuka Kvaratskehlia, and Kathleen Boris-Lawrie, unpublished). We have characterized the process to generate FLAG-RHA complexes for mass spectrometry. Comparison of four preparative approaches are summarized in Fig. 19.1.

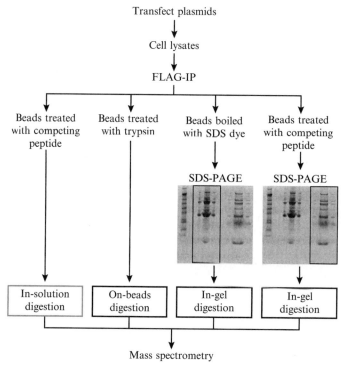

Figure 19.1 Scheme to elute and process samples following epitope immunoprecipitation for mass spectrometry. The approach of epitope immunoprecipitation is coupled with mass spectrometry to identify cofactors that coprecipitate with FLAG-tagged RHA. Immunoprecipitated protein can be trypsin-digested directly on sepharose-conjugated beads (On-beads digestion) or digested postelution with a competing peptide (In-Solution Digestion). Alternatively, beads can be treated with competing peptide or directly lysed in SDS buffer followed by SDS-PAGE. Specific band of interest or entire lane can then be trypsin digested.

2.3. Polysome association of RNA helicase with target RNA or protein cofactors

It is well appreciated that RNA–nuclear protein interaction induces RNP rearrangements that are necessary to produce a translation-competent mRNA template in the cytoplasm. RNA-binding proteins and *cis*-acting viral RNA sequences have the potential to modulate the function and fate of export and translation RNP.

In a typical ribosomal profile analysis, nontranslating, free mRNPs are separated from higher molecular mass of 40S and 60S ribosomal subunits, 80S monosomes and polyribosomes. The analysis of protein and RNA complexes in each gradient fraction is a potent technique to characterize cellular RNA helicase and viral protein(s) and RNAs that interact with the translation apparatus of the cell. The 5′ UTR of all retroviruses and selected complex cellular mRNAs contain a *cis*-acting PCE that is necessary for polysome loading and efficient translation of viral mRNA. The technique of ribosomal profile analysis demonstrated the specific interaction of PCE–RNA with host cofactor RHA as absolutely necessary for polysome association and translation of the PCE-containing viral and cellular mRNAs (Bolinger *et al.*, 2007; Hartman *et al.*, 2006).

2.3.1. Ribosomal profile analysis

1. Typically two 150-mm plates of HEK293 or COS7 cells are utilized to generate one profile. Incubate overnight four 150-mm plates of 5×10^6 COS7 cells and transfect with siRNA and/or expression plasmids of interest (15 µg total plasmid amount per plate) as described in Section 2.1.1 for 48 h.
2. Prior to harvesting, add cycloheximide at a final concentration of 0.1 µg/ml to the culture medium and incubate at 37 °C for 20 min.
3. Harvest cells from each plate in 2 ml of 1 × PBS. Centrifuge at $500 \times g$ for 5 min at 4 °C.
4. Resuspend the cells in 450 µl gradient buffer (10 mM HEPES, 10 mM NaCl, 3 mM CaCl$_2$, 7 mM MgCl$_2$, 0.5% NP40, 1 µl 1 M DTT, 80 U RNaseOUT (Invitrogen) and 1 µl/ml cycloheximide). Incubate on ice for 15 min; vortex gently at 5 min intervals.
5. Centrifuge at $3000 \times g$ for 5 min at 4 °C.
6. Layer the clarified lysate on a 15–47.5% sucrose gradient. Centrifuge at 36,000 rpm at 4 °C for 2.25 h (Beckman L-80 ultracentrifuge, SW41 rotor).
7. Decant gradient fractionates and measure A$_{254}$ absorbance to generate the ribosomal RNA profile. Fractions can be combined as mRNPs, 40S and 60S, 80S, light polyribosomes, and heavy polyribosomes, respectively.

2.3.2. Efficient isolation of protein from gradient fractions

1. Incubate 1 ml of each fraction with 50 μg BSA in TCA (final concentration 20%). Mix by inverting several times and incubate on ice for 30–60 min. Alternatively, fractions can be stored at $-20\,°C$ overnight.
2. Centrifuge at 12,000 rpm at 4 °C for 20 min. Discard supernatant and resuspend pellet in 500 μl acetone (combine selected fractions at this point).
3. Centrifuge at 12,000 rpm at 4 °C for 5 min. Discard supernatant and repeat acetone washes twice.
4. Spin the pellets in a SpeedVac for 20–30 min. *Note:* At this point, pellets may be stored at $-20\,°C$ for analysis later.
5. Resuspend pellets in 35 μl water and 35 μl 2× SDS loading dye, boil for 5 min, and subject to SDS-PAGE and/or immunoblot analysis. Some samples may be chunky or yellowish in color but will resolve once loaded onto gel.

2.3.3. Isolating RNA from gradient fractions

1. Decant 1 ml of each fraction to a fresh tube and add equivalent volume of 100% ethanol, 10 μg glycogen, or 1 μg tRNA; incubate overnight at $-80\,°C$.
2. Centrifuge at 12,000 rpm for 15–20 min at 4 °C.
3. Extract RNA from the pellet using TRIzol (Invitrogen) or RNeasy MinElute Kit (Qiagen) according to manufacturer's instructions.

RNA prepared from the fractions is suitable for several analyses, including native gel and Northern blotting, RT-PCR, quantitative real-time PCR, RNase protection assay, and analysis of polyA tail length.

2.3.3.1. Approach to distinguish ribosomes from nonribosomal RNPs
In order to confirm that the gradient fractions represent polysome, EDTA supplementation will dissociate polysomes and generate an upward shift of ribosomal RNA in the A_{254} profile. Any polysome-associated protein and/or RNA will also shift toward the left of the gradient. For EDTA treatment, we recommend supplementation of 30 mM EDTA to the PBS used to harvest the cells and the cell lysis buffer.

A complementary approach to EDTA is inhibition of translation puromycin, a chain terminator. Puromycin (Sigma) supplementation of culture medium (400 μM) is initiated 40 min prior to addition of cycloheximide for a total incubation time of 1 h. Puromycin treatment eliminates polysomes. By contrast, nontranslating RNP complexes or virus particles will remain intact in the heavy sucrose fractions.

2.3.3.2. Approach to distinguish ribosomes from intracellular virus-like particles The EDTA-mediated dissociation of ribosomes will produce a shift in polysome-associated proteins and RNAs. The failure to shift indicates that the RNPs in question do not represent polysomes. In particular, a lack of shift may be due to virus particles that cosediment with light and heavy polysomes. This differential provides a tool to differentiate between polysome-associated and virion-associated cellular and viral proteins and RNAs, respectively. Puromycin treatment will not disrupt the sedimentation of virus particles in the sucrose gradient.

3. Biochemical and Biophysical Methods to Study RNA Helicase

3.1. RNA-affinity chromatography for cofactor identification

RNA-affinity chromatography coupled with proteomic analysis is a powerful tool to identify host factors that specifically interact with the subject RNA bait. The differential between functional and nonfunctional RNAs has robustly identified host RNA helicase as a necessary effector protein of proto-oncogene junD PCE and retroviral PCE (spleen necrosis virus, human T-cell leukemia type 1 virus, HIV-1). The proteomic analysis was followed by RNA–protein coimmunoprecipitation that demonstrated selective interaction of RHA with structural features of PCE. RHA coprecipitate with HIV-1 RU5 more abundantly when compared to equimolar amounts of HIV-1 R or U5 RNA alone. The combination of RNA-affinity chromatography, RHA immunoblot analysis, and PCE activity assays determined HIV-1 R and U5 RNA elements interact synergistically with RHA to facilitate efficient gag mRNA translation (Bolinger *et al.*, 2010).

3.1.1. Biotinylation of RNA

1. Generate DNA templates for *in vitro* transcription by standard PCR with primers containing T7 promoter sequence.
2. Generate biotinylated *in vitro* transcripts using the MEGAshortscript™ Kit (Ambion) in the presence of 15 m*M* biotinylated UTP or CTP (ENZO) and 135 m*M* UTP (Promega), using T7 RNA polymerase.
3. Treat biotinylated transcripts with 1 unit DNase (Promega) at 37 °C for 30 min. Stop the reaction by bringing up the volume to 500 μl with DEPC-treated water and then add 500 μl of acid phenol. Shake and spin at 12,000 rpm for 2–3 min.
4. Take upper layer carefully into a new tube and add equal volume of chloroform. Shake and then vortex at 12,000 rpm for 2–3 min. Take upper layer into a new tube and add 1 μl glycogen, 2½ times the volume

ice-cold ethanol and 10% 3 M NaOAc. Shake and keep at $-80\ °C$ for at least 15 min.
5. Centrifuge 12,000 rpm for 10 min. Wash pellet with 500 μl of 70% ethanol. Spin at 12,000 rpm for 10 min. Dry pellet for 2 min and resuspend in 30 μl DEPC-treated water.
6. Separate 1 μl RNA on a 2% agarose gel and stain with ethidium bromide to verify the presence of the expected transcription product. Subject the RNA to the Quick Spin Sephadex G25 exclusion column (Roche) to remove nonincorporated dNTPs using the manufacturer protocol.

3.1.2. Isolation of cofactors associated with biotinylated RNA

1. Mix the slurry of streptavidin-coated beads by inverting several times. The 100 μl aliquot of beads is suitable for analysis of a candidate protein by immunoblot. The 200 μl aliquot of beads is suitable for reactions to identify RNA-interactive proteins by mass spectrometry.
2. Supplement the beads with 5 volumes of binding buffer (10 mM HEPES, pH 7.6, 5 mM EDTA, pH 8.0, 3 mM MgCl$_2$, 40 mM KCl, 5% glycerol, 1% NP40, 2 mM DTT). Gently mix by inverting 3–5 times. Centrifuge at $1200 \times g$ for 1 min. Repeat these washes twice.
3. Supplement the reaction with biotinylated RNA. For candidate analysis by immunoblot, use 15 μM of RNA. For detection of RNA-interactive proteins by mass spectrometry use ∼8 μg.
4. Incubate for 1 h at 4 °C with gentle rocking. Do not vortex. Centrifuge at $1200 \times g$ for 1 min and discard supernatant.
5. Add 1 ml of 2 mM Biotin blocking solution. Invert to mix, incubate at room temperature for 5 min. Centrifuge at $1200 \times g$ for 1 min and discard supernatant. Add 1 ml binding buffer, spin, and discard. Repeat this blocking step once more.
6. Add HeLa nuclear or total cellular extract. Use 100–500 μg for the small-scale candidate analysis, and ∼3 mg for the larger scale mass spectrometry-based screen. Supplement with 200 μl of binding buffer and incubate for 2 h at 4 °C with gentle rocking. Do not vortex. Incubation with lysate may be repeated to enrich isolation of low-abundance proteins.
7. Centrifuge at $1200 \times g$ for 1 min and discard supernatant.
8. Add 1 ml of binding buffer, mix well by inverting, and incubate at room temperature for 1 min. Centrifuge at $1200 \times g$ for 1 min and discard supernatant. Repeat washing for three additional times.
9. Wash four times with 200 μl elution buffer (10 mM HEPES, pH 7.6, 3 mM MgCl$_2$, 5 mM EDTA, pH 8.0, 0.2% glycerol, 2 mM DTT) with 40 mM, 100 mM, 200 mM, and 2 M KCl, respectively. Mix well by inverting; incubate at room temperature for 5 min, centrifuge at $1200 \times g$ for 1 min and save supernatants in separate prechilled tubes.

10. Dialyze the 2 M KCl elutants against binding buffer overnight at 4 °C. Concentrate on Millipore concentrator columns or by SpeedVac; Analyze by SDS-PAGE.

Downstream processing analyses are: staining with Coomassie or silver stain; in-gel trypsin digestion and mass spectrometry; or immunoblotting with antisera against candidate proteins. If cellular helicases are identified to interact with a specific RNA bait, additional experiments are warranted to confirm specificity of interaction. Electrophoretic mobility shift assays (EMSAs) and RNA-immunoprecipitation assays can be used to examine the specificity of interaction of the identified helicase with the RNA. The RNA–protein complexes isolated by RNA-affinity chromatography will be selectively competed by excess RNA in an RNA EMSA. A specificity control is necessary for RNA-affinity chromatography and typical controls are c-myc, gapdh, and nonfunctional mutant RNAs. An additional control of streptavidin beads without RNA bait is required to measure nonspecific enrichment of protein of interest. The RNA-affinity chromatography with total, nuclear, and/or cytoplasmic extracts provides interference to the subcellular compartment of an RNA–host protein interaction.

3.2. Determination of RNA-binding activity using EMSA

This approach can be used to evaluate the binding activity of an RNA helicase to viral and/or cellular RNA elements. The component of the experiment can include full-length-purified or individual domains of the helicase. An important consideration is to establish the specificity of interaction by comparing functional and nonfunctional RNA elements with generic control-like double-stranded RNAs (dsRNAs). EMSAs are useful for initial screening of RNA-binding activity of helicases. The quantitative assessment of RNA binding by fluorescence anisotropy (FA) is warranted to measure the binding affinity.

The results of EMSAs and FA experiments showed that the N-terminal domain of RHA exhibits higher binding affinity for SNV and junD PCE than for nonfunctional mutant PCE RNA or nonspecific control RNAs (Ranji et al., 2011). By comparison, the isolated DEIH domain lacks detectable binding to the SNV and junD PCE RNAs, and the C-terminal RG-rich domain bound nonspecifically, as designated by interaction with nonspecific control RNAs.

3.2.1. Methods for measuring RNA-binding affinity using EMSA
3.2.1.1. In vitro transcription

1. Generate *in vitro* transcripts using the RiboMAX™ large-scale RNA production system (Promega) in the presence of [α-^{32}P]UTP/[α-^{32}P]CTP (PerkinElmer Life Sciences), using T7 RNA polymerase.

2. Treat transcription reactions with 1 unit DNase (Promega) at 37 °C for 30 min, separate on 8% denaturing urea gels, and elute in probe gel elution buffer (0.5 M NH$_4$OAc, 1 mM EDTA, 0.1% SDS) for 30 min at 37 °C and then overnight at 4 °C.
3. Precipitate RNAs in 95% ethanol and 0.3 M NaOAc at -80 °C for 20 min in the presence of glycogen (20 μg), collect by centrifugation at 12,000 rpm for 10 min, wash in 500 μl of 75% ethanol. Resuspend in DEPC-treated water and measure the scintillation counts per microliter of RNA.
4. Treat RNA for 2 min in boiling water bath and move to room temperature water bath for 2 to yield native RNA conformation. Dilute the RNA to 100,000 cpm in DEPC-treated water.

3.2.1.2. Expression and purification of recombinant proteins

The preparation of expression plasmids for the RHA N-terminal (N-term), DEIH, and C-terminal (C-term) domains are described in Ranji et al. (2011). Proteins were expressed in BL21-CodonPlus optimized cells (Stratagene). The cells were treated with 1 mM isopropyl-β-D-thiogalactopyranoside (United States Biochemical Corp.) for 2.5 h at 33 °C. Cell pellets were resuspended in 1× PBS with 10 μl/ml protease inhibitor mixture (Sigma) and 1 μl/ml 1 M dithiothreitol, and subjected to 5000 units of pressure in an Aminco French pressure cell. Soluble proteins were harvested by centrifugation in a Sorvall RC5C SS-34 rotor at 12,000 rpm for 20 min at 4 °C.

1. Recombinant RNA helicase domains are purified from the soluble protein lysate on glutathione-Sepharose beads (Pierce). Incubate the lysate and beads overnight, wash four times with 50 mM HEPES and 150 mM NaCl (pH 7.0), wash once with 50 mM HEPES and 500 mM NaCl (pH 7.0), and once with thrombin cleavage buffer (20 mM HEPES, pH 8, 0.15 M NaCl, and 2.5 M CaCl$_2$). Biotinylated thrombin (10 units) is useful to release N-term RHA from the GST tag and can be further removed from the solution by incubation with streptavidin.
2. For purification of the DEIH domain of RHA, incubate soluble protein lysate with 2 ml of glutathione-Sepharose (GE Healthcare) for 30 min at room temperature. Wash the beads with 20 ml of buffer containing 50 mM HEPES, pH 7.3, 500 mM NaCl, 5 mM DTT, and 1 mM EDTA. Elute the DEIH domain from the beads using buffer containing 50 mM HEPES, pH 7.3, 25 mM reduced glutathione, 500 mM NaCl, 5 mM DTT, and 1 mM EDTA. Dialysis of protein fractions in the same buffer without glutathione is important to remove glutathione.
3. His-tagged RNA helicase was purified from the soluble protein lysate on nickel-Sepharose (GE Healthcare). Incubate the lysate with 2 ml of nickel beads at room temperature for 1 h. Remove nonspecifically bound proteins by extensive washing of the beads with 20 ml of buffer

containing 50 mM HEPES, pH 7.4, 500 mM NaCl, 7.5 mM CHAPS, 20 mM imidazole, and 4 mM β-mercaptoethanol. Elute the C-term domain in 50 mM HEPES, pH 7.4, 500 mM NaCl, 7.5 mM CHAPS, 500 mM imidazole, 5 mM EDTA, and 4 mM β-mercaptoethanol and dialyze against the same buffer without imidazole.
4. Evaluation of each protein preparation for size and purity is necessary. The results of SDS-PAGE are followed by assessment of the concentration of recombinant proteins by Bio-Rad DC protein assay.

3.2.1.3. Gel electrophoresis

1. Incubate recombinant protein and 100,000 cpm of *in vitro* transcribed α-^{32}P-labeled RNA (above) in EMSA buffer (2% glycerol, 0.8 mM EGTA, 0.2 mM EDTA, 2 mM Tris, pH 7.6, 14 mM KCl, and 0.2 mM Mg(OAc)$_2$) for 30 min on ice.
2. Electrophoresis is performed at 4 °C and using 5% native Tris borate/EDTA-acrylamide gels. Fix, dry, and expose overnight in a Phosphor-Imager cassette.

3.3. Determination of RNA-binding activity using FA

FA assays using synthetic 5′-fluorescein-tagged SNV PCE RNAs verify EMSA RNA-binding trends and determine the RNA-binding affinity of the RNA helicase domains (a related approach is described in this volume in Chapter 11). The length limit of chemical RNA synthesis is < 100 nt. For RHA-SNV-PCE studies 96- and 98-nt RNAs were labeled. The specificity and selectivity of functional PCE RNA was determined relative to nonfunctional controls. The SNV-PCE^{AC} and SNV-PCE^{AB}, which are necessary for RHA translation activity and for precipitation of epitope-tagged RHA in cells, were chosen for FP analysis and are described in Ranji *et al.* (2011).

3.3.1. Method for fluorescence anisotropy measurements

1. Synthetic RNA oligonucleotides labeled at the 5′-nucleotide with fluorescein (Dharmacon) are resuspended in DEPC-treated water (20 μM RNA).
2. Treat RNA at 80 °C for 2 min and then 60 °C for 2 min to facilitate native conformation. Incubate 3 μl of 20 μM RNA in 5.25 μl of DEPC-treated water, 3.75 μl of 100 mM HEPES, pH 7.5, and 1.5 μl of 1 M NaCl. Complete the incubation by addition of 1.5 μl of 100 mM MgCl$_2$.
3. Add DEPC-treated water to bring final volume of RNA solution to 300 μl. *Note:* For folded RNA, add 285 μl DEPC-treated water; for

RNA that does not require folding, add 297 μl DEPC-treated water. This final working solution is 200 nM.
4. Begin mixing reactions according to Table 19.2. Add DEPC water to the tubes first, then the 5× FP buffer (2% glycerol, 0.8 mM EGTA, 0.2 mM EDTA, 2 mM Tris, pH 7.6, 14 mM KCl, and 0.2 mM Mg(OAc)$_2$), then the protein and finally, the RNA. Vortex each tube briefly.
5. Incubate reactions for 30 min at room temperature in the dark to allow samples to reach equilibrium.
6. Perform FA measurements in triplicate by loading wells in Corning 3676 low volume 384-well black nonbinding surface polystyrene plates.
7. Scan plate in MD SpectraMax M5 fluorimeter (Molecular Devices). For fluorescein, use excitation wavelength of 485 nm and emission wavelength of 525 nm. Measure both anisotropy and intensity.

Table 19.2 Template for FA measurements is constructed to calculate of equilibrium dissociation constant by using variable amounts of recombinant N-terminal RHA domain with constant amount of 5′-fluorescein-labeled RNA

RHA N-term (nM)	DEPC water (μl)	5× FP buffer (μl)	RHA N-term 2 μM (μl)	RHA N-term 10 μM (μl)	RNA 200 nM (μl)
Blank	80.00	20	0.00	0	0
0	70.00	20	0.00	0	10
25	68.75	20	1.25	0	10
50	67.50	20	2.50	0	10
75	66.25	20	3.75	0	10
100	65.00	20	5.00	0	10
125	63.75	20	6.25	0	10
150	62.50	20	7.50	0	10
175	61.25	20	8.75	0	10
200	60.00	20	10.00	0	10
250	57.50	20	12.50	0	10
300	55.00	20	15.00	0	10
350	52.50	20	17.50	0	10
400	50.00	20	20.00	0	10
450	47.50	20	22.50	0	10
500	45.00	20	25.00	0	10
600	64.00	20	0.00	6	10
800	62.00	20	0.00	8	10
1000	60.00	20	0.00	10	10
1400	56.00	20	0.00	14	10
1600	54.00	20	0.00	16	10
2000	50.00	20	0.00	20	10

8. Plot anisotropy as a function of increasing protein concentration. Obtain the equilibrium dissociation constants (K_d) by fitting the binding curves to a single-binding site model on KaleidaGraph as described (Stewart-Maynard et al., 2008). Calculate weighted averages and SD as described (Taylor, 1997).

Equilibrium binding conditions should be determined by varying the incubation temperature and time. For RHA and SNV or junD PCE RNA, incubation for 30 or 60 min produces similar binding patterns indicating equilibrium binding by 30 min. Binding temperature and time should be empirically established for each helicase and its corresponding nucleic acid.

In both EMSA- and FA-based approaches, control dsRNAs should be used to compare RNA binding and establish a baseline for RNA-binding activity. Nonfunctional SNV-PCE dsRNA stem loops (termed mutAC which is a structural mutant of SNV PCE^{AC}), which lacks translation activity and does not coprecipitate RHA in cells, was used as control to establish a baseline of RHA-binding affinity. The minihelix^{Lys} 35-nt RNA derived from the acceptor-TΨC stem of human tRNA$_3$Lys and the human 7SL 27-nt hairpin RNA further provided generic dsRNA controls that lack RHA translation activity. In order to generate maximum power for statistical analysis for binding affinities, we recommend that EMSAs and FA measurements should be repeated with at least three independent preparations of protein and RNA.

4. Methods Used to Study Virion-Associated RNA Helicase in Cultured Mammalian Cells

4.1. Detection of RNA helicase in virion preparations

Viruses package host proteins in the virus particle that promote replication of virus in the subsequent generation. The analysis of whether an RNA helicase is incorporated into progeny virus particles is limited by sensitivity of immunoblotting and the purity of virus particle preparation. Given these caveats, the analysis distinguishes the role of the helicase in the producer cell (in which the virus particles are produced) and the target cells (in which the nascent virus infects and replicates).

RHA is a necessary cellular cofactor for HIV-1 replication and infectivity on primary lymphocytes. Early in the HIV-1 lifecycle, cell-associated RHA is necessary for translation of viral mRNA. The molecular basis of RHA translational stimulation involves the specific and selective interaction with structural features of the 5′ UTR via the amino-terminal residues of RHA and tethering of the ATPase-dependent helicase activity that facilitates ribosome access to the open reading frame. A minority of cell-intrinsic RHA is assembled into virus particles and the virion-associated RHA (~2 mol/particle)

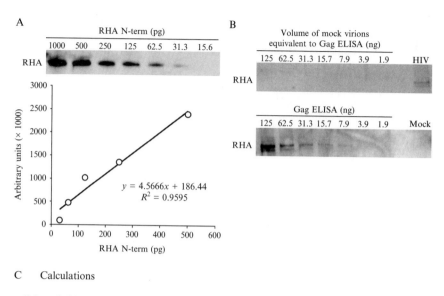

C Calculations

62.5 ng of p24 = 862 (× 1000) arbitrary units = 148 pg of RHA (based of N-term RHA standard curve)

148 pg of RHA/62.5 ng of Gag

148 pg of RHA = ? molecules

Molecular weight of RHA = 142 kDa

$= 6.27 \times 10^8$ mol/62.5 ng of Gag

62.5 ng of Gag (p24) = ? molecules

Molecular weight of Gag (p24) = 24 kDa

$= 1.56 \times 10^{12}$ mol

6.27×10^8 molecules of RHA/1.56×10^{12} mol of Gag

4.01×10^{-4} molecules of RHA/mol of Gag

1 virion = 5000 mol of Gag

$4.01 \times 10^{-4} \times 5000 = 2$ mol of RHA/virion

Figure 19.2 RHA is incorporated into HIV-1 particles. (A) Indicated mass (pg) of purified N-terminal domain (aa 1–300) of RHA was evaluated by RHA immunoblot and a standard curve was generated. Graph summarizes densitometry of N-term RHA. (B) HEK 293 cells were transfected with HIV-1^{NL4-3} or left nontransfected for 48 h. Cell-free medium from indicated cells was isolated on 25% sucrose pad and particles were lysed in RIPA buffer. Gag p24 ELISA on cell-free medium was performed indicated ng of particles was evaluated by RHA immunoblot. (C) Based on the N-term RHA standard curve from (A), amount of RHA for 62.5 ng of Gag p24 was calculated. Number of RHA molecule per virion was calculated assuming 1 virion equals 5000 Gag p24 molecules (Briggs *et al.*, 2004).

(Fig. 19.2) promotes infectivity by a mechanism that remains poorly elucidated. Downregulation of cell-intrinsic RHA using the approach described in Section 2.1 impairs the infectivity of progeny virions on primary human lymphocytes and reporter cell lines.

4.1.1. Method for quantitative detection of virion-associated RNA helicase

1. Incubate 1×10^6 HEK293 cells in a 100-mm dish overnight, transfect with 10 μg of pNL4-3 HIV-1 provirus and 30 μl FuGene6 (Roche) for 48 h.
2. Collect the virus-containing tissue culture supernatant in a Falcon tube. Pellet the debris and broken cells by centrifugation at 2000 rpm for 5 min at room temperature. Collect the supernatant through a 0.45 μm filter.
3. Ultracentrifuge the filtered supernatant over 0.5–1 ml of 25% sucrose pad at 35,000 rpm (100,000 × g) for 2 h at 4 °C in a Beckman ultracentrifuge using SW-41 rotor or comparable ultracentrifuge and rotor (e.g., Sorvall).
4. Remove supernatant by decanting, then wipe the inside wall of the tube with rolled paper towel to remove as much supernatant as possible (do not touch the bottom of the tube).
5. Add 100 μl of RIPA cell lysis buffer (50 mM Tris, pH 8.0, 0.1% SDS, 1% Triton-X, 150 mM NaCl, 1% deoxycholic acid, 2 mM PMSF). Leave the tube at 37 °C for 15 min and pipette several times to resuspend the viral pellet. Briefly vortex the pellet and collect in an eppendorf.
6. In parallel, harvest cells from plate in 1 ml of 1 × PBS, resuspend in 500 μl of RIPA cell lysis buffer on ice for 15 min with an intermittent vortex.
7. Use 25 μl of virion lysate to perform Gag p24 ELISA (Zeptometrix) with appropriate range of standard curve (7.8–125 pg/ml Gag).
8. Quantify the Gag units in the viral lysate and load a range of 2–125 ng/ml Gag units for SDS-PAGE. Immunoblot with RHA and HIV-1 Gag antisera.
9. In parallel, immunoblot serial dilutions of the recombinant RHA (15.7–1000 ng). Quantify the signal intensity of the bands; generate a standard curve and compare the abundance of RNA helicase from the virion immunoblot (Fig. 19.2).

A potential caveat is the association during ultracentrifugation of culture medium of virions with extracellular microvesicle contaminants. To overcome this pitfall, use conditioned medium from equivalent cultures of noninfected cells. The immunoblot will ascertain possible contaminants. An additional step is to perform a second ultracentrifugation (100,000 × g) through a 15%/65% sucrose step gradient. Purified virus particles are collected at the 15%/65% sucrose interface and may be concentrated at

100,000 × g. Herein, the virus preparation is placed at the top of a 20–70% continuous sucrose gradient, centrifuged at 100,000 × g at 4 °C for 16 h. Fractions are then collected and subjected to immunoblotting with antisera against viral core and the RNA helicase under investigation.

The viral RNP is disrupted and the envelope solubilized in 1% Triton X-100 at room temperature for 5 min in a wash buffer (10 mM Tris–HCl, pH 8.0, and 100 mM NaCl). The treated samples are centrifuged at 10,000 × g for 10 min and the pellet is washed three times with wash buffer. This approach will ensure that the cellular helicase is incorporated into viral cores and is not merely sticking to the outside of viral envelope. The majority of the HIV-1 matrix (MA) and capsid (CA) proteins are removed following the treatment with 1% Triton X-100, whereas substantial amounts of reverse transcriptase and nucleocapsid proteins, together with viral genomic RNA, are recovered. This preparation is designated the viral RNP complex.

To probe whether the RNA helicase requires specific interaction with the viral RNA for incorporation in the assembling HIV-1 virions, mutant proviruses are a suitable tool. The transfected provirus is used to provide mutant virion precursor RNA. A negative control is expressed by a provirus with nonfunctional mutation of the viral RNA packaging signal (Ψ).

4.2. Detection of virion-associated protein and RNA: Host and viral factors

RNA helicases are composed of domains that are interchangeable among RNA-interactive proteins: C-terminal arginine and glycine-rich domain; central DEIH helicase domain; and N-terminal double-stranded RNA-binding domains (dsRBDs) with conserved α-β-β-β-α topology. These domains provide a modular structure for multiple protein–protein and protein–RNA interactions. Such interactions can result in selective incorporation of certain host factors into assembling virions. Furthermore, interactions of RNA helicase with viral core proteins and viral nucleic acids can contribute to productive morphogenesis of the virions.

Using the approaches described in Sections 2.1.1 and 4.1.1, HIV-1 virions produced from cells treated with RHA siRNA or nonsilencing control siRNA were examined for packaging of viral and cellular factors in virions. Equivalent quantities of virions were subjected to immunoblot or RNA extraction to examine virion-associated RNA. Typically, 5–100 ng of HIV-1 Gag (as measured by Gag ELISA) is sufficient for immunoblotting and 25–500 ng HIV-1 Gag is sufficient for RNA extraction. Using the methods described below, it was shown that virions produced from cells treated with RHA siRNA are deficient in RHA but not deficient in another cellular cofactor LysRS. The results of RT-real-time PCR with

HIV-1 gag primers (described below) demonstrated viral RNA packaging efficiency is not reduced by RHA downregulation.

4.2.1. Method to screen candidate virion-associated protein(s) and RNA

1. Incubate 1×10^6 HEK293 cells/100-mm dish and culture overnight.
2. Treat with 20 µM RHA siRNAs or nonsilencing Sc RNAs as described in Section 2.1.1.
3. After second siRNA treatment, transfect cells with 10 µg pNL4-3 HIV-1 provirus in 1:3 DNA to FuGene6 (Roche) ratio.
4. Harvest cells from plate in 1 ml of 1× PBS and lyse in 250–500 µl of RIPA cell lysis buffer. Perform immunoblotting to verify downregulation of RHA from siRNA treatment. Also, immunoblotting with viral protein antisera measures the effect of RHA downregulation on viral gene expression.
5. Collect the virus-containing tissue culture supernatant in a sterile tube. Pellet the debris and broken cells by centrifugation at 2000 rpm for 5 min at room temperature. Collect the supernatant and pass it through 0.45 µm filter.
6. Ultracentrifuge the filtered supernatant over 0.5–1 ml of 25% sucrose pad at 35,000 rpm ($100,000 \times g$) for 2 h at 4 °C in a Beckman ultracentrifuge using SW-41 rotor or use a comparable ultracentrifuge and rotor (e.g., Sorvall).
7. Remove supernatant by decanting, then wipe the inside wall of the tube with rolled paper towel to remove as much supernatant as possible (do not touch the bottom of the tube).
8. Add 100 µl of RIPA cell lysis buffer (50 mM Tris, pH 8.0, 0.1% SDS, 1% Triton-X, 150 mM NaCl, 1% deoxycholic acid, 2 mM PMSF). Leave the tube at 37 °C for 15 min and pipette several times to resuspend the viral pellet. Briefly vortex the pellet and collect in an eppendorf. Use 25 µl of virion lysate to perform Gag p24 ELISA (Zeptometrix) with appropriate range of standard curve (7.8–125 pg/ml p24). Use equivalent p24 units for Sci/RHAi virions for immunoblotting.
9. Mix 50 µl virion preparation in 0.5 ml of TRIzol LS (Invitrogen) and isolate the RNA, treat with DNaseI (Ambion), extract with acid phenol (Ambion), and precipitate in 100% ethanol and 0.3 M NaOAc. Resuspend in DEPC-treated water.
10. Treat total virion RNA with Omniscript reverse transcriptase (RT, Qiagen) and random hexamer primer for 1 h at 37 °C. Use 10% of the cDNA preparation for *gag* and *β-actin* real-time PCR, respectively with HIV-1 *gag* primers KB1614 (GTAAGAAAAAGGCACAGCAAG-CAGC) and KB1615 (CATTTGCCCCTGGAGGTTCTG) or

β-*actin* primers KB1252 (TCACCCACACTGTGCCCATCTACGA) and KB1253 (CAGCGGAACCGCTCATTGCCAATGG) and Lightcycler480 SYBR Green master mix (Roche) in a Lightcycler480 (Roche, Germany). Generate standard curves to determine RNA copy numbers on pHIV-1^{NL4-3} or β-*actin* plasmid in the range of 10^2 to 10^8 copies.

RNA prepared from the virus particles is suitable for native gel electrophoresis, Northern blot, RNase protection assay, polyA length analysis.

HIV-1 virions package two copies of viral genomic RNA, which dimerize via noncovalent linkage. The approach of native gel electrophoresis and Northern blots on virion RNA preparations is useful to ascertain sustained HIV-1 RNA dimerization during downregulation of a candidate RNA helicase. A caveat is redundant functional activity of other RNA helicase superfamily members and viral RNA chaperones. In addition, the folding of viral RNA detected in this assay is not comprehensive and alternative folds may not recapitulate efficient viral replication.

4.3. Measurement of infectivity of RNA helicase-deficient virions

It is critical to investigate whether the candidate RNA helicase is important to sustain viral infectivity. If a cellular helicase is necessary for viral infection, downregulation of endogenous protein is expected to decrease virion-intrinsic RNA helicase. Herein, the outcome is production of progeny virions that are poorly infectious. Alternatively, the RNA helicase can act as a restriction factor. Herein, the downregulation may result in enhanced infectivity.

To examine the effect of RHA downregulation on HIV-1 infectivity, equivalent cell-free virion preparations were used for infection of human PBMCs or a HeLa-based Luciferase reporter cell line, TZM-bl. TZM-bl is a genetically engineered HeLa cell line that expresses high levels of HIV-1 receptor and coreceptors: CD4, CXCR4, and CCR5 (Platt *et al.*, 1998; Wei *et al.*, 2002). These cells contain reporter cassettes of Luciferase and β-galactosidase that are each expressed from Tat-inducible HIV-1 LTR. Expression of these reporter genes is directly dependent on production of HIV-1 Tat postinfection. HEK293 cells were used as producer cells and were treated with RHA siRNA or Sc siRNA as described in Section 2.1.1 (Fig. 19.3A). Infections of TZM-bl cells were performed with 2 ng Gag aliquots of virions deficient in RHA or containing RHA as described below. RHA-deficient virions are twofold less infectious on TZM-bl cells compared to virions containing RHA (Fig. 19.3B). Further, expression of siRNA-resistant RHA in the producer cells is sufficient to rescue HIV-1 infectivity on TZM-bl cells (Fig. 19.3B).

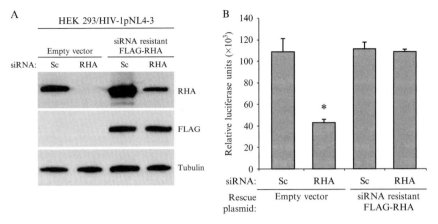

Figure 19.3 RHA downregulation in producer cells reduces HIV-1 infectivity on target cells. (A) HEK 293 cells were transfected with scrambled (Sc) or RHA (RHA) siRNAs, and then second dose of siRNA with either empty vector or siRNA-resistant FLAG-RHA and VSV-G pseudotyped HIV-1$^{NL4-3\Delta Env}$ for 48 h. Immunoblot of total cell protein with indicated antiserum verified RHA downregulation, expression of siRNA-resistant FLAG-RHA and equal protein loading, respectively. (B) Cell-free virus equivalent to 2 ng Gag was used to infect TZM-bl cells and Luciferase activity determined at 48 h ($n=3$). Asterisk indicates statistically significant difference from Sc siRNA control was observed ($P \leq 0.0005$).

4.3.1. Method for measuring infectivity of RHA-deficient HIV-1 virions

1. Incubate 1×10^6 HEK293 cells/100-mm dish and culture overnight.
2. Treat with 20 μM RHA siRNAs or nonsilencing Sc RNAs as described in Section 2.1.1.
3. After second siRNA treatment, transfect cells with 10 μg pNL4-3 HIV-1 provirus in 1:3 DNA to FuGene6 (Roche) ratio. For rescue experiments, cotransfect 5 μg of siRNA-resistant pcDNA-FLAG-RHA or pcDNA-FLAG with 10 μg HIV-1 provirus.
4. Harvest cells from plate in 1 ml of 1 × PBS and lyse in 250–500 μl of RIPA cell lysis buffer. Perform immunoblotting to verify downregulation of RHA from siRNA treatment.
5. Collect the virus-containing tissue culture supernatant in a Falcon tube. Pellet the debris and broken cells by centrifugation at 2000 rpm for 5 min at room temperature. Collect the supernatant and pass it through 0.45 μm filter.
6. Quantify the HIV-1 Gag p24 units in the supernatant. Use 25 μl of supernatant to perform Gag p24 ELISA (Zeptometrix) with appropriate range of standard curve (7.8–125 pg/ml p24).

7. Incubate 2×10^5 HeLa TZM-bl cells per well in a 6-well plate and culture overnight.
8. Infect HeLa TZM-bl cells with 0.2, 2, and 20 ng HIV-1 p24 respectively in 6-well plates by spinoculation at $1500 \times g$ for 1 h at 32 °C. Adjust the volumes of HIV-1 containing supernatant to 500 µl with complete DMEM media.
9. Harvest the HeLa TZM-bl cultures 48 h postinfection and extract in 75 µl NP40 lysis buffer (20 mM Tris–HCL, pH7.4, 150 mM NaCl, 2 mM EDTA, 1% NP40, and 2 mM PMSF).
10. Perform Luciferase assay (Promega) on 10 µg cell lysate (determined in DCA Bradford assay). The mock-infected cell lysate is a requisite control for background luminescence, and is subtracted from the sample wells.

In order to further characterize the role of cell-associated RHA in promoting or restricting viral infectivity in target cells, equivalent amounts of RHA-deficient virions and virions containing RHA is used to infect target cells (like TZM-bl, MAGI, and PBMCs) that have experienced either control or RHA siRNA treatment. If cell-intrinsic RHA acts as an antiviral factor, then an increase in viral infectivity will be observed when RHA is downregulated in target cells. On the contrary, a reduction in viral infectivity is expected if cell-intrinsic RHA acts to promote viral infection in target cells.

For reporter assays, including the TZM-bl infectivity assay, the linear range of detectable infection is determined in advance. This parameter addresses pitfalls of infections with high virus input that results in cell killing. We recommend titration of the infection parameters using range of 0.2–20 ng HIV-1 Gag p24 ELISA units. Alternatively, the virus multiplicity of infection is utilized in the range of 0.1–10. Infection or transduction via spinoculation at $1500 \times g$ for 1 h at 32 °C results in uniform infection outcome. A further alternative to spinoculation is infection at 37 °C for 2 h in the presence of 8 µg/µl polybrene or 40 µg/ml DEAE-dextran.

5. Concluding Remarks

Besides their cellular functions, RNA helicases also perform roles that are beneficial to viruses that infect plant and animal cells. A complete understanding of targets, mechanisms, and redundancy among RNA helicases is critical for new avenues toward efficacious antiviral therapies. The methods described in this chapter focused on RHA, a necessary host cofactor for retrovirus replication. However, these approaches can readily be adapted for other cellular RNA helicases with pro- or antiviral roles.

ACKNOWLEDGMENTS

We gratefully acknowledge Mr. Tim Vojt for illustrations, Dr. Amy Hayes, and Dr. Priya Kannian for helpful comments on this chapter, and Wei Jing for technical contributions in Fig. 19.1. The authors appreciate the support of Pelotonia Predoctoral Fellowship, Comprehensive Cancer Center, Ohio State University to A. S. and NIH NCI P30CA100730 and RO1CA108882 to K. B. L.

REFERENCES

Abudu, A., Wang, X., Dang, Y., Zhou, T., Xiang, S. H., and Zheng, Y. H. (2011). Identification of molecular determinants from moloney Leukemia virus 10 homolog (MOV10) protein for virion packaging and anti-human immunodeficiency virus type 1 (HIV-1) activity. *J. Biol. Chem.* **287,** 1220–1228.

Angus, A. G., Dalrymple, D., Boulant, S., McGivern, D. R., Clayton, R. F., Scott, M. J., Adair, R., Graham, S., Owsianka, A. M., Targett-Adams, P., Li, K., Wakita, T., *et al.* (2010). Requirement of cellular DDX3 for hepatitis C virus replication is unrelated to its interaction with the viral core protein. *J. Gen. Virol.* **91,** 122–132.

Ariumi, Y., Kuroki, M., Abe, K., Dansako, H., Ikeda, M., Wakita, T., and Kato, N. (2007). DDX3 DEAD-box RNA helicase is required for hepatitis C virus RNA replication. *J. Virol.* **81,** 13922–13926.

Ariumi, Y., Kuroki, M., Kushima, Y., Osugi, K., Hijikata, M., Maki, M., Ikeda, M., and Kato, N. (2011). Hepatitis C virus hijacks P-body and stress granule components around lipid droplets. *J. Virol.* **85,** 6882–6892.

Bolinger, C., Yilmaz, A., Hartman, T. R., Kovacic, M. B., Fernandez, S., Ye, J., Forget, M., Green, P. L., and Boris-Lawrie, K. (2007). RNA helicase A interacts with divergent lymphotropic retroviruses and promotes translation of human T-cell leukemia virus type 1. *Nucleic Acids Res.* **35,** 2629–2642.

Bolinger, C., Sharma, A., Singh, D., Yu, L., and Boris-Lawrie, K. (2010). RNA helicase A modulates translation of HIV-1 and infectivity of progeny virions. *Nucleic Acids Res.* **38,** 1686–1696.

Briggs, J. A., Simon, M. N., Gross, I., Krausslich, H. G., Fuller, S. D., Vogt, V. M., and Johnson, M. C. (2004). The stoichiometry of Gag protein in HIV-1. *Nat. Struct. Mol. Biol.* **11,** 672–675.

Burdick, R., Smith, J. L., Chaipan, C., Friew, Y., Chen, J., Venkatachari, N. J., Delviks-Frankenberry, K. A., Hu, W. S., and Pathak, V. K. (2010). P body-associated protein Mov10 inhibits HIV-1 replication at multiple stages. *J. Virol.* **84,** 10241–10253.

Chable-Bessia, C., Meziane, O., Latreille, D., Triboulet, R., Zamborlini, A., Wagschal, A., Jacquet, J. M., Reynes, J., Levy, Y., Saib, A., Bennasser, Y., and Benkirane, M. (2009). Suppression of HIV-1 replication by microRNA effectors. *Retrovirology* **6,** 26–36.

Chang, P. C., Chi, C. W., Chau, G. Y., Li, F. Y., Tsai, Y. H., Wu, J. C., and Wu Lee, Y. H. (2006). DDX3, a DEAD box RNA helicase, is deregulated in hepatitis virus-associated hepatocellular carcinoma and is involved in cell growth control. *Oncogene* **25,** 1991–2003.

Chen, J. Y., Chen, W. N., Poon, K. M., Zheng, B. J., Lin, X., Wang, Y. X., and Wen, Y. M. (2009). Interaction between SARS-CoV helicase and a multifunctional cellular protein (Ddx5) revealed by yeast and mammalian cell two-hybrid systems. *Arch. Virol.* **154,** 507–512.

Cocude, C., Truong, M. J., Billaut-Mulot, O., Delsart, V., Darcissac, E., Capron, A., Mouton, Y., and Bahr, G. M. (2003). A novel cellular RNA helicase, RH116, differentially regulates cell growth, programmed cell death and human immunodeficiency virus type 1 replication. *J. Gen. Virol.* **84,** 3215–3225.

Edgcomb, S. P., Carmel, A. B., Naji, S., Ambrus-Aikelin, G., Reyes, J. R., Saphire, A. C., Gerace, L., and Williamson, J. R. (2011). DDX1 is an RNA-dependent ATPase involved in HIV-1 Rev function and virus replication. *J. Mol. Biol.* **415,** 61–74.

Fang, J., Kubota, S., Yang, B., Zhou, N., Zhang, H., Godbout, R., and Pomerantz, R. J. (2004). A DEAD box protein facilitates HIV-1 replication as a cellular co-factor of Rev. *Virology* **330,** 471–480.

Fang, J., Acheampong, E., Dave, R., Wang, F., Mukhtar, M., and Pomerantz, R. J. (2005). The RNA helicase DDX1 is involved in restricted HIV-1 Rev function in human astrocytes. *Virology* **336,** 299–307.

Fujii, R., Okamoto, M., Aratani, S., Oishi, T., Ohshima, T., Taira, K., Baba, M., Fukamizu, A., and Nakajima, T. (2001). A role of RNA helicase A in cis-acting transactivation response element-mediated transcriptional regulation of human immunodeficiency virus type 1. *J. Biol. Chem.* **276,** 5445–5451.

Fuller-Pace, F. V. (2006). DExD/H box RNA helicases: Multifunctional proteins with important roles in transcriptional regulation. *Nucleic Acids Res.* **34,** 4206–4215.

Furtak, V., Mulky, A., Rawlings, S. A., Kozhaya, L., Lee, K., Kewalramani, V. N., and Unutmaz, D. (2010). Perturbation of the P-body component Mov10 inhibits HIV-1 infectivity. *PLoS One* **5,** e9081.

Garbelli, A., Radi, M., Falchi, F., Beermann, S., Zanoli, S., Manetti, F., Dietrich, U., Botta, M., and Maga, G. (2011). Targeting the human DEAD-box polypeptide 3 (DDX3) RNA helicase as a novel strategy to inhibit viral replication. *Curr. Med. Chem.* **18,** 3015–3027.

Goh, P. Y., Tan, Y. J., Lim, S. P., Tan, Y. H., Lim, S. G., Fuller-Pace, F., and Hong, W. (2004). Cellular RNA helicase p68 relocalization and interaction with the hepatitis C virus (HCV) NS5B protein and the potential role of p68 in HCV RNA replication. *J. Virol.* **78,** 5288–5298.

Greer, A. E., Hearing, P., and Ketner, G. (2011). The adenovirus E4 11 k protein binds and relocalizes the cytoplasmic P-body component Ddx6 to aggresomes. *Virology* **417,** 161–168.

Hartman, T. R., Qian, S., Bolinger, C., Fernandez, S., Schoenberg, D. R., and Boris-Lawrie, K. (2006). RNA helicase A is necessary for translation of selected messenger RNAs. *Nat. Struct. Mol. Biol.* **13,** 509–516.

Haussecker, D., Cao, D., Huang, Y., Parameswaran, P., Fire, A. Z., and Kay, M. A. (2008). Capped small RNAs and MOV10 in human hepatitis delta virus replication. *Nat. Struct. Mol. Biol.* **15,** 714–721.

He, Q. S., Tang, H., Zhang, J., Truong, K., Wong-Staal, F., and Zhou, D. (2008). Comparisons of RNAi approaches for validation of human RNA helicase A as an essential factor in hepatitis C virus replication. *J. Virol. Methods* **154,** 216–219.

Ishaq, M., Hu, J., Wu, X., Fu, Q., Yang, Y., Liu, Q., and Guo, D. (2008). Knockdown of cellular RNA helicase DDX3 by short hairpin RNAs suppresses HIV-1 viral replication without inducing apoptosis. *Mol. Biotechnol.* **39,** 231–238.

Isken, O., Grassmann, C. W., Sarisky, R. T., Kann, M., Zhang, S., Grosse, F., Kao, P. N., and Behrens, S. E. (2003). Members of the NF90/NFAR protein group are involved in the life cycle of a positive-strand RNA virus. *EMBO J.* **22,** 5655–5665.

Jangra, R. K., Yi, M., and Lemon, S. M. (2010). DDX6 (Rck/p54) is required for efficient hepatitis C virus replication but not for internal ribosome entry site-directed translation. *J. Virol.* **84,** 6810–6824.

Jankowsky, E., and Bowers, H. (2006). Remodeling of ribonucleoprotein complexes with DExH/D RNA helicases. *Nucleic Acids Res.* **34,** 4181–4188.

Jankowsky, E., and Jankowsky, A. (2000). The DExH/D protein family database. *Nucleic Acids Res.* **28,** 333–334.

Jeang, K. T., and Yedavalli, V. (2006). Role of RNA helicases in HIV-1 replication. *Nucleic Acids Res.* **34,** 4198–4205.

Jong, J. E., Park, J., Kim, S., and Seo, T. (2010). Kaposi's sarcoma-associated herpesvirus viral protein kinase interacts with RNA helicase a and regulates host gene expression. *J. Microbiol.* **48,** 206–212.

Kalverda, A. P., Thompson, G. S., Vogel, A., Schroder, M., Bowie, A. G., Khan, A. R., and Homans, S. W. (2009). Poxvirus K7 protein adopts a Bcl-2 fold: Biochemical mapping of its interactions with human DEAD box RNA helicase DDX3. *J. Mol. Biol.* **385,** 843–853.

Kawaguchi, A., Momose, F., and Nagata, K. (2011). Replication-coupled and host factor-mediated encapsidation of the influenza virus genome by viral nucleoprotein. *J. Virol.* **85,** 6197–6204.

Kim, T., Pazhoor, S., Bao, M., Zhang, Z., Hanabuchi, S., Facchinetti, V., Bover, L., Plumas, J., Chaperot, L., Qin, J., and Liu, Y. J. (2010). Aspartate-glutamate-alanine-histidine box motif (DEAH)/RNA helicase A helicases sense microbial DNA in human plasmacytoid dendritic cells. *Proc. Natl. Acad. Sci. U.S.A.* **107,** 15181–15186.

Lawrence, P., and Rieder, E. (2009). Identification of RNA helicase A as a new host factor in the replication cycle of foot-and-mouth disease virus. *J. Virol.* **83,** 11356–11366.

Li, J., Tang, H., Mullen, T. M., Westberg, C., Reddy, T. R., Rose, D. W., and Wong-Staal, F. (1999). A role for RNA helicase A in post-transcriptional regulation of HIV type 1. *Proc. Natl. Acad. Sci. U.S.A.* **96,** 709–714.

Linder, P. (2006). Dead-box proteins: A family affair—Active and passive players in RNP-remodeling. *Nucleic Acids Res.* **34,** 4168–4180.

Linder, P., and Jankowsky, E. (2011). From unwinding to clamping—The DEAD box RNA helicase family. *Nat. Rev. Mol. Cell Biol.* **12,** 505–516.

Linder, P., Tanner, N. K., and Banroques, J. (2001). From RNA helicases to RNPases. *Trends Biochem. Sci.* **26,** 339–341.

Liu, J., Henao-Mejia, J., Liu, H., Zhao, Y., and He, J. J. (2011). Translational regulation of HIV-1 replication by HIV-1 Rev cellular cofactors Sam68, eIF5A, hRIP, and DDX3. *J. Neuroimmune Pharmacol.* **6,** 308–321.

Ma, J., Rong, L., Zhou, Y., Roy, B. B., Lu, J., Abrahamyan, L., Mouland, A. J., Pan, Q., and Liang, C. (2008). The requirement of the DEAD-box protein DDX24 for the packaging of human immunodeficiency virus type 1 RNA. *Virology* **375,** 253–264.

Majerciak, V., Deng, M., and Zheng, Z. M. (2010). Requirement of UAP56, URH49, RBM15, and OTT3 in the expression of Kaposi sarcoma-associated herpesvirus ORF57. *Virology* **407,** 206–212.

Miyaji, K., Nakagawa, Y., Matsumoto, K., Yoshida, H., Morikawa, H., Hongou, Y., Arisaka, Y., Kojima, H., Inoue, T., Hirata, I., Katsu, K., and Akao, Y. (2003). Over-expression of a DEAD box/RNA helicase protein, rck/p54, in human hepatocytes from patients with hepatitis C virus-related chronic hepatitis and its implication in hepatocellular carcinogenesis. *J. Viral Hepat.* **10,** 241–248.

Miyashita, M., Oshiumi, H., Matsumoto, M., and Seya, T. (2011). DDX60, a DEXD/H box helicase, is a novel antiviral factor promoting RIG-I-like receptor-mediated signaling. *Mol. Cell. Biol.* **31,** 3802–3819.

Momose, F., Basler, C. F., O'Neill, R. E., Iwamatsu, A., Palese, P., and Nagata, K. (2001). Cellular splicing factor RAF-2p48/NPI-5/BAT1/UAP56 interacts with the influenza virus nucleoprotein and enhances viral RNA synthesis. *J. Virol.* **75,** 1899–1908.

Oshiumi, H., Ikeda, M., Matsumoto, M., Watanabe, A., Takeuchi, O., Akira, S., Kato, N., Shimotohno, K., and Seya, T. (2010). Hepatitis C virus core protein abrogates the DDX3 function that enhances IPS-1-mediated IFN-beta induction. *PLoS One* **5,** e14258.

Owsianka, A. M., and Patel, A. H. (1999). Hepatitis C virus core protein interacts with a human DEAD box protein DDX3. *Virology* **257,** 330–340.

Platt, E. J., Wehrly, K., Kuhmann, S. E., Chesebro, B., and Kabat, D. (1998). Effects of CCR5 and CD4 cell surface concentrations on infections by macrophagetropic isolates of human immunodeficiency virus type 1. *J. Virol.* **72,** 2855–2864.

Ranji, A., and Boris-Lawrie, K. (2010). RNA helicases: Emerging roles in viral replication and the host innate response. *RNA Biol.* **7,** 775–787.

Ranji, A., Shkriabai, N., Kvaratskhelia, M., Musier-Forsyth, K., and Boris-Lawrie, K. (2011). Features of double-stranded RNA-binding domains of RNA helicase A are necessary for selective recognition and translation of complex mRNAs. *J. Biol. Chem.* **286,** 5328–5337.

Robertson-Anderson, R. M., Wang, J., Edgcomb, S. P., Carmel, A. B., Williamson, J. R., and Millar, D. P. (2011). Single-molecule studies reveal that DEAD box protein DDX1 promotes oligomerization of HIV-1 Rev on the Rev response element. *J. Mol. Biol.* **410,** 959–971.

Roy, B. B., Hu, J., Guo, X., Russell, R. S., Guo, F., Kleiman, L., and Liang, C. (2006). Association of RNA helicase a with human immunodeficiency virus type 1 particles. *J. Biol. Chem.* **281,** 12625–12635.

Sadler, A. J., Latchoumanin, O., Hawkes, D., Mak, J., and Williams, B. R. (2009). An antiviral response directed by PKR phosphorylation of the RNA helicase A. *PLoS Pathog.* **5,** e1000311.

Scheller, N., Mina, L. B., Galao, R. P., Chari, A., Gimenez-Barcons, M., Noueiry, A., Fischer, U., Meyerhans, A., and Diez, J. (2009). Translation and replication of hepatitis C virus genomic RNA depends on ancient cellular proteins that control mRNA fates. *Proc. Natl. Acad. Sci. U.S.A.* **106,** 13517–13522.

Schroder, M., Baran, M., and Bowie, A. G. (2008). Viral targeting of DEAD box protein 3 reveals its role in TBK1/IKKepsilon-mediated IRF activation. *EMBO J.* **27,** 2147–2157.

Stewart-Maynard, K. M., Cruceanu, M., Wang, F., Vo, M. N., Gorelick, R. J., Williams, M. C., Rouzina, I., and Musier-Forsyth, K. (2008). Retroviral nucleocapsid proteins display nonequivalent levels of nucleic acid chaperone activity. *J. Virol.* **82,** 10129–10142.

Sunden, Y., Semba, S., Suzuki, T., Okada, Y., Orba, Y., Nagashima, K., Umemura, T., and Sawa, H. (2007). DDX1 promotes proliferation of the JC virus through transactivation of its promoter. *Microbiol. Immunol.* **51,** 339–347.

Tang, H., and Wong-Staal, F. (2000). Specific interaction between RNA helicase A and Tap, two cellular proteins that bind to the constitutive transport element of type D retrovirus. *J. Biol. Chem.* **275,** 32694–32700.

Taylor, J. R. (1997). An Introduction to Error Analysis: The Study of Uncertainties in Physical Measurements. Second Edition, University Science Books, Sausalito, CA, USA pp. 173–180.

Wang, H., and Ryu, W. S. (2010). Hepatitis B virus polymerase blocks pattern recognition receptor signaling via interaction with DDX3: Implications for immune evasion. *PLoS Pathog.* **6,** e1000986.

Wang, H., Kim, S., and Ryu, W. S. (2009). DDX3 DEAD-Box RNA helicase inhibits hepatitis B virus reverse transcription by incorporation into nucleocapsids. *J. Virol.* **83,** 5815–5824.

Wang, X., Han, Y., Dang, Y., Fu, W., Zhou, T., Ptak, R. G., and Zheng, Y. H. (2010). Moloney leukemia virus 10 (MOV10) protein inhibits retrovirus replication. *J. Biol. Chem.* **285,** 14346–14355.

Ward, A. M., Bidet, K., Yinglin, A., Ler, S. G., Hogue, K., Blackstock, W., Gunaratne, J., and Garcia-Blanco, M. A. (2011). Quantitative mass spectrometry of DENV-2 RNA-interacting proteins reveals that the DEAD-box RNA helicase DDX6 binds the DB1 and DB2 3′ UTR structures. *RNA Biol.* **8,** 1173–1186.

Wei, X., Decker, J. M., Liu, H., Zhang, Z., Arani, R. B., Kilby, J. M., Saag, M. S., Wu, X., Shaw, G. M., and Kappes, J. C. (2002). Emergence of resistant human immunodeficiency virus type 1 in patients receiving fusion inhibitor (T-20) monotherapy. *Antimicrob. Agents Chemother.* **46,** 1896–1905.

Westberg, C., Yang, J. P., Tang, H., Reddy, T. R., and Wong-Staal, F. (2000). A novel shuttle protein binds to RNA helicase A and activates the retroviral constitutive transport element. *J. Biol. Chem.* **275,** 21396–21401.

Wisskirchen, C., Ludersdorfer, T. H., Muller, D. A., Moritz, E., and Pavlovic, J. (2011). The cellular RNA helicase UAP56 is required for prevention of double-stranded RNA formation during influenza A virus infection. *J. Virol.* **85,** 8646–8655.

Xu, L., Khadijah, S., Fang, S., Wang, L., Tay, F. P., and Liu, D. X. (2010). The cellular RNA helicase DDX1 interacts with coronavirus nonstructural Protein 14 and enhances viral replication. *J. Virol.* **84,** 8571–8583.

Xu, Z., Anderson, R., and Hobman, T. C. (2011). The capsid-binding nucleolar helicase DDX56 is important for infectivity of West Nile virus. *J. Virol.* **85,** 5571–5580.

Yedavalli, V. S., Neuveut, C., Chi, Y. H., Kleiman, L., and Jeang, K. T. (2004). Requirement of DDX3 DEAD box RNA helicase for HIV-1 Rev-RRE export function. *Cell* **119,** 381–392.

Yu, S., Chen, J., Wu, M., Chen, H., Kato, N., and Yuan, Z. (2010). Hepatitis B virus polymerase inhibits RIG-I- and Toll-like receptor 3-mediated beta interferon induction in human hepatocytes through interference with interferon regulatory factor 3 activation and dampening of the interaction between TBK1/IKKepsilon and DDX3. *J. Gen. Virol.* **91,** 2080–2090.

Yu, S. F., Lujan, P., Jackson, D. L., Emerman, M., and Linial, M. L. (2011). The DEAD-box RNA helicase DDX6 is required for efficient encapsidation of a retroviral genome. *PLoS Pathog.* **7,** e1002303.

Zhang, Z., Yuan, B., Bao, M., Lu, N., Kim, T., and Liu, Y. J. (2011). The helicase DDX41 senses intracellular DNA mediated by the adaptor STING in dendritic cells. *Nat. Immunol.* **12,** 959–965.

Zhou, Y., Ma, J., Bushan, R. B., Wu, J. Y., Pan, Q., Rong, L., and Liang, C. (2008). The packaging of human immunodeficiency virus type 1 RNA is restricted by overexpression of an RNA helicase DHX30. *Virology* **372,** 97–106.

CHAPTER TWENTY

Inhibitors of Translation Targeting Eukaryotic Translation Initiation Factor 4A

Regina Cencic,* Gabriela Galicia-Vázquez,* and Jerry Pelletier*,†

Contents

1. Introduction	438
1.1. Helicases implicated in general translation initiation—eIF4AI (DDX2A), eIF4AII (DDX2B), and DHX29	438
1.2. Helicases as druggable targets	441
2. Small Molecule Inhibitors of RNA Helicases	441
2.1. Hippuristanol	442
2.2. Pateamine A	443
2.3. Rocaglates	443
2.4. NTPase/helicase inhibitors	444
3. Small Molecule Inhibitors of eIF4A ATPase Activity	444
3.1. Material and supplies	445
3.2. Production of recombinant protein	445
3.3. Screening for eIF4AI ATPase inhibitors	446
3.4. UV-induced ATP cross-linking	447
3.5. RNA pull-down assay	449
3.6. RNA helicase assay	453
4. Conclusions	455
Acknowledgments	456
References	456

Abstract

The RNA helicases eIF4AI and eIF4AII play key roles in recruiting ribosomes to mRNA templates during eukaryotic translation initiation. Small molecule inhibitors of eIF4AI and eIF4AII have been useful for chemically dissecting their role in translation *in vitro* and *in vivo*. Here, we describe a screen performed on a small

* Department of Biochemistry, McGill University, Montreal, Quebec, Canada
† The Rosalind and Morris Goodman Cancer Research Center, McGill University, Montreal, Quebec, Canada

focused library of kinase inhibitors to identify a novel helicase inhibitor. We describe assays that have been critical for characterizing novel RNA helicase inhibitors.

1. INTRODUCTION

1.1. Helicases implicated in general translation initiation—eIF4AI (DDX2A), eIF4AII (DDX2B), and DHX29

Helicases utilize the energy derived from NTP hydrolysis to rearrange nucleic acids, nucleic acid–protein interactions, or protein complexes. They have been grouped into several superfamilies (SF) depending on the number and type of conserved motifs, with RNA helicases being residents of either SF1 or SF2 (Caruthers and McKay, 2002; Gorbalenya and Koonin, 1993). Those RNA helicases implicated in eukaryotic protein synthesis are members of the DEAD and DEAH/RHA subfamilies of SF2.

Ribosome recruitment to mRNA templates during eukaryotic translation initiation is a highly orchestrated event, requiring the interplay of at least 12 initiation factors, Met-tRNA$_i^{Met}$ and 40S and 60S ribosomes (Hinnebusch, 2011; Pestova et al., 2007). It is the rate-limiting step of eukaryotic translation, and several RNA helicases have been implicated in this process. Notably, eIF4AI, eIF4AII, and DHX29 have been identified as core translation factors that are required by most, if not all eukaryotic mRNAs, whereas other RNA helicases (Ded1p, Vasa, DHX9, DDX19, DDX25) may act in a more mRNA-selective manner or affect processes such as mRNA stability or localization that impact on translational output (Hilliker et al., 2011; Parsyan et al., 2011). Translation initiation is often deregulated in human cancers due to the altered activity of regulatory signal transduction pathways (PI3K/Akt/mTOR and MAPK/ERK) with tumor cells often becoming "addicted" to these higher translation rates (Malina et al., 2011). Hence, there has been significant interest over the past decade in identifying small molecule inhibitors of translation initiation and in assessing their potential as therapeutic agents. One set of promising compounds identified from chemical biology efforts to target eukaryotic translation initiation are inhibitors of eIF4A (Malina et al., 2011).

EIF4A. EIF4A is the archetypal member of the DEAD-box family of RNA helicases, and the properties of the enzyme have been extensively reviewed (Rogers et al., 2002). It is one of the more abundant translation factors—present at 3 copies/ribosome (Duncan et al., 1987). There exist two highly related isoforms of eIF4A, eIF4AI, and eIF4AII that share 90% identity (Nielsen and Trachsel, 1988). The bulk of the biochemical experiments characterizing eIF4A have been performed with eIF4AI and it is thought that eIF4AI and eIF4AII are functionally interchangeable (Yoder-Hill et al., 1993). The relative abundance of the two eIF4A isoforms differs

(Nielsen and Trachsel, 1988; Yoder-Hill et al., 1993) with eIF4AI being expressed in a greater range of tissues (Nielsen and Trachsel, 1988; Nielsen et al., 1985; Sudo et al., 1995) (see also www.genecards.org). eIF4A exhibits ATP-dependent RNA binding, RNA-stimulated ATPase, and RNA helicase activities (Grifo et al., 1984; Rogers et al., 1999). Its activity is stimulated by two RNA chaperones, eIF4B and eIF4H (Richter-Cook et al., 1998; Rogers et al., 1999, 2001).

eIF4A is delivered to the mRNA template in the vicinity of the 5′ m^7GpppN cap structure as a part of eIF4F, a heterotrimeric complex that also harbors the eIF4E cap-binding protein and a large scaffolding protein, eIF4G (Fig. 20.1A). The helicase activity of eIF4A is ∼20-fold more efficient when it is a subunit of eIF4F (referred herein as $eIF4A_c$) compared to when in its free form ($eIF4A_f$) (Pause and Sonenberg, 1992; Rogers et al., 1999). Structural analyses have revealed that the middle domain of eIF4G binds to the C-terminal domain of eIF4A to a region that lies adjacent to residues involved in RNA binding, ATP hydrolysis, and RNA unwinding (Oberer et al., 2005; Schutz et al., 2008). eIF4G stabilizes the closed form of eIF4A by acting as a clamp. ATP is thought to assist in juxtapositioning the N-terminal and C-terminal domains, and kinetic/thermodynamic analyses of eIF4A have indicated that binding of RNA and ATP are coupled (Lorsch and Herschlag, 1998a,b).

Several macromolecules have been described as inhibitors of eIF4A activity and provided insight into eIF4A's role and regulation in translation. The tumor suppressor PDCD4 inhibits translation by binding to eIF4A and competing with eIF4G binding (LaRonde-LeBlanc et al., 2007; Suzuki et al., 2008; Waters et al., 2007; Yang et al., 2004) (Fig. 20.1A). PDCD4: eIF4A interaction is regulated by the mTOR-S6 Kinase signaling pathway highlighting the importance of this regulation in normal cellular homeostasis (Dorrello et al., 2006). In addition, a nonprotein coding RNA, BC1, has been reported to interact with eIF4A and uncouples ATP hydrolysis from helicase activity (Eom et al., 2011; Lin et al., 2008; Wang et al., 2002, 2005). Mutational analysis of eIF4AI and extensive characterization of the resultant mutants lead to the generation of several dominant-negative mutants (Pause et al., 1994), one of which inhibited translation initiation and was found to reduce the amount of wild-type eIF4A in eIF4F, suggesting that eIF4A recycles through the eIF4F complex during initiation (Svitkin et al., 2001). As well, inhibition of eIF4A using an RNA aptamer that blocked ATPase activity suppressed cap-dependent translation *in vitro* (Oguro et al., 2003). Taken together, these results reveal an essential role for eIF4A in the ribosome recruitment phase of translation initiation, and indicate that this activity is regulated in cells. Although the exact role of eIF4A in translation has not been elucidated, eIF4A may participate in unwinding 5′-proximal secondary structure to facilitate 40S ribosome recruitment, participate in preinitiation complex (PIC) scanning of the 5′ UTR, and/or may hydrolyze ATP to rearrange protein–protein or protein–mRNA interactions.

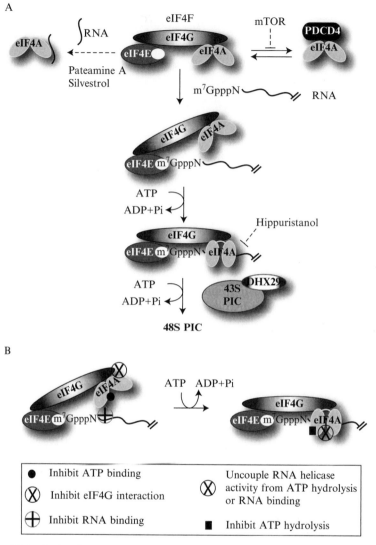

Figure 20.1 Role of eIF4A in eukaryotic translation initiation. (A) Schematic diagram of eIF4F-dependent 48S preinitiation complex (PIC) formation in eukaryotes. (B) Potential small molecule interaction sites on eIF4A. See text for details.

DHX29. Recently, the RNA helicase DHX29 has been implicated in translation initiation (Pisareva *et al.*, 2008). DHX29 does not have specificity for ATP and can hydrolyze all four NTPs. It has been shown *in vitro* to stimulate translation of mRNAs with highly structured 5′ UTRs by increasing 48S complex formation in the presence of eIF4F and eIF4B, either through remodeling of the 40S subunit (Pisareva *et al.*, 2008) or directly by unwinding 5′ UTR secondary structure (Abaeva et al., 2011). Further evaluation of DHX29's effects *in vivo* supports a role in translation initiation as silencing of DHX29 inhibits translation of mRNAs with structured 5′ UTRs (Parsyan et al., 2009).

1.2. Helicases as druggable targets

Over the past 20 years, helicases have attracted increasing attention as it has become clear that they are involved in all aspects of RNA metabolism, ranging from transcription, splicing, nucleo-cytoplasmic RNA export, to translation and turnover (Jankowsky, 2011). Moreover, many medically important RNA viruses carry their own helicase (e.g., Hepatitis C virus) (Kwong et al., 2005) or usurp cellular helicases (e.g., Influenza virus) (Brass et al., 2009; Karlas et al., 2010; Konig et al., 2010; Shapira et al., 2009) in order to replicate their genetic material. The diversity of biological activities that this class of enzymes participates in is truly wide-ranging and in many instances there is a clear rationale for exploring their potential as drug targets (Abdelhaleem, 2004; Kwong et al., 2005). The finding that small molecule metabolites can regulate cellular helicases (Montpetit et al., 2011), as well as the identification of compounds that target specific family members (Lindqvist et al., 2008), has provided proof of concept that regulation of specific helicases can be achieved. Additionally, helicases offer many more druggable opportunities than most enzymes. Not only are their surface area, ATP-binding sites, and nucleic acid binding sites potential targets for small molecular interdiction, but because their mechanism of action involves movement between intermolecular domains, additional drug targeting possibilities exist among the different conformation states, as well as between helicase–nucleic acid interaction sites (Kwong et al., 2005) (Fig. 20.1B). Here, we describe our experience in small molecule targeting of eIF4A.

2. SMALL MOLECULE INHIBITORS OF RNA HELICASES

Three natural products that target eIF4A have been identified from a high-throughput screen monitoring translational efficiency (Bordeleau et al., 2005, 2006a,b, 2008; Novac et al., 2004). The detailed setup of this forward chemical genetic screen, as well as experiments designed to identify the steps of translation targeted by these compounds have been previously reported (Cencic et al., 2007). In sum, using an *in vitro* translation assay in Krebs extracts and a bicistronic mRNA reporter, a screen of 300,000 compounds identified three natural products with very different molecular scaffolds; hippuristanol, pateamine A (Pat A), and silvestrol (Fig. 20.2A) that

A

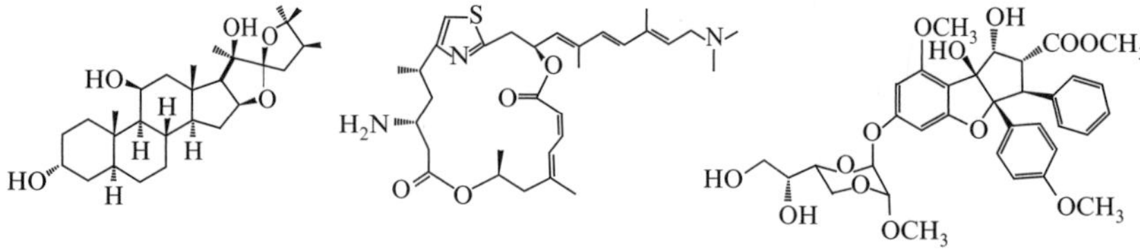

Hippuristanol Pateamine A Silvestrol

B

		Conserved Motif V		Conserved Motif VI	
DDX2A	328	-TTDLLARGIDVQQVSLVIN-	(11)	-HRIGRGGRFGRKGVAINM-	375
DDX2B	329	-TTDLLARGIDVQQVSLVIN-	(11)	-HRIGRGGRFGRKGVAINF-	376
DDX48	333	-STDVWARGLDVPQVSLIIN-	-----	-HRIGRSGRYGRKGVAINF-	380
DDX19	388	-TTNVCARGIDVEQVSVVIN-	(17)	-HRIGRTGRFGKRGLAVNM-	460
DDX25	394	-TTNVCARGIDVKQVTIVVN-	(17)	-HRIGRTGRFGKKGLAFNM-	447
DDX3	497	-ATAVAARGLDISNVKHVIN-	(11)	-HRIGRTGRVGNLGLATSF-	544
Ded1p	455	-ATAVAARGLDIPNVTHVIN-	-----	-HRIGRTGRAGNTGLATAF-	502

C

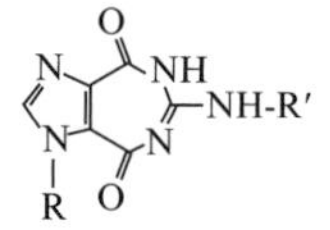

R= C_1-C_{18}-alkyl, aryl, C_1-C_4-aralkyl
R′= α/β-D-ribosyl, α/β-2′-deoxyribosyl, H, aryl, aralkyl

Ring expanded nucleoside

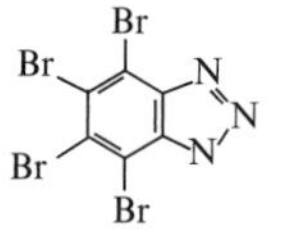

4,5,6,7-Tetrabromobenzotriazole

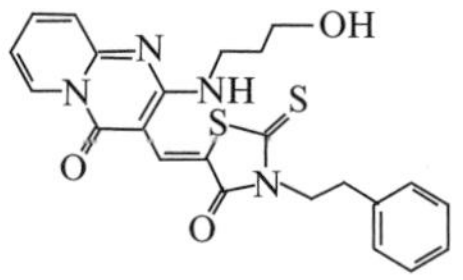

Rhodanine derivative

Figure 20.2 Small molecule RNA helicase inhibitors. (A) Structure of eIF4A activity modulators. (B) Sequence alignment of the hippuristanol binding site among chosen DDX family members. Residues in eIF4AI shown to directly interact with hippuristanol are shown in bold. Gray boxes indicate residues that lie within 5 Å. (C) Small molecule inhibitors of NTPase/helicase activity.

impact on eIF4A's helicase activity in different ways (Bordeleau *et al.*, 2005, 2006a,b, 2008).

2.1. Hippuristanol

Hippuristanol is a natural product first isolated from the coral *Isis hippuris* (Higa and Tanaka, 1981). It acts as an allosteric inhibitor of RNA interaction by binding to the C-terminal domain of eIF4A (Bordeleau *et al.*,

2006b; Lindqvist *et al.*, 2008). Both eIF4A$_f$ and eIF4A$_c$ are inhibited by hippuristanol. NMR and mutational studies have identified the binding site of hippuristanol on eIF4A to a region including and flanking conserved motifs V and VI (Fig. 20.2B) (Lindqvist *et al.*, 2008). Since the flanking region is not conserved among DEAD-box family members, hippuristanol does not inhibit the ATPase activity of other RNA helicases harboring a closely related hippuristanol binding site (i.e., DDX52 or DDX19), does not inhibit Ded1p activity (Lindqvist *et al.*, 2008), and does not affect nuclear splicing indicating that helicases involved in that process are not targets for hippuristanol (Bordeleau *et al.*, 2006b; Lindqvist *et al.*, 2008).

2.2. Pateamine A

Pat A was first isolated from the marine sponge *Mycale* sp. (Northcote *et al.*, 1991) and is an irreversible inhibitor of translation—the likely consequence of a reactive Michael addition site on the molecule (Bordeleau *et al.*, 2005). It appears to act as a chemical inducer of dimerization, stimulating nonsequence specific binding of eIF4A to RNA (Bordeleau *et al.*, 2006a). By inducing eIF4A: RNA complexes, Pat A causes eIF4A to be depleted from the eIF4F complex—resulting in inhibition of cap-dependent translation (Bordeleau *et al.*, 2006a).

Pat A stimulates the ATPase activity of eIF4A$_f$ but not of eIF4A$_c$ (Bordeleau *et al.*, 2005), suggesting that Pat A's binding site is not accessible on eIF4A$_c$. The binding site of Pat A on eIF4A$_f$ has not been mapped, but Pat A's activity is dependent on the nature of the linker region between the N- and C-terminal domains (Low *et al.*, 2007). It is possible that Pat A may interact with both domains to favor formation of the closed (high RNA affinity) conformation. In addition to eIF4AI and eIF4AII, Pat A can also interact with eIF4AIII (Bordeleau *et al.*, 2005), a third isoform of eIF4A that shares 65% identity to eIF4AI and eIF4AII and which has been implicated in playing a role in nonsense-mediated mRNA decay (Ferraiuolo *et al.*, 2004; Shibuya *et al.*, 2004). Through this interaction, Pat A inhibits nonsense-mediated mRNA decay (Dang *et al.*, 2009). Pat A does not inhibit the helicase activity of Ded1p nor does it appear to target any RNA helicases involved in mRNA splicing (Bordeleau *et al.*, 2005).

2.3. Rocaglates

Rocaglates, or cyclopenta[*b*]benzofuran flavaglines, are a family of compounds that exhibit a number of biological activities, including inhibition of cell growth (Bohnenstengel *et al.*, 1999; Hwang *et al.*, 2004; Kim *et al.*, 2007; Lee *et al.*, 1998; Ohse *et al.*, 1996; Wu *et al.*, 1997), antifungal, and insecticidal activity (Greger *et al.*, 2001; Güssregen *et al.*, 1997; Schneider *et al.*, 2000). Among the rocaglates, the most potent inhibitor of translation to date identified is silvestrol (Fig. 20.2A). Similar to Pat A, silvestrol also appears to function as a chemical inducer of dimerization, forcing an engagement between eIF4A$_f$ or eIF4A$_c$ and RNA (Bordeleau et al., 2008; Cencic et al., 2009). Silvestrol does not inhibit in vitro splicing reactions and appears specific for eIF4A (Bordeleau et al., 2008). In contrast to Pat A, the inhibitory effects of silvestrol on translation are reversible (Bordeleau et al., 2008).

2.4. NTPase/helicase inhibitors

A few additional classes of chemical compounds have also been reported to inhibit RNA helicases. A series of "ring-expanded" nucleoside analogs (Zhang et al., 2003a,b) have been extensively studied as candidate viral (West Nile virus, Hepatitis C virus, and Japanese encephalitis virus) NTPase/helicase inhibitors and DDX3 inhibitors by the Hosmane lab (Yedavalli et al., 2008; Zhang et al., 2003a,b) (Fig. 20.2C). As well, in screening for NTPase/helicase inhibitors of the Hepatitis C virus enzyme, Borowski et al. (2003) identified a series of benzimidazole and benzotriazole analogs that appear to behave as allosteric nucleoside/nucleotide binding site inhibitors (Fig. 20.2C). One compound (4,5,6,7-tetrabromobenzotriazole) was identified as an inhibitor of HCV helicase activity and is also a known selective inhibitor of protein kinase 2 (Borowski et al., 2003). Recent pharmacophore modeling and molecular docking simulations have lead to the identification of rhodanine-based compounds as inhibitors of DDX3 (Maga et al., 2008, 2011) (Fig. 20.2C). DDX3 has been implicated in HIV replication (Yedavalli et al., 2004) and these inhibitors behave as uncompetitive inhibitors and block HIV replication in vitro.

A chemical proteomics study investigating the interaction of kinase inhibitors with the human proteome reported on the capture of protein kinases to several immobilized kinase inhibitors (labeled as kinobeads). Bantscheff et al. (2007) showed that a number of ATP- and purine-binding proteins, including chaperones, helicases, ATPases, and metabolic enzymes were also retained on the kinobeads. eIF4AI was identified as a weakly binding target of Dasatinib (Bantscheff et al., 2007). Given that fairly selective inhibition of kinases can be crafted from nucleoside analogs (Yedavalli et al., 2008; Zhang et al., 2003a,b), and the possibility of using these as tools to elucidate protein function in vivo (Dar and Shokat, 2011), we investigated the potential of kinase inhibitors to inhibit eIF4AI.

3. SMALL MOLECULE INHIBITORS OF eIF4A ATPASE ACTIVITY

We have previously described a high-throughput assay designed to identify inhibitors of eukaryotic translation (Cencic et al., 2007). Here, we utilize a low-throughput, but kinetically rich assay to screen for inhibitors of eIF4A ATPase activity from a small focused collection of kinase inhibitors.

3.1. Material and supplies

Expression vectors: The expression vectors for eIF4AI (pET15b/4AI) and eIF4AII (pET28a/4AII) have been previously described (Bordeleau *et al.*, 2005; Lindqvist *et al.*, 2008).

Affinity resins: For purification of His$_6$-tagged proteins, Ni^{2+}-NTA agarose (Qiagen, Toronto) and Q-Sepharose fast flow matrix (GE Healthcare, Baie d'Urfe) are utilized.

Recombinant protein storage buffer (Buffer A): Recombinant His$_6$-eIF4AI and His$_6$-eIF4AII are stored in 20 mM Tris–HCl (pH 7.5), 10% glycerol, and 0.1 mM EDTA.

Sonication buffer: 20 mM Tris–HCl (pH 7.5), 10% glycerol, 200 mM KCl, 0.1 mM EDTA, 0.1% Triton X-100, and 3.4 mM β-mercaptoethanol [added fresh].

Buffer A100: 20 mM Tris–HCl (pH 7.5), 10% glycerol, 100 mM KCl, 0.1 mM EDTA, and 2 mM DTT [added fresh].

Buffer A300: 20 mM Tris–HCl (pH 7.5), 10% glycerol, 300 mM KCl, 0.1 mM EDTA, and 20 mM imidazole.

Buffer A800: 20 mM Tris–HCl (pH 7.5), 10% glycerol, 800 mM KCl, 0.1 mM EDTA, and 20 mM imidazole.

Radioactive isotope: γ-^{32}P-ATP (10 Ci/mmol) (Perkin Elmer, Woodbridge).

10× ATPase assay reaction buffer: 25 mM MgCl$_2$, 10 mM DTT, 10% glycerol, 200 mM MES–KOH (pH 6), and 100 mM KOAc.

Thin layer chromatography (TLC): PEI cellulose F TLC plates (20 × 20 cm; Merck) are developed in 0.3 M NaH$_2$PO$_4$ and 1 M LiCl.

RNA and nucleotides: 31.4 μM poly (U) RNA stock (GE Healthcare); 1 mM AMP-PNP stock (Sigma, Oakville).

Quantification: Quantification is performed using a BAS-1800II phosphoimager (Fuji) and an imaging screen.

3.2. Production of recombinant protein

Recombinant His$_6$-tagged murine eIF4AI and eIF4AII proteins are produced in *Escherichia coli* strain BL21 (DE3) pLysS. An overnight culture of a bacterial clone transformed with a pET-based eIF4A vector is inoculated into 1 l LB containing ampicillin (100 μg/ml) and chloramphenicol (30 μg/ml), and grown at 37 °C to an OD$_{600}$ of 0.6. Induction is performed with 1 mM IPTG for 3 h at 37 °C. Cells are harvested by centrifugation and quick frozen once. After thawing, the pellet is resuspended in 50 ml sonication buffer and sonicated utilizing a Sonic Dismembrator (Fisher Scientific, Ottawa) nine times at 50% intensity on ice for 20 s intervals with 1 min resting periods. Once complete, the suspension is centrifuged twice at 27,000 × g for 30 min at 4 °C. After the last centrifugation, imidazole is added to the clarified supernatants to a final concentration of 20 mM. All subsequent purification steps are performed at 4 °C. Recombinant eIF4A proteins are loaded onto a Ni^{2+}-NTA agarose column preequilibrated with 10 column volumes of sonication buffer. Columns containing bound recombinant eIF4AI or eIF4AII protein are washed with 10 column volumes of buffer A800 and 10 column volumes of buffer A300, followed by elution with 2.5 column volumes of buffer A300 containing 200 mM imidazole and 2 mM DTT.

Eluted eIF4AI and eIF4AII proteins are dialyzed using Spectra/Por molecular membrane tubing (Spectrum Laboratories Inc.; cut off 12,000–14,000 Da) against buffer A100 overnight at 4 °C. The eluted recombinant proteins are analyzed by SDS-PAGE followed by Coomassie Blue staining. Recombinant eIF4AI and eIF4AII proteins are further purified on a Q-Sepharose fast flow matrix. Proteins are loaded onto a column preequilibrated with 10 column volumes of buffer A100. Bound proteins are washed with 10 column volumes of buffer A100, and eluted in buffer A containing 0.1 mM DTT, using a linear salt gradient (100–500 mM KCl). Washes and elutions are performed utilizing a peristaltic pump P-1 (GE Healthcare) at a flow rate of 15 ml/h. An aliquot of the eluted fractions is analyzed by SDS-PAGE for the presence of recombinant protein. Fractions containing recombinant proteins are dialyzed separately against buffer A overnight, aliquoted, and stored at -80 °C.

3.3. Screening for eIF4AI ATPase inhibitors

The ATPase assay we utilize is a modification of that described by Lorsch and Herschlag (1998a) utilizing their condition "B."

1. Essentially, in a reaction volume of 20 μl; 2.5 μM poly (U), 1 μM γ-^{32}P-ATP (10 Ci/mmol; 0.01 μCi), and 0.5 μg recombinant proteins (His$_6$-eIF4AI or His$_6$-cIF4AII) are incubated in 1× ATPase reaction buffer at 25 °C. A 10 μM γ-^{32}P-ATP working solution is prepared right before addition to the reaction by diluting the stock 1:10 in deionized, double-distilled RNAse-free water (20 μM) and mixing it 1:1 with 20 μM MgCl$_2$.
2. Control reactions include an incubation without addition of recombinant protein, a reaction without addition of RNA, and an incubation in the presence of 50 μM AMP-PNP, a nonhydrolyzable ATP analog. When piloting these assays, we find it reassuring to purify and test an eIF4A mutant that is defective for ATP hydrolysis to evaluate the presence of potential contaminating NTPases present in the protein preparations (Pause et al., 1994). Aliquots (2 μl) are taken from the ATPase reactions at various time points and terminated by the addition of 2 μl of 25 mM EDTA (pH 8) and immediately put on ice. Inorganic phosphate (^{32}P$_i$) and γ-^{32}P-ATP are resolved by TLC. TLC plates are cut into 10 × 20 cm pieces, prerun in 100 ml of double-distilled deionized water, and dried using a hairdryer. Samples are spotted 3 cm above

the bottom edge of the TLC plate and placed 1.5 cm apart from each other. A total of 1.5 μl of the reaction is spotted onto the TLC plate with the volume spotted twice, each time as 0.75 μl to minimize sample spreading. The TLC plate is developed until the front migrates ¾ the length of the plate, dried, wrapped in Saran wrap and either exposed to X-ray film or quantitated using a BAS-1800II phosphoimager (Fuji).

3. Utilizing this assay, we have screened a small molecule collection from Biomol International consisting of 80 kinase inhibitors (Fig. 20.3A; Table 20.1). The primary screen was performed utilizing recombinant eIF4AI and the aforementioned controls. Kinase inhibitors were tested at a final concentration of 50 μM. The rate of ATP hydrolysis was set relative to the rate of hydrolysis obtained in the presence of vehicle (0.5% DMSO) and compounds that inhibited ATPase activity by at least 50% were considered primary hits (Fig. 20.3A). Using these criteria and this small focused collection, hypericin (Fig. 20.3B) was identified as a potent inhibitor of eIF4AI ATPase activity and was validated in follow-up experiments using commercially available compound (Figs. 20.3C and D).

Hypericin is an aromatic polycyclic dione isolated from plants of the *Hypericum* family (Takahashi *et al.*, 1989). It has been characterized as an inhibitor of protein kinase C (PKC) and has been proposed to inhibit PKC activity by binding to a regulatory domain containing a tandem repeat of Cys-rich, zinc finger structures (Kocanova *et al.*, 2006). Hypericin is a photoreactive compound with antiproliferative activity against glioma cells (Chen *et al.*, 2002; Chen *et al.*, 2003; Vantieghem *et al.*, 1998). Calphostin C (Fig. 20.3B) is structurally similar to hypericin and is also a PKC inhibitor showing 10-fold more potent activity than hypericin (Takahashi *et al.*, 1989; Tamaoki *et al.*, 1990). However, when tested as an eIF4AI ATPase inhibitor, it shows much weaker activity than hypericin (Fig. 20.3C and D).

4. To determine the specificity of candidate inhibitors, we assess their ability to inhibit the closely related eIF4AII isoform, as well as a more distant member of the DEAD-box family, Ded1p (Fig. 20.4A). Hypericin inhibited ATP hydrolysis of all RNA helicases tested in this assay, suggesting a broad range of activity (Fig. 20.4A and B). In contrast, Calphosin C did not show significant inhibition of ATP hydrolysis by eIF4AII or Ded1p (Fig. 20.4A). As previously reported, hippuristanol was selective for eIF4AII (and eIF4AI) but not Ded1p (Fig. 20.4A) (Bordeleau *et al.*, 2006b).

3.4. UV-induced ATP cross-linking

To determine if a compound inhibits ATP hydrolysis by blocking substrate binding, we perform UV-induced cross-linking experiments to trap α-^{32}P-ATP to eIF4AI. The UV cross-linking assay is adapted from Sarkar *et al.* (1985) and performed as follows.

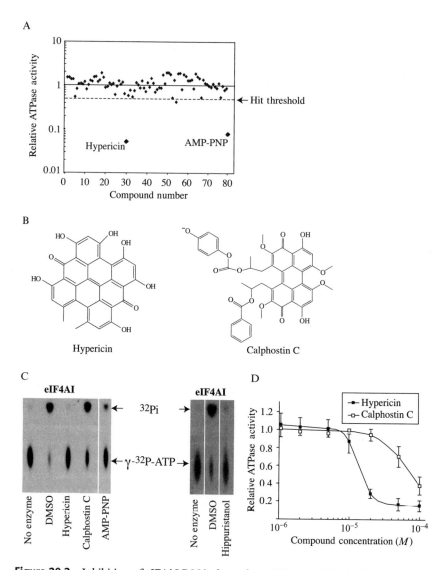

Figure 20.3 Inhibition of eIF4AI RNA-dependent ATPase activity by Hypericin. (A) Results of an ATPase-based screen utilizing a small focused kinase inhibitor library to identify inhibitors of eIF4AI ATPase activity. ATPase activity is set relative to vehicle controls and the Log_{10} of the relative ATP hydrolysis is plotted. Values represent the average of two to four reactions. (B) Chemical structure of Hypericin and Calphostin C. (C) TLC analysis of reaction products following eIF4AI-dependent ATPase assay. The enzymes and compounds tested are indicated above and below the TLC plate, respectively. DMSO was present at a final concentration of 0.5%, whereas compounds were present at 50 μM. The position of migration of P_i and ATP is indicated. (D) Titration of Hypericin and Calphostin C on eIF4AI ATPase activity. ATPase assays were performed as described for 30 min, at which point aliquots were taken and

1. A 20 µl reaction containing 1–5 µg eIF4AI in 1× ATP cross-linking buffer (50 mM Tris–HCl, pH 7.5, 50 mM KCl, 20 mM Mg(OAc)$_2$, 1 mM DTT) supplemented with 6% glycerol and 7.5 µM poly(U) RNA (from a 31.4 µM poly(U) RNA stock; GE Healthcare) is preincubated in the presence of vehicle or compound at 30 °C for 5 min. After preincubation, 0.5 µl α^{32}-P ATP (5 µCi; 3000 Ci/mmol; Perkin Elmer) is added to the reaction. Samples are spotted onto a layer of parafilm that is positioned on a prechilled glass plate placed on an ice bucket and irradiated at a distance of 2–4 cm using a germicidal UV lamp (G8T5, 254 nm, Sankyo Denki) for 20 min in a cold room. After irradiation, the samples are transferred to eppendorf tubes, unlabelled dATP added to a final concentration of 3.5 mM and RNAse A treatment (20 µg/reaction) performed for 10 min at 37 °C.
2. The samples are resolved by 10% SDS-PAGE, the polyacrylamide gel is stained with Coomassie Blue to visualize the levels of eIF4A in the reactions, followed by drying of the gels and autoradiography.
3. In this assay, hypericin inhibited UV-induced ATP cross-linking in a dose-dependent manner (Fig. 20.5A).

3.5. RNA pull-down assay

Since ATP and RNA binding to eIF4A are coupled, it is important to assess whether a small molecule that impacts on eIF4A ATPase activity is indirectly doing so by inhibiting RNA binding. We describe an RNA pull-down assay that has been informative in many cases where we have had to assess the RNA binding properties of eIF4A.

1. Poly (U) pull-down assays are performed utilizing Krebs-2 extracts (Kerr *et al.*, 1966; Martin *et al.*, 1961) which have been preincubated in the presence of vehicle (0.5% DMSO), 10 µM Pat A (as a positive control reaction for stimulating eIF4AI binding to RNA) and the indicated concentrations of hypericin for 10 min at 30 °C. Reactions are performed in 500 µl volumes. After preincubation, 50 µl of a 50% slurry of Poly (U) Sepharose beads (GE Healthcare) preequilibrated in 1× binding buffer (20 mM HEPES–KOH, pH 7.5, 250 mM KOAc, 0.1% NP-40, 1 mM DTT) is added to each sample. Reactions are incubated end-over-end for 1 h at 4 °C and beads washed three times with 10 volumes of binding buffer.

---

analyzed by TLC. The amount of P$_i$ released was quantitated and used to determine the relative ATPase activity. Values represent the average of three experiments performed in duplicates. Error bars denote the standard error of the mean.

Table 20.1 List of kinase inhibitors tested for activity in the eIF4AI ATPase assay

Compound	Target	Compound	Target
LY 294002	PI3K	H-8	PKA, PKG
Worthmannin	PI3K	HA-1004	PKA, PKG
Quercetin dihydrate	PI3K	HA-1077	PKA, PKG
Tricirbine	Akt signaling pathway	Palmitoyl-DL-carnitine Cl	PKC
BML-257	Akt	Rottlerin	PKC delta
Rapamycin	mTOR	GF 109203X	PKC
		Hypericin	PKC
SP 600125	JNK	Ro-31-8220	PKC
AG-490	JAK-2	Sphingosine	PKC
ZM 449829	JAK-3	H-89	PKA
SB-202190	p38, MAPK	HBDDE	PKC alpha, PKC gamma
SB-203580	p38, MAPK	H-7	PKA, PKG MLCK, PKC
5-Iodotubercidin	ERK2, CK1, CK2, adenosine kinase	H-9	PKA, PKG MLCK, PKC
PD-98059	MEK	ML-7	MLCK
U-0126	MEK	ML-9	MLCK
ZM 336372	cRAF	Y-27632	ROCK
GW 5074	cRAF	AG-825	HER1-2
BAY 11-7082	IKK pathway	LFM-A13	BTK
SC-514	IKK2	Terreic acid	BTK
		HMNPA	IRK

Table 20.1 (Continued)

Compound	Target	Compound	Target
Kenpaullone	GSK-3beta	AG-126	IRAK
Indirubin-3′-monoxime	GSK-3beta	SU1498	Flk1
Indirubin	GSK-3beta, CDK5	SU 4312	Flk1
BML-259	CDK5/p25	PP1	Src family
Roscovitine	CDK	PP2	Src family
Olomoucine	CDK	Piceatannol	Syk
N9-isopropyl-olomoucine	CDK	Damnacanthal	p56lck
iso-Olomoucine	Negative control for Olomoucine		
2-Aminopurine	p58 PITSLRE beta 1		
Erbstatin analog	EGFRK	Tyrphostin 9	PDGFRK
Lavendustin A	EGFRK	AG-370	PDGFRK
AG-494	EGFRK, PDGFRK	AG-879	NGFRK
RG-14620	EGFRK	HBDA	EGFRK, CaMK II
Tyrphostin 23	EGFRK	KN-62	CaMK II
Tyrphostin 25	EGFRK	KN-93	CaMK II
Tyrphostin 46	EGFRK, PDGFRK	AG-1296	PDGFRK
Tyrphostin 47	EGFRK	BML-265 (Erlotinib analog)	EGFRK
Tyrphostin 51	EGFRK		
Tyrphostin 1	Negative control for kinase inhibitors	Apigenin	CK II
Tyrphostin AG 1228	Tyrosine kinases	DRB	CK II
Tyrphostin AG 1478	EGFRK		
Tyrphostin AG 1295	Tyrosine kinases	Staurosporin	Pan-specific
Genistein	Tyrosine kinases		
Daidzein	Negative control for Genistein		

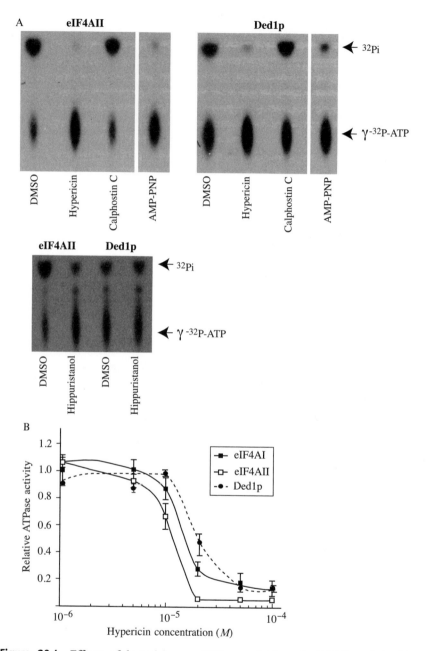

Figure 20.4 Effects of hypericin on ATPase activities of eIF4AII and Ded1p. (A) ATPase assays with the indicated recombinant proteins (0.5 μg eIF4AII and 1 μg Ded1p, respectively) were performed for 30 min (eIF4AII) or 2 h (Ded1p) in the presence of 50 μM compound, followed by TLC analysis. (B) Hypericin inhibits ATPase activity of eIF4AII and Ded1p in a concentration dependent fashion.

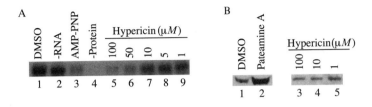

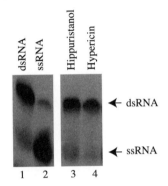

Figure 20.5 Effects of hypericin on ATP-binding, RNA-binding, and helicase activity of eIF4AI. (A) Hypericin inhibits UV-induced γ-^{32}P-ATP cross-linking to recombinant eIF4AI in a dose-dependent fashion. (B) Poly (U)-dependent RNA pull-down assays of eIF4AI in the presence of the indicated compounds. (C) Hypericin inhibits eIF4AI helicase activity. Helicase assays were performed with recombinant eIF4AI, utilizing RNA-1/RNA-11 duplexes (sequences indicated above the panel). The position of migration of RNA-1/RNA-11 duplex or single-stranded RNA-11 is indicated to the right.

2. The bound proteins are eluted in Laemmli sample buffer and resolved by SDS-PAGE, followed by Western blot analyses. Endogenous eIF4AI is detected using an anti-eIF4AI antibody (Abcam: ab31217).
3. Hypericin does not significantly inhibit eIF4AI–RNA binding at the concentrations tested (Fig. 20.5B, compare lanes 3–5 to 1), whereas Pat A significantly increased this interaction (compare lane 2–1), as previously reported (Bordeleau et al., 2006a).

3.6. RNA helicase assay

To monitor eIF4A RNA helicase activity, we perform assays essentially as described by Rogers et al. (1999). Here, a duplex RNA containing two RNAs of different lengths (RNA-1 and RNA-11; Rogers et al., 1999) is prepared as substrate (Fig. 20.5C).

1. *In vitro* transcription of RNA-1. The longer of the two RNA strands (RNA-1) is generated by *in vitro* transcription using T7 RNA polymerase (New England Biolabs [NEB]) and two DNA oligonucleotides, (1) 5′-GAATTTAATACGACTCACTATAG-3′ and (2) 3′-CTTAAAT-TATGCTGAGTGATAT★CCCCTCT(TTTTG)$_5$ATCGTGG-CATTTCGTGCG-5′ as templates (★indicates start of transcription site). The two DNA oligonucleotides are annealed by incubating 10 pmol of each template in a total volume of 20 μl in 1× hybridization buffer (10 mM Tris, pH 8, 1 mM EDTA, 100 mM KCl) for 5 min at 95 °C and the reaction allowed to slowly cool to 4 °C. Two picomoles of the oligonucleotide 1/2 duplex is then used in *in vitro* transcription reactions with 150 U of T7 RNA polymerase in 1× transcription buffer (NEB, 40 mM Tris–HCl, pH 7.9, 6 mM MgCl$_2$, 2 mM spermidine, 1 mM DTT) containing 0.5 mM of each ATP, CTP, GTP, and UTP (GE Healthcare), as well as 100 U RNase inhibitor (NEB). The reaction is incubated for 2 h at 37 °C, followed by a 10 min incubation at 37 °C with 8 U DNase I (NEB) to remove template DNA. The reaction is phenol:chloroform extracted, passed through a Sephadex G50 spin column, and precipitated with 1/5 volume of 10.5 M NH$_4$OAc and 2.5 volumes of ethanol. The RNA is recovered by centrifugation, washed with 75% ethanol, and resuspended in sterile T.E. (10 mM Tris, pH 8.0, and 1 mM EDTA).

2. ^{32}P-labeling of RNA-11. RNA-11 is purchased (Integrated DNA Technology) and used as template for ^{32}P-labeling. The labeling reaction is set up with 40 pmol of RNA, 60 μCi γ-^{32}P-ATP (stock 6000 Ci/mmol, 10 μCi/μl; Perkin Elmer) and 10 U T4 PNK (NEB) in 1× T4 PNK buffer. The reaction is incubated at 37 °C for 1 h, followed by inactivation of the enzyme for 10 min at 65 °C. To purify the ^{32}P-labeled RNA, 20 μl ddH$_2$O and 70 μl deionized formamide (Sigma) are added to the sample, incubated for 5 min at 65 °C, and the samples resolved on an 8 M urea, 12% polyacrylamide gel. The gel is prerun in 1× TBE for 30 min at 1800 V/55 mA/60 W before loading of the samples and continuing the run for another 1 h. Following electrophoresis, one glass plate is removed, the gel covered with saran wrap and exposed to X-ray film to determine the position of the ^{32}P-labeled RNA-11 on the gel (if the kinase reaction worked well, an exposure lasting only a few seconds is required). The gel slice containing RNA-11 is excised with a razor blade and RNA-11 eluted in a 50 ml conical tube (BD Biosciences, Mississauga) into 5 ml of 0.5 M NH$_4$OAc and 1 mM EDTA at 4 °C overnight. The eluted RNA-11 is purified on a DE-52 column (GE Healthcare). To this end, 1 cm of DE-52 matrix is packed in a sterile Pasteur pipette plugged with silanized glass wool (Chromatographic Specialties Inc., Brockville). Ten milliliters of ddH$_2$O are added to RNA-11 (thus diluting the NH$_4$OAc concentration to

0.17 M), and the RNA solution is loaded onto the DE-52 column. The column is washed with 1 ml of wash buffer (50 mM Tris, pH 7.5, 50 mM NH$_4$OAc, 1 mM EDTA) and the RNA is eluted in 3 × 500 μl elution buffer (50 mM Tris, pH 7.5, 2.5 M NH$_4$OAc, 1 mM EDTA, 50% deionized formamide) preheated to 65 °C. Twenty micrograms of glycogen or 5 μg calf liver tRNA and 2 volumes of 100% EtOH are added to each fraction and the RNA is precipitated overnight at −80 °C. RNA-11 is recovered by centrifugation, washed with 75% EtOH, and resuspended in 10 μl ddH$_2$O (~2 pmol/μl).

3. Generation of RNA duplexes. Ten picomoles of RNA-1 and radiolabeled RNA-11 are incubated in a total volume of 20 μl in 1 × hybridization buffer (10 mM Tris, pH 8, 1 mM EDTA, 100 mM KCl) for 5 min at 95 °C and the reaction is allowed to cool slowly to 4 °C. The RNA-1/RNA-11 duplex is diluted to 20 nM with 1 × hybridization buffer and is stored for a week at −80 °C.

4. Helicase assay. A 10 μl reaction containing 2 μM RNA-1/RNA-11 duplex, 0.4 μM recombinant eIF4AI, 1 mM ATP, 20 μg acetylated BSA (Ambion), and either vehicle (0.5% DMSO) or compound is incubated in 1 × helicase assay buffer (20 mM HEPES, pH 7.5, 70 mM KCl, 2 mM DTT, 1 mM Mg(OAc)$_2$) for 15 min at 35 °C. Reactions are stopped by the addition of 5 μl stop solution (50% glycerol, 2% SDS, 20 mM EDTA, Bromophenol Blue and Xylene Cyanol dyes) and immediately loaded onto a 12% polyacrylamide gel prerun (30 min) at 200 V in 1 × TBE at 4 °C. Electrophoresis is performed in the cold room for 2–2.5 h, the gel dried, and subjected to autoradiography (Kodak X-OMat).

5. Hypericin, like hippuristanol, inhibits unwinding of RNA-1/RNA-11 by eIF4AI (Fig. 20.5C, compare lane 4–3).

4. Conclusions

The approaches described herein have been invaluable in characterizing a number of translation initiation inhibitors that target eIF4A (Bordeleau et al., 2005, 2006a,b, 2008). We report on a novel RNA helicase inhibitor, hypericin, identified from a small collection of kinase inhibitors. Although identified in an eIF4AI-based ATPase assay, hypericin inhibited other SF2 family members (e.g., Ded1p) (Fig. 20.4A). The results indicate that kinases inhibitors may represent a potential source of pharmacophores that could be crafted into more selective helicase inhibitors by medical chemistry efforts. Improving the selectivity of hypericin will be a future challenge, and assays of the type described herein will be useful in elucidating structure–function relationships.

ACKNOWLEDGMENTS

Recombinant Ded1p protein was a kind gift from Dr. Eckhard Jankowsky (Case Western Reserve University, Cleveland, Ohio). G. G. -V. is funded by a Cole Foundation Fellowship. Work in the author's laboratory (J. P.) is supported by grants from the Canadian Institutes of Health Research and Canadian Cancer Society Research Institute.

REFERENCES

Abaeva, I. S., Marintchev, A., Pisareva, V. P., Hellen, C. U., and Pestova, T. V. (2011). Bypassing of stems versus linear base-by-base inspection of mammalian mRNAs during ribosomal scanning. *EMBO J.* **30,** 115–129.

Abdelhaleem, M. (2004). Do human RNA helicases have a role in cancer? *Biochim. Biophys. Acta* **1704,** 37–46.

Bantscheff, M., Eberhard, D., Abraham, Y., Bastuck, S., Boesche, M., Hobson, S., Mathieson, T., Perrin, J., Raida, M., Rau, C., Reader, V., Sweetman, G., et al. (2007). Quantitative chemical proteomics reveals mechanisms of action of clinical ABL kinase inhibitors. *Nat. Biotechnol.* **25,** 1035–1044.

Bohnenstengel, F. I., Steube, K. G., Meyer, C., Quentmeier, H., Nugroho, B. W., and Proksch, P. (1999). 1H-cyclopenta[b]benzofuran lignans from Aglaia species inhibit cell proliferation and alter cell cycle distribution in human monocytic leukemia cell lines. *Z. Naturforsch. C* **54,** 1075–1083.

Bordeleau, M. E., Matthews, J., Wojnar, J. M., Lindqvist, L., Novac, O., Jankowsky, E., Sonenberg, N., Northcote, P., Teesdale-Spittle, P., and Pelletier, J. (2005). Stimulation of mammalian translation initiation factor eIF4A activity by a small molecule inhibitor of eukaryotic translation. *Proc. Natl. Acad. Sci. U. S. A.* **102,** 10460–10465.

Bordeleau, M. E., Cencic, R., Lindqvist, L., Oberer, M., Northcote, P., Wagner, G., and Pelletier, J. (2006a). RNA-mediated sequestration of the RNA helicase eIF4A by Pateamine A inhibits translation initiation. *Chem. Biol.* **13,** 1287–1295.

Bordeleau, M. E., Mori, A., Oberer, M., Lindqvist, L., Chard, L. S., Higa, T., Belsham, G. J., Wagner, G., Tanaka, J., and Pelletier, J. (2006b). Functional characterization of IRESes by an inhibitor of the RNA helicase eIF4A. *Nat. Chem. Biol.* **2,** 213–220.

Bordeleau, M. E., Robert, F., Gerard, B., Lindqvist, L., Chen, S. M., Wendel, H. G., Brem, B., Greger, H., Lowe, S. W., Porco, J. A., Jr., and Pelletier, J. (2008). Therapeutic suppression of translation initiation modulates chemosensitivity in a mouse lymphoma model. *J. Clin. Invest.* **118,** 2651–2660.

Borowski, P., Deinert, J., Schalinski, S., Bretner, M., Ginalski, K., Kulikowski, T., and Shugar, D. (2003). Halogenated benzimidazoles and benzotriazoles as inhibitors of the NTPase/helicase activities of hepatitis C and related viruses. *Eur. J. Biochem.* **270,** 1645–1653.

Brass, A. L., Huang, I. C., Benita, Y., John, S. P., Krishnan, M. N., Feeley, E. M., Ryan, B. J., Weyer, J. L., van der Weyden, L., Fikrig, E., Adams, D. J., Xavier, R. J., et al. (2009). The IFITM proteins mediate cellular resistance to influenza A H1N1 virus, West Nile virus, and dengue virus. *Cell* **139,** 1243–1254.

Caruthers, J. M., and McKay, D. B. (2002). Helicase structure and mechanism. *Curr. Opin. Struct. Biol.* **12,** 123–133.

Cencic, R., Robert, F., and Pelletier, J. (2007). Identifying small molecule inhibitors of eukaryotic translation initiation. *Methods Enzymol.* **431,** 269–302.

Cencic, R., Carrier, M., Galicia-Vazquez, G., Bordeleau, M. E., Sukarieh, R., Bourdeau, A., Brem, B., Teodoro, J. G., Greger, H., Tremblay, M. L., Porco, J. A., Jr., and Pelletier, J. (2009). Antitumor activity and mechanism of action of the cyclopenta[b]benzofuran, silvestrol. *PLoS One* **4**, e5223.

Chen, B., Roskams, T., Xu, Y., Agostinis, P., and de Witte, P. A. (2002). Photodynamic therapy with hypericin induces vascular damage and apoptosis in the RIF-1 mouse tumor model. *Int. J. Cancer* **98**, 284–290.

Chen, T. C., Su, S., Fry, D., and Liebes, L. (2003). Combination therapy with irinotecan and protein kinase C inhibitors in malignant glioma. *Cancer* **97**, 2363–2373.

Dang, Y., Low, W. K., Xu, J., Gehring, N. H., Dietz, H. C., Romo, D., and Liu, J. O. (2009). Inhibition of nonsense-mediated mRNA decay by the natural product pateamine A through eukaryotic initiation factor 4AIII. *J. Biol. Chem.* **284**, 23613–23621.

Dar, A. C., and Shokat, K. M. (2011). The evolution of protein kinase inhibitors from antagonists to agonists of cellular signaling. *Annu. Rev. Biochem.* **80**, 769–795.

Dorrello, N. V., Peschiaroli, A., Guardavaccaro, D., Colburn, N. H., Sherman, N. E., and Pagano, M. (2006). S6K1- and betaTRCP-mediated degradation of PDCD4 promotes protein translation and cell growth. *Science* **314**, 467–471.

Duncan, R., Milburn, S. C., and Hershey, J. W. (1987). Regulated phosphorylation and low abundance of HeLa cell initiation factor eIF-4F suggest a role in translational control. Heat shock effects on eIF-4F. *J. Biol. Chem.* **262**, 380–388.

Eom, T., Berardi, V., Zhong, J., Risuleo, G., and Tiedge, H. (2011). Dual nature of translational control by regulatory BC RNAs. *Mol. Cell. Biol.* **31**, 4538–4549.

Ferraiuolo, M. A., Lee, C. S., Ler, L. W., Hsu, J. L., Costa-Mattioli, M., Luo, M. J., Reed, R., and Sonenberg, N. (2004). A nuclear translation-like factor eIF4AIII is recruited to the mRNA during splicing and functions in nonsense-mediated decay. *Proc. Natl. Acad. Sci. U. S. A.* **101**, 4118–4123.

Gorbalenya, A. E., and Koonin, E. V. (1993). Helicases: Amino acid sequence comparisons and structure-function relationships. *Curr. Opin. Struct. Biol.* **3**, 419–429.

Greger, H., Pacher, T., Brem, B., Bacher, M., and Hofer, O. (2001). Insecticidal flavaglines and other compounds from Fijian *Aglaia* species. *Phytochemistry* **57**, 57–64.

Grifo, J. A., Abramson, R. D., Satler, C. A., and Merrick, W. C. (1984). RNA-stimulated ATPase activity of eukaryotic initiation factors. *J. Biol. Chem.* **259**, 8648–8654.

Güssregen, R., Fuhr, M., Nugroho, B. W., Wray, V., Witte, L., and Proksch, P. (1997). New Insecticidal rocaglamide derivatives from flowers of Aglaia odorata. *Z. Naturforsch.* **52C**, 334–339.

Higa, T., and Tanaka, J. (1981). 18-oxygenated polyfunctional steroids from the gorgonian Isis hippuris. *Tetrahedron Lett.* **22**, 2777–2780.

Hilliker, A., Gao, Z., Jankowsky, E., and Parker, R. (2011). The DEAD-box protein Ded1 modulates translation by the formation and resolution of an eIF4F-mRNA complex. *Mol. Cell* **43**, 962–972.

Hinnebusch, A. G. (2011). Molecular mechanism of scanning and start codon selection in eukaryotes. *Microbiol. Mol. Biol. Rev.* **75**, 434–467.

Hwang, B. Y., Su, B. N., Chai, H., Mi, Q., Kardono, L. B., Afriastini, J. J., Riswan, S., Santarsiero, B. D., Mesecar, A. D., Wild, R., Fairchild, C. R., Vite, G. D., *et al.* (2004). Silvestrol and episilvestrol, potential anticancer rocaglate derivatives from *Aglaia silvestris*. *J. Org. Chem.* **69**, 3350–3358.

Jankowsky, E. (2011). RNA helicases at work: Binding and rearranging. *Trends Biochem. Sci.* **36**, 19–29.

Karlas, A., Machuy, N., Shin, Y., Pleissner, K. P., Artarini, A., Heuer, D., Becker, D., Khalil, H., Ogilvie, L. A., Hess, S., Maurer, A. P., Muller, E., *et al.* (2010). Genome-wide RNAi screen identifies human host factors crucial for influenza virus replication. *Nature* **463**, 818–822.

Kerr, I. M., Cohen, N., and Work, T. S. (1966). Factors controlling amino acid incorporation by ribosomes from krebs II mouse ascites-tumour cells. *Biochem. J.* **98,** 826–835.

Kim, S., Hwang, B. Y., Su, B. N., Chai, H., Mi, Q., Kinghorn, A. D., Wild, R., and Swanson, S. M. (2007). Silvestrol, a potential anticancer rocaglate derivative from Aglaia foveolata, induces apoptosis in LNCaP cells through the mitochondrial/apoptosome pathway without activation of executioner caspase-3 or -7. *Anticancer Res.* **27,** 2175–2183.

Kocanova, S., Hornakova, T., Hritz, J., Jancura, D., Chorvat, D., Mateasik, A., Ulicny, J., Refregiers, M., Maurizot, J. C., and Miskovsky, P. (2006). Characterization of the interaction of hypericin with protein kinase C in U-87 MG human glioma cells. *Photochem. Photobiol.* **82,** 720–728.

Konig, R., Stertz, S., Zhou, Y., Inoue, A., Hoffmann, H. H., Bhattacharyya, S., Alamares, J. G., Tscherne, D. M., Ortigoza, M. B., Liang, Y., Gao, Q., Andrews, S. E., *et al.* (2010). Human host factors required for influenza virus replication. *Nature* **463,** 813–817.

Kwong, A. D., Rao, B. G., and Jeang, K. T. (2005). Viral and cellular RNA helicases as antiviral targets. *Nat. Rev. Drug Discov.* **4,** 845–853.

LaRonde-LeBlanc, N., Santhanam, A. N., Baker, A. R., Wlodawer, A., and Colburn, N. H. (2007). Structural basis for inhibition of translation by the tumor suppressor Pdcd4. *Mol. Cell. Biol.* **27,** 147–156.

Lee, S. K., Cui, B., Mehta, R. R., Kinghorn, A. D., and Pezzuto, J. M. (1998). Cytostatic mechanism and antitumor potential of novel 1H-cyclopenta[b]benzofuran lignans isolated from *Aglaia elliptica*. *Chem. Biol. Interact.* **115,** 215–228.

Lin, D., Pestova, T. V., Hellen, C. U., and Tiedge, H. (2008). Translational control by a small RNA: Dendritic BC1 RNA targets the eukaryotic initiation factor 4A helicase mechanism. *Mol. Cell. Biol.* **28,** 3008–3019.

Lindqvist, L., Oberer, M., Reibarkh, M., Cencic, R., Bordeleau, M. E., Vogt, E., Marintchev, A., Tanaka, J., Fagotto, F., Altmann, M., Wagner, G., and Pelletier, J. (2008). Selective pharmacological targeting of a DEAD box RNA helicase. *PLoS One* **3,** e1583.

Lorsch, J. R., and Herschlag, D. (1998a). The DEAD box protein eIF4A. 1. A minimal kinetic and thermodynamic framework reveals coupled binding of RNA and nucleotide. *Biochemistry* **37,** 2180–2193.

Lorsch, J. R., and Herschlag, D. (1998b). The DEAD box protein eIF4A. 2. A cycle of nucleotide and RNA-dependent conformational changes. *Biochemistry* **37,** 2194–2206.

Low, W. K., Dang, Y., Bhat, S., Romo, D., and Liu, J. O. (2007). Substrate-dependent targeting of eukaryotic translation initiation factor 4A by pateamine A: Negation of domain-linker regulation of activity. *Chem. Biol.* **14,** 715–727.

Maga, G., Falchi, F., Garbelli, A., Belfiore, A., Witvrouw, M., Manetti, F., and Botta, M. (2008). Pharmacophore modeling and molecular docking led to the discovery of inhibitors of human immunodeficiency virus-1 replication targeting the human cellular aspartic acid-glutamic acid-alanine-aspartic acid box polypeptide 3. *J. Med. Chem.* **51,** 6635–6638.

Maga, G., Falchi, F., Radi, M., Botta, L., Casaluce, G., Bernardini, M., Irannejad, H., Manetti, F., Garbelli, A., Samuele, A., Zanoli, S., Este, J. A., *et al.* (2011). Toward the discovery of novel anti-HIV Drugs. Second-generation inhibitors of the cellular ATPase DDX3 with improved anti-HIV activity: Synthesis, structure-activity relationship analysis, cytotoxicity studies, and target validation. *ChemMedChem* **6,** 1371–1389.

Malina, A., Cencic, R., and Pelletier, J. (2011). Targeting translation dependence in cancer. *Oncotarget* **2,** 76–88.

Martin, E. M., Malec, J., Coote, J. L., and Work, T. S. (1961). Studies on protein and nucleic acid metabolism in virus-infected mammalian cells. 3. Methods for the disruption of Krebs II mouse-ascites-tumour cells. *Biochem. J.* **80,** 606–611.

Montpetit, B., Thomsen, N. D., Helmke, K. J., Seeliger, M. A., Berger, J. M., and Weis, K. (2011). A conserved mechanism of DEAD-box ATPase activation by nucleoporins and InsP6 in mRNA export. *Nature* **472,** 238–242.

Nielsen, P. J., McMaster, G. K., and Trachsel, H. (1985). Cloning of eukaryotic protein synthesis initiation factor genes: Isolation and characterization of cDNA clones encoding factor eIF-4A. *Nucleic Acids Res.* **13,** 6867–6880.

Nielsen, P. J., and Trachsel, H. (1988). The mouse protein synthesis initiation factor 4A gene family includes two related functional genes which are differentially expressed. *EMBO J.* **7,** 2097–2105.

Northcote, P. T., Blunt, J. W., and Munro, M. H. G. (1991). Pateamine: A potent cytotoxin from the new zealand marine sponge, mycale Sp. *Tetrahedron Lett.* **32,** 6411–6414.

Novac, O., Guenier, A. S., and Pelletier, J. (2004). Inhibitors of protein synthesis identified by a high throughput multiplexed translation screen. *Nucleic Acids Res.* **32,** 902–915.

Oberer, M., Marintchev, A., and Wagner, G. (2005). Structural basis for the enhancement of eIF4A helicase activity by eIF4G. *Genes Dev.* **19,** 2212–2223.

Oguro, A., Ohtsu, T., Svitkin, Y. V., Sonenberg, N., and Nakamura, Y. (2003). RNA aptamers to initiation factor 4A helicase hinder cap-dependent translation by blocking ATP hydrolysis. *RNA* **9,** 394–407.

Ohse, T., Ohba, S., Yamamoto, T., Koyano, T., and Umezawa, K. (1996). Cyclopentabenzofuran lignan protein synthesis inhibitors from *Aglaia odorata*. *J. Nat. Prod.* **59,** 650–652.

Parsyan, A., Shahbazian, D., Martineau, Y., Petroulakis, E., Alain, T., Larsson, O., Mathonnet, G., Tettweiler, G., Hellen, C. U., Pestova, T. V., Svitkin, Y. V., and Sonenberg, N. (2009). The helicase protein DHX29 promotes translation initiation, cell proliferation, and tumorigenesis. *Proc. Natl. Acad. Sci. U. S. A.* **106,** 22217–22222.

Parsyan, A., Svitkin, Y., Shahbazian, D., Gkogkas, C., Lasko, P., Merrick, W. C., and Sonenberg, N. (2011). mRNA helicases: The tacticians of translational control. *Nat. Rev. Mol. Cell Biol.* **12,** 235–245.

Pause, A., and Sonenberg, N. (1992). Mutational analysis of a DEAD box RNA helicase: The mammalian translation initiation factor eIF-4A. *EMBO J.* **11,** 2643–2654.

Pause, A., Methot, N., Svitkin, Y., Merrick, W. C., and Sonenberg, N. (1994). Dominant negative mutants of mammalian translation initiation factor eIF-4A define a critical role for eIF-4F in cap-dependent and cap-independent initiation of translation. *EMBO J.* **13,** 1205–1215.

Pestova, T. V., Lorsch, J. R., and Hellen, C. U. (2007). The mechanism of translation initiation in eukaryotes. *In* "Translational Control in Biology and Medicine," (M. B. Mathews, N. Sonenberg, and J. W. B. Hershey, eds.), pp. 87–128. Cold Spring Harbor Laboratory Press, Cold Spring Harbor, New York.

Pisareva, V. P., Pisarev, A. V., Komar, A. A., Hellen, C. U., and Pestova, T. V. (2008). Translation initiation on mammalian mRNAs with structured 5'UTRs requires DExH-box protein DHX29. *Cell* **135,** 1237–1250.

Richter-Cook, N. J., Dever, T. E., Hensold, J. O., and Merrick, W. C. (1998). Purification and characterization of a new eukaryotic protein translation factor. Eukaryotic initiation factor 4H. *J. Biol. Chem.* **273,** 7579–7587.

Rogers, G. W., Jr., Richter, N. J., and Merrick, W. C. (1999). Biochemical and kinetic characterization of the RNA helicase activity of eukaryotic initiation factor 4A. *J. Biol. Chem.* **274,** 12236–12244.

Rogers, G. W., Jr., Richter, N. J., Lima, W. F., and Merrick, W. C. (2001). Modulation of the helicase activity of eIF4A by eIF4B, eIF4H, and eIF4F. *J. Biol. Chem.* **276,** 30914–30922.

Rogers, G. W., Jr., Komar, A. A., and Merrick, W. C. (2002). eIF4A: The godfather of the DEAD box helicases. *Prog. Nucleic Acid Res. Mol. Biol.* **72,** 307–331.

Sarkar, G., Edery, I., and Sonenberg, N. (1985). Photoaffinity labeling of the cap-binding protein complex with ATP/dATP. Differential labeling of free eukaryotic initiation factor 4A and the eukaryotic initiation factor 4A component of the cap-binding protein complex with [alpha-32P]ATP/dATP. *J. Biol. Chem.* **260**, 13831–13837.

Schneider, C., Bohnenstengel, F. I., Nugroho, B. W., Wray, V., Witte, L., Hung, P. D., Kiet, L. C., and Proksch, P. (2000). Insecticidal rocaglamide derivatives from *Aglaia spectabilis* (Meliaceae). *Phytochemistry* **54**, 731–736.

Schutz, P., Bumann, M., Oberholzer, A. E., Bieniossek, C., Trachsel, H., Altmann, M., and Baumann, U. (2008). Crystal structure of the yeast eIF4A-eIF4G complex: An RNA-helicase controlled by protein-protein interactions. *Proc. Natl. Acad. Sci. U. S. A.* **105**, 9564–9569.

Shapira, S. D., Gat-Viks, I., Shum, B. O., Dricot, A., de Grace, M. M., Wu, L., Gupta, P. B., Hao, T., Silver, S. J., Root, D. E., Hill, D. E., Regev, A., et al. (2009). A physical and regulatory map of host-influenza interactions reveals pathways in H1N1 infection. *Cell* **139**, 1255–1267.

Shibuya, T., Tange, T. O., Sonenberg, N., and Moore, M. J. (2004). eIF4AIII binds spliced mRNA in the exon junction complex and is essential for nonsense-mediated decay. *Nat. Struct. Mol. Biol.* **11**, 346–351.

Sudo, K., Takahashi, E., and Nakamura, Y. (1995). Isolation and mapping of the human EIF4A2 gene homologous to the murine protein synthesis initiation factor 4A-II gene Eif4a2. *Cytogenet. Cell Genet.* **71**, 385–388.

Suzuki, C., Garces, R. G., Edmonds, K. A., Hiller, S., Hyberts, S. G., Marintchev, A., and Wagner, G. (2008). PDCD4 inhibits translation initiation by binding to eIF4A using both its MA3 domains. *Proc. Natl. Acad. Sci. U. S. A.* **105**, 3274–3279.

Svitkin, Y. V., Pause, A., Haghighat, A., Pyronnet, S., Witherell, G., Belsham, G. J., and Sonenberg, N. (2001). The requirement for eukaryotic initiation factor 4A (elF4A) in translation is in direct proportion to the degree of mRNA 5' secondary structure. *RNA* **7**, 382–394.

Takahashi, I., Nakanishi, S., Kobayashi, E., Nakano, H., Suzuki, K., and Tamaoki, T. (1989). Hypericin and pseudohypericin specifically inhibit protein kinase C: Possible relation to their antiretroviral activity. *Biochem. Biophys. Res. Commun.* **165**, 1207–1212.

Tamaoki, T., Takahashi, I., Kobayashi, E., Nakano, H., Akinaga, S., and Suzuki, K. (1990). Calphostin (UCN1028) and calphostin related compounds, a new class of specific and potent inhibitors of protein kinase C. *Adv. Second Messenger Phosphoprotein Res.* **24**, 497–501.

Vantieghem, A., Assefa, Z., Vandenabeele, P., Declercq, W., Courtois, S., Vandenheede, J. R., Merlevede, W., de Witte, P., and Agostinis, P. (1998). Hypericin-induced photosensitization of HeLa cells leads to apoptosis or necrosis. Involvement of cytochrome c and procaspase-3 activation in the mechanism of apoptosis. *FEBS Lett.* **440**, 19–24.

Wang, H., Iacoangeli, A., Popp, S., Muslimov, I. A., Imataka, H., Sonenberg, N., Lomakin, I. B., and Tiedge, H. (2002). Dendritic BC1 RNA: Functional role in regulation of translation initiation. *J. Neurosci.* **22**, 10232–10241.

Wang, H., Iacoangeli, A., Lin, D., Williams, K., Denman, R. B., Hellen, C. U., and Tiedge, H. (2005). Dendritic BC1 RNA in translational control mechanisms. *J. Cell Biol.* **171**, 811–821.

Waters, L. C., Veverka, V., Bohm, M., Schmedt, T., Choong, P. T., Muskett, F. W., Klempnauer, K. H., and Carr, M. D. (2007). Structure of the C-terminal MA-3 domain of the tumour suppressor protein Pdcd4 and characterization of its interaction with eIF4A. *Oncogene* **26**, 4941–4950.

Wu, T. S., Liou, M. J., Kuoh, C. S., Teng, C. M., Nagao, T., and Lee, K. H. (1997). Cytotoxic and antiplatelet aggregation principles from *Aglaia elliptifolia*. *J. Nat. Prod.* **60**, 606–608.

Yang, H. S., Cho, M. H., Zakowicz, H., Hegamyer, G., Sonenberg, N., and Colburn, N. H. (2004). A novel function of the MA-3 domains in transformation and translation suppressor Pdcd4 is essential for its binding to eukaryotic translation initiation factor 4A. *Mol. Cell. Biol.* **24,** 3894–3906.

Yedavalli, V. S., Neuveut, C., Chi, Y. H., Kleiman, L., and Jeang, K. T. (2004). Requirement of DDX3 DEAD box RNA helicase for HIV-1 Rev-RRE export function. *Cell* **119,** 381–392.

Yedavalli, V. S., Zhang, N., Cai, H., Zhang, P., Starost, M. F., Hosmane, R. S., and Jeang, K. T. (2008). Ring expanded nucleoside analogues inhibit RNA helicase and intracellular human immunodeficiency virus type 1 replication. *J. Med. Chem.* **51,** 5043–5051.

Yoder-Hill, J., Pause, A., Sonenberg, N., and Merrick, W. C. (1993). The p46 subunit of eukaryotic initiation factor (eIF)-4F exchanges with eIF-4A. *J. Biol. Chem.* **268,** 5566–5573.

Zhang, N., Chen, H. M., Koch, V., Schmitz, H., Liao, C. L., Bretner, M., Bhadti, V. S., Fattom, A. I., Naso, R. B., Hosmane, R. S., and Borowski, P. (2003a). Ring-expanded ("fat") nucleoside and nucleotide analogues exhibit potent in vitro activity against flaviviridae NTPases/helicases, including those of the West Nile virus, hepatitis C virus, and Japanese encephalitis virus. *J. Med. Chem.* **46,** 4149–4164.

Zhang, N., Chen, H. M., Koch, V., Schmitz, H., Minczuk, M., Stepien, P., Fattom, A. I., Naso, R. B., Kalicharran, K., Borowski, P., and Hosmane, R. S. (2003b). Potent inhibition of NTPase/helicase of the West Nile Virus by ring-expanded ("fat") nucleoside analogues. *J. Med. Chem.* **46,** 4776–4789.

CHAPTER TWENTY-ONE

IDENTIFICATION AND ANALYSIS OF INHIBITORS TARGETING THE HEPATITIS C VIRUS NS3 HELICASE

Alicia M. Hanson, John J. Hernandez, William R. Shadrick, *and* David N. Frick

Contents

1. Introduction	464
2. The Need for Additional HCV Drug Targets	465
3. Targeting the NS3 Helicase	466
4. HTS for HCV Helicase Inhibitors	467
5. Expression and Purification of NS3h	469
6. The Molecular Beacon-Based Helicase Assay (MBHA)	470
6.1. MBHA-based high-throughput screens	472
6.2. Examining effects of inhibitors on the kinetics of DNA strand separation	473
6.3. Evaluating alternate MBHAs for HTS	476
7. An RNA-Based Split Beacon Helicase Assay (SBHA)	477
8. Discussion	479
Acknowledgments	480
References	480

Abstract

This chapter describes two types of FRET-based fluorescence assays that can be used to identify and analyze compounds that inhibit the helicase encoded by the hepatitis C virus (HCV). Both assays use a fluorescently labeled DNA or RNA oligonucleotide to monitor helicase-catalyzed strand separation, and they differ from other real-time helicase assays in that they do not require the presence of other nucleic acids to trap the reaction products. The first assay is a molecular beacon-based helicase assay (MBHA) that monitors helicase-catalyzed displacement of a hairpin-forming oligonucleotide with a fluorescent moiety on one end and a quencher on the other. DNA-based MBHAs have been used extensively for high-throughput screening (HTS), but RNA-based MBHAs are

Department of Chemistry and Biochemistry, University of Wisconsin-Milwaukee, Milwaukee, Wisconsin, USA typically less useful because of poor signal to background ratios. In the second assay discussed, the fluorophore and quencher are split between two hairpin-forming oligonucleotides annealed in tandem to a third oligonucleotide. This split beacon helicase assay can be used for HTS with either DNA or RNA oligonucleotides. These assays should be useful to the many labs searching for HCV helicase inhibitors in order to develop new HCV therapies that are still desperately needed.

1. INTRODUCTION

Specific helicase inhibitors of viral RNA helicases are needed for two reasons. First, they are valuable chemical probes needed to understand the roles that RNA helicases play in biology. Second, inhibitors of viral helicases may be valuable as antiviral agents. Most RNA viruses that replicate outside the cell's nucleus encode an RNA helicase. If such a virus lacks a functional helicase, neither can it replicate (Kolykhalov et al., 2000) nor can it synthesize its RNA genome (Lam and Frick, 2006). RNA helicases provide medicinal chemists many targets because helicase inhibitors could, at any one of several clearly defined ligand binding sites (or other critical motifs), block ATP binding, ATP hydrolysis, RNA binding, strand separation, or protein translocation. Once a binding site is clearly defined, the many available RNA helicase crystal structures could be used to rationally design more potent derivatives.

Interest in helicases as drug targets peaked about 10 years ago when two classes of compounds targeting a helicase encoded by herpes simplex virus (HSV) were shown to elicit potent antiviral effects in animal models. These novel HSV antiviral drugs target a DNA helicase that coordinates DNA replication and the action of DNA primase (Crute et al., 2002; Katsumata et al., 2011; Kleymann et al., 2002). Inspired by the success of the HSV compounds, several teams have led extensive searches for inhibitors of human helicases (Aggarwal et al., 2011; Yedavalli et al., 2008) and helicases encoded by important human pathogens (Frick, 2006; Kwong et al., 2005; Tuteja, 2007). One of the most frequently targeted RNA helicases is the one encoded by the hepatitis C virus (HCV). The methods discussed in this chapter were specifically designed for use with HCV helicase but they could be used with related helicases with relatively minor changes. Medically relevant RNA helicases related to HCV helicase include enzymes encoded by the flaviviruses (e.g., Dengue virus, Yellow fever virus, and West Nile virus) and the human DEAD-box proteins. The procedures below have been developed and implemented to screen over 290,000 compounds as part of the National Institutes of Health's Molecular Libraries Probe

Production Centers Network (MLPCN). All screening results from this project are posted regularly on PubChem BioAssay (http://www.ncbi.nlm.nih.gov/pcassay).

2. THE NEED FOR ADDITIONAL HCV DRUG TARGETS

HCV infects nearly one in every 50 people alive today causing fibrosis, cirrhosis, and ultimately, liver failure. There are no approved HCV vaccines, but there are effective HCV treatments that all use the broad-acting drugs ribavirin and pegylated recombinant human interferon alpha (INFα). Current HCV drug combinations have an impressive impact on viral proliferation, typically curing more than half of patients, but they are expensive and their considerable side effects make HCV therapy difficult to tolerate (Edlin, 2011). Development of HCV vaccines and less toxic HCV drugs has been slow because it was not possible to study wild-type HCV in the lab until the recent advent of robust cell culture systems and small animal models (Murray and Rice, 2011). For years, HCV drug development focused almost entirely on recombinant HCV proteins that were expressed in model organisms and used to develop assays suitable for high-throughput screening (HTS). "Hits" in these HTS assays were then developed by rationally designing better compounds using high-resolution protein structures of the HCV targets. Eventually, this process led to the discovery of numerous direct acting antivirals (DAAs), which are now being developed to replace INFα and ribavirin in HCV therapy. It is hoped that DAAs will cause fewer side effects because, unlike INFα and ribavirin, DAAs are not designed to modulate the host response to viruses.

The HCV genome contains a single open reading frame encoding an approximately 3000 amino acid long polypeptide. Host and viral proteases cleave the polyprotein into 10 mature HCV proteins. Three HCV proteins are structural, forming the virus particle, and seven are nonstructural (NS) proteins. NS3 is the HCV helicase, but it also has several additional important functions. Upon translation, NS3 combines with NS2 to form an autocatalytic protease that cleaves the NS2/NS3 junction. Processed NS3 then contains another protease active site that is activated after newly translated NS4A binds to the NS3 N-terminal protease domain. This second HCV protease cleaves itself in *cis* and other HCV and cellular proteins in *trans*. The most advanced HCV DAAs attack this NS3/NS4A protease. Currently, the most advanced protease inhibitors are the recently approved drugs Telaprevir (Zeuzem et al., 2011) and Boceprevir (Bacon et al., 2011). Triple therapy with INFα, ribavirin, and a protease inhibitor cures up to 88% of patients who have failed prior therapies. There are, however, still several problems with this state-of-the-art HCV therapy. First, the protease inhibitors are only effective when administered with interferon and ribavirin because of a low resistance barrier. In other words, single point mutations confer resistance to Telaprevir and Boceprevir. These mutations evolve rapidly and have only relatively minor effects on HCV fitness or viability (Hiraga et al., 2011). Second, Telaprevir and Boceprevir are only effective against specific HCV strains and genotypes, mainly ones common in North America. Third, additional side effects are associated with the protease inhibitors. Fourth, triple therapy is even more costly, making the new therapy even less accessible to most patients. New drugs are therefore still needed to make HCV treatment more accessible and better tolerated so that it might start to impact the global HCV burden.

It will not likely be possible to decrease HCV therapy cost and toxicity unless drugs are found to replace, rather than supplement, INFα. To this end, other DAAs are being tested alone and in combination with protease inhibitors. None of these drug cocktails contain helicase inhibitors, but, in theory, helicase inhibitors would be particularly attractive additions to DAA cocktails since they would target the same protein as the protease inhibitors already in use. Two drugs targeting the same protein could interact synergistically so that combined they are more effective. The accumulating evidence that the NS3 helicase and protease depend on one another supports the notion that helicase and protease inhibitors might act synergistically (Beran et al., 2007, 2009; Frick et al., 2004).

3. Targeting the NS3 Helicase

HCV helicase was one of the first HCV targets identified with its activity first characterized shortly after HCV was discovered (Choo et al., 1989; Kim et al., 1995; Porter et al., 1998; Preugschat et al., 1996; Suzich et al., 1993). The HCV helicase was also the first RNA helicase crystallized (Yao et al., 1997) and NS3 has been studied extensively both as a model helicase and as a drug target ever since. As discussed extensively in other reviews, tremendous progress has since been made to understand exactly how the helicase unwinds DNA and RNA in an ATP-fueled reaction (Frick, 2007; Pyle, 2008; Raney et al., 2010).

Early HCV helicase studies were performed mainly with truncated NS3 lacking the protease domain (referred to here as NS3h) because such proteins express in *Escherichia coli* at higher levels than full-length NS3 and they are more stable. In NS3h proteins, NS3 is truncated at a linker connecting the helicase to the NS3/NS4A protease by deleting between 166 and 190 amino acids from the NS3 N-terminus. The protease is then replaced with an affinity tag, or an affinity tag is fused to the C-terminus of NS3h. Most early studies used NS3h as a surrogate for full-length NS3, but more recent studies tend to focus on full-length NS3. Direct comparisons of NS3h to full-length NS3 have revealed that the protease domains and NS4A influence the helicase, and *vice versa*, suggesting that the NS3 helicase and protease functions do not act independently but instead they are tightly coordinated (Beran *et al.*, 2007, 2009; Frick *et al.*, 2004). The protocols below have been used with a variety of recombinant NS3 and NS3h proteins isolated from a wide array of HCV strains and genotypes (Belon and Frick, 2009b; Belon *et al.*, 2010). We find that results are most consistent with an NS3h lacking the first 166 NS3 amino acids with a His-tag attached to the NS3 C-terminus. The NS3h protein we use most often in screens is the one isolated from the Con1 strain of HCV genotype 1b. The Con1 strain forms the backbone for many common HCV replicons used to study HCV replication in cells (Lohmann *et al.*, 1999).

All NS3 and NS3h proteins unwind both RNA and DNA. This robust DNA helicase activity facilitates *in vitro* analysis, but it is unusual because HCV has no DNA stage and related proteins act only on RNA. It has been speculated that the activity of NS3 on DNA is somehow related to the fact that HCV infection correlates with high rates of hepatocellular carcinoma. However, only two indirect lines of evidence link NS3 to a role in liver cancer. The first is the observation that, when HCV helicase is overexpressed in human cells, some of the protein has been observed in the nucleus where it might affect host gene expression or transforms cells to a cancerous phenotype (Muramatsu *et al.*, 1997). The second is the biochemical observation that NS3h can catalyze strand exchange reactions, which hints toward a possible role for NS3 in genetic recombination (Rypma *et al.*, 2009).

Regardless of why NS3h unwinds DNA, DNA has already been used in many screens for HCV helicase inhibitors. The major concern with such assays is that compounds inhibiting HCV helicase-catalyzed DNA unwinding might not inhibit the action of NS3 on its natural RNA substrates. The procedures below address this concern by first providing a readout as to whether or not the compound interacts with the DNA substrate and, second, by using a second assay as a counterscreen that uses an RNA-based substrate.

4. HTS for HCV Helicase Inhibitors

Standard helicase assays and some early HTS assays monitor helicase action using radioactive oligonucleotides to observe strand displacement (Kyono *et al.*, 1998). In order to avoid using hazardous radioisotopes, simplify protocols, and provide real-time readouts, newer helicase assays often use fluorescently labeled oligonucleotides and Förster resonance energy transfer (FRET) to monitor strand separation. In most FRET-based helicase assays, one nucleic acid strand is labeled with a donor fluorophore and the complementary oligonucleotide is labeled with an accepter moiety. When a helicase separates the two oligonucleotides upon ATP addition, donor fluorophore fluorescence increases because it is separated from the FRET acceptor (Bjornson et al., 1994; Houston and Kodadek, 1994). FRET-based assays have been used extensively for HCV assays (Boguszewska-Chachulska et al., 2004; Frick et al., 2007; Tani et al., 2009), but we have found they are not ideal for HTS because well-to-well variation in apparent reaction rates and extent makes hit identification difficult. Rypma et al. demonstrated that some of this variability stems from the fact that nucleic acid traps added to the above FRET-based assays to prevent the two labeled strands from reannealing heavily influence observed reaction rates and their extent (Rypma et al., 2009).

To help facilitate HTS for HCV helicase inhibitors, our lab developed a real-time helicase assay that did not require nucleic acid traps to observe helicase action on duplexes. This second-generation FRET-based helicase assay uses a helicase substrate made with a molecular beacon (Tyagi and Kramer, 1996) annealed to a longer DNA oligonucleotide such that a $3'$ single-stranded region is available for the helicase to load (Belon and Frick, 2008). Using protocols below (Section 6), this molecular beacon-based helicase assay (MBHA) has been used in mechanistic analyses (Belon and Frick, 2009b), HTS (Belon and Frick, 2010), and for the analysis of known HCV helicase inhibitors (Belon et al., 2010).

One serious limitation of the MBHA is that, when the DNA oligonucleotides are substituted with RNA oligonucleotides, the signal to background (S/B) ratio decreases to a level where the assay is no longer appropriate for HTS (Belon and Frick, 2008). Because the natural substrate for HCV NS3h is most likely RNA, a screen with RNA is needed to identify compounds that act only when DNA is used as a substrate. Compounds that inhibit HCV helicase-catalyzed DNA unwinding but not RNA unwinding might be useful chemical probes to understand if the HCV helicase action on DNA plays any role in HCV biology. However, it is also possible that only compounds that inhibit HCV helicase action on RNA will be effective antivirals. To improve S/B ratios with RNA, we have recently developed a split beacon assay where the fluorophore and quenching moieties are present on separate hairpin-forming oligonucleotides. Unlike this MBHA, this split beacon helicase assay (SBHA) performs similarly when HCV helicase acts on DNA or RNA (Section 7), with Z' factors in ranges appropriate for HTS (Zhang et al., 1999).

5. Expression and Purification of NS3h

Purified full-length NS3 is typically more active than NS3h, and these differences are most apparent in assays where long stretches of DNA or RNA must be unwound to detect activity. However, full-length NS3 is also notably less stable in solution, losing activity after only a few freeze–thaw cycles and sometimes even during prolonged storage at −80 °C. In contrast, most NS3h proteins we have tested retain activity longer at room temperature or after repeated freeze–thaw cycles. NS3h proteins with fusion tags at the N-terminus, such that the fusion partner replaces the protease domain, generally behave more like full-length NS3 in that they are initially more active (Frick et al., 2004), but they also lose activity more rapidly upon prolonged storage. The assays described here have, therefore, been optimized using an NS3h (isolated from the Con1 strain of HCV genotype 1b; Heck et al., 2008) with a C-terminal His-tag that has been purified using the protocol below. NS3h from other HCV genotypes has also been purified with the below protocol (Lam et al., 2003; Neumann-Haefelin et al., 2008). Our lab's method to purify full-length NS3 has been published elsewhere (Frick et al., 2010).

1. Streak *E. coli* Rosetta (DE3) cells (EMD Biosciences) harboring the plasmid pET24-Hel-Con1 (Heck et al., 2008) on LB-agar containing kanamycin (50 μg/ml) and tetracycline (50 μg/ml) to isolate single colonies. After overnight incubation at 37 °C, inoculate 5 ml of LB broth containing kanamycin (50 μg/ml) and tetracycline (50 μg/ml) with single colonies. Shake vigorously at 37 °C until slightly turbid. Transfer culture to 1 l of LB containing the same antibiotics. Shake vigorously at 37 °C and periodically monitor OD_{600}. When OD_{600} is approximately 1.0, add isopropyl β-D-1-thiogalactopyranoside to a final concentration of 1 mM. Incubate 2–3 h at room temperature and harvest cells by centrifugation, wash cells with phosphate buffered saline, and store pellet at −80 °C.

   Perform all subsequent steps at 4 °C or with all tubes on ice.
2. Suspend frozen cells in 10 ml 20 mM Tris, pH 8, 0.5 M NaCl, and 5 mM Imidazole (buffer A). Lyse cells using French press or Sonifier Cell disrupter (Branson). Centrifuge at 10,000 g, discard pellet, and filter the supernatant through a 0.8-μm glass fiber filter (Fraction I).
3. Load Fraction I onto a 5-ml Ni-NTA column (GE Healthcare) equilibrated with buffer A. Wash with buffer A containing 40 mM Imidazole. Elute with buffer A containing an imidazole gradient from 40 to 500 mM Imidazole. NS3h should elute when the imidazole approaches 100 mM. Analyze fractions using 10% SDS-PAGE to identify fractions containing 53 kDa NS3h protein. Combine fractions containing NS3h (Fraction II).

4. Precipitate NS3h from fraction II by slowly adding solid $(NH_4)_2SO_4$ to 60% saturation (0.361 g/ml). Centrifuge at 12,000 g for 20 min. Discard supernatant. Dissolve pellet in 2 ml storage buffer (20 mM Tris, pH 8, 50 mM NaCl, 1 mM EDTA, 0.1 mM DTT, 25% glycerol) (Fraction III).
5. Load Fraction III onto a 100-ml gel filtration column (Sephacryl S-300 HR, GE Healthcare) that has been previously equilibrated with 20 mM Tris, pH 8, 50 mM NaCl, 1 mM EDTA, and 0.1 mM DTT (GF buffer). Elute protein with GF buffer by collecting 2 ml fractions at 0.1 ml/min. Analyze fractions using a 10% SDS-PAGE. Combine fractions containing NS3h (fraction IV).
6. Load fraction IV on a 1-ml DEAE Sepharose FF column (GE Healthcare) that has been equilibrated with GF buffer. After washing with GF buffer, elute with a GF buffer containing a gradient of NaCl from 0 to 500 mM. NS3h should elute around 150 nM NaCl. Analyze the fractions with a 10% SDS-PAGE, and combine fractions containing NS3h (Fraction V).
7. Dialyze protein with GF buffer (1 l). Protein may be concentrated at this point by sprinkling dialysis tubing with polyethylene glycol (average molecular weight > 20,000) and allowing liquid to absorb at 4 °C. Let desired buffer absorb and return bag to GF buffer. After two changes of GF buffer, dialyze with storage buffer (prepared in step 4).
8. Determine protein concentration from absorbance at 280 nm using an extinction coefficient calculated from the protein sequence (51,890 M^{-1} cm^{-1} for NS3h_1b(Con1)). Store aliquots at -80 °C.

6. The Molecular Beacon-Based Helicase Assay (MBHA)

The MBHA most commonly used in our laboratory employs a substrate designed to mimic a hairpin-forming region at the 3' end of the HCV polyprotein reading frame (Fig. 21.1A). The shorter (top) oligonucleotide in this substrate is modified by attaching a cyanine 5 (Cy5) at the 5' end and attaching an Iowa black RQ (IAbRQ) to its 3' end. The longer DNA oligonucleotide is complementary to the shorter strand and also has a 20-nucleotide long 3' tail. The assay works by monitoring the fluorescence of the DNA probe before and after the addition of ATP, which is needed to fuel helicase movement. As the reaction proceeds, the helicase rearranges the nucleic acids such that the complementary nucleotides near the ends of the hybridization probe bind, forming a hairpin loop structure (Fig. 21.1A). The hairpin loop allows the fluorophore and quencher molecule to come into contact, with the result being a reduction in Cy5 fluorescence. This drop in fluorescence is plotted versus time to determine both the rate and extent of the reaction (Fig. 21.1B). For screening, each MBHA needs to be read only twice, before ATP addition (F_0) and at the completion of the

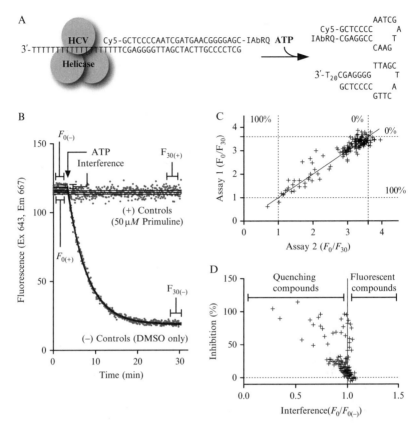

Figure 21.1 The molecular beacon-based helicase assay and its use in HTS. (A) A Cy5-labeled MBHA substrate based on the sequence of the HCV genome. (B) Four MBHAs performed in the absence of test compounds and four performed with a saturating concentration (50 μM) of a known HCV helicase inhibitor (primuline). (C) Reaction extents (F_{30}/F_0) for duplicate reactions performed with the same concentration (20 μM) of 125 different test compounds. (D) Average compound interference and inhibition obtained from the duplicate reactions shown in panel C. All reactions contained in 25 mM MOPS, pH 6.5, 1.25 mM MgCl2, 5 nM MBHA substrate, 2% (v/v) DMSO, 12.5 nM NS3h_1b(con1), and 1 mM ATP.

reaction (F_{30}). The F_0/F_{30} ratio reflects the extent of the reaction. An inhibitor will cause a decrease in the F_0/F_{30} ratio to a limit of one, where, by definition, no reaction takes place.

Another important parameter obtained with an MBHA concerns compound interference. Three classes of compounds tend to interfere with the MBHA. The first interfering class contains fluorescent compounds that absorb and emit light at wavelengths similar to Cy5. The second quenches Cy5 fluorescence by absorbing light at the Cy5 excitation or emission wavelength, and the third class alters substrate fluorescence by binding the MBHA substrate and changing the orientation of Cy5 relative to IAbRQ.

Simply scanning compound absorbance and fluorescence in the absence of the MBHA substrate can identify compounds in the first two classes. Examining the interaction of the compounds and DNA using other assays can identify compounds in the third class. All such compounds would alter both F_0 and F_{30} in such a way that they might be misidentified as helicase inhibitors because they lower the F_0/F_{30} ratio. To identify interfering compounds, the F_0 of each particular compound is compared to the F_0 of DMSO-only controls ($F_{0(-)}$). A compound with an $F_0/F_{0(-)}$ near 1 generally does not interfere, quenching compounds and DNA-binders have $F_0/F_{0(-)} < 1$, and fluorescent compounds have $F_0/F_{0(-)} > 1$.

6.1. MBHA-based high-throughput screens

The following protocol is routinely used in our lab to monitor helicase activity in either low-volume 96-well microplates or standard 384-well microplates. The protocol is optimized for NS3h_con(1b), but it can be used with other NS3 proteins by simply changing the amount of NS3h in each assay, using less of more active helicases and more of less active helicases. We typically use the yellow dye primuline (MP Biochemicals) as a positive control in these assays. Belon and Frick previously reported that a similar dye called thioflavine S is a HCV helicase inhibitor (Belon and Frick, 2010). Thioflavine S is a more heterogeneous mixture of compounds than is primuline, which we have since found to give more consistent results in MBHAs. The protocol below is for 60 μl reactions, but the reaction can be scaled down to conserve reagents. Precision at low volumes is limited only by sensitivity of the micro plate reader used and the precision of available liquid handlers. A protocol for 5 μl reactions suitable for ultra-high-throughput screening (uHTS) performed in 1536-well plates is available on PubChem BioAssay (AID #1800).

1. Anneal substrate by combining oligonucleotide as shown in Fig. 21.1A (Integrated DNA Technologies, Coralville, IA) at 50 μM each in 10 mM Tris–HCl, pH 8.0. Heat to 95 °C, and cool slowly to room temperature. Store concentrated substrate in the dark at −20 °C.
2. Prepare 0.5 M 3-(N-morpholino)propanesulfonic acid (MOPS) adjusting pH to 6.5 with NaOH. Also prepare 25 mM MOPS, pH 6.5, 25 mM MgCl$_2$, 10 mM ATP, 1 mM Primuline (MP Biochemicals) in dimethyl sulfoxide (DMSO), and NS3h dilution buffer (25 mM MOPS, pH 6.5, 1 mM DTT, 0.2% Tween20, 0.1 mg/mL BSA).
3. Dilute the Cy5-MBHA substrate to 100 nM in 25 mM MOPS.
4. Dilute NS3h to 250 nM in NS3h dilution buffer.
5. Assemble enough reaction mixture for the desired number of assays. For each assay, combine 3 μl of 0.5 M MOPS, 3 μl of 25 mM MgCl$_2$, 3 μl of 100 nM Cy5-MBHA substrate, 3 μl of 250 nM NS3h, and 39 μl of nuclease free water.

6. Dispense 51 μl of the reaction mix to each well of a polystyrene white low-volume 96-well microplate (Corning) at 23 °C. Add 3 μl of DMSO or compounds dissolved in DMSO. (*Note*: The assay tolerates up to 35% DMSO, so more or less DMSO can be added depending on compound solubility.)
7. Read Cy5 fluorescence (Ex 643 Em 667) of each well.
8. Add 6 μl of 10 mM ATP, and read fluorescence until values in negative control reactions remain constant (typically 20–30 min) (Fig. 21.1B). Record endpoint (F_{30}). *Note*: After ATP addition each reaction will contain 25 mM MOPS, pH 6.5, 1.25 mM MgCl$_2$, 0.05 mM DTT, 5% DMSO, 0.01% Tween20, 5 μg/mL BSA, 5 nM MBHA substrate, 12.5 nM NS3h, and 1 mM ATP.
9. Calculate F_0/F_{30} ratios for each reaction. Figure 21.1C shows results of duplicate MBHAs that were performed with a library of 150 known or suspected HCV helicase inhibitors and DNA binding compounds that we use to evaluate HCV helicase assays in our lab. The duplicate assays are plotted against each other to evaluate reproducibility. Dotted lines mark average F_0/F_{30} ratios for negative controls (DMSO only) and positive control reactions (50 μM primuline).
10. Calculate percent inhibition by normalizing F_0/F_{30} ratios for each reaction to the ratio obtained with positive (+) and negative (−) controls (Eq. 21.1):

$$\text{Inhibition}(\%) = \frac{(F_0/F_{30}) - (F_{0(-)}/F_{30(-)})}{(F_{0(+)}/F_{30(+)}) - (F_{0(-)}/F_{30(-)})} \times 100. \qquad (21.1)$$

11. Calculate compound interference by calculating the ratio of F_0 for each compound and F_0 for the negative controls ($F_0/F_{0(-)}$). Figure 21.1D shows the average percent inhibition and interference ratio values of the duplicate assays performed with our helicase inhibitor library, where all library compounds were tested at 10 μM.

6.2. Examining effects of inhibitors on the kinetics of DNA strand separation

Once a compound is identified to influence the MBHA, the next step we take is to examine the apparent affinity of the compound for the MBHA substrate and the apparent affinity of the compound for the unwinding complex. Affinity of the compound for the substrate is estimated by first examining the effect of compound concentration on substrate fluorescence before ATP is added (if there is an effect, direct DNA binding is measured with another assay). The apparent affinity of the compound for the unwinding complex is estimated from effect of the compound on the initial rates of unwinding reactions. Both parameters can be estimated from time courses obtained in the presence of various compound concentrations, by first calculating the average fluorescence of the substrate during the time before and after ATP is added, and then fitting data obtained after ATP addition to a rate equation. There are many commercially available software packages, including some supplied with more sophisticated plate readers, which will fit kinetic readouts directly to first order rate equations. However, the protocol below uses the program Graphpad Prism (La Jolla, CA) software because it can be used with data exported from any reader. As demonstrated with primuline (Fig. 21.2), the protocol below fits time course data to calculate F_0 and initial velocities (Fig. 21.2A), export the values, and fit them directly to a dose response equation (Fig. 21.2B).

1. Add 2 mM solutions of each inhibitor to the first tube in an eight-tube strip and serially dilute the compound 1:2 into DMSO in remaining tubes. This creates an eight-step 1:2 dilution series beginning at 2 mM. To a second eight-tube strip, add DMSO to four tubes and a positive control inhibitor (e.g., 1 mM primuline in DMSO) to the other four tubes.
2. Assemble MBHA reactions as described above, except add diluted compounds and positive/negative control series instead of the chemical library.
3. Perform the assay using a microplate reader equipped with a reagent dispenser capable of precisely adding the 6 μl of ATP to each well needed to start the reaction. Collect fluorescence data as rapidly as the plate reader allows for ~2 min before ATP injection, and then until fluorescence no longer changes in the negative control reactions.
4. Export time course data to Prism, and fit fluorescence time courses to the following Prism nonlinear regression equation series that calculates F_0

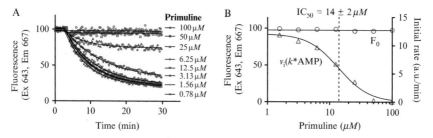

Figure 21.2 Effect of various primuline concentrations on a HCV helicase-catalyzed MBHA. (A) MBHAs were performed with the indicated concentration of the known helicase inhibitor primuline. Solid lines show nonlinear regression fits described in text. (B) Secondary plot of initial velocities (triangles) and F_0s (circles) fit to a dose response equation as described in text. Dotted line indicates the IC_{50} value determined by nonlinear regression.

and uses the calculated F_0 as a starting fluorescence that decays according to a first order rate constant (Fig. 21.2A).

$$Y_1 = F_0$$
$$Y_2 = \mathrm{AMP}(\exp(-K \times (X - \mathrm{START}))) + (F_0 - \mathrm{AMP})$$
$$Y = \mathrm{IF}[(X < \mathrm{START}), Y_1, Y_2]$$

In the above equations, X is time, Y is observed fluorescence, Y_1 is the data before ATP injection, Y_2 is data collected after ATP injection, START is the time that ATP was added, AMP is reaction amplitude, and K is a first order rate constant describing the loss of fluorescence. Constrain "START" to the time ATP was added, and AMP, F_0, and K so they remain greater than zero. Set initial estimates for F_0, AMP, and K to Y_{max}, Y_{min}, and the 1/value of X at Y_{mid}, respectively. When entering this equation, set Prism to calculate initial velocities for each reaction by defining a "V" under "transforms to report" as $K \times \mathrm{AMP}$. While setting parameters for nonlinear regression, set Prism to output F_0, V, and AMP to a summary table and plot each on an XY graph. *Notes*: if a program other than Prism is used to fit reaction time courses, calculate initial rates for each reaction by multiplying observed first order rate constant by the reaction amplitudes. If Prism or another nonlinear regression software is not available, estimates of initial velocities can be obtained from the slopes of the initial linear phase of each reaction time course.

5. Fit plots of initial velocities versus compound concentration to a dose response equation such as the Prism equation below:

$$Y = V_0 - \left(\frac{V_0 X^h}{\mathrm{IC}_{50}^h + X^h} \right),$$

where X is compound concentration, Y is velocity in the presence of inhibitor, V_0 is velocity in the absence of inhibitor, h is the Hill coefficient, and IC_{50} is the concentration of compound that inhibits the unwinding reaction by 50%. *Note*: several other dose response equations are available in Prism, but data must first be transformed to the LOG of inhibitor concentration.

6. Examine plots of F_0's versus compound concentration. A negative slope to the F_0 graph indicates that the compounds quench the MBHA substrate fluorescence. Data obtained with compounds that interact with DNA, or quench by inner filter effects, will sometimes also fit the equation in step 5 and yield IC_{50} values that mirror those obtained if velocities are plotted. Such compounds likely exert their effect by binding DNA rather than the helicase itself. Compounds that do not affect F_0 in a dose response manner are assumed not to significantly interfere with the assay (Fig. 21.2B); however, this does not mean that

they do not interact with the DNA substrate. To identify all potential DNA binding compounds, we subject all hits to a counterscreen that monitors DNA binding potential. One such assay monitors a compound's ability to displace a fluorescent DNA intercalator like ethidium bromide (Boger and Tse, 2001; Boger et al., 2001).

6.3. Evaluating alternate MBHAs for HTS

The brief protocol below is designed to evaluate the quality of an MBHA or related assay. When analyzing hits from a helicase HTS, we typically examine a compound's behavior in helicase assays that use a variety of DNA sequences and fluorophores because it is possible that hits might act only by interacting with certain DNA sequences or the fluorophore used to monitor the reaction (Belon et al., 2010). One alternate MBHA substrate we use is shown in Fig. 21.3A, which differs from the one in Fig. 21.1 in that the bottom strand does not form a hairpin when separated from the top strand. Figure 21.3A–D show results obtained when the sequence is labeled with fluorophores with different chemistry. Although primuline inhibits the observed reactions with all four beacons, S/B varies among the assays. All but one assay has a Z' factor in the range appropriate for HTS.

1. Assemble an eight-tube strip containing four tubes of DMSO and four tubes of a known helicase inhibitor (e.g., 1 mM primuline).
2. Assemble reaction mixtures as described in Section 6.1 and dispense 51 μl into all wells of a low-volume white 96-well plate. Add 3 μl of the controls from the eight-tube strip to each column.
3. Monitor DNA fluorescence at appropriate wavelength (F_0); dispense 6 μl of 10 mM ATP to each well to start the reactions. Continue to monitor fluorescence of each well until negative control reaction fluorescence no longer changes. Record F_{30} for each well.
4. Calculate F_0/F_{30} ratios. Plot data versus well number and inspect. A slope to the negative control data will result if all the reactions have not gone to completion.
5. Average the ratios for the positive control reactions (M_+) and negative control reaction (M_-), and the standard deviation for positive controls (SD_+) and negative controls (SD_-).
6. Calculate a Z' factor (Zhang et al., 1999) from the means and standard deviations (Eq. 21.2).

$$Z' = 1 - \left(\frac{3SD_+ + 3SD_-}{M_+ - M_-}\right). \tag{21.2}$$

A Z' value of 1.0 indicates an ideal assay, excellent assays yield values between 0.5 and 1.0, a value between 0.5 and 0 indicates a marginal assay

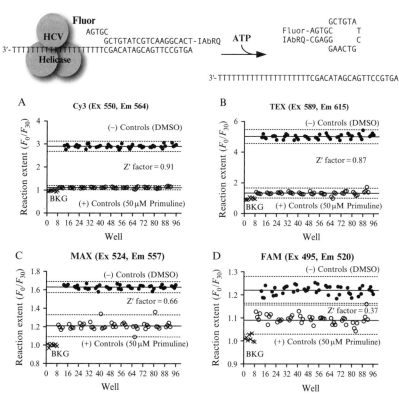

Figure 21.3 Evaluation of alternate fluorophores for use in MBHA-based screens. Each substrate was labeled with Iowa Black RQ (IAbRQ) and one of the four indicated fluorophores available from Integrated DNA technologies (Coralville, IA). Each panel shows results from a single 96-well low-volume plate containing 8 reactions without DNA (BKG), 44 reactions with 2% DMSO, and 44 reactions containing the HCV helicase inhibitor primuline at a final concentration of 50 μM in 2% DMSO. Solid lines show means of positive and negative controls and the dotted lines show three times the standard deviations.

needing improvement, and assays with Z' factors below 0 are generally not useful for HTS.

7. An RNA-Based Split Beacon Helicase Assay (SBHA)

We have performed many of the above assays using substrates where RNA forms one or both strands of the MBHA substrate. While such assays provide valuable results when performed carefully under analytical conditions, their usefulness in HTS is limited. Specifically, the F_{30}/F_0 ratios obtained from negative control reactions (no inhibitors) with RNA substrates is usually two to four times lower than it is for the same negative control reactions performed with DNA substrates. As a result, difference is smaller between the positive and negative controls in RNA assays and Z' factors are usually only in the marginal range at best.

To monitor HCV helicase action on RNA in HTS, we instead use an RNA-based assay where the fluorophore and quencher are split between two different hairpin-forming oligonucleotides that both anneal to a third strand at adjacent positions. In this SBHA, the two oligonucleotides that the helicase must separate for a signal change are made of RNA, while the third oligonucleotide containing the quenching moiety is made of DNA (Fig. 21.4A). In SBHAs, F_0 and F_{30} signals are similar when either DNA or RNA is used as a substrate (Fig. 21.4B), but more enzyme is needed with RNA because the same amount of NS3h unwinds the RNA substrate more slowly. Similar rates are obtained if five times more NS3h (i.e., 62.5 nM) is added to the reactions

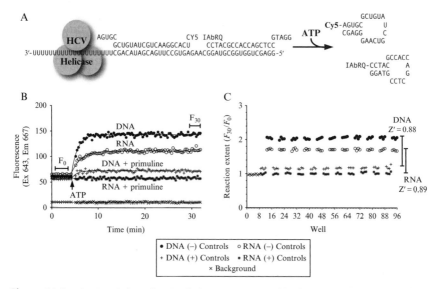

Figure 21.4 An RNA-based HCV helicase assay suitable for HTS. (A) Design of the split beacon helicase assay (SBHA). In the RNA substrate, the bottom strand and the Cy5-labeled strand are RNA, while the IAbRQ-labeled strand is composed of DNA. In the DNA-labeled substrate, all three strands are composed of DNA, with Ts replacing Us in the bottom strand and the Cy5-labeled strand. (B) Time courses for sample negative control reactions using the DNA (closed circles) and the RNA (open circles) substrate, positive control reactions containing DNA plus 50 μM primuline (+), and the RNA substrate with 50 μM primuline (★). Lines show fit to the equations described in the text. The fluorescence observed in a well without substrate (x) is shown for comparison. (C) Reaction extent for two 96-well plates, one of which contained an RNA substrate and one a DNA substrate.

with RNA SBHA substrate than is added to DNA SBHA substrates (i.e., 5 nM NS3h). Unlike the MBHA, fluorescence increases when the helicase unwinds an SBHA substrate, and reaction extent is therefore calculated as an F_{30}/F_0 ratio (rather than an F_0/F_{30} ratio). Reaction extents and Z' factors are similar for RNA- and DNA-based SBHA (Fig. 21.4C). The following protocol can be used to perform RNA-based SBHAs.

1. Anneal substrate by combining three oligonucleotides shown in Fig. 21.1A (Integrated DNA Technologies, Coralville, IA) at 50 µM each in RNAse-free 10 mM Tris–HCl, pH 8.5. Heat to 95 °C. Cool slowly to room temperature. Store concentrated substrate in the dark at −20 °C.
2. Assemble reaction mixtures, controls, and inhibitors as described above for the MBHA (Section 6.1), except that NS3h_1b(Con1) should be included at 60 nM in each reaction. Add 51 µl of reaction mix and 3 µl of either control compounds or dilutions of compounds to be tested.
3. Monitor Cy5 fluorescence before ATP addition. Add 6 µl of ATP. Record final fluorescence (F_{30}) after fluorescence in control reactions no longer changes. Unlike the MBHA, fluorescence will increase upon ATP addition. Calculate F_{30}/F_0 ratios and normalize data with ratios obtained with positive and negative control reactions to determine percent inhibition.
4. For kinetic analyses, exported data can be analyzed as discussed in Section 6.2, except that the data obtained after ATP addition should be fit a rate equation describing substrate appearance rather than decay. The following set of equations can be used to fit SBHA data using Prism.

$$Y_1 = F_0$$
$$Y_2 = F_0 + \text{AMP}(1 - \exp(-K(X - \text{START})))$$
$$Y = \text{IF}[(X < \text{START}), Y_1, Y_2]$$

In the above equations, X is time, Y is observed fluorescence, Y_1 is the data before ATP injection, Y_2 is data collected after ATP injection, START is the time that ATP was added, AMP is reaction amplitude, and K is a first order rate constant describing the gain of fluorescence. Constrain "START" to the time ATP was added, and AMP, F_0, and K so they remain greater than zero. Set initial estimates for F_0, AMP, and K to $Y_{\min}$, $Y_{\max}$, and the 1/the value of X at Y_{mid}, respectively.

8. Discussion

The above protocols are being used to analyze HCV helicase inhibitors reported by others (Belon *et al.*, 2010), to screen compound libraries for HCV helicase inhibitors, and to analyze hits (Belon and Frick, 2010). These assays, however, are only the first step in the process of finding a compound that could be used as a molecular probe or as a lead compound for drug discovery. After identifying compounds that inhibit HCV helicase activity on both DNA and RNA, we next examine whether or not they can inhibit HCV replication in cells. The assay most useful in this regard employs an HCV replicon in which a reporter gene is linked to a marker that can be used to select cells in which HCV RNA replicates (Hao *et al.*, 2007). Although we have studied helicase inhibitors that are not active against the HCV replicon (Belon *et al.*, 2010), our lab now primarily focuses on inhibitors that display some antiviral potency in HCV replicon-based assay. These compounds are now being studied using a variety of techniques to elucidate how they inhibit the unwinding reaction on a molecular level and how they effect HCV replication in cells. It should be appreciated that many other assays are available to identify inhibitors of HCV helicase, which have been the subject of other recent reviews (Belon and Frick, 2009a, 2011). There are also other resources available that summarize compounds that have been discovered to inhibit HCV helicase (Belon and Frick, 2009a, 2011; Borowski *et al.*, 2000; Briguglio *et al.*, 2011; Lemon *et al.*, 2010) and compounds that target other helicases (Frick, 2003; Frick and Lam, 2006; Kwong *et al.*, 2005).

ACKNOWLEDGMENTS

This work was supported by NIH grant RO1 AI088001 and a Research Growth Initiative Award (101X219) from the UWM Research Foundation. We would also like to thank all our collaborators at the MLPCN for their help with this project, particularly Dmitriy Minond of the Scripps Research Institute for helpful advice with assay design and Kelin Li, Kevin J. Frankowski, and Ben Neuenswander of the University of Kansas for providing the compounds used in our helicase inhibitor library.

REFERENCES

Aggarwal, M., Sommers, J. A., Shoemaker, R. H., and Brosh, R. M. J. (2011). Inhibition of helicase activity by a small molecule impairs Werner syndrome helicase (WRN) function in the cellular response to DNA damage or replication stress. *Proc. Natl. Acad. Sci. USA* **108,** 1525–1530.

Bacon, B. R., Gordon, S. C., Lawitz, E., Marcellin, P., Vierling, J. M., Zeuzem, S., Poordad, F., Goodman, Z. D., Sings, H. L., Boparai, N., Burroughs, M., Brass, C. A., *et al.* (2011). Boceprevir for previously treated chronic HCV genotype 1 infection. *N. Engl. J. Med.* **364,** 1207–1217.

Belon, C. A., and Frick, D. N. (2008). Monitoring helicase activity with molecular beacons. *Biotechniques* **45**(433–40), 442.

Belon, C. A., and Frick, D. N. (2009a). Helicase inhibitors as specifically targeted antiviral therapy for hepatitis C. *Future Virol.* **4,** 277–293.

Belon, C. A., and Frick, D. N. (2009b). Fuel specificity of the hepatitis C virus NS3 helicase. *J. Mol. Biol.* **388,** 851–864.

Belon, C., and Frick, D. N. (2010). Thioflavin S inhibits hepatitis C virus RNA replication and the viral helicase with a novel mechanism. *FASEB J.* **24,** lb202.

Belon, C. A., and Frick, D. N. (2011). NS3 helicase inhibitors. In "Hepatitis C: Antiviral Drug Discovery and Development," (Y. He and S. L. Tan, eds.), pp. 327–356. Caister Academic Press, Norfolk, UK.

Belon, C. A., High, Y. D., Lin, T. I., Pauwels, F., and Frick, D. N. (2010). Mechanism and specificity of a symmetrical benzimidazolephenylcarboxamide helicase inhibitor. *Biochemistry* **49,** 1822–1832.

Beran, R. K., Serebrov, V., and Pyle, A. M. (2007). The serine protease domain of hepatitis C viral NS3 activates RNA helicase activity by promoting the binding of RNA substrate. *J. Biol. Chem.* **282,** 34913–34920.

Beran, R. K., Lindenbach, B. D., and Pyle, A. M. (2009). The NS4A protein of hepatitis C virus promotes RNA-coupled ATP hydrolysis by the NS3 helicase. *J. Virol.* **83,** 3268–3275.

Bjornson, K. P., Amaratunga, M., Moore, K. J., and Lohman, T. M. (1994). Single-turnover kinetics of helicase-catalyzed DNA unwinding monitored continuously by fluorescence energy transfer. *Biochemistry* **33,** 14306–14316.

Boger, D. L., and Tse, W. C. (2001). Thiazole orange as the fluorescent intercalator in a high resolution fid assay for determining DNA binding affinity and sequence selectivity of small molecules. *Bioorg. Med. Chem.* **9,** 2511–2518.

Boger, D. L., Fink, B. E., Brunette, S. R., Tse, W. C., and Hedrick, M. P. (2001). A simple, high-resolution method for establishing DNA binding affinity and sequence selectivity. *J. Am. Chem. Soc.* **123,** 5878–5891.

Boguszewska-Chachulska, A. M., Krawczyk, M., Stankiewicz, A., Gozdek, A., Haenni, A. L., and Strokovskaya, L. (2004). Direct fluorometric measurement of hepatitis C virus helicase activity. *FEBS Lett.* **567,** 253–258.

Borowski, P., Mueller, O., Niebuhr, A., Kalitzky, M., Hwang, L. H., Schmitz, H., Siwecka, M. A., and Kulikowsk, T. (2000). ATP-binding domain of NTPase/helicase as a target for hepatitis C antiviral therapy. *Acta Biochim. Pol.* **47,** 173–180.

Briguglio, I., Piras, S., Corona, P., and Carta, A. (2011). Inhibition of RNA Helicases of ssRNA+ Virus Belonging to Flaviviridae, Coronaviridae and Picornaviridae Families. *Int. J. Med. Chem.* 2011. Article ID 213135, 22 pages.

Choo, Q. L., Kuo, G., Weiner, A. J., Overby, L. R., Bradley, D. W., and Houghton, M. (1989). Isolation of a cDNA clone derived from a blood-borne non-A, non-B viral hepatitis genome. *Science* **244,** 359–362.

Crute, J. J., Grygon, C. A., Hargrave, K. D., Simoneau, B., Faucher, A. M., Bolger, G., Kibler, P., Liuzzi, M., and Cordingley, M. G. (2002). Herpes simplex virus helicase-primase inhibitors are active in animal models of human disease. *Nat. Med.* **8,** 386–391.

Edlin, B. R. (2011). Perspective: Test and treat this silent killer. *Nature* **474,** S18–S19.

Frick, D. N. (2003). Helicases as antiviral drug targets. *Drug News Perspect.* **16,** 355–362.

Frick, D. N. (2006). HCV helicase: Structure, function, and inhibition. In "Hepatitis C Viruses: Genomes and Molecular Biology," (S. L. Tan, ed.).Horizon Bioscience, Norfolk (UK) Chapter 7.

Frick, D. N. (2007). The hepatitis C virus NS3 protein: A model RNA helicase and potential drug target. *Curr. Issues Mol. Biol.* **9,** 1–20.

Frick, D. N., and Lam, A. M. (2006). Understanding helicases as a means of virus control. *Curr. Pharm. Des.* **12,** 1315–1338.

Frick, D. N., Rypma, R. S., Lam, A. M., and Gu, B. (2004). The nonstructural protein 3 protease/helicase requires an intact protease domain to unwind duplex RNA efficiently. *J. Biol. Chem.* **279,** 1269–1280.

Frick, D. N., Banik, S., and Rypma, R. S. (2007). Role of divalent metal cations in ATP hydrolysis catalyzed by the hepatitis C virus NS3 helicase: Magnesium provides a bridge for ATP to fuel unwinding. *J. Mol. Biol.* **365,** 1017–1032.

Frick, D. N., Ginzburg, O., and Lam, A. M. (2010). A method to simultaneously monitor hepatitis C virus NS3 helicase and protease activities. *Methods Mol. Biol.* **587,** 223–233.

Hao, W., Herlihy, K. J., Zhang, N. J., Fuhrman, S. A., Doan, C., Patick, A. K., and Duggal, R. (2007). Development of a novel dicistronic reporter-selectable hepatitis C virus replicon suitable for high-throughput inhibitor screening. *Antimicrob. Agents Chemother.* **51,** 95–102.

Heck, J. A., Lam, A. M., Narayanan, N., and Frick, D. N. (2008). Effects of mutagenic and chain-terminating nucleotide analogs on enzymes isolated from hepatitis C virus strains of various genotypes. *Antimicrob. Agents Chemother.* **52,** 1901–1911.

Hiraga, N., Imamura, M., Abe, H., Nelson Hayes, C., Kono, T., Onishi, M., Tsuge, M., Takahashi, S., Ochi, H., Iwao, E., Kamiya, N., Yamada, I., et al. (2011). Rapid emergence of telaprevir resistant hepatitis C virus strain from wild type clone *in vivo*. *Hepatology* **54,** 781–788.

Houston, P., and Kodadek, T. (1994). Spectrophotometric assay for enzyme-mediated unwinding of double-stranded DNA. *Proc. Natl. Acad. Sci. USA* **91,** 5471–5474.

Katsumata, K., Chono, K., Sudo, K., Shimizu, Y., Kontani, T., and Suzuki, H. (2011). Effect of ASP2151, a herpesvirus helicase-primase inhibitor, in a guinea pig model of genital herpes. *Molecules* **16,** 7210–7223.

Kim, D. W., Gwack, Y., Han, J. H., and Choe, J. (1995). C-terminal domain of the hepatitis C virus NS3 protein contains an RNA helicase activity. *Biochem. Biophys. Res. Commun.* **215,** 160–166.

Kleymann, G., Fischer, R., Betz, U. A., Hendrix, M., Bender, W., Schneider, U., Handke, G., Eckenberg, P., Hewlett, G., Pevzner, V., Baumeister, J., Weber, O., et al. (2002). New helicase-primase inhibitors as drug candidates for the treatment of herpes simplex disease. *Nat. Med.* **8,** 392–398.

Kolykhalov, A. A., Mihalik, K., Feinstone, S. M., and Rice, C. M. (2000). Hepatitis C virus-encoded enzymatic activities and conserved RNA elements in the 3' nontranslated region are essential for virus replication *in vivo*. *J. Virol.* **74,** 2046–2051.

Kwong, A. D., Rao, B. G., and Jeang, K. T. (2005). Viral and cellular RNA helicases as antiviral targets. *Nat. Rev. Drug Discov.* **4,** 845–853.

Kyono, K., Miyashiro, M., and Taguchi, I. (1998). Detection of hepatitis C virus helicase activity using the scintillation proximity assay system. *Anal. Biochem.* **257,** 120–126.

Lam, A. M., and Frick, D. N. (2006). Hepatitis C virus subgenomic replicon requires an active NS3 RNA helicase. *J. Virol.* **80,** 404–411.

Lam, A. M., Keeney, D., Eckert, P. Q., and Frick, D. N. (2003). Hepatitis C virus NS3 ATPases/helicases from different genotypes exhibit variations in enzymatic properties. *J. Virol.* **77,** 3950–3961.

Lemon, S. M., McKeating, J. A., Pietschmann, T., Frick, D. N., Glenn, J. S., Tellinghuisen, T. L., Symons, J., and Furman, P. A. (2010). Development of novel therapies for hepatitis C. *Antiviral Res.* **86,** 79–92.

Lohmann, V., Korner, F., Koch, J., Herian, U., Theilmann, L., and Bartenschlager, R. (1999). Replication of subgenomic hepatitis C virus RNAs in a hepatoma cell line. *Science* **285,** 110–113.

Muramatsu, S., Ishido, S., Fujita, T., Itoh, M., and Hotta, H. (1997). Nuclear localization of the NS3 protein of hepatitis C virus and factors affecting the localization. *J. Virol.* **71,** 4954–4961.

Murray, C. L., and Rice, C. M. (2011). Turning hepatitis C virus into a real virus. *Annu. Rev. Microbiol.* **65,** 307–327.

Neumann-Haefelin, C., Frick, D. N., Wang, J. J., Pybus, O. G., Salloum, S., Narula, G. S., Eckart, A., Biezynski, A., Eiermann, T., Klenerman, P., Viazov, S., Roggendorf, M., et al. (2008). Analysis of the evolutionary forces in an immunodominant CD8 epitope in hepatitis C virus at a population level. *J. Virol.* **82,** 3438–3451.

Porter, D. J., Short, S. A., Hanlon, M. H., Preugschat, F., Wilson, J. E., Willard, D. H. J., and Consler, T. G. (1998). Product release is the major contributor to kcat for the hepatitis C virus helicase-catalyzed strand separation of short duplex DNA. *J. Biol. Chem.* **273,** 18906–18914.

Preugschat, F., Averett, D. R., Clarke, B. E., and Porter, D. J. (1996). A steady-state and pre-steady-state kinetic analysis of the NTPase activity associated with the hepatitis C virus NS3 helicase domain. *J. Biol. Chem.* **271,** 24449–24457.

Pyle, A. M. (2008). Translocation and unwinding mechanisms of RNA and DNA helicases. *Annu. Rev. Biophys.* **37,** 317–336.

Raney, K. D., Sharma, S. D., Moustafa, I. M., and Cameron, C. E. (2010). Hepatitis C virus non-structural protein 3 (HCV NS3): A multifunctional antiviral target. *J. Biol. Chem.* **285,** 22725–22731.

Rypma, R. S., Lam, A. M., and Frick, D. N. (2009). Effect of substrate traps on hepatitis C virus NS3 helicase catalyzed DNA unwinding: Evidence for enzyme catalyzed strand exchange. *In* "Bacterial DNA, DNA Polymerase and DNA Helicases," (W. D. Knudsen and S. S. Bruns, eds.), pp. 389–407. Nova Science Publishers, Inc., New York.

Suzich, J. A., Tamura, J. K., Palmer-Hill, F., Warrener, P., Grakoui, A., Rice, C. M., Feinstone, S. M., and Collett, M. S. (1993). Hepatitis C virus NS3 protein polynucleotide-stimulated nucleoside triphosphatase and comparison with the related pestivirus and flavivirus enzymes. *J. Virol.* **67,** 6152–6158.

Tani, H., Akimitsu, N., Fujita, O., Matsuda, Y., Miyata, R., Tsuneda, S., Igarashi, M., Sekiguchi, Y., and Noda, N. (2009). High-throughput screening assay of hepatitis C virus helicase inhibitors using fluorescence-quenching phenomenon. *Biochem. Biophys. Res. Commun.* **379,** 1054–1059.

Tuteja, R. (2007). Helicases—Feasible antimalarial drug target for Plasmodium falciparum. *FEBS J.* **274,** 4699–4704.

Tyagi, S., and Kramer, F. R. (1996). Molecular beacons: Probes that fluoresce upon hybridization. *Nat. Biotechnol.* **14,** 303–308.

Yao, N., Hesson, T., Cable, M., Hong, Z., Kwong, A. D., Le, H. V., and Weber, P. C. (1997). Structure of the hepatitis C virus RNA helicase domain. *Nat. Struct. Biol.* **4,** 463–467.

Yedavalli, V. S., Zhang, N., Cai, H., Zhang, P., Starost, M. F., Hosmane, R. S., and Jeang, K. T. (2008). Ring expanded nucleoside analogues inhibit RNA helicase and intracellular human immunodeficiency virus type 1 replication. *J. Med. Chem.* **51,** 5043–5051.

Zeuzem, S., Andreone, P., Pol, S., Lawitz, E., Diago, M., Roberts, S., Focaccia, R., Younossi, Z., Foster, G. R., Horban, A., Ferenci, P., Nevens, F., *et al.* (2011). Telaprevir for retreatment of HCV infection. *N. Engl. J. Med.* **364,** 2417–2428.

Zhang, J. H., Chung, T. D., and Oldenburg, K. R. (1999). A simple statistical parameter for use in evaluation and validation of high throughput screening assays. *J. Biomol. Screen.* **4,** 67–73.

Author Index

Note: Page numbers followed by "*f*" indicate figures, and "*t*" indicate tables.

A

Abaan, O. D., 340
Abaeva, I. S., 440–441
Abdelhaleem, M., 30, 441
Abe, H., 465–466
Abe, K., 407*t*
Abelson, J. N., 76–78, 94–95, 292*t*, 295
Abrahams, J. P., 177–178
Abraham, Y., 444
Abramoff, M. D., 335
Abramson, R. D., 215*t*, 438–439
Abudu, A., 407*t*
Acheampong, E., 407*t*
Achsel, T., 329–330, 330*t*, 337–338
Adair, R., 407*t*
Adam, H., 86
Adams, D. J., 441
Adams, P. D., 186–187
Adelman, J. L., 174–175
Adolph, K. W., 387
Aebi, U., 240–241
Afonine, P. V., 186–187
Afriastini, J. J., 443–444
Agarwal, S., 291–294
Aggarwal, M., 464–465
Agostinis, P., 447
Agrawal, S., 8
Ahamed, E., 348–349, 363–365
Aigueperse, C., 326*t*, 327, 329–331, 330*t*, 334, 337–338
Aitchison, J. D., 312–313
Akao, Y., 30, 407*t*
Akimitsu, N., 467–468
Akimoto, C., 363–365
Akinaga, S., 447
Akira, S., 140–141, 407*t*
Akman, S. A., 325–327
Alain, T., 440–441
Alamares, J. G., 441
Alcaraz, L. D., 370
Alcazar-Roman, A. R., 214–217, 215*t*, 218, 241–242
Aldea, M., 304
Alexandrov, A., 214–217
Alexeyev, M. F., 392
Alexis, J. D., 30
Ali, S., 348–349, 363–365
Allen, M. M., 387–388

Allison, T. J., 173–174
Aloy, P., 311–312
Altmann, K. H., 164–165
Altmann, M., 80, 99, 100–102, 101*f*, 208*f*, 214–217, 439, 441, 442–443, 445
Altman, R. B., 126
Amaratunga, M., 467–468
Amberg, D. C., 326*t*, 327–328
Ambrus-Aikelin, G., 407*t*
Amrani, N., 256
Anantharaman, V., 15
Andersen, C. B. F., 76, 78, 95, 99–100, 201
Andersen, G. R., 34–35, 41, 76, 78, 95, 99–100, 191–212
Andersen, J. B., 378
Andersen, J. S., 291–294
Anderson, B. J., 4–5, 9–10, 12
Anderson, J. S., 291–294
Anderson, J. T., 2
Anderson, K., 337
Anderson, P., 324–325, 330–331, 334–335
Anderson, R., 135–137, 407*t*
Anderson, V. E., 126
Andersson, S., 10
Andre, B., 337
Andrecka, J., 98–99
Andre, E., 311–312
Andrei, M. A., 329–330, 330*t*, 337–338
Andrenyak, D. M., 67
Andreone, P., 465–466
Andreou, A. Z., 75–109
Andresen, K., 126
Andreu, J. M., 198–199
Andrews, C. L., 173–174
Andrews, S. E., 441
Angus, A. G., 407*t*
Antao, V. P., 36
Anthonycahill, S., 84–85
Antonik, M., 91, 93
Apostolou, V., 388–389
Appel, B., 98–99
Appleby, T. C., 135–137
Arabshahi, A., 152
Arani, R. B., 429–430
Arao, Y., 348–349
Aratani, S., 407*t*
Aravind, L., 15, 172
Aregger, R., 83–84, 86, 96, 97–98, 97*f*
Arenas, C. D., 265

485

Arenas, J. E., 295
Ares, M. Jr., 292t, 294, 295, 296–297
Arisaka, Y., 407t
Ariumi, Y., 407t
Aronoff, R., 256
Aronova, A., 215t, 217–218, 219
Arrowsmith, C. H., 185–186
Arslan, S., 9–10, 12–13
Artarini, A., 441
Arunakumara, K. K., 391
Asai, K., 102–103
Asayama, M., 397
Ascano, M. Jr., 276–277, 286, 338
Ashburner, M., 370
Ashe, M. P., 324–325, 331
Assefa, Z., 447
Astromoff, A., 337
Atoom, A. M., 134
Attri, A. K., 36
Atz, J., 348–349
Auerbach, D., 219
Averett, D. R., 466
Awrey, D. E., 311–312
Ayodele, M., 324–325, 330–331
Azzalin, C. M., 325, 338

B

Babajko, S., 325
Baba, M., 407t
Babcock, H. P., 126
Babu, M., 311–312
Bacher, M., 443–444
Bachi, A., 241
Bacon, B. R., 465–466
Badeaux, F., 326t, 328
Baguet, A., 214–217
Bahr, G. M., 407t
Baird, C. L., 67
Baker, A. R., 439
Baker, L., 349
Balasubramanian, S., 325–327
Ballut, L., 76, 78, 95, 99–100, 138, 201, 214–217, 256, 260, 260f, 261f, 264, 267f
Bangham, R., 337
Banik, S., 467–468
Bannwarth, W., 98–99
Banroques, J., 30–31, 241, 307–308, 369–370, 408–409
Bantscheff, M., 444
Bao, M., 407t
Baran, M., 348–349, 407t
Barbee, S., 325
Barberato, C., 198–199, 201, 202b, 203b
Barber-Rotenberg, J. S., 340
Barr, E. W., 174–175
Barshop, B. A., 56
Barta, A., 30, 112–113

Barten, R., 391
Bartenschlager, R., 466–467
Barthelmebs, L., 371
Bartley, L. E., 126
Baserga, R., 312–313
Baserga, S. J., 214–217, 219, 289–322
Basler, C. F., 407t
Basquin, C., 76–78, 138, 214–218, 256, 264, 265f, 269f, 270
Bassler, J., 290–291, 296–297, 304, 309
Bastuck, S., 311–312, 444
Basu, S., 325–327
Bates, G. J., 348–349, 363–365
Bauch, A., 312–313, 348–349
Baudin, A., 305–306, 310
Bauer, A., 311–312
Bauer, R., 196b
Baumann, U., 80, 99, 100–102, 101f, 208f, 214–217, 439
Baumeister, J., 464–465
Bear, D. G., 173–174
Beasley, S., 185–186
Beattie, B. K., 309
Becker, D., 441
Bedu, S., 397
Beermann, S., 407t
Beggs, J. D., 214–217, 215t, 220, 286, 292t, 294
Behlen, L. S., 165
Behlke, M. A., 9–10, 12
Behrens, M. A., 191–212
Behrens, S. E., 407t, 409
Beisenherz, G., 244
Belfiore, A., 340, 444
Bell, A. W., 326t, 328
Belli, G., 304
Bellsolell, L., 208f
Belon, C. A., 466–467, 468, 472–473, 476–477, 480
Belsham, G. J., 340, 439, 441–443, 447, 455–456
Benard, L., 302t
Bender, W., 464–465
Bendridi, N., 377–378
Benes, V., 374
Benita, Y., 441
Benito, R., 337
Benito, Y., 377–378
Benkirane, M., 407t
Benkovic, S. J., 3
Bennasser, Y., 407t
Bennett, K. L., 312–313, 348–349
Bennett, M., 140–141
Ben-Yishay, R., 218
Benz, J., 214–217
Beran, R. K., 135–137, 370–371, 466–467
Berardi, V., 439

Author Index

Berger, J. M., 2, 41, 76–78, 95, 99, 102, 133, 138, 157*f*, 171–190, 214–217, 218, 241–244, 246, 248, 249*f*, 441
Berges, T., 302*t*
Bergman, N., 173–174
Bernadó, P., 205–206
Bernard, C., 392–393, 393*f*
Bernardini, M., 444
Bernheimer, A. W., 381
Berninger, P., 276–277, 286, 338
Bernstein, E., 140–141
Bernstein, J., 12
Bernstein, K. A., 290–291, 292*t*, 297, 308, 312
Bertrand, E., 326*t*, 327, 334, 340
Betz, U. A., 464–465
Beyer, A. L., 309–310, 312
Bhadti, V. S., 444
Bhaskaran, H., 15, 79, 113–114, 123, 124
Bhat, S., 443
Bhattacharya, A., 256
Bhattacharyya, S. N., 326*t*, 328, 441
Bidet, K., 407*t*
Bieniossek, C., 80, 99, 100–102, 101*f*, 208*f*, 214–217, 439
Biezynski, A., 469–470
Bigay, J., 96
Billaut-Mulot, O., 407*t*
Bisognano, C., 375–377
Bi, X., 215*t*
Bizebard, T., 15, 35
Bjorkroth, B., 241
Bjornson, K. P., 467–468
Black, J. D., 256
Blackledge, M., 205–206
Blackstock, W., 407*t*
Blackwell, T. K., 325
Blandford, V., 326*t*, 328
Bleichert, F., 290–291, 292*t*, 304, 312
Blobel, G., 76–78, 79
Blom, T., 56
Blume, A. C., 276–277
Blume, J. E., 295
Blum, S., 290
Blunt, J. W., 443
Bock, R., 326*t*, 328
Boeke, J. D., 337
Boesche, M., 311–312, 444
Bogden, C. E., 173–174
Boger, D. L., 476
Boguszewska-Chachulska, A. M., 467–468
Bohm, M., 439
Bohnenstengel, F. I., 443–444
Bohnsack, M. T., 220, 275–288, 290–291, 292*t*, 295
Boismenu, D., 326*t*, 328
Boisset, S., 377–378
Boisvenue, S., 325–327, 326*t*
Bokinsky, G., 126
Boldt, K., 312–313

Bolger, G., 464–465
Bolger, T. A., 241
Bolinger, C., 407*t*, 409–410, 415, 417
Bonneau, F., 138, 214–217, 218, 255–274
Bono, F., 76–78, 95, 99–100
Boparai, N., 465–466
Bordeleau, M. E., 325–327, 334–335, 340, 441–444, 445, 447, 453, 455–456
Boris-Lawrie, K., 405–436
Bornard, I., 371
Borner, T., 397
Borowski, P., 444, 480
Bosche, M., 311–312
Boscheron, C., 241
Botlagunta, M., 348–349
Botta, L., 444
Botta, M., 340, 407*t*, 444
Boudvillain, M., 149–169, 273
Boulant, S., 407*t*
Bourdeau, A., 443–444
Bourdon, J. C., 348–349, 356, 363–365
Bousquet-Antonelli, C., 292*t*
Bouveret, E., 310, 312
Bover, L., 407*t*
Bowers, H. A., 76–78, 112–113, 135–137, 214, 408–409
Bowie, A. G., 348–349, 407*t*
Boyer, P. D., 66–67, 69, 70–71
Brachat, A., 280, 304, 305–306
Bradley, D. W., 466
Bradley, M. J., 29–63, 67, 241–242
Bragado-Nilsson, E., 310, 312
Braig, K., 176
Brame, C. J., 309
Brand, L., 8
Brass, A. L., 441
Brass, C. A., 465–466
Braun, P., 315
Brautigam, C. A., 165
Bray, S. E., 348–349, 363–365
Breenblatt, J., 309–310, 312
Brem, B., 441–442, 443–444, 455–456
Brendza, K. M., 135–137
Brengues, M., 324–325, 332–333
Brennan, C. A., 173–174
Brenowitz, M., 112, 117, 126
Breslauer, K. J., 10, 12–13
Bretner, M., 444
Breukink, E., 80
Briber, R. M., 126
Briercheck, D. M., 173–174
Briggs, J. A., 424*f*
Briguglio, I., 480
Brillard, J., 371
Briolat, V., 371
Brochon, J. C., 91
Brogna, S., 256

Brosh, R. M. J., 464–465
Broussolle, V., 371
Brown, A. H., 215t, 256
Brown, E. L., 372, 380–381
Browne, R. J., 174–175
Brown, M. L., 340
Bruckner, A., 219
Bruckner, F., 98–99
Brunel, J., 140–141, 143
Brune, M., 52, 53
Brunette, S. R., 476
Brunger, A. T., 98–99
Bruno, M. M., 135–137
Brzozka, K., 141
Buchan, J. R., 324–328, 326t, 330–334, 330t, 335–336
Bucher, P., 141
Bücher, T., 244
Bud, L. O., 214–217
Bugaut, A., 325–327
Buhot, A., 36
Buikema, W. J., 392–393
Bujalowski, W., 37, 46–51, 172–173
Bujnicki, J. M., 215t
Bumann, M., 80, 99, 100–102, 101f, 208f, 214–217, 439
Bunkoczi, G., 186–187
Burckstummer, T., 312–313, 348–349
Burdick, R., 407t
Burger, F., 292t
Burger, L., 276–277, 286, 338
Burgess, B. R., 173–174
Burns, C. M., 173–174
Burroughs, M., 465–466
Burton, A. S., 265
Buser, P., 214–217, 290
Bushan, R. B., 407t
Bussey, H., 337
Bustamante, C., 135
Bustelo, X., 309
Bustin, S. A., 374
Buttner, K., 137
Byrd, A. K., 156

C

Cable, M., 135, 466
Cagney, G., 309, 311–312, 315–316
Cai, H., 444, 464–465
Caizergues-Ferrer, M., 192, 214–218, 290, 292t, 295, 296–297, 309–310, 312
Calabrese, J. M., 334
Callahan, S. M., 392–393
Calle, A., 291–294
Cambie, D., 186–187
Cameron, C. E., 135, 328–329, 466
Campbell, S. G., 324–325, 331
Canadien, V., 309
Canaves, J. M., 185–186

Cantor, C. R., 247–248
Cao, D., 407t
Cao, W., 3, 31–32, 34, 35, 36–37, 38–42, 40f, 44, 46–56, 57–58, 67, 86, 97–98, 103, 214–217, 241–242, 328–329
Capeyrou, R., 192, 214–218, 295
Capron, A., 407t
Cardullo, R. A., 8
Carell, T., 140–141
Caretti, G., 348–349
Carmel, A. B., 407t
Carmo-Fonseca, M., 241
Carpousis, A. J., 370–371
Carrier, M., 443–444
Carrington, M., 328–329
Carr, M. D., 439
Carroll, A. S., 291–294, 331–332
Carroll, J. S., 328–330, 330t
Carroll, K. S., 118, 302t
Carta, A., 480
Caruthers, J. M., 76–78, 80, 94–95, 100–102, 201, 205, 438
Casaluce, G., 444
Caspary, F., 310, 312
Castelli, L. M., 324–325, 331
Castenholz, R. W., 387–388
Cate, J. H., 166–167
Caudy, A. A., 140–141
Cavalcanti, V. L., 30
Cavin, J. F., 371
Cech, T. R., 114, 118
Celestre, R. S., 186–187
Cencic, R., 437–461
Chable-Bessia, C., 407t
Chabre, M., 96
Chacón, P., 198–199
Chadee, A. B., 113–114
Chai, H., 443–444
Chaipan, C., 407t
Chait, B. T., 312–313
Chakrabarti, S., 138, 214–217, 218, 256, 264, 265f, 269f, 270
Chalupnikova, K., 325–327, 326t, 328–329, 340
Chamberlin, M. J., 16–17
Chamieh, H., 76, 78, 95, 99–100, 138, 201, 256, 260, 260f, 261f, 264, 267f
Chamot, D., 5, 386, 387, 393f, 395, 396f
Chamot, S., 371
Champion, E. A., 214–217, 219, 309–310
Chance, M. R., 126
Chan, C. Y., 36
Chanfreau, G., 118
Chang, J. H., 101f
Chang, P. C., 407t
Chang, T.-H., 241, 292t, 370
Chang, W., 67
Chang, Y. F., 256

Author Index

Chao, C. H., 30, 338, 340, 348–349
Chaperot, L., 407t
Chard, L. S., 340, 441–443, 447, 455–456
Charette, J. M., 219, 309–310, 315
Charollais, J., 219
Charter, N. W., 41–42
Chau, G. Y., 407t
Chavanikamannil, F., 292t
Chemla, D. S., 90–91
Chen, B., 447
Chen, C. M., 30, 348–349
Chen, D. S., 135
Cheng, P. L., 30, 348–349
Cheng, S. C., 215t, 217–218
Cheng, W., 135
Cheng, Z., 76–78, 138, 249f, 329, 339
Chen, H. M., 407t, 444
Chen, J. H., 348–349
Chen, J. Y.-F., 370, 407t
Chen, P. Y., 325
Chen, S. M., 441–442, 443–444, 455–456
Chen, T. C., 447
Chen, V. B., 186–187
Chen, W. N., 407t
Chen, Y. F., 15, 35, 55, 97–98
Chen, Y. L., 192, 214–218, 295
Chepetan, A., 265
Chesebro, B., 429–430
Cheung, K. H., 291–294
Chevalier, C., 377–378
Chiba, T., 315
Chi, C. W., 407t
Chin, K., 118
Chi, S. W., 276–277, 295
Chi, W. K., 135
Chi, Y. H., 407t, 444
Choe, J., 466
Choi, J. M., 101f
Choi, N. S., 325–327
Cho, M. H., 439
Chono, K., 464–465
Choong, P. T., 439
Choo, Q. L., 466
Cho-Park, P. F., 208f
Chorvat, D., 447
Chou, T. H., 340
Cho, Y. H., 101f
Christendat, D., 185–186
Christensen, B. B., 378
Chua, G., 337
Chu, A. M., 337
Chua, P. K., 141–142
Chung, T. D., 468, 477
Chu, S., 98–99, 126
Civril, F., 140–141
Clarke, B. E., 466
Clark, E. L., 348–349, 362
Clark, P. K., 67

Clark, T. A., 276–277, 295
Clayton, C., 328–329
Clayton, R. F., 407t
Clegg, R. M., 8
Clerici, M., 138, 256
Clima, L., 98–99
Clodi, E., 114
Cocude, C., 407t
Coelho, P. S., 291–294
Cohen-Bazire, G., 387
Cohen, F. E., 198–199
Cohen, N., 449
Colburn, N. H., 99–100, 439
Cole, C. N., 214–217, 215t, 218, 241–242, 326t, 327–328
Coller, J., 329–330, 338
Collett, M. S., 466
Colley, A., 292t, 294
Collins, R., 34–35, 76–78, 79, 95, 243–244
Colognori, D., 214–217
Colot, H. V., 214–217, 241
Colvin, K. R., 5, 387
Coman, M. M., 31–32, 34, 35, 36–37, 38–42, 40f, 44, 46–54, 55–56, 57–58, 67, 241–242
Combet, C., 83–84, 84f
Combs, D. J., 292t, 294, 295, 296–297
Compagnone-Post, P. A., 309, 312
Conaway, R., 173–174
Confalonieri, F., 102–103
Connelly, C., 337
Connors, K. A., 250
Consler, T. G., 466
Conti, E., 76–78, 95, 99–100, 138, 214–218, 256, 264, 265f, 269f, 270
Conzelmann, K. K., 141
Cook, N. J., 138
Coombes, R. C., 348–349, 363–365
Coote, J. L., 449
Coppolecchia, R., 214–217
Corbett, A. H., 241
Cordingley, M. G., 464–465
Cordin, O., 30–31, 241, 307–308, 369–370
Cork, C. W., 186–187
Cornish-Bowden, A., 41, 44
Cornish, P. V., 12–13, 141–142
Corona, P., 480
Corrie, J. E. T., 52, 53
Costa-Mattioli, M., 443
Costes, S., 99–100
Cote, J., 325–327, 326t
Cougot, N., 325, 326t, 327, 334, 340
Coulson, A., 348–349, 362
Courtois, S., 447
Coute, Y., 291–294
Cramer, P., 98–99
Crawford, P. A., 348–349
Creacy, S. D., 325–327
Criscuolo, A., 386

Croughan, T. P., 392
Cruceanu, M., 423
Cruciat, C. M., 311–312
Crute, J. J., 464–465
Cui, B., 443–444
Cui, S., 12–13, 141–142
Culin, C., 305–306, 310
Cunha-Neto, E., 30
Curk, T., 276–277
Cusack, S., 138, 140–141, 143, 256
Cuzic, S., 150–152, 166–167
Czaplinski, K., 256, 270

D

Dahan, M., 90–91
Dahlgren, L. G., 34–35, 76–78, 79, 95, 243–244
Dalbadie-McFarland, G.
Dale, M. P., 66–67, 70, 72
Dalgliesh, C., 348–349, 362
Dalrymple, D., 407t
Daneholt, B., 240–241
Dang, Y., 325–327, 334–335, 407t, 443
Dansako, H., 407t
Dar, A. C., 444
Darcissac, E., 407t
Darnell, J. C., 276–277, 295
Darnell, R. B., 276–277, 295, 338
Das, R., 113–114
D'Assoro, A. B., 362
Datta, K., 10
Datta, N., 311–312
Daugeron, M. C., 292t, 294, 308
Dautry, F., 326t, 327, 329–331, 330t, 334, 337–338
Dave, R., 407t
David, M. D., 329
Davidson, L. A., 114
Davierwala, A. P., 237
Davis, I. W., 186–187
Day, R. S. III., 394–395
Debler, E. W., 76–78, 79
Decatur, W. A., 290
Decker, J. M., 429–430
Decker, T., 348–349
Declais, A. C., 102–103
Declercq, W., 447
De Francesco, R., 16–17
de Grace, M. M., 441
DeGrasse, J. A., 312–313
Deimling, T., 140–141
Deinert, J., 444
Dekker, C., 102–103
Dekker, N. H., 102–103
de Kruijff, B., 80
De La Cruz, E. M., 3, 29–63, 67, 79, 86, 95, 97–98, 99, 103, 214–217, 228, 230–231, 241–242, 328–329

de la Cruz, J., 290–291, 292t, 296–297, 302t
DeLano, W. L., 187
De Las Rivas, J., 309
de La Tour, C. B., 102–103
Del Campo, M., 5, 15, 20, 34–35, 55, 76–78, 79, 97–98, 99, 112–113, 116–117, 118, 120–121, 121f, 123, 124–126, 125f, 214
Deleage, G., 83–84, 84f
de Lencastre, A., 166–167
Deloche, O., 292t, 296–297
del Rayo Sanchez-Carbente, M., 326t, 328
Delsart, V., 407t
del Toro Duany, Y., 102–103
Delviks-Frankenberry, K. A., 407t
Demarini, D. J., 280, 304, 305–306
Demuth, J. P., 371
Deng, M., 407t
Deniz, A. A., 90–91
Denman, R. B., 439
Denouel, A., 305–306, 310
Deon, C., 291–294
Deplazes, A., 325, 338
Deruelles, J., 387–388
Desclozeaux, M., 348–349
DesGroseillers, L., 326t, 328
Desplancq, D., 392–393, 393f
Deterre, P., 96
Deuring, R., 326t, 328
Deval, J., 141–142
Dever, T. E., 215t, 438–439
de Witte, P. A., 447
Diago, M., 465–466
Díaz, J. F., 198–199
Diaz, J. J., 291–294
Dietrich, U., 407t
Dietz, H. C., 256, 443
Diges, C. M., 34–35, 55, 78–79, 86, 370–371
Dignam, J. D., 356
Dillingham, M. S., 14–15, 133, 138, 172
Dilworth, F. J., 348–349
Ding, S. C., 31–32, 34, 35, 36–37, 38–42, 40f, 44, 46–54, 55–56, 57–58, 67, 131–147, 241–242
Ding, Y., 36
Di Padova, M., 348–349
Dittmann, E., 397
Doan, C., 480
Dockter, M. E., 7
Doere, M., 369–370
Dohmae, N., 325–327, 326t, 328
Dolan, J. W., 173–174
Doma, M. K., 30
Dombroski, A. J., 173–174
Domcke, S., 138, 214–217, 218, 256, 264, 265f, 269f, 270
Dominguez, D., 214–217
Domning, E. E., 186–187
Dong, G., 386

Dong, S., 256
Doniach, S., 198–199
Doolittle, W. F., 394–395
Dorn, C., 348–349
Dorrello, N. V., 439
Dosil, M., 309
Doye, V., 240–241
Dragon, F., 309, 312
Drakas, R., 312–313
Drawid, A., 291–294
Dreyfus, M., 15, 35, 55, 79, 214, 219, 370–371
Dreyfuss, G., 273
Dricot, A., 441
Drigo, S. A., 30
Duarte, C. M., 114, 115f
Duarte, R. M., 186–187
Duggal, R., 480
Duguet, M., 102–103
Dumont, S., 135
Dumpelfeld, B., 311–312
Dunbar, D., 309
Duncan, R., 438–439
Dunman, P. M., 371, 372, 380–381
Durand, S., 340
Dutta, A., 328–329

E

Earnest, T. N., 186–187
Easton, J. B., 172–173
Eaton-Rye, J. J., 391
Eberhard, D., 444
Ebert, J., 76–78, 95, 99–100
Echols, N., 186–187
Eckart, A., 469–470
Eckenberg, P., 464–465
Eckert, P. Q., 469–470
Eckstein, F., 150, 153
Edelmann, A., 311–312
Edersen, J. S., 198
Edery, I., 215t, 449
Edgcomb, S. P., 407t
Edlin, B. R., 465
Edmonds, K. A., 41, 80, 439
Eglen, R. M., 41–42
Ehira, S., 397
Eiermann, T., 469–470
Eisenacher, K., 141
Eleaume, H., 375–377
El-Fahmawi, B., 387, 397, 398–400, 399f
Elhai, J., 390, 391
Eliason, W. K., 31–32, 34, 36–37, 79, 99, 228, 230–231
Elie, C., 102–103
Ellegast, J., 141
Elles, L. M., 370–371
Ellington, A. D., 276

Elliott, D. J., 348–349, 362
Ellis, L., 328–329
Ellis, N. A., 102–103
Ellman, J., 84–85
Elvira, G., 326t, 328
Emerick, V. L., 114
Emerman, M., 407t
Emery, B., 292t, 296–297, 308
Endoh, H., 348–349, 363–365
Enemark, E. J., 138, 184–185
Engelhardt, M. A., 113–114
Eom, T., 439
Eperon, I. C., 152, 154–155
Eritja, R., 10, 12–13
Erkizan, H. V., 340
Erlacher, K., 198
Erzberger, J. P., 41, 214–217, 241, 242–243, 246
Este, J. A., 444
Etcheverry, T., 304
Evdokimova, E., 185–186

F

Fabrizio, P., 214–217, 215t
Facchinetti, V., 407t
Faé, K. C., 30
Fagotto, F., 441, 442–443, 445
Fahrenkrog, B., 214–217, 240–241
Fairchild, C. R., 443–444
Fairman, M. E., 2, 5, 15, 20, 214
Fairman-Williams, M. E., 14–15, 112–113, 133, 214, 307–308, 325
Fak, J. J., 276–277, 295
Falchi, F., 340, 407t, 444
Falvo, J. V., 291–294, 331–332
Fam, B. C., 172–173
Fang, J., 407t
Fang, Y., 378
Fan, J. S., 76–78, 249f
Fan, X., 326t, 328
Farabaugh, P., 348–349
Fass, D., 173–174
Fattom, A. I., 444
Faucher, A. M., 464–465
Fauth, M., 327–328
Fechter, P., 377–378
Fedorova, O., 114, 135–137, 150–152, 155t, 156, 158, 166–167
Feeley, E. M., 441
Feinstone, S. M., 464, 466
Felekyan, S., 91, 93
Ferenci, P., 465–466
Ferlenghi, I., 15, 35
Fernandez, S., 407t, 409–410, 415
Ferraiuolo, M. A., 443
Ficner, R., 214–217
Fields, S., 276, 314, 315–316
Fikrig, E., 441

Filion, C., 329
Filipowicz, W., 326t, 328
Fink, B. E., 476
Fiorini, F., 138, 214–217, 218, 255–274
Fire, A. Z., 407t
Fischer, C. J., 9, 18, 20, 268–270
Fischer, R., 464–465
Fisher, A. J., 96
Fisher, P. B., 141–142
Flannery, B. P., 165
Fleming, S., 348–349, 362
Flick, J. S., 41, 214–217, 241, 242–243, 246
Florencio, F. J., 397
Flores, C., 8
Flores, E., 390
Focaccia, R., 465–466
Folkmann, A. W., 218
Förster, T., 80–82
Forterre, P., 102–103
Foster, G. R., 465–466
Fourney, R. M., 394–395
Fournier, M. J., 290–291, 292t
Fox, A. H., 291–294
Foy, E., 140–141
Franceschi, F., 46–51, 78–79
Francois, P., 375–377
Franke, D., 199, 209–210
Franklin, S., 312–313
Franks, T. M., 256, 328–329, 339
Fredslund, F., 99–100
Freed, E. F., 300
Freier, S. M., 164–165
Frey, P. A., 152
Frick, D. N., 463–485
Frieden, C., 56
Friedman, S. L., 348–349
Friew, Y., 407t
Fritzler, M. J., 324–325, 330–331
Froment, C., 192, 214–218, 295, 309–310, 312
Fromont-Racine, M., 295, 309–310, 312
Fry, D., 447
Frydlova, I., 324–325, 331, 332–333, 340–341
Fuhr, M., 443–444
Fuhrman, S. A., 480
Fujii, R., 407t
Fujiki, Y., 325–327, 326t, 328–329, 340
Fujita, O., 467–468
Fujita, T., 140–141, 467
Fujiyama, S., 363–365
Fukamizu, A., 407t
Fukuda, T., 363–365
Fuller, M. T., 326t, 328
Fuller-Pace, F., 407t
Fuller-Pace, F. V., 78–79, 214, 347–367, 370–371, 408–409
Fuller, S. D., 424f

Funk, M., 304
Fu, Q., 407t
Furch, M., 41–42
Fu, R. H., 215t, 217–218
Furman, P. A., 480
Furtak, V., 407t
Futai, E., 325–327, 326t
Futcher, A. B., 305–306, 310
Fu, W., 407t

G

Gack, M. U., 12–13, 141–142
Gadal, O., 309
Gaiduk, A., 91, 93
Gale, M. Jr., 140–141, 143
Galicia-Vázquez, G., 437–461
Gallagher, J. E., 302t, 309, 312
Gallinari, P., 16–17
Gallouzi, I., 325–327, 334–335
Galmozzi, C. V., 397
Ganesan, R., 256
Ganguly, A., 102–103
Gao, Q., 441
Gao, Z., 2, 325–327, 326t, 327f, 328–330, 330t, 332, 338, 339, 340, 341, 438
Garbade, K. H., 244
Garbelli, A., 340, 407t, 444
Garces, R. G., 439
Garcia-Blanco, M. A., 407t
Gareau, J. R., 358
Gari, E., 304
Garson, J. A., 374
Garvis, S., 378, 380
Gaspin, C., 377–378
Gat-Viks, I., 441
Gaughan, L., 348–349, 362
Gavin, A. C., 311–312
Gebreab, F., 315
Gee, P., 141–142
Geeves, M. A., 41–42
Gehring, N. H., 138, 256, 443
Geiselmann, J., 172–173
Geisinger, E., 378
Geissmann, T., 377–378
Gelugne, J. P., 292t
Genevaux, P., 371, 377–378
Georgopoulos, C., 371, 377–378
Geourjon, C., 83–84, 84f
Gerace, L., 407t
Gerard, B., 441–442, 443–444, 455–456
Gerke, L. C., 291–294, 331–332
Gerlier, D., 140–141, 143
Gersappe, A., 256
Gerstein, M., 185–186
Gerus, M., 290, 296–297
Gevorkyan, J., 141–142
Ghosh, S., 256

Author Index

Gillian, A. L., 348–349
Gill, S. J., 34, 36–37
Gimple, O., 158
Ginalski, K., 444
Gindorf, C., 348–349
Ginzburg, O., 469–470
Giot, L., 315
Giraud, G., 9–10, 12–13
Gish, G., 150
Givskov, M., 378
Gkogkas, C., 325–327, 438
Glatter, O., 193b, 195, 196b, 197–198
Glenn, J. S., 480
Glesias, N., 240
Glieden, A., 348–349
Gloeckner, C. J., 312–313
Gocheva, V., 157f, 159
Godbout, R., 30, 407t
Goel, A., 362
Goh, P. Y., 407t
Golan, D. E., 324–325, 330–331
Goldberg, A. C., 30
Goldeck, M., 141
Golden, B. L., 118, 126
Golden, J. W., 386
Golden, S. S., 386
Gold, L., 276
Goldstein, A. L., 241
Gollnick, P., 214
Goncalves, A., 348–349
Goodman, M. F., 10, 12–13
Goodman, Z. D., 465–466
Goody, R. S., 37
Gopalkrishnan, R. V., 141–142
Gopich, I., 91
Gorbalenya, A. E., 438
Gordon, M. S., 67
Gordon, S. C., 465–466
Gordus, A., 126
Gorelick, R. J., 423
Gorospe, M., 334–335
Goss, D. J., 215t
Goto, M., 348–349
Goulet, I., 325–327, 326t
Gozdek, A., 467–468
Graham, S., 407t
Grainger, R. J., 9–10, 12–13
Grakoui, A., 466
Grande, J. P., 348–349, 358, 362
Grandi, P., 309, 311–312
Granneman, S., 214–217, 220, 275–288, 290–291, 292t, 295, 296–297, 309, 312
Grant, R. A., 173–174
Graslund, A., 10
Grasser, F. A., 348–349
Grassmann, C. W., 407t, 409
Greco, A., 291–294

Greenblatt, J. F., 311–312
Green, J. B., 116–117
Green, R., 114
Greer, A. E., 407t
Greger, H., 441–442, 443–444, 455–456
Gregory, D. J., 348–349, 356, 363–365
Grierson, D. S., 340
Griffiths, A. D., 152, 154–155
Grifo, J. A., 438–439
Grigorieva, G., 391
Grigoriev, M., 192, 214–218, 295
Grigorov, B., 140–141, 143
Grimsley, G. R., 32
Groebe, D. R., 222
Grohman, J. K., 79
Gross, C. H., 134, 135
Grosse, F., 407t, 409
Grosse-Kunstleve, R. W., 186–187
Grosshans, C. A., 114
Gross, I., 424f
Grousl, T., 324–325, 331, 332–333, 340–341
Gruber, G., 135
Grubmuller, H., 98–99
Grundhoff, A. T., 348–349
Grygon, C. A., 464–465
Grzechnik, S. K., 185–186
Guardavaccaro, D., 439
Gu, B., 466–467, 469–470
Gubaev, A., 81f, 83–85, 86, 89–90, 92, 94, 95, 97f, 100–102, 101f, 214–217
Guenier, A. S., 441–442
Guenther, U. P., 2, 14–15, 112–113, 133, 214, 307–308, 325
Guex, N., 84f
Guilligay, D., 256
Guinier, A., 195b
Guja, K., 4–5, 9–10, 12
Gu, M., 14–15
Gunaratne, J., 407t
Guo, D., 407t
Guo, F., 407t
Guo, J., 348–349
Guo, L., 34, 76–78, 79, 126
Guo, S., 214–217, 241
Guo, X., 309, 311–312, 407t
Guo, Y., 309
Gupta, M., 331
Gupta, P. B., 441
Gu, Q., 7
Gury, J., 371
Güssregen, R., 443–444
Gutell, R. R., 161
Gutfreund, H., 48
Guthrie, C., 214–217, 215t, 241, 370
Gutsche, I., 138, 256
Gwack, Y., 466

H

Haas, E., 92
Hackney, D. D., 3, 31–32, 34, 35, 36–37, 38–42, 40*f*, 44, 46–56, 57–58, 65–74, 86, 97–98, 103, 214–217, 241–242, 328–329
Haenni, A. L., 467–468
Hafner, M., 276–277, 286, 338
Hagan, K. W., 256, 270
Hagemann, M., 386, 397
Haghighat, A., 439
Hahn, D., 286
Hall, K. B., 3, 9, 10
Halls, C., 5, 112–113, 124
Halperin, A., 36
Hamadani, K. M., 91, 93
Hamborg, K., 99–100
Hammarström, M., 34–35, 76–78, 79, 95, 243–244
Hammond, S. M., 140–141
Hanabuchi, S., 407*t*
Handke, G., 464–465
Haney, S., 372, 380–381
Han, J. H., 466
Hanlon, M. H., 466
Hannemann, D. E., 50, 52
Hannon, G. J., 140–141
Hansen, S., 196*b*
Hanson, A. M., 463–485
Hantschel, O., 312–313, 348–349
Han, Y., 407*t*
Hao, T., 315, 441
Hao, W., 480
Hardin, J. W., 98–99
Hargrave, K. D., 464–465
Harnpicharchai, P., 309
Hartmann, R. K., 78–79, 150–152, 166–167
Hartman, T. R., 407*t*, 409–410, 415
Hasek, J., 324–325, 331, 332–333, 340–341
Haselkorn, R., 387, 392–393
Hashimoto, S., 348–349
Ha, T. J., 7–8, 9–10, 12–13, 14–15, 90–91, 126, 137, 141–142
Hattori, M., 315
Haussecker, D., 407*t*
Hausser, J., 276–277, 286, 338
Hawkes, D., 407*t*
Haw, R., 309
Hayano, T., 309–310
Hayden, E. J., 265
Haynes, J., 337
Hay, R. T., 348–349, 358, 363
Hazlett, T. L., 53
Headd, J. J., 186–187
Hearing, P., 407*t*
Heath, C. V., 326*t*, 327–328
Heck, J. A., 469–470
Hedrick, M. P., 476

Heer, R., 348–349, 362
He, F., 215*t*, 256
Hegamyer, G., 439
Heidtman, M., 291–294
Heintzmann, R., 329–330, 330*t*, 337–338
Heitkamp, S. E., 3, 31–32, 35, 37, 38–39, 41–42, 55–56, 67, 86, 241–242, 328–329
He, J. J., 407*t*
Helfer, A. C., 377–378
Hellemans, J., 374
Hellen, C. U., 80, 438, 439, 440–441
Hellwig, A., 240–241
Helmke, K. J., 2, 41, 76–78, 95, 99, 102, 133, 214–217, 218, 241–242, 243–244, 248, 249*f*, 441
Henao-Mejia, J., 407*t*
Hendrickson, E., 41, 80
Hendrix, M., 464–465
Henn, A., 3, 30–32, 34, 35, 36–37, 38–42, 40*f*, 44, 46–56, 57–58, 67, 79, 86, 97–98, 99, 103, 214–217, 228, 230–231, 241–242, 328–329
Henras, A. K., 290, 296–297
Henry, Y., 192, 214–218, 290, 295, 296–297
Hensold, J. O., 438–439
Hentze, M. W., 138, 256
He, Q. S., 407*t*, 409
Herdman, M., 387–388
Herdy, B., 41, 80
Herian, U., 466–467
Herlihy, K. J., 480
Hermann-Le Denmat, S., 219
Hernandez, J. J., 463–485
Herrero, E., 304
Herschlag, D., 30, 35, 95, 97–98, 112, 113–114, 118, 123, 126, 439, 446–447
Hershey, J. W., 438–439
Herzberg, C., 371
Hesson, T., 135, 466
Hess, S., 441
Hetzer, M. W., 240–241
Heuer, D., 441
Heurtier, M. A., 311–312
Hewlett, G., 464–465
He, Y., 191–212
Heyman, J. A., 291–294
Hickman, T., 334–335
Hieter, P., 297–298
Higa, T., 340, 441–443, 447, 455–456
Higgins, C. F., 370–371
High, Y. D., 466–467, 468, 476–477, 480
Hijikata, M., 407*t*
Hilbert, M., 30–31, 34–35, 38, 76, 89–90, 94, 100–102, 101*f*, 112–113, 214–217
Hiley, S. L., 292*t*, 294, 295, 296–297, 309–310, 312
Hill, D. E., 441
Hillebrand, J., 325

Hiller, S., 439
Hilliker, A., 2, 438
Hilliker. A., 323–346
Hingorani, M. M., 172–173
Hinnebusch, A. G., 192, 438
Hiraga, N., 465–466
Hirata, I., 407t
Hiratsuka, T., 32
Hir, H. L., 201
Hirokawa, N., 325–327, 326t, 328
Hirozane-Kishikawa, T., 315
Hislop, R. G., 348–349, 358, 363
Hitomi, M., 241
Hoang, T., 309–310, 312
Hobman, T. C., 407t
Hobson, S., 444
Hochstrasser, D., 291–294
Hodge, C. A., 214–217, 215t, 218, 241–242, 326t, 327–328
Hoelz, A., 76–78, 79
Hofer, O., 443–444
Hofert, C., 311–312
Hoffman, E. P., 348–349
Hoffmann, H. H., 441
Hofmann, K., 141
Hogue, K., 407t
Hoheisel, J. D., 328–329
Hohng, S., 7–8, 9–10
Holden, D. W., 378, 380
Holden, H. M., 96
Holmes, L. E., 324–325, 331
Holzbaur, E. L. F., 67
Homans, S. W., 407t
Homma, K., 309–310
Hong, F., 348–349
Hongou, Y., 407t
Hong, W., 407t
Hong, Z., 135, 466
Hopfner, K. P., 12–13, 137, 140–142
Horban, A., 465–466
Horiguchi, T., 173–174
Hornakova, T., 447
Hornung, V., 141
Horsey, E., 309
Hoshino, S., 273
Hosmane, R. S., 444, 464–465
Hotta, H., 467
Houghton, M., 466
Houston, P., 467–468
Houston, S., 185–186
Howard, J., 38
Howell, S. A., 53
Howells, M., 186–187
Howson, R. W., 291–294, 331–332
Hoyle, N. P., 324–325, 331
Hritz, J., 447
Hsu, J. L., 443
Huang, H. R., 112–113, 116–117

Huang, I. C., 441
Huang, Y. H., 153, 215t, 217–218, 388–389, 407t
Huber, J., 240–241
Hubner, C. G., 6–7
Huggett, J., 374
Hughes, T. R., 292t, 294, 295, 296–297, 309–310, 312
Hug, N., 325, 338
Huh, W. K., 291–294, 331–332
Hu, J., 407t
Humbert, O., 217–218
Hung, L. W., 186–187
Hung, P. D., 443–444
Hunke, C., 135
Hunt, D. F., 309, 312
Hunter, J. L., 52, 53
Huntzinger, E., 377–378
Huot, M. E., 329
Huppert, J. L., 325–327
Hurt, E., 240–241, 290–291, 296–297, 304, 309
Hurt, E. C., 302t
Huson, L., 348–349, 363–365
Hutton, R. L., 67
Hu, W. S., 407t
Hu, Y. X., 34–35, 78–79, 98–99
Hwang, B. Y., 443–444
Hwang, L. H., 135, 480
Hyberts, S. G., 439

I

Iacoangeli, A., 439
Ianni, B., 30
Igarashi, M., 467–468
Iglesias, N., 240
Ignatchenko, A., 311–312
Ikeda, M., 407t
Imai, B., 309
Imaizumi, T., 140–141
Imam, J. S., 256
Imamura, M., 465–466
Imamura, S., 397
Imataka, H., 439
Ingelfinger, D., 329–330, 330t, 337–338
Ingraham, H. A., 348–349
Inn, K. S., 141
Inoue, A., 441
Inoue, T., 407t
Iost, I., 15, 35, 55, 79, 214, 219, 370–371
Iqbal, A., 9–10, 12–13
Irannejad, H., 444
Irving, T., 126
Ishaq, M., 407t
Ishida, K., 397
Ishido, S., 467
Ishikawa, H., 309–310
Ishiura, S., 325–327, 326t
Isken, O., 256, 273, 407t, 409

Isobe, T., 309–310
Itoh, M., 467
Ito, T., 315
Ivanov, K. A., 138
Ivanov, P., 324–325, 331, 332–333, 340–341
Iwamatsu, A., 407t
Iwamoto, F., 325–327, 326t, 328–329, 340
Iwao, E., 465–466
Iyer, K. S., 7
Iyer, L. M., 172
Izaurralde, E., 138, 240, 325
Izumikawa, K., 309–310

J

Jabbouri, S., 375–377
Jackrel, M. E., 219, 309–310
Jackson, D. L., 407t
Jackson, R. J., 80
Jackups, R. R. Jr., 370
Jacobs, A. M., 348–349, 356, 358, 363–365
Jacobson, A., 215t, 256
Jacq, C., 219
Jacquet, J. M., 407t
Jacquier, A., 118, 377–378
Jacquinot, F., 150–151, 153, 154–155, 156, 157f, 158, 159–161, 160f, 162–165, 166–167
Jaffray, E. G., 348–349, 358, 363
Jagessar, K. L., 370
Jahnke, W., 135, 137
Jain, C., 370–371
Jain, D., 328–329
Jakovljevic, J., 309
Jalal, C., 5, 363–365
Jambon, M., 83–84, 84f
Jamison, E., 126
Jancura, D., 447
Janda, F., 324–325, 331, 332–333, 340–341
Jangra, R. K., 407t
Jang, S. K., 101f
Janknecht, R., 348–349, 358, 362
Jankowsky, A., 406
Jankowsky, E., 1–27, 30, 34–35, 38, 55–56, 76–78, 96, 97–98, 112–113, 116–117, 124, 133, 135–137, 141–142, 192, 214–217, 232–233, 241, 256, 276, 290–291, 304, 307–308, 325–327, 326t, 327f, 328–330, 330t, 332, 338, 339, 340, 341, 369–370, 406, 408–409, 438, 441–442, 443, 445, 455–456
Janoskova, D., 324–325, 331, 332–333, 340–341
Jansen, A. P., 99–100
Jansen, J. A., 166–167
Jansen, R., 291–294
Jarmoskaite, I., 30–31, 34, 38, 76–78, 79, 112–113
Jaskiewicz, L., 326t, 328
Jawhari, A., 98–99
Jayachandran, U., 138, 214–217, 218, 256, 264, 265f, 269f, 270
Jeang, K. T., 406, 441, 444, 464–465, 480
Jean, J. M., 9, 10
Jenkins, G. S., 325–327
Jensen, E. D., 348–349
Jensen, K. B., 276–277, 295
Jensen, L. J., 311–312
Jens, M., 276–277
Jeong, Y. J., 174–175
Jezewska, M. J., 37, 46–51
Jia, H., 2, 112–113, 116–117, 214
Jiang, F., 140–141, 143
Jiang, J., 348–349, 363–365
Jiang, Y., 112–113, 116–117
Jin, S. B., 241
Johansen, J. S., 76, 78, 95, 99–100, 201
Johnson, E. R., 76–78, 80, 94–95, 100–102, 201, 205
Johnson, J., 172–173, 174–175
Johnson, K. A., 46, 48, 51–52, 56, 67
Johnson, M. C., 424f
Johnson, N. P., 10
John, S. P., 441
Johnstone, O., 326t, 328
Jones, D. M., 134
Jong, J. E., 407t
Jonikas, M. A., 126
Jordan, L. B., 349
Jose, D., 10
Joshua-Tor, L., 138, 184–185
Jovin, S. M., 215t
Joyner-Butt, C., 325–327
Judson, R. S., 315
Jung, A., 141
Jungblut, S. P., 102–103
Jung, J. U., 12–13, 141–142, 240–241
Jungkamp, A. C., 276–277, 286, 338
Ju, Q., 298, 312
Juranek, S., 141

K

Kabat, D., 429–430
Kadlec, J., 138, 256
Kagawa, R., 176
Kalicharran, K., 444
Kalil, J., 30
Kalinin, S., 94, 98–99
Kalitzky, M., 480
Kalverda, A. P., 407t
Kamiya, N., 465–466
Kanai, Y., 325–327, 326t, 328
Kanehisa, M., 386
Kanesaki, Y., 386
Kang, D. C., 141–142
Kann, M., 407t, 409
Kantardjieff, K. A., 178–180

Kantor, J., 214–217, 241
Kao, C., 154f
Kao, P. N., 407t, 409
Kaplan, A., 397
Kappes, J. C., 429–430
Kapral, G. J., 186–187
Karbstein, K., 118, 213–237
Kardono, L. B., 443–444
Karginov, F. V., 34–35, 78–79, 98–99
Karlas, A., 441
Karlberg, T., 34–35, 76–78, 79, 95, 243–244
Karow, A. R., 30–31, 34–35, 38, 76, 77f, 79, 81f, 83–85, 84f, 86, 87, 89–90, 92, 94, 95, 96, 97f, 98–99, 112–113, 228, 230–231
Karpel, R. L., 114
Karunatilaka, K. S., 103–104, 114, 126
Kashima, I., 273
Kassube, S. A., 76–78, 79
Katahira, J., 240–241
Kataoka, N., 273
Katchalskikatzir, E., 92
Kato, H., 141
Katoh, A., 166
Katoh, M., 270
Kato, N., 407t
Kato, S., 348–349
Katsu, K., 407t
Katsumata, K., 464–465
Kauffman, L., 41–42
Kaufman, R. J., 324–327, 330–331, 334–335
Kawaguchi, A., 407t
Kawaguchi, Y., 392–393
Kawaoka, J., 135, 150–151, 155t, 156, 158, 163, 166–167
Kayikci, M., 276–277, 295
Kay, M. A., 407t
Kazantsev, A. V., 166–167
Kebbel, F., 94, 100–102, 101f, 214–217
Kedersha, N. L., 324–325, 330–331, 334–335
Keeney, D., 469–470
Kehr, J. C., 397
Kelez, N., 186–187
Kelley, W. L., 371, 375–378
Kennedy, D., 329
Kenny, J. E., 7
Kerr, I. M., 449
Kervestin, S., 256
Ketner, G., 407t
Kewalramani, V. N., 407t
Khalil, H., 441
Khan, A. R., 407t
Khandjian, E. W., 329
Khemici, V., 370–371
Khorshid, M., 276–277, 286, 338
Khosla, M., 114
Kibler, P., 464–465
Kieffer, B., 392–393, 393f
Kienzler, A., 98–99

Kiet, L. C., 443–444
Kikuchi, A., 102–103
Kikuchi, M., 140–141
Kilburn, J. D., 126
Kilby, J. M., 429–430
Kim, A., 101f
Kimber, M. S., 185–186
Kim, D. E., 172–173, 174–175
Kim, D. W., 466
Kimmins, S., 326t, 328
Kim, S., 141, 407t, 443–444
Kim, T., 407t
Kim, Y. C., 101f
Kindbeiter, K., 291–294
Kinghorn, A. D., 443–444
King, P. A., 126
Kino, Y., 325–327, 326t
Kirchhofer, A., 12–13, 141–142
Kistler, A. L., 370
Kitagawa, H., 348–349
Klee, W. A., 7
Kleiman, L., 407t, 444
Klempnauer, K. H., 439
Klenerman, P., 469–470
Kleymann, G., 464–465
Klonowska, M. M., 172–173
Klostermeier, D., 30–31, 34–35, 38, 75–109, 112–113, 214–217, 228, 230–231
Klumpp, K., 141–142
Knight, J. R., 315
Knop, M., 310
Kobayashi, E., 447
Kobayashi, K., 326t, 327–328
Kobayashi, S., 34–35, 76–78, 95
Kobayashi, Y., 348–349
Kocanova, S., 447
Koch, J., 466–467
Koch, M. H. J., 198–200, 201, 202b, 203b, 204, 209–210
Koch, V., 444
Kodadek, T., 467–468
Koegl, M., 315
Kohara, Y., 325
Kohler, J., 81f, 83–85, 86, 89–90, 92, 94, 95, 97f
Kohlway, A., 140–141, 143
Kohn, H., 172–173, 174–175
Kojima, H., 407t
Kokaram, A., 325
Kolykhalov, A. A., 464
Komar, A. A., 80, 99–100, 438–439, 440–441
Komatsu, W., 309–310
Kong, K. F., 377–378, 380–381
Kong, Y., 340
Konig, J., 276–277
Konig, R., 441
Kono, T., 465–466
Kontani, T., 464–465
Kool, E. T., 164–165

Koonin, E. V., 15, 172, 438
Kopp, J., 84f
Korner, F., 466–467
Koshida, I., 363–365
Kos, M., 290–291, 292t, 308
Kossen, K., 34–35, 78–79, 133
Kotaja, N., 326t, 328
Kottilil, S., 134
Kovchegov, Y., 91, 93
Kowalinski, E., 140–141, 143
Koyano, T., 443–444
Kozhaya, L., 407t
Kozin, M. B., 198–199, 200, 209–210
Kraemer, B., 276
Kramer, F. R., 468
Kramer, S., 328–329
Kratky, O., 193b, 195, 196b, 197–198
Krause, R., 311–312
Krausslich, H. G., 424f
Krawczyk, M., 467–468
Kremmer, E., 348–349
Kressler, D., 290–291, 292t, 296–297, 302t, 304
Kress, M., 326t, 327, 329–331, 330t, 334, 337–338
Kreusch, A., 185–186
Kriedemann, P. E., 390
Krisch, H. M., 370–371
Krishnakumar, S., 326t, 327–328
Krishnan, M. N., 441
Krogan, N. J., 309–310, 311–312
Krug, A., 141
Krupnick, J., 186–187
Kubista, M., 374
Kubota, S., 407t
Kudla, G., 277, 280–281, 285, 286, 295
Kuhmann, S. E., 429–430
Kuhn, I., 397
Kuhn, P., 185–186
Kujat-Choy, S. L., 5, 387
Kujat, S. L., 386, 395
Kulikowski, T., 444
Kulikowsk, T., 480
Kulozik, A., 138, 256
Kumar, A., 291–294
Kumari, S., 325–327
Kumar, K., 386
Kunisawa, R., 387
Kuo, G., 466
Kuoh, C. S., 443–444
Kuo, L. Y., 114
Kuroki, M., 407t
Kushima, Y., 407t
Kuster, B.
Kutach, A. K., 215t
Kvaratskhelia, M., 407t, 411, 419–421
Kwok, L. W., 126
Kwong, A. D., 135, 441, 464–465, 466, 480
Kyono, K., 467–468

L

Labelle, Y., 329
Lacaille, J. C., 326t, 328
Lacroute, F., 305–306, 310
Laederach, A., 112, 113–114, 117, 126
Lafontaine, D. L., 292t, 294, 309–310, 312
Lai, M. C., 325–328, 326t, 330, 330t, 338
Lai, S. Y., 36
Lakowicz, J. R., 37, 247–248
Lalev, A., 309
Lam, A. M., 464, 466–468, 469–470, 480
Lamanna, A. C., 226
Lamb, J. S., 126
Lambowitz, A. M., 5, 15, 20, 34–35, 55, 76–78, 79, 97–98, 99, 112–113, 116–117, 118, 120–121, 121f, 123, 124–126, 125f, 214, 378
Lammens, A., 141
Lammens, K., 141
Lamond, A. I., 291–294
Lam, Y. W., 291–294
Landthaler, M., 276–277, 286, 325, 338
Lane, B. H., 219, 309–310
Lane, D. P., 78–79, 214, 348–349, 356, 363–365, 370–371
Larkin, C., 4–5, 9–10, 12
LaRonde-LeBlanc, N., 439
Larson, B. C., 265
Larsson, O., 440–441
Lasko, P. F., 325–327, 326t, 328, 370, 438
Last, R. L., 214–217
Latchoumanin, O., 407t
Latreille, D., 407t
Lattmann, S., 325–327, 326t, 328–329, 340
Latus, L. J., 370
Laurence, T. A., 91, 93
Lawitz, E., 465–466
Lawrence, C. E., 36
Lawrence, P., 407t, 409
Law, S. M., 10, 12–13
Lebaron, S., 192, 214–218, 290, 295, 296–297, 309–310, 312
Lebedeva, L. A., 348–349
Lebedeva, S., 276–277
Leblanc, P., 388–389
Lebovitz, R. M., 356
Lebowitz, J., 36
Lechner, J., 309
LeCuyer, K. A., 165
Lee, A. V., 157f, 159, 290–291, 292t, 312
Lee, C. S., 443
Leeds, N. B., 292t, 294, 295, 296–297
Lee, K. H., 407t, 443–444
Lee, M. B., 348–349
Lee, S. K., 443–444
Lee, U., 348–349
Lee, W. C., 296–297

Author Index

Lee, Y. H., 30, 325–328, 326t, 330, 330t, 338, 348–349
Le Hir, H., 34–35, 41, 76, 78, 95, 99–100, 138, 214–217, 218, 255–274
Lehman, N., 265
Lehnik-Habrink, M., 371
Lehtiö, L., 34–35, 76–78, 79, 95, 243–244
Le, H. V., 135, 466
Leipe, D. D., 172
Lejeune, F., 340
Lemke, E. A., 84–85
Lemmens, I., 315
Lemon, S. M., 134, 407t, 480
Lentze, N., 219
Ler, L. W., 443
Le Roy, F., 329–331, 330t, 337–338
Leroy, P., 370
Ler, S. G., 407t
Lescar, J., 135
Lesley, S. A., 185–186
Leslie, A. G., 176, 177–178
Lessmann, T., 309
Leulliot, N., 217–218
Leung, A. K., 291–294, 334
Leung, H. Y., 348–349, 362
Levy, Y., 407t
Lew, D. P., 375–377
Lewis, M. S., 36
Liang, C., 141, 407t
Liang, H., 337
Liang, M. P., 126
Liang, X. H., 290–291, 292t
Liang, Y., 134, 441
Liao, C. L., 444
Liao, J. C., 174–175
Li, C., 256
Licatalosi, D. D., 276–277, 295
Licciardello, N., 3, 31–32, 35, 37, 38–39, 41–42, 55–56, 67, 86, 241–242, 328–329
Liebes, L., 447
Li, F. Y., 407t
Li, H., 76–78, 94–95, 141
Li, J., 311–312, 407t
Li, K., 407t
Li, L., 30
Liljestrom, P., 141
Lilley, D. M., 8, 9–10, 12–13
Lill, H., 391
Lima, C. D., 358
Lima, W. F., 12, 15, 41, 215t, 438–439
Lim, C. L., 348–349
Lim, M. K., 138, 329, 339
Lim, S. G., 407t
Lim, S. P., 135, 137, 407t
Lim, Y. H., 315
Li, N., 315
Lin, C., 214–217
Lin, D., 439

Lindenbach, B. D., 140–141, 143, 466–467
Linden, M. H., 78–79
Linder, P., 2, 30–31, 34–35, 38, 55–56, 76, 79, 112–113, 133, 214–217, 241, 276, 290–291, 292t, 294, 296–297, 302t, 307–308, 326t, 328, 369–383, 406, 408–409
Lindner, P., 193b, 196b
Lindqvist, L., 340, 441–444, 445, 447, 453, 455–456
Lindsley, J. E., 67
Lingner, J., 325, 338
Linial, M. L., 407t
Lin, R. J., 214–217
Lin, T. I., 466–467, 468, 476–477, 480
Lin, X., 407t
Lin, Y. C., 215t, 217–218
Liou, M. J., 443–444
Lisok, A., 348–349
Liu, B., 391
Liu, F., 2, 15, 55, 96, 97–98, 214–217, 290, 328–329, 339
Liu, H., 407t, 429–430
Liu, J. O., 325–327, 334–335, 407t, 443
Liu, Q., 407t
Liu, R. Z., 30
Liu, S., 126, 391
Liu, X. Q., 84–85, 135–137
Liu, Y. J., 241, 291–294, 407t
Liuzzi, M., 464–465
Li, W., 331
Li, Y., 312–313, 315
Lockett, S. J., 99–100
Lockshon, D., 315
Lohmann, V., 466–467
Lohman, T. M., 3, 8, 9, 12–13, 18, 20, 23, 55, 268–270, 467–468
Loke, J., 348–349
Lomakin, I. B., 439
Longtine, M. S., 280, 304, 305–306
Loo, Y. M., 140–141
Lopez, A., 337
Lopez-Ramirez, V., 370
Lorentzen, E., 76–78, 95, 99–100
Lorsch, J. R., 35, 95, 97–98, 112–113, 438, 439, 446–447
Louber, J., 140–141, 143
Loverix, S., 165
Lowery, C., 174
Lowery-Goldhammer, C., 174
Lowe, S. W., 441–442, 443–444, 455–456
Low, W. K., 325–327, 334–335, 443
Lu, C., 114
Lucius, A. L., 3, 8, 9, 12–13, 18, 20, 46–51, 268–270
Ludersdorfer, T. H., 407t
Ludwig, J., 141
Luhrmann, R., 214–217, 215t, 240–241, 325, 329–330, 330t, 337–338

Lührmann, R., 290
Lujan, P., 407t
Luke, B., 325, 338
Lu, N., 407t
Lunardi, T., 140–141, 143
Lund, M. K., 241
Luo, D., 135, 137, 140–141, 143
Luo, M. J., 443
Luscombe, N. M., 276–277
Lutter, R., 177–178
Lykke-Andersen, J., 256, 324–325, 328–329, 330–331, 339
Lyon, C. E., 291–294
Lyubimov, A. Y., 172
Lzumi, N., 273

M

MacDowell, A. A., 186–187
Machuy, N., 441
Mackie, G. A., 370–371
Maddock, J. R., 214–217
Madix, R. A., 265
Mady, C., 30
Maeda, A., 214–217
Maeda, S., 392–393
Maeder, C., 214–217, 215t
Maga, G., 340, 407t, 444
Magalhães, P. J., 335
Magde, D., 92
Magee, W. C., 386
Mahuteau-Betzer, F., 340
Mailer, R., 304
Ma, J., 407t
Majerciak, V., 407t
Maki, M., 407t
Mak, J., 407t
Malcova, I., 324–325, 331, 332–333, 340–341
Malec, J., 449
Malina, A., 438
Malinska, K., 324–325, 331, 332–333, 340–341
Mallam, A. L., 34, 76–78, 79
Maluf, N. K., 9, 18, 20, 268–270
Mandel, M., 387
Manetti, F., 340, 407t, 444
Mangus, D. A., 256
Manickman, S., 290–291, 292t, 312
Mann, D. A., 348–349
Mann, M., 240–241, 257, 291–294, 312
Mansfield, T. A., 315
Manstein, D. J., 41–42
Maquat, L. E., 256, 273
Marcellin, P., 465–466
Marchadier, B., 214–217
Marcotrigiano, J., 140–141, 143
Marecek, J. F., 67
Margeat, E., 149–169

Marintchev, A., 41, 80, 439, 440–441, 442–443, 445
Marintcheva, B., 41, 80
Maroney, P. A., 214
Marsault, J., 102–103
Marshall, N. F., 173–174
Martin, A., 217–218
Martineau, Y., 440–441
Martin, E. M., 449
Martinez, A., 173–174
Martinez-Murillo, F., 256
Martin, R., 220, 277, 280, 295
Martin, S. R., 53
Maruyama, K., 348–349
Marzioch, M., 311–312
Masuhiro, Y., 348–349
Mateasik, A., 447
Mathews, D., 36
Mathieson, T., 444
Mathonnet, G., 440–441
Matson, S., 291–294
Matsuda, Y., 467–468
Matsumoto, K., 140–141, 407t
Matsumoto, M., 407t
Matsumoto, T., 363–365
Matsumura, A., 270
Matsuura, M., 116–117
Matthews, B. W., 178–180
Matthews, E. E., 31–32, 34, 36–37, 79, 99, 228, 230–231
Matthews, J., 441–442, 443, 445, 455–456
Maurer, A. P., 441
Maurizot, J. C., 447
Ma, Y., 134
Mayer, O., 114
Mazroui, R., 325–327, 326t, 329, 334–335
McAdam, R., 326t, 328
McCarthy, A. A., 140–141, 143
McCarthy, T. J., 166–167
McCoy, A. J., 186–187
McGivern, D. R., 407t
McKay, D. B., 34–35, 76–79, 80, 94–95, 98–99, 100–102, 201, 205, 438
McKeating, J. A., 480
McKenzie, A. III., 280, 304, 305–306
McKinney, W., 186–187
McLaughlin, L. W., 10
McMaster, G. K., 438–439
McNamara, P. J., 375–377
McPherson, A., 173–174
McPherson, P. S., 326t, 328
Meacher, D., 309
Medalia, O., 46–51
Meerovitch, K., 215t
Mehta, R. R., 443–444
Mei, J. M., 378, 380
Meissner, S., 397
Meister, G., 325

Author Index

Mele, A., 276–277, 286, 295
Mele, M., 276–277
Melese, T., 296–297
Mella-Herrera, R. A., 386
Mendel, D., 84–85
Mendell, J. T., 256
Menz, R. I., 176
Merchant, M., 340
Merlevede, W., 447
Merrick, W. C., 12, 15, 35, 41, 76, 80, 99–100, 215t, 325–327, 438–439, 446, 454–455
Mesecar, A. D., 443–444
Methot, N., 208–209, 439, 446
Metzger, D., 348–349
Meyer, C., 443–444
Meyers, J. L., 256
Meziane, O., 407t
Michaelis, J., 98–99
Michalet, X., 91, 93
Michaud, C., 371
Michon, A. M., 311–312
Micura, R., 141
Middleton, E. R., 31–32, 34, 35, 36–37, 38–42, 40f, 44, 46–54, 55–56, 57–58, 67, 241–242
Miguet, L., 392–393, 393f
Mihalik, K., 464
Mihara, M., 363–365
Mikami, K., 386
Mikkat, S., 397
Milburn, S. C., 438–439
Miles, T., 309
Millar, D. P., 3, 92, 407t
Miller, I., 331
Miller, L. C., 326t, 328
Miller, M. T., 140–141, 143
Miller, P., 291–294
Milligan, J. F., 222
Minczuk, M., 444
Mindrinos, M., 326t, 327–328
Minor, W., 186–187
Minton, A. P., 36
Mi, Q., 443–444
Mironchik, Y., 348–349
Miskovsky, P., 447
Misra, V. K., 126
Mitchell, B. M., 309, 312
Mitchell, D., 113, 122–123
Mitchess, B., 309
Mitra, S., 112, 117, 126
Miura, Y., 309–310
Miwa, Y., 173–174
Miyagishi, M., 140–141
Miyaji, K., 407t
Miyakoshi, J., 394–395
Miyashiro, M., 467–468
Miyashita, M., 407t
Miyata, R., 467–468
Miyazawa, N., 309–310

Mizushima, S., 80
Mnaimneh, S., 337
Modrak, D., 173–174
Moffat, J., 337
Mohr, G., 214
Mohr, S., 5, 112–113, 116–117, 124, 126, 214
Mokas, S., 325–327, 326t
Moldt, M., 140–141
Molin, S., 378
Mollet, S., 326t, 327, 334
Momose, F., 407t
Moncoeur, E., 166–167
Monsarrat, B., 192, 214–218, 295, 309–310, 312
Montgomery, M. G., 176
Montpetit, B., 2, 41, 76–78, 95, 99, 102, 133, 214–217, 218, 239–254, 441
Mooney, S. M., 362
Moore, H. C., 349, 362, 363–365
Moore, K. J., 467–468
Moore, M. J., 240, 443
Mooseker, M. S., 67
Morán, F., 198–199
Moreira, B. G., 9–10, 12
Moreland, R. B., 388–389
Moreno-Hagelsieb, G., 370
Morgan, W. D., 173–174
Mori, A., 340, 441–443, 447, 455–456
Moriarty, N. W., 186–187
Morikawa, H., 407t
Morishita, R., 273
Morita, T., 325–327, 326t
Moritz, E., 407t
Morlang, S., 78–79
Morris, M. J., 325–327
Morse, D., 337
Mortimer, S. A., 126
Mouffok, S., 217–218
Mougin, A., 290, 296–297
Mouillet, J. F., 348–349
Mourão, A., 138, 256
Moustafa, I. M., 135, 466
Mouton, Y., 407t
Mueller-Lantzsch, N., 348–349
Mueller, O., 480
Mueller, R., 374
Mueller, T. J., 7
Muhlemann, O., 328–329
Muhlrad, D., 138, 324–325, 326t, 327–328, 329, 330–331, 330t, 332–333, 334, 339
Mukadam, S., 348–349
Mukhtar, M., 407t
Mulky, A., 407t
Mullen, T. M., 407t
Muller, D. A., 407t
Muller, E., 441
Müller, N. G., 30
Muller, S., 98–99
Mumberg, D., 304

Munchel, S. E., 328–330, 330t
Munro, M. H. G., 443
Munschauer, M., 276–277, 286, 338
Muramatsu, S., 467
Murata, N., 386
Muro-Pastor, A. M., 390
Muro-Pastor, M. I., 397
Murphy, E., 372, 380–381
Murphy, R. M., 36
Murray, C. L., 465
Muschielok, A., 98–99
Musier-Forsyth, K., 407t, 411, 419–421, 423
Muskett, F. W., 439
Muslimov, I. A., 439
Mylonas, E., 205–206
Myong, S., 12–13, 14–15, 137, 141–142

N

Nadal, M., 102–103
Nagamine, Y., 325–327, 326t, 328–329, 340
Nagao, T., 443–444
Nagashima, K., 407t
Nagata, K., 407t
Nagel, R. J., 292t, 294, 295, 296–297
Naitou, M., 363–365
Najera, I., 141–142
Najima, T., 407t
Naji, S., 407t
Nakagawa, Y., 407t
Nakamura, A., 34–35, 76–78, 95
Nakamura, H., 166
Nakamura, T., 363–365
Nakamura, Y., 99–100, 438–439
Nakamuta, H., 270
Nakanishi, S., 447
Nakano, H., 447
Nakano, S., 166, 270
Nakasugi, K., 391
Nandineni, M. R., 214–217
Napetschnig, J., 76–78, 79
Narayanan, N., 469–470
Narayan, V., 315
Narula, G. S., 469–470
Naslund, T. I., 141
Nasmyth, K., 310
Naso, R. B., 444
Natsukawa, T., 140–141
Navarro, R. E., 325
Nayak, R. K., 3
Necakov, A., 185–186
Nehring, S., 137
Neilan, B. A., 391
Nelson, F. K., 291–294
Nelson Hayes, C., 465–466

Neuhauser, D., 91, 93
Neumann-Haefelin, C., 469–470
Neuveut, C., 407t, 444
Nevens, F., 465–466
Newman, J., 178–180
Nguyen, H., 98–99
Nguyen-the, C., 371
Ng, W. O., 391
Nichols, M. D., 173–174
Nicol, S. M., 78–79, 214, 347–367, 370–371
Niebuhr, A., 480
Nielsen, A. T., 378
Nielsen, K. H., 76, 78, 95, 99–100, 191–212
Nielsen, P. J., 370, 438–439
Niepel, M., 240–241
Nikonowicz, E. P., 165
Nilsen, T. W., 214
Nilsson, L., 10
Nir, E., 91, 93
Nishi, K., 370
Nissan, T. A., 309, 325–327, 331–332, 333–334, 335–336
Niu, L., 348–349
Nix, J., 214–217
Noble, C., 76–78, 249f
Noble, K. N., 215t, 218, 241–242
Noda, N., 467–468
Nolan, T., 374
Nollmann, M., 166
Nolte, A., 118
Nordlund, T. M., 10
Noren, C. J., 84–85
Norman, D. G., 9–10, 12–13
Northcote, P. T., 325–327, 334–335, 441–442, 443, 445, 453, 455–456
Nouwen, N., 80
Novac, O., 441–442, 443, 445, 455–456
Novakova, L., 324–325, 331, 332–333, 340–341
Nover, L., 327–328
Novick, R. P., 378
Nowatzke, W., 153
Nugroho, B. W., 443–444
Nureki, O., 34–35, 76–78, 95
Nyen, C. H., 340

O

Oberer, M., 41, 80, 340, 439, 441–443, 445, 447, 453, 455–456
Oberholzer, A. E., 80, 99, 100–102, 101f, 208f, 214–217, 439
Ochi, H., 465–466
Ochocka, A. M., 348–349, 363–365
O'Day, C. L., 292t
Odenwalder, P., 214–217
Oeffinger, M., 312–313

Oesterhelt, F., 98–99
Offmann, S. V., 200, 201f, 206, 208f
Ogawa, S., 348–349
Ogilvie, L. A., 441
Ogryzko, V., 348–349
Oguro, A., 99–100, 439
Ohba, S., 443–444
Ohe, T. A., 392–393
Ohmichi, T., 166, 270
Ohmori, M., 397
Ohno, M., 273
Ohno, S., 273
Ohse, T., 443–444
Ohshima, T., 407t
Ohtsu, T., 99–100, 439
Oihi, T., 407t
Okada, Y., 407t
Okamoto, M., 407t
Okumus, B., 9–10, 12–13
Oldenburg, K. R., 468, 477
Olivares, A. O., 67
Oliveira, C. L. P., 76, 78, 95, 99–100, 191–212
Olmedo-Alvarez, G., 370
Omata, T., 392–393
O'Neill, R. E., 407t
Onishi, H., 325–327, 326t
Onishi, M., 465–466
Opperman, T., 173–174
Orba, Y., 407t
Orlowski, J., 215t
Ortigoza, M. B., 441
Ortoleva-Donnelly, L., 161, 166–167
O'Shea, E. K., 291–294, 331–332
Osheim, Y. M., 309–310, 312
Oshiumi, H., 407t
Ostap, E. M., 41–42
Ostap, M. E., 41–42
Oster, G., 174–175
Osugi, K., 407t
Otegui, M. S., 326t, 327–328
Otsu, N., 336
Otto, M., 377–378, 380–381
Otwinowski, Z., 186–187
Otzen, D., 198
Ou, Q., 348–349
Overby, L. R., 466
Overgaard, M. T., 34–35, 78–79, 98–99
Owczarzy, R., 9–10, 12
Owsianka, A. M., 407t
Owttrim, G. W., 5, 371, 385–403
Oyelere, A. K., 166–167
Ozawa, R., 315
Ozier-Kalogeropoulos, O., 305–306, 310

P

Pace, C. N., 32
Pacher, T., 443–444
Padilla, R., 153
Pagano, M., 439
Page, R., 185–186
Pajunen, M., 371, 377–378
Pakrasi, H. B., 391
Palacios, D., 372, 380–381
Palese, P., 407t
Palmer-Hill, F., 466
Palmer, R. J. Jr., 378
Pan, C., 30, 117, 329
Pandiani, F., 371
Pan, J., 114, 123
Pan, K., 325
Pannell, L. K., 84–85
Pan, Q., 407t
Pante, N., 241
Pantos, E., 198–199
Paolini, C., 16–17
Papin, C., 192, 214–218, 295
Parameswaran, P., 407t
Parker, C. A., 92
Parker, R., 2, 30, 138, 324–334, 326t, 327f, 330t, 335–336, 337–338, 339, 340, 341, 438
Park, H. Y., 126
Park, J., 407t
Parsyan, A., 325–327, 438, 440–441
Parvinen, M., 326t, 328
Patel, A. H., 348–349
Patel, G., 174–175
Patel, S. S., 140–141, 143, 155, 172–173, 174–175
Paterson, M. C., 394–395
Pathak, V. K., 407t
Patick, A. K., 480
Patterson, D. N., 12
Pauli, S., 325–327
Pause, A., 76, 290, 438–439, 446
Pauwels, F., 466–467, 468, 476–477, 480
Pavlovic, J., 407t
Pazhoor, S., 407t
Peck, M. L., 97–98
Pedersen, J. S., 76, 78, 95, 99–100, 191–212
Peersen, O. B., 3
Peitsch, M. C., 84f
Pelletier, J., 325–327, 334–335, 340, 437–461
Peltz, S. W., 256, 270
Pena, V., 215t
Peng, L., 397
Peng, W. T., 309–310, 312, 337
Peregrin-Alvarez, J. M., 311–312
Pereira, G., 310
Pérez-Fernández, J., 309
Periyasamy, M., 348–349, 363–365
Perkins, N. D., 348–349, 356, 363–365

Perlman, P. S., 112–113, 116–117
Perrin, J., 444
Peschiaroli, A., 439
Pestova, T. V., 80, 438, 439, 440–441
Peter, M., 325, 338
Peters, G., 375–377
Peters, L., 325
Petfalski, E., 277, 280–281, 285, 295, 302t, 309
Petoukhov, M. V., 199–200, 203–206, 207, 209–210
Petroulakis, E., 440–441
Pevzner, V., 464–465
Pezzuto, J. M., 443–444
Pfaffl, M. W., 374
Pfeffer, S. R., 16–17
Pfister, C., 96
Pfleiderer, G., 244
Pflieger, D., 219
Pfortner, H., 371
Philippsen, P., 280, 304, 305–306
Piccirillo, S., 291–294
Pichlmair, A., 141
Pico, S., 114
Piedrafita, L., 304
Pietack, N., 371
Pietschmann, T., 480
Piras, S., 480
Pisarev, A. V., 440–441
Pisareva, V. P., 440–441
Plate, D. W., 186–187
Platt, E. J., 429–430
Platt, T., 173–174
Platzmann, F., 214–217
Pleiss, J. A., 214–217
Pleissner, K. P., 441
Plumas, J., 407t
Pochart, P., 276, 315
Podtelejnikov, A., 240–241
Poeck, H., 141
Polach, K. J., 35, 36–37
Polge, C., 219
Poljak, L., 370–371
Pollack, L., 126
Pollard, T. D., 33
Pol, S., 465–466
Pomerantz, R. J., 407t
Poon, K. M., 407t
Poordad, F., 465–466
Pootoolal, J., 337
Popp, S., 439
Porco, J. A. Jr., 441–442, 443–444, 455–456
Porra, R. J., 390
Porter, D. J., 466
Porwancher, K. A., 309, 312
Possedko, M., 377–378
Potier, N., 392–393, 393f
Potratz, J. P., 15, 35, 55, 97–98, 111–130, 155

Potter, B. V., 152, 154–155
Poulin, F., 208f
Powell, B., 138
Preradovic, A., 312–313
Press, W. H., 165
Preugschat, F., 466
Pringle, J. R., 280, 304, 305–306
Prisco, M., 312–313
Proctor, R. A., 375–377
Projan, S. J., 372, 380–381
Proksch, P., 443–444
Proux, F., 219
Prud'homme-Genereux, A., 370–371
Ptak, R. G., 407t
Puig, O., 310, 312
Punna, T., 311–312
Purdie, C. A., 349
Pu, S., 311–312
Putnam, A., 1–27, 55, 96, 97–98, 214–217, 232–233, 290, 328–329, 339
Py, B., 370–371
Pybus, O. G., 469–470
Pyle, A. M., 14–15, 18, 31–32, 34, 35, 36–37, 38–42, 40f, 44, 46–54, 55–56, 57–58, 67, 103–104, 114, 115f, 116–117, 118, 126, 131–147, 150–152, 155t, 156, 158, 163, 166–167, 241–242, 290, 466–467
Pyronnet, S., 439

Q

Qian, S., 407t, 409–410, 415
Qin, J., 407t
Qin, P. Z., 114
Queiroz, R., 328–329
Quentmeier, H., 443–444
Quinlan, P. R., 349
Qureshi-Emili, A., 315

R

Rabhi, M., 149–169
Radi, M., 407t, 444
Radzikowska, A., 220
Rahmouni, A. R., 150–151, 154, 156–158, 157f, 159, 160f, 162–165, 166–167, 273
Raida, M., 444
Raings, S. A., 407t
Rain, J., 295, 309–310, 312
Rajagopala, S. V., 315–316
Rajan, P., 348–349, 362
Rajewsky, N., 276–277
Rajkowitsch, L., 114
Rajyaguru, P., 329
Raman, A., 348–349
Ramanathan, A., 140–141, 143
Raman, V., 348–349
Ramasawmy, R., 30
Ramaswami, M., 325
Ramey, C. S., 370–371

Author Index

Ram, S. J., 335
Raney, K. D., 3, 135, 156, 466
Ranji, A., 406, 407t, 408–409, 411–412, 419–421
Rao, B. G., 441, 464–465, 480
Rastinejad, F., 173–174
Rau, C., 311–312, 444
Ravelli, R. B., 256
Rawling, D. C., 289–322
Rayment, I., 96
Reader, V., 444
Read, R. J., 186–187
Rebbapragada, I., 256
Redder, P., 369–383
Reddy, T. R., 407t
Reed, R., 443
Reese, J. C., 328–329
Rees, W. T., 92
Refregiers, M., 447
Regan, L., 219, 309–310
Regev, A., 441
Reibarkh, M., 441, 442–443, 445
Reid, A. D., 78–79, 214, 370–371
Reinstein, J., 266
Reise Sousa, C., 141
Remor, M., 311–312
Rempeters, L., 371
Renge, I., 6–7
Ren, J., 215t
Renn, A., 6–7
Reuter, J., 36
Reyes, J. R., 407t
Reynes, J., 407t
Reysset, G., 371
Rhoades, E., 31–32, 34, 35, 36–37, 38–42, 40f, 44, 46–54, 55–56, 57–58, 67, 241–242
Rhodes, M. M., 126
Rice, C. M., 14–15, 464, 465, 466
Richards, D. P., 309
Richardson, J. P., 153, 173–174
Richardson, L., 153
Richter-Cook, N. J., 438–439
Richter, N. J., 12, 15, 35, 41, 76, 215t, 438–439, 454–455
Rick, J. M., 311–312
Rieder, E., 407t, 409
Rigaut, G., 241, 257, 310, 312
Rigler, R., 10
Riley, C. A., 265
Ripmaster, T. L., 292t, 296–297
Rippka, R., 387–388, 393
Risuleo, G., 439
Riswan, S., 443–444
Rivera-Pomar, R., 329–330, 330t, 337–338
Robblee, J. P., 50, 52
Robert, F., 441–442, 443–444, 455–456
Roberts, J. W., 173–174
Robertson-Anderson, R. M., 407t
Roberts, S., 465–466

Robinson, M. D., 309
Robson, C. N., 348–349, 362
Rocak, S., 241, 292t, 296–297, 308
Rodrigues, J. P., 241
Roeder, R. G., 356
Roemer, T., 291–294
Rogers, G. W. Jr., 12, 15, 35, 41, 76, 80, 215t, 438–439, 454–455
Rogers, R., 312–313
Roggendorf, M., 469–470
Roh, J. H., 126
Rojo, M., 290–291, 292t, 296–297, 302t
Román, A., 309
Roman, J., 309
Romo, D., 325–327, 334–335, 443
Rong, L., 407t
Root, D. E., 441
Rosbach, M., 214–217, 241
Rosen, G., 67
Roskams, T., 447
Rossler, O. G., 5, 6
Ross-Macdonald, P., 291–294
Rot, G., 276–277
Rothballer, A., 276–277, 286, 338
Rothwell, P. J., 94
Routh, E. D., 325–327
Rout, M. P., 240–241, 309, 312–313
Roux, C. M., 371
Rouzina, I., 423
Rowe, C. E., 112–113, 116–117
Roy, B. B., 407t
Roy, K., 30
Roy, R., 7–8, 9–10
Roy, S., 7
Rozen, F., 215t
Rual, J. F., 315
Rudisser, S., 154f
Rudolph, M. G., 34, 78–79, 99, 102–103
Rueda, D., 103–104, 114, 126
Ruffing, A. M., 386
Ruggiu, M., 276–277
Ruiz-Albert, J., 378, 380
Rule, G. S., 173–174
Rupp, B., 178–180
Ruprecht, M., 220, 277, 280, 295
Russell, D. R., 374
Russell, D. W., 394–395
Russell, R. S., 15, 30–31, 34, 35, 38, 55, 76–78, 79, 97–98, 111–130, 134, 155, 329, 407t
Rust, M. J., 126
Rutz, B., 257, 310, 312
Ryan, B. J., 441
Rychner, E., 214–217, 241
Ryder, S. P., 166–167
Rypma, R. S., 466–468, 469–470
Ryu, W. S., 407t

S

Saag, M. S., 429–430
Sadler, A. J., 407t
Sadovsky, Y., 348–349
Saelices, L., 397
Sagi, I., 31–32, 34, 36–37, 46–51, 79, 99, 228, 230–231
Sahalie, J., 315
Sahni, A., 30
Saib, A., 407t
Saikrishnan, K., 138
Sakaki, Y., 315
Salisbury, J. L., 348–349, 358, 362
Salloum, S., 469–470
Sambrook, J., 374, 394–395
Sampath, A., 135, 137
Samuele, A., 444
Sanchez-Carbente, M. R., 326t, 328
Sanchez, J. C., 291–294
Santarsiero, B. D., 443–444
Santhanam, A. N., 439
Saphire, A. C., 407t
Sarisky, R. T., 407t, 409
Sarkar, G., 449
Sartorelli, V., 348–349
Sasagawa, N., 325–327, 326t
Sasaki, M., 166, 270
Sassone-Corsi, P., 326t, 328
Satler, C. A., 438–439
Sattler, M., 138, 256
Savchenko, A., 185–186
Savilahti, H., 371, 377–378
Sawa, H., 407t
Scarcelli, J. J., 218, 326t, 327–328
Schalinski, S., 444
Scherer-Becker, D., 135, 137
Scherl, A., 291–294
Scheuner, D., 324–325, 330–331
Schiebel, E., 310
Schiesser, S., 140–141
Schildbach, J. F., 4–5, 9–10, 12
Schiltz, R. L., 348–349
Schimmel, P. R., 247–248
Schlatterer, J. C., 126
Schlattner, U., 219
Schlee, M., 141
Schleiff, E., 220, 277, 280, 295
Schlichting, I., 266
Schlottmann, S., 340
Schmedt, T., 439
Schmid, S. R., 290
Schmidt, A. S., 102–103
Schmitt, C., 241
Schmitz, H., 444, 480
Schmitzova, J., 214–217
Schneider, B. L., 305–306, 310
Schneider, C., 443–444

Schneider-Poetsch, T., 325–327
Schneider, S., 217–218
Schneider, U., 464–465
Schnier, J., 370
Schoenberg, D. R., 407t, 409–410, 415
Schön, A., 158
Schrader, J. W., 329
Schrenzel, J., 371, 375–378
Schroder, G. F., 98–99
Schroder, M., 348–349, 407t
Schroeder, R., 30, 112–113, 114
Schubach, W. H., 348–349
Schuberth, C., 141
Schubert, P., 329
Schuck, P., 36
Schuler, B., 84–85
Schüler, H., 34–35, 76–78, 79, 95, 243–244
Schultz, J., 311–312
Schultz, P. G., 84–85, 90–91
Schulz, O., 141
Schumacher, A., 312–313
Schutze, G., 312–313
Schutz, P., 76–78, 79, 80, 95, 99, 100–102, 101f, 208f, 214–217, 243–244, 439
Schütz, P., 34–35
Schwanhausser, B., 276–277
Schwartz, A., 149–169
Schwede, T., 84f
Schweitzer, A. C., 276–277, 295
Schwer, B., 2, 215t, 217–218, 219
Sclavi, B., 126
Scott, M. J., 407t
Seaphin, B., 312
Seeliger, M. A., 2, 41, 76–78, 95, 99, 102, 133, 214–217, 218, 239–254, 441
Segref, A., 240–241
Seidel, C. A. M., 91, 93, 94, 98–99
Seidel, R., 102–103
Seifert, S., 34, 76–78, 79
Seifried, S. E., 172–173
Sekiguchi, Y., 467–468
Selak, N., 325–327, 326t, 328–329, 340
Selbach, M., 276–277
Semba, S., 407t
Semrad, K., 30, 112–113, 114
Sengoku, T., 34–35, 76–78, 95
SenGupta, D. J., 276
Seol, Y., 36
Seo, T., 407t
Séraphin, B., 76, 78, 95, 99–100, 201, 214–217, 241, 257, 310, 312, 325
Serebrov, V., 135, 466–467
Serin, G., 256
Serman, A., 329–331, 330t, 337–338
Serrano, A., 326t, 328
Servant, F., 326t, 328
Settlage, R. E., 309, 312
Seufert, W., 305–306, 310

Author Index

Seya, T., 407t
Seybold, P. G., 92
Shabanowitz, J., 309, 312
Shadrick, W. R., 463–485
Shahbazian, D., 325–327, 438, 440–441
Shah, N. G., 280, 304, 305–306
Shanks, J., 214–217
Shapira, S. D., 441
Sharifi, N. A., 256
Sharma, A., 405–436
Sharma, K., 240–241
Sharma, S. D., 135, 466
Sharpe Elles, L. M., 78–79, 214
Sharp, P. A., 334
Shav-Tal, Y., 218
Shaw, G. M., 429–430
Shcherbakova, I., 112, 117, 126
Sheehan, A., 291–294
She, M., 329
Sheng, G., 141
Sherman, N. E., 439
Shestakov, S., 391
Sheth, U., 325, 326t, 328–329, 330t, 331, 332–333, 337–338, 339
Shevchenko, A., 257, 312
Shibuya, T., 443
Shigesada, K., 172–174
Shih, J. W., 30, 338, 340, 348–349
Shim, E. Y., 325
Shimizu, Y., 464–465
Shimotohno, K., 407t
Shinobu, N., 140–141
Shin, Y. C., 141, 441
Shipley, G. L., 374
Shirai, M., 397
Shi, S. P., 46–51
Shi, Z., 325–327
Shkriabai, N., 407t, 411, 419–421
Shlaes, D., 372, 380–381
Shoemaker, D. D., 337
Shoemaker, R. H., 464–465
Shokat, K. M., 444
Shokolenko, I. N., 392
Short, S. A., 466
Shousha, S., 348–349, 363–365
Shrma, A., 407t, 409–410, 417
Shugar, D., 444
Shuman, S., 134, 135
Shum, B. O., 441
Sibler, A. P., 392–393, 393f
Siegers, K., 310
Sigel, R. K., 114
Sikorski, R. S., 297–298
Silverman, E. J., 215t
Silver, S. J., 441
Simoneau, B., 464–465
Simonis, N., 315
Simon, M. N., 424f

Simons, R. W., 370–371
Simpson, Z. B., 56
Sinan, S., 114, 121f, 123, 126
Sindbert, S., 98–99
Singer, J. D., 173–174
Singh, D., 407t, 409–410, 417
Singh, G., 256, 328–329, 339
Singh, R., 41–42
Singleton, M. R., 14–15, 133, 172
Sings, H. L., 465–466
Singson, A., 325
Sisamakis, E., 94
Siwecka, M. A., 480
Skarina, T., 185–186
Skinner, G. M., 36
Skordalakes, E., 157f, 173–174, 176, 180–182, 186–187
Sleep, J. A., 66–67, 70–71
Slonimski, P. P., 370
Small, E. C., 292t, 294, 295, 296–297
Smith, C. A., 96
Smith, H., 126
Smith, J. L., 407t
Smith, J. S., 165
Smith, P., 215t
Smith, R., 96
Snay-Hodge, C. A., 241
Sninsky, J. J., 348–349
Sohn, S. Y., 101f
Solem, A., 103–104, 114, 118, 126, 150–151, 158, 163, 166–167
Solomon, S., 329
Sommers, J. A., 464–465
Somoza, J. R., 135–137
Sonenberg, N., 41, 76, 80, 99–100, 192, 208–209, 208f, 215t, 290, 325–327, 438–439, 440–442, 443, 445, 446, 449, 455–456
Song, H., 76–78, 138, 249f, 329, 339
Song, M., 208–209
Song, O., 314
Sossin, W. S., 326t, 328
Sottrup Jensen, L., 200, 201f, 206, 208f
Soudet, J., 290, 296–297
Soukup, G. A., 150–151, 166–167
Soukup, J. K., 150–151, 166–167
Soulat, D., 348–349
Soultanas, P., 138
Sousa, R., 153
Sowers, L. C., 3
Spatrick, P., 256
Spraggon, G., 185–186
Srinivasan, M., 315
Srinivasan, S., 2
Staal, F., 407t
Staely, J. P., 292t, 294, 295, 296–297
Stafford, P., 215t, 241
Stahl, H., 5, 6, 363–365

Stahl, S. J., 16–17
Stalder, L., 328–329
Staley, J. P., 370
Stands, L., 370
Stanier, R. Y., 387–388
Stankiewicz, A., 467–468
Stanley, L. K., 102–103
Stan, R., 102–103
Stark, H., 215t
Starost, M. F., 444, 464–465
Steen, H., 291–294
Stefanovic, A., 348–349
Steinberg, I. Z., 92
Steinberg, M., 46–51
Steiner, B., 305–306, 310
Steiner, M., 114
Stein, G. S., 348–349
Steitz, J. A., 215t, 313–314
Steitz, T. A., 165
Stempel, K. E., 66–67, 69, 70
Stepien, P., 444
Sternberg, C., 378
Stertz, S., 441
Steube, K. G., 443–444
Stevens, R. C., 185–186
Stevens, S. W., 292t, 294, 295, 296–297
Stewart, B., 388–389
Stewart, M., 240–241
Stewart-Maynard, K. M., 423
Steyaert, J., 165
Stitt, B. L., 172–173, 174–175
Stoecklin, G., 324–325, 330–331
Storoni, L. C., 186–187
Story, R. M., 76–78, 94–95
Stotz, A., 215t
Strahm, Y., 215t, 241
Strahs, D., 126
Straka, A., 5, 6
Strambio-De-Castillia, C., 240–241
Strasser, K., 240–241
Strobel, S. A., 150–152, 154–155, 155t, 158, 161, 162, 163, 166–167
Strokovskaya, L., 467–468
Stromberg, R., 165
Strongin, A., 135, 137
Strop, P., 98–99
Struve, W. G., 7
Strycharska, M., 172
Stryker, J. M., 112–113
Stuhrmann, H. B., 198–199
Stulke, J., 371
Stutz, F., 215t, 240, 241
Su, B. N., 443–444
Subramanya, H. S., 138
Suchanek, M., 220
Sudo, K., 438–439, 464–465
Sugimoto, N., 166, 270
Suh, H., 113–114

Sukarieh, R., 325–327, 334–335, 443–444
Su, L. J., 114, 115f
Sullivan, M., 126
Sun, C., 292t
Sunden, Y., 407t
Superti-Furga, G., 312–313
Superti-Furga, G., 348–349
Su, S., 447
Sutoh, K., 96
Suydam, I. T., 150–152, 154–155, 155t, 158, 161, 162, 163, 166–167
Suzawa, M., 348–349
Suzich, J. A., 466
Suzuki, C., 41, 80, 439
Suzuki, H., 464–465
Suzuki, I., 386
Suzuki, K., 447
Suzuki, T., 407t
Svaren, J., 348–349
Svenson, C. J., 391
Svergun, D., 198–199, 201, 202b, 203b
Svergun, D. I., 196b, 198–200, 198b, 203–205, 206, 207, 208–210
Svitkin, Y. V., 99–100, 325–327, 438, 439, 440–441, 446
Swanson, S. M., 443–444
Sweeney, H. L., 41–42
Sweetman, G., 444
Swiatkowska, A., 277, 285
Swinney, D. C., 141–142
Swisher, J., 114, 115f
Swisher, K. D., 326t, 327, 332, 333–334
Sykes, M. T., 78–79, 214
Symoniatis, D., 291–294
Symons, J., 480
Szabo, A., 91
Szczelkun, M. D., 102–103
Szewczak, A. A., 161, 166–167
Szewczak, L. B., 150–151
Szostak, J. W., 7–8, 276

T

Tachikawa, H., 309–310
Taguchi, I., 467–468
Tai, C. L., 135
Tai, H., 348–349
Taira, K., 140–141, 407t
Takahashi, E., 438–439
Takahashi, I., 447
Takahashi, N., 309–310
Takahashi, S., 465–466
Takeuchi, O., 407t
Takeyama, K., 348–349
Talavera, M. A., 31–32, 34, 36–37, 46–51, 79, 95, 99, 228, 230–231
Tamaoki, T., 447
Tamura, J. K., 466

Tanaka, J., 325–327, 334–335, 340, 441–443, 445, 447, 455–456
Tanaka, K., 397
Tanaka, N., 2, 215t, 217–218, 219
Tan, C. P., 141
Tanenr, N. K., 307–308
Tange, T. O., 443
Tang, G. Q., 140–141, 143
Tang, H., 407t, 409
Tani, H., 467–468
Tanner, M., 114
Tanner, N. K., 2, 30–31, 241, 290, 308, 369–370, 408–409
Tan, Y. H., 407t
Tan, Y. J., 407t
Tapscott, S. J., 348–349
Targett-Adams, P., 407t
Tarn, W. Y., 325–328, 326t, 330, 330t, 338
Tartakoff, A. M., 241
Taylor, J. R., 423
Taylor, J. S., 391
Taylor, S. D., 150–151, 158, 163, 166–167
Tazi, J., 340
Teesdale-Spittle, P., 441–442, 443, 445, 455–456
Teixeira, D., 324–325, 329–330, 330t, 332–333, 337–338
Tellinghuisen, T. L., 480
Teng, C. M., 443–444
Teodoro, J. G., 443–444
Teplyuk, N., 348–349
Tetenbaum-Novatt, J., 240–241
Tettweiler, G., 440–441
Teukolsky, S. A., 165
Theil, K., 276–277
Theilmann, L., 466–467
Theissen, B., 81f, 83–85, 86, 87, 89–90, 92, 94, 95, 97f, 228, 230–231
Thiele, C., 220
Thiel, T., 391
Thoden, J. B., 96
Thomä, N., 37
Thomas, R. S., 348–349, 363–365
Thompson, A. M., 349
Thompson, G. S., 407t
Thompson, K. C., 9–10, 12–13
Thompson, W. A., 390
Thomsen, N. D., 2, 41, 76–78, 95, 99, 102, 133, 138, 157f, 171–190, 215t, 218, 241–242, 243–244, 248, 249f, 441
Thorner, J., 41, 215t, 241, 242–243, 246
Tiboulet, R., 407t
Tiedge, H., 439
Tijerina, P., 15, 34, 35, 55, 76–78, 79, 97–98, 112–113, 123, 126, 155
Tikuisis, A. P., 311–312
Tiller, G. E., 7
Tinoco, I. Jr., 36, 135

Tisdale, S., 334–335
Toesca, I., 370–371
Tollervey, D., 220, 275–288, 290–291, 292t, 294, 295, 296–297, 302t, 308, 309
Tomasetto, C., 215t
Tomko, E. J., 23, 55
Tommassen, J., 80
Tong, X. K., 326t, 328
Torchet, C., 219
Toretsky, J. A., 340
Toth, E. A., 12
Trachsel, H., 80, 99, 100–102, 101f, 208f, 214, 215t, 290, 438–439
Tran, E. J., 215t, 218, 241–242
Travers, K. J., 113–114
Treiber, D. K., 112
Tremblay, M. L., 443–444
Tremblay, S., 329
Trifillis, P., 256
Trinkle-Mulcahy, L., 291–294
Trochesset, M., 337
Truong, K., 407t, 409
Truong, M. J., 407t
Tsai, R. T., 215t, 217–218
Tsai, T. Y., 338, 340
Tsai, Y. H., 407t
Tscherne, D. M., 441
Tschopp, J., 141
Tseng, C. K., 215t, 217–218
Tseng, S. S., 241
Tse, W. C., 476
Tsou, A. P., 30, 348–349
Tsu, C. A., 34–35, 41–42, 133
Tsuchihashi, Z., 114
Tsuge, M., 465–466
Tsuneda, S., 467–468
Tucker-Kellogg, G., 372, 380–381
Tuerk, C., 276
Tuma, R., 150–151, 156, 157f
Tu Quoc, P. H., 371, 377–378
Tureci, O., 348–349
Turner, D. J., 276–277
Tuschl, T., 141, 325
Tuteja, R., 464–465
Tyagi, S., 468

U

Ueffing, M., 312–313
Uetz, P., 315–316
Uhlenbeck, O. C., 34–35, 36–37, 41–42, 55, 78–79, 86, 98–99, 114, 133, 165, 214, 222, 370–371
Uhlmann-Schiffler, H., 5, 363–365
Uhrin, D., 9–10, 12–13
Ule, A., 276–277
Ule, J., 276–277, 295

Ulicny, J., 447
Ulrich, A., 276–277, 286, 338
Umansky, L., 291–294
Umemura, T., 407t
Umezawa, K., 443–444
Unterholzner, L., 325
Unutmaz, D., 407t
Uren, A., 340
Urlaub, H., 215t, 325

V

Vaidya, N., 265
Valasek, L., 324–325, 331, 332–333, 340–341
Valdez, B. C., 5
Valencia-Sanchez, M. A., 332–333
Valeri, A., 94
Vallee, F., 185–186
Valle, R. C., 302t
van Brabant, A. J., 102–103
Vandenabeele, P., 447
van den Berg, S., 34–35, 76–78, 79, 95, 243–244
Vandenesch, F., 377–378
Vandenheede, J. R., 447
van der Scheer, C., 102–103
van der Weyden, L., 441
Vandesompele, J., 374
Van Diest, P., 348–349
Van Dorsselaer, A., 392–393, 393f
van Nues, R. W., 215t, 220
Van Orden, A., 3
van Raalte, A., 80
Vanrobays, E., 309–310, 312
Vantieghem, A., 447
van Tilbeurgh, H., 217–218
van Wijnen, A. J., 348–349
Vanzo, N. F., 370–371
Vasicova, P., 324–325, 331, 332–333, 340–341
Vasudevan, S. G., 135
Vaudaux, P., 375–377
Vaughn, G. P., 292t, 296–297
Vaughn, J. P., 325–327
Vavilin, D., 387–388
Vedadi, M., 185–186
Vela, A., 140–141, 143
Velankar, S. S., 138
Venema, J., 292t, 296–297, 302t
Venkatachari, N. J., 407t
Venkatesan, K., 315
Vepritskiy, A., 390
Vermaas, W., 387–388
Vesuna, F., 348–349
Vetterling, W. T., 165
Veverka, V., 439
Viazov, S., 469–470
Vierling, J. M., 465–466
Viggiano, S., 326t, 327–328
Vinh, J., 219
Vinnemeier, J., 386
Vioque, A., 395
Visscher, K., 36
Vite, G. D., 443–444
Vogel, A., 407t
Vogt, E., 441, 442–443, 445
Vogt, V. M., 424f
Vojtova, J., 324–325, 331, 332–333, 340–341
Volkmann, G., 84–85
Volkov, V. V., 198–199, 200, 204–205
Vo, M. N., 423
Vondriska, T. M., 312–313
Von Eiff, C., 375–377
von Hippel, P. H., 10, 172–174
VonKobbe, C., 241
von Moeller, H., 76–78, 215t
Vrljic, M., 98–99
Vuong, C., 377–378, 380–381

W

Wach, A., 280, 304, 305–306
Wadekar, S. A., 348–349
Wagner, G., 41, 80, 340, 439, 441–443, 445, 447, 453, 455–456
Wagschal, A., 407t
Wahl, M. C., 215t, 290
Wakita, T., 407t
Waldsich, C., 112–113, 114, 150–152, 162, 163
Walker, G. T., 164–165
Walker, J. E., 176, 177–178
Walmacq, C., 153, 154–155, 156–158, 159–161, 164, 273
Walsh, M. J., 348–349
Walstrum, S. A., 114
Walter, N. G., 126
Walther, D., 198–199
Wang, B., 329
Wang, C. H., 135, 137
Wang, F., 423
Wang, H., 407t, 439
Wang, J. J., 31–32, 34, 36–37, 79, 99, 228, 230–231, 407t, 469–470
Wang, L., 9–10, 12–13
Wang, N., 30
Wang, R., 135–137
Wang, S., 34–35, 78–79, 98–99, 164–165
Wang, W., 214
Wang, X., 2, 276–277, 295, 407t
Wang, Y. X., 141, 174, 391, 407t
Wan, Y., 113–114, 122–123
Warburg, O. H., 244
Ward, A. M., 407t
Wardle, G. S., 276–277, 286, 338
Wardrop, J., 348–349, 356, 363–365
Warkocki, Z., 215t
Warner, J. R., 298, 312
Warrener, P., 466

Author Index

Wasiak, S., 326t, 328
Watanabe, A., 407t
Watanabe, M., 348–349
Waterbury, J. B., 387–388
Waters, L. C., 439
Watson, R. P., 135, 137
Weaver, P. L., 241, 292t
Webb, M. R., 52, 53, 138, 174–175
Weber, C., 327–328
Weber, F., 141
Weber, O., 464–465
Weber, P. C., 135, 466
Weeks, K.M., 126
Weglohner, W., 78–79
Wehner, K. A., 302t, 309, 312
Wehrly, K., 429–430
Weigelt, J., 34–35, 76–78, 79, 95, 243–244
Wei, J., 215t
Wei, K. E., 312–313
Weil, D., 326t, 327, 329–331, 330t, 334, 337–338
Weiner, A. J., 466
Weirich, C. S., 41, 214–217, 215t, 241, 242–243, 246
Wei, R. R., 173–174
Weis, K., 2, 41, 76–78, 95, 99, 102, 133, 215t, 218, 239–254, 328–330, 330t, 441
Weiss, E., 392–393, 393f
Weissman, J. S., 291–294, 331–332
Weiss, S., 90–91, 93
Wei, X., 429–430
Wendel, H. G., 441–442, 443–444, 455–456
Weng, Y., 256, 270
Weninger, K. R., 98–99
Wen, J., 256
Wente, S. R., 215t, 218, 240–242
Wen, Y. M., 407t
Westberg, C., 407t
Westendorf, J. J., 348–349
Westermayer, S., 348–349
Weyer, J. L., 441
Wickens, M., 276
Wickner, R. B., 302t
Widger, W. R., 172–173, 174–175
Wieslander, L., 241
Wigley, D. B., 14–15, 133, 138, 172
Wilczynska, A., 326t, 327, 334
Wild, R., 443–444
Wild, U. P., 6–7
Wilkinson, M. F., 256
Willard, D. H. J., 466
Will, C. L., 290
Williams, B. R., 407t
Williams, K., 439
Williams, M. C., 423
Williamson, J. R., 78–79, 112, 214, 407t

Wilm, M., 241, 257, 310, 312
Wilson, B. J., 348–349, 356, 363–365
Wilson, D. S., 7–8
Wilson, G. M., 12
Wilson, J. E., 466
Wilson, T. J., 8, 9–10, 12–13
Winnard, P. Jr., 348–349
Winn, M. D., 186–187
Winquist, A., 165
Winsor, B., 310
Winzeler, E. A., 337
Wisskirchen, C., 407t
Witherell, G. W., 222, 439
Witte, G., 140–141
Witte, L., 443–444
Wittwer, C. T., 374
Witvrouw, M., 340, 444
Wlodawer, A., 439
Wlotzka, W., 280, 285
Wojnar, J. M., 441–442, 443, 445, 455–456
Wolf, D. E., 8
Wolf, R. Z., 116–117, 118, 120–121, 121f, 123, 124–126, 125f
Wolk, C. P., 390, 391
Wong, C. J., 3, 8, 9, 12–13, 18
Wong, R. W., 76–78, 79, 92
Wong-Staal, F., 407t, 409
Woodson, S. A., 30, 114, 118, 123, 126
Wood, T. C., 173–174
Wooldford, J. L. Jr., 215t, 292t, 296–297
Work, T. S., 449
Wormsley, S., 309, 312
Wortham, N. C., 348–349, 363–365
Wozniak, A. K., 98–99
Wray, V., 443–444
Wrenn, R. F., 56
Wu, C. G., 23, 55
Wu, C. W., 173–174
Wu, J. C., 407t
Wu, J. Y., 407t
Wu, L., 441
Wu Lee, Y. H., 338, 340, 407t
Wu, M., 407t
Wu, P., 8
Wu, Q., 141–142
Wu, S., 372, 380–381
Wu, T. S., 443–444
Wu, X., 429–430
Wyatt, J. R., 164–165
Wyman, J., 34, 36–37

X

Xavier, R. J., 441
Xiang, S. H., 407t
Xiao, M., 114
Xu, J., 443

Xu, T., 135, 137, 215t
Xu, Y., 172–173, 174–175, 329, 447
Xu, Z., 407t

Y

Yacono, P., 324–325, 330–331
Yamada, I., 465–466
Yamagata, K., 363–365
Yamamoto, T., 443–444
Yamamoto, Y., 363–365
Yamashita, A., 273
Yamauchi, Y., 309–310
Yanagisawa, J., 348–349
Yang, B., 407t
Yang, D., 76–78, 249f
Yang, H. S., 99–100, 439
Yang, J. P., 407t
Yang, Q. H., 5, 7, 12, 15, 20, 35, 55, 76–78, 112–113, 124, 214, 305–306, 310
Yang, X., 337
Yang, Y., 407t
Yano, T., 348–349
Yan, X., 348–349
Yao, D., 387–388
Yao, J., 378
Yao, N., 135, 466
Yates, J., 134
Yea, S., 348–349
Yeates, T. O., 172–173
Yedavalli, V. S., 406, 407t, 444, 464–465
Yeh, F. L., 215t, 217–218
Yeong, S. S., 135, 137
Yildirim, M. A., 315
Yi, M., 134, 407t
Yinglin, A., 407t
Yoder-Hill, J., 438–439
Yokoyama, S., 34–35, 76–78, 95
Yoneyama, M., 140–141, 166, 270
Yoon, J. H., 326t, 331, 332–334
Yoshida, H., 407t
Yoshida, M., 315
Yoshikawa, H., 309–310, 348–349
Yoshimura, K., 363–365
Young, C., 213–237
Younossi, Z., 465–466
You, Y., 9–10, 12
Yuan, B., 407t
Yuan, L., 340
Yuan, X., 114, 121f, 123, 126
Yuan, Z., 407t
Yu, E., 386, 387
Yu, H., 311–312, 315
Yu, L., 407t, 409–410, 417
Yu, S. F., 407t

Z

Zabetakis, D., 296–297
Zachariae, W., 310
Zagursky, R. J., 372, 380–381
Zakowicz, H., 439
Zamborlini, A., 407t
Zambryski, P., 326t, 327–328
Zamecnik, P. C., 8
Zang, X., 391
Zanoli, S., 407t, 444
Zarnack, K., 276–277
Zaug, A. J., 114
Zemb, T., 193b, 196b
Zemora, G., 112–113
Zenklusen, D., 215t, 241
Zentella, R., 391
Zeuzem, S., 465–466
Zhang, B., 276
Zhang, C. C., 338, 397
Zhang, H., 407t
Zhang, J., 76–78, 249f, 407t, 409
Zhang, J. H., 468, 477
Zhang, L., 114, 215t
Zhang, M. J., 312–313
Zhang, N. J., 444, 464–465, 480
Zhang, P., 444, 464–465
Zhang, S., 407t, 409
Zhang, W., 337
Zhang, X., 134, 309, 391
Zhang, Y., 114
Zhang, Z., 429–430
Zhao, J., 241
Zhao, P., 348–349
Zhao, R., 215t
Zhao, Y., 407t
Zheng, B. J., 407t
Zheng, M., 154f
Zheng, S., 328–329
Zheng, X., 99–100
Zheng, Y. H., 407t
Zheng, Z. M., 407t
Zhong, G., 309, 311–312
Zhong, J., 378, 439
Zhou, D., 407t, 409
Zhou, N., 407t
Zhou, T., 407t
Zhou, Y., 241, 407t, 441
Zhou, Z., 76–78, 249f
Zhuang, X., 126
Ziebuhr, J., 138
Zilliges, Y., 397
Zimmerle, C. T., 56
Zingler, N., 118
Zivarts, M. V., 166–167
Zolinger, C., 407t, 409–410, 415
Zorca, C., 215t
Zupan, B., 276–277

Subject Index

Note: Page numbers followed by "*f*" indicate figures, and "*t*" indicate tables.

A

Advanced light source (ALS), 186–187
Anabaena sp. strain PCC 7120 transformation
  cargo plasmids, 390
  E. coli parental cells preparation, 391
Atomic resolution models
  crystal structure, Prp43, 203
  eIF4A solution structure, 201
  scattering data, Prp43 and eIF4A, 201–203, 202*f*
  solution structures
    constraints, 204
    CORAL program, 204
    DAMAVER program, 205
    dummy residues, 204
    macromolecules, 205–206
    Prp43, 204–205, 205*f*
    rigid body modeling, 203, 203*b*
    SASREF program, 203–204
    spherical harmonics, 203–204
  structure, 201, 202*b*
ATP hydrolysis. *See* Oxygen isotopic exchange probes, ATP hydrolysis

B

Biochemical and biophysical methods
  RNA-affinity chromatography, 417–419
  RNA-binding activity using EMSA, 419–424
Briggs-Haldane equation
  observed ATPase rates, 44
  quadratic form, 44
  steady-state, 44

C

Catalytic RNA. *See* RNA Folding
Cell-associated RNA helicase, cultured mammalian cells
  polysome association
    efficient isolation of protein, 416
    isolation from gradient fractions, 416–417
    ribosomal profile analysis, 415–416
  siRNA downregulation
    advantages, 410
    RHA silencing, 410–411
    transfection with siRNA complementary, 409–410
  with target RNA/protein cofactors
    epitope immunoprecipitation, 412–413
    genome-wide or candidate target mRNA, 413–414
    protein extraction for proteomic analysis, 414–415
    RHA translational control, 412
Cellular ultrastructure, 397–398
Cofactors, RNA helicases
  effects, 220
  enzymatic activity
    biochemical and structural data, 218
    crystal structures, Dbp5 and Upf1, 218
  identification
    co-immunoprecipitation assays, 219–220
    genetic interactions, 219
    helicase–cofactor interaction, 220
    in vitro effects, 219
    photoactivatable and chemical cross linkers, 220
    yeast two-hybrid assays, 219
  in vivo, 217–218
  kinetic cycle, 214–217, 217*f*
  protocols (*see* Protocols, RNA helicases)
  Prp43 and Upf1, 214
  reagents
    gels and loading dyes, 221
    purification, nucleotides (*see* Nucleotide purification)
    purification, RNAs, 223–224
    reaction buffers, 221
    RNAs labeling, ^{32}P, 221–223
  S. cerevisiae, 214, 215*t*
  tandem RecA-like domains, 214
Co-immunoprecipitation (co-IP), 309–310, 355–356, 357–358
Conformational cycle, DEAD-box proteins
  full-length YxiN structural model (*see* Full-length YxiN structural model)
  guidance mechanism, eIF4A activation
    crystal structure, comparison, 100–102
    inspection, eIF4G-stabilized conformation, 102
    open and closed state, 99–100
    primary and secondary interface, 100–102
    regulation, helicase core, 99–100, 101*f*
    "stopper function", 100–102
  helicase modules, 102–103
  YxiN helicase core (*see* YxiN helicase core)
CRAC. *See* Cross-linking and analysis of cDNA (CRAC)

513

Subject Index

Cross-linking
  CLIP technique, 276–277
  conditions
    growing cultures, 281
    living cells, 280–281
    protein-RNA interactions, 280
    yeast, culture medium, 281
  CRAC (see Cross-linking and analysis of cDNA (CRAC))
  mutants, 287
  UV cross-linking
    growing cells, 282
    in vitro, 281–282
    resuspended cells, 282
  weak, 286–287
Cross-linking and analysis of cDNA (CRAC)
  average length, 278
  data analysis, 285, 286
  His6 tag, 279–280
  in vitro experiments, 281–282
  RNA helicase, 282
  TAP protocol, 277
Crystallization, RNA and adenosine nucleotides
  crystal-form I
    data processing, 178–180
    Matthew's analysis, 178–180
    substrate matrix, 178–180, 178t, 179f
  crystal-form II
    data collection and partial refinement, 180–182, 180t
    r(CU)$_n$ polymers, 180–182
    structure solution and ligand binding, 180–182, 181f
    substrate matrix, 178t, 180–182
  crystal-form III
    "aging" screen, 184–185
    cryoprotectant, 182–183
    description, 182–183
    diffraction images, 183
    nonmerohedral twinning, 183–184, 184f
    pseudo-merohedral twinning, 184–185
Cyanobacteria
  CrhR and CrhC proteins, 387
  description, 386
  growth
    cautionary note, 388
    freshwater, 387–388
    preservation, 388
  RNA helicases (see RNA helicases)
Cytoplasmic mRNP granules, RNA helicases
  ATPase domain mutations, 339–340
  genetic depletion/deletion
    benefits, 337–338
    null phenotype, 337
  localization
    Dbp5/Rat8 and Mex67, 327–328
    Ded1 accumulation, 325–327, 327f
    Dhh1/RCK, Upf1 and Mov10, 325
    ISE2 and *Arabidopsis thaliana*, 327–328
    mammalian cell culture, 334–335
    mechanism, 328–329
    neuronal transport granules, 328
    neurons and germ cell granules, 328
    P-bodies and SGs, 325, 326t
    quantification, 335–336
    S. cerevisiae (see *Saccharomyces cerevisiae*, cytoplasmic mRNP granules)
    yeast and mammalian cells, 325–327, 327f
  N and C terminal mutations, 340
  overexpression, 338
  small molecule inhibitors, 340

D

DAMAVER, 200, 204–205
DAMFILT, 200
DAMMIF, 199, 209–210, 209f
DEAD-box helicases
  chaperones (see RNA chaperones)
  RNAs folding, 113–114
  RNAs folds
    3-D structures, 112
    "kinetically trapped" folding, 112
  self-splicing (see Self-splicing)
  smFRET (see Single molecule fluorescence resonance energy transfer (smFRET))
  substrate cleavage, native state formation, 118–124
DEAD-box proteins (DBP), mRNA export
  cargo, 240–241
  Dbp5, Gle1 and Nup159, purification
    concentration and storage, 243t
    culture growth and induction, 242t
    expression constructs, 242–243
    freezing and thawing, 243–244
    protein, 242t
  description, Dbp5, 241
  fluorescence polarization
    ATPase measurements, Dbp5, 247f, 248
    fluorophores, 247–248
    measuring RNA and nucleotide release rates, Dbp5, 251–252
    monochromators, 247–248
    reagents and instrumentation setup, 248–250
    RNA and adenosine nucleotide, 250–251
    separation process, 246–247
  NPC (see Nuclear pore complex (NPC))
  production and translation, 240
  recycling, enzyme, 241–242
  RNA–protein interactions, 241
  steady-state ATPase assay (see Steady-state ATPase assay)
DEAD-box proteins (DBP) preparation
  activities, 86
  donor and acceptor dyes, 87

Subject Index

fluorophores, 84–85
hairpin 92, 86
ligation, 87
procedure, 84–85, 85f
RecA domains, 83–84
statistical labeling reaction, 87
strand displacement, 86
structure, YxiN, 83–84, 84f
surface-exposed cysteines, 86
DEAD box RNA helicases, transcriptional regulation
  cellular processes, 348
  innate immune signaling pathways, 348–349
  interactions, *in vitro* and mammalian cell lines
    co-IP, 355–356, 357–358
    GST-tagged proteins (*see* GST-tagged proteins)
    His-tagged proteins (*see* His-tagged proteins)
    NaCl concentration, 356
    nuclear extracts, preparation, 356–357
    ^{35}[S]-labeled proteins, 353
  p68 and p72 proteins
    description, 348–349
    luciferase assays, 363, 364f
    RNA helicase activity, 362
    sumoylation, cell lines (*see* p68/p72 sumoylation, cell lines)
    transfection, cells, 362–363
  transcriptional coregulators, 348
Dengue virus, 464–465
Deoxyribonucleic acid (DNA)
  helicase, 464–465
  inhibitors effects, strand separation
    apparent affinity, compound, 474–475
    prism equation, 475–476
    protocol, 474–475
DExD/H-box helicases, 304, 307–308
Discontinuous assay
  footprinting data, 126
  RNA catalytic activity, 126
  SAXS, 126
  unfolding native structure
    chaperone protein, 124–126
    misfolded states, 124
    reversible unfolding, 124–126
DNA. *See* Deoxyribonucleic acid (DNA)
Double-stranded translocation
  ATP binding and hydrolysis, 143–144
  CARD1 and CARD2, 143
  central dogma, 141
  dsRNA and ssRNA, 140–141
  dsRNA, HEL2 and HEL2i, 144
  dynamic motions, RIG-I, 141–142, 142f
  HEL1 and HEL2, 142–143
  immune signaling and RNA interference, 140–141
  N and C-termini, 140–141
  RecA-like domains, 140–141
  recognizing and binding dsRNA, 140–141
  RNA duplex, 143
  SAXS, 143
  single-molecule fluorescence, 141–142
  triphosphorylated duplex substrates, 144
Duplex concentration
  RNA, 4–5
  strand annealing
    fluorescence measurements, 5
    rate constants, pseudo first-order, 5
  strand exchange
    ATP-independent, 6
    concentration change, 5–6
    scavenger RNA/DNA, 5
Duplex substrate design and preparation
  2-AP, RNA labeled
    description, 10
    preparation and purification, 11t
    radiolabeled, 10–12
    strands, purification, 10t
  Cy dyes, RNA labeled, 9–10
  fluorophor assessment, unwinding reaction
    competition experiments, 13
    EMSA-based methods, 13
    protein-induced fluorescence enhancement, 12–13
    stability changes, 12
    stacking, 12
  labeling techniques, 7–8
  principles, 7
  RNA labeling approaches
    description, 8–9, 8f
    FRET, 8–9
    quenching techniques, 9

E

eIF4AI
  DHX29, 440–441
  eIF4A isoforms, 438–439
  eIF4A role, eukaryotic translation initiation, 439, 440f
  helicases, 438
  macromolecules, 439
  ribosome recruitment, 438
  screening, ATPase inhibitors
    control reactions, 446
    hypericin effects, 447, 452f
    kinase inhibitors, 447, 450t
    reaction volume, 446
    RNA-dependent ATPase activity, hypericin, 447, 448f
eIF4AII
  helicases, translation initiation, 438–441
  isoform, 447
eIF4F
  eIF4A, 439

eIF4F (cont.)
and eIF4B, 440–441
Electrophoretic mobility shift assays (EMSA), RNA-binding activity
  expression and purification, recombinant proteins, 420–421
  gel electrophoresis, 421
  in vitro transcription, 420
Eukaryotic ribosome assembly, RNA helicases
  Mtr4, 290–291
  NTP-dependent nucleic acid remodeling enzymes, 290
  pre-rRNA processing, *Saccharomyces cerevisiae*, 291f
  pre-SSU and pre-LSU, 290
  protein–protein interactions
    Co-IP, tagged proteins, 313–314
    pre-rRNA transcript, 309
    TAP analysis, 309–313
    Y2H analysis, 314–316
  RB helicases evaluation
    HITS-CLIP, 295
    nucleolar localization, 291–294
    protein, nucleolus, 294
    putative helicase protein, 296f
    RNP factors identification, 294
    yeast, 291–294
    yeast DEAH-box protein Prp43, 295
  RB pathway
    evaluation, rRNA production, 297–304
    perturbation, 304–308
    pre-rRNA trimming, 297
    RNA polymerase I-mediated transcription, 296–297
Eukaryotic translation initiation factor 4A
  eIF4AI, eIF4AII and DHX29, 438–441
  helicases, druggable targets, 441
  molecule inhibitors, eIF4A ATPase activity
    material and supplies, 445
    production, recombinant protein, 445–446
    RNA pull-down assay, 449–453
    screening, eIF4AI, 446–448
    UV-induced ATP cross-linking, 449
  molecule inhibitors, RNA helicases
    hippuristanol, 442–443
    in vitro translation assay, 441–442, 442f
    NTPase, 444
    pateamine A, 443
    rocaglates, 443–444

F

Fluorescence anisotropy (FA), RNA-binding activity
  measurements, 422–424
  template, 422t
Fluorescent-labeled ATP and ADP analysis
  ADP dissociation and isomerization steps, 49
  biphasic time courses, 49
  fluorescence signal, 49
  parameters, 49
  phases, 49
  description, 46–51
  method, 48t
  methylanthraniloyl (mant)
    ADP dissociation, 48t
    ATP/ADP association, 48t
  Förster mechanism, 80–82
  Förster resonance energy transfer (FRET), 467–468
Full-length YxiN structural model
  approaches, 98–99
  distance restraints, 98–99
  mean distances, 98–99
  orientations, RNA-binding domain (RBD), 99
  protein-interaction site, 99

G

GASBOR, 199–200
Gel-shift RNA binding assay, 225
Global analysis, NAIM
  duplex unwinding, 164
  helicase and nonbridging pro-R oxygens, 165
  helicase translocation, 164
  Rho interactions, 164
  RNA:DNA helix, 164–165
Gram-positive RNA decay, DEAD-box RNA helicases
  CsdA and SrmB, 370–371
  cyanobacteria, 371
  dissociation activity, 370
  eukaryotic genomes, 370
  motifs, 369–370
  mRNA decay measurement (*see* mRNA decay measurement)
  phenotypic readouts
    alpha and delta hemolysin, 378
    biofilm assay, 378–379
    cshA gene disruption, 377–378
    degradosome role, 377–378, 379f
    hemolysis assay, 380–381
    targetron system, 378
  quorum-sensing system, 371
  RhlB and RhlE protein, 370–371
  Tn916 mutagenesis, 371
  translation initiation and ribosome biogenesis, 370
GST-tagged proteins
  bacterial
    expression, 349t
    lysis buffer, 350t
    purification, 350t

Subject Index

binding
  beads preparation, 350–351
  in vitro translated proteins, 353–355

H

Helicases. *See also* Target sites identification, RNA helicase
  activity selection
    accurate measurements, 159–161
    autoradiography, 159–161
    high-salt concentrations, 158
    NPH-II helicase, 158
    rate, duplex unwinding, 159
    Rho helicase, 159, 160*f*
    Rho hexamers, 159–161
    RNA:RNA:DNA substrates, 159
  substrates
    architecture, 156
    DEAD-box, 155
    duplex region, 156
    lack, redundant/stabilizing components, 156
    PBS and SBS, 156
    Rho helicase, 156–158, 157*f*
    RNA arm, 156–158
    RNA:DNA/RNA:RNA, 155
  UPF1 (*see* RNA helicase UPF1, biochemical characterization)
Hepatitis C virus (HCV)
  drug targets
    helicase inhibitors, 466
    HTS, 465
    NS3, 465–466
    Telaprevir and Boceprevir, 465–466
  HSV, 464–465
  HTS, helicase inhibitors, 467–468
  inhibitors, viral RNA helicases, 464
  MBHA, 470–477
  NS3h, 466–467, 469–470
  replicon-based assay, 480
  RNA-based SBHA, 478–480
Herpes simplex virus (HSV), 464–465
High-throughput screening (HTS)
  evaluation, HTS, 476–477, 477*f*
  HCV helicase inhibitors
    MBHA, 468
    and standard helicase assays, 467–468
  MBHA-based, 472–473
Hill binding isotherm, 23
Hippuristanol, 442–443
His-tagged proteins
  binding
    in vitro translated proteins, 353–355
    Ni^{2+} beads preparation, 352–353
    expression and purification, 351–352

HSV. *See* Herpes simplex virus (HSV)
Hypericin
  concentrations, 449
  defined, 447
  identification, 455–456

I

IFT. *See* Indirect Fourier transformation (IFT)
In vitro transcription
  NαS and dNαS analogs, 154–155, 155*t*
  PAGE, 154–155
  PCR, 154
  synthetic DNA templates, 154, 154*f*
  T7 RNAP variant, 154–155
Illumina Solexa sequencing, 278, 285
Indirect Fourier transformation (IFT)
  description, 195
  particle oligomerization and aggregation, 198
  scattering data, 196*b*, 197*f*
Inducible RNA helicase expression, 392–393
Influenza virus, 441
Inhibitors identification and analysis. *See* Hepatitis C virus (HCV)
In situ RNA helicase localization
  Cryo-EM, *Synechocystis* sp. strain PCC 6803, 399*f*
  harvesting cyanobacterial cells, 398–400
Intereference signals
  correlation function, 165
  physicochemical parameters, 165
  power spectrum analysis, 165
  QSAR, 166
  Rho chemomechanical cycle, 166
  specific value, 166

K

Kinetic and equilibrium methods, DBP ATPase
  activity, 31
  cycle reaction scheme, 30–31
  binding stoichiometry, RNA-DBP, 36–37
  description, 30
  equilibrium RNA binding affinity
    ADP-bound states, DBPs, 38
    affinity measurement, ADP, 38
    binding/dissociation rate constants, ADP, 39
    DBP-bound nucleotide, 38
    DBP-RNA equilibrium binding, 38–39
    density values, 37–38
    description, 37
    titrations, 37
  oligomeric state and stability/aggregation, 34
  quantitative analysis, DBP ATPase cycle
    expressions, 57–58
    ^{18}O-isotope exchange, 56–57
    qualitative agreement, 58

Kinetic and equilibrium methods, DBP (cont.)
  rate constants, 57–58
  steady-state kinetic constants, 57
  reagents and equipment
    ATP, ADP, and mant-labeled nucleotides, 32
    DBP, 32
    fluorimeter, 33
    quench-flow apparatus, 33
    RNA substrate(s), 32
    solution conditions and temperature, 31–32
    stopped-flow apparatus, 33
    UV-visible spectrophotometer, 33
  RNA substrate selection/design, 34–36
  RNA unwinding assays and ATPase coupling
    description, 55
    fast-mixing techniques, 55–56
    strand displacement, 56
  simulations, 56
  steady-state ATPase measurements
    DBP ATPase activity, 39–41
    real-time enzyme-coupled assay, 41–45
  transient kinetic analysis (see Transient kinetic analysis, DBP ATPase cycle)

M

MBHA. See Molecular beacon-based helicase assay (MBHA)
Michaelis constant (K_m), 41
Model-independent analysis
  experimental scattering data, 197–198, 197f
  IFT, 195, 196b
  Kratky plot, 197–198
  particle oligomerization and aggregation, 198
  p(r) function, 195–197
  radius, 194, 195, 195b
  SAXS data, 194
  scattering data, 195–197, 197f
Molecular beacon-based helicase assay (MBHA)
  compound interference, 471–472
  HTS, 470–471, 471f, 472–473, 476–477
  inhibitors effects, 474–476
mRNA decay measurement
  data analyses
    reference RNA, choosing, 374, 375–377
    relative half-life, 374–375, 375f
    northern blotting, 373f, 374
    qRT-PCR, 372–374
  RNA preparation
    growth phases, 372
    rifampicin treatment, 371–372
    RNeasy Mini Kit, 372
    sample preparation, 372, 373f
mRNA protein complexes (mRNPs). See Cytoplasmic mRNP granules, RNA helicases

N

NAIM. See Nucleotide analog interference mapping (NAIM)
NanoDrop, 285
NMD. See Nonsense-mediated mRNA decay (NMD)
Nonsense-mediated mRNA decay (NMD)
  core components, 256
  description, 256
  RNA helicase UPF1 (see RNA helicase UPF1, biochemical characterization)
Northern analysis, transcript levels
  electrophoresis and blotting, 394–395
  RNA loading control, 395
  transcript detection, 395
Northern blotting, mRNA decay measurement, 373f, 374
NPC. See Nuclear pore complex (NPC)
NS3 helicase (NS3h). See also Hepatitis C virus (HCV)
  expression and purification, 469–470
  HCV, 466
  NS3, liver cancer, 467
  NS3/NS4A protease, 466–467
NS3/NPH-II family. See SF2 proteins
Nuclear pore complex (NPC), 240–241
Nucleotide analog interference mapping (NAIM)
  materials and reagents
    buffers, 153
    chemicals, 152
    enzymes, 153
    equipment, 152–153
  methods
    global analysis, 164–165
    helicase activity selection, 158–161
    helicase substrates, 155–158
    in vitro transcription, 154–155
    intereference signals, 165–166
    preparation and purification, tripartite, 158
    quantitation and normalization, NAIM signals, 161–163
    sequence effects, 163–164
    sequencing, selected populations, 161
  RNA folding, 150–151
  RNA–helicase interactions, 150
  transcripts, 150, 151f
Nucleotide purification
  ^{32}P-labeled ATP, 223
  unlabeled ATP, 223

O

Oligonucleotides, RNA helicase target sites identification
  complex genomes, 278
  5′DNA/RNA hybrid adapter sequences, 278
Oxygen isotopic exchange probes, ATP hydrolysis

Subject Index

exchange approaches, 67
fitting intermediate exchange data
  description, 70–71
  mass spectroscopic analysis, 71–72
  unenriched ATP, hydrolysis, 70–71, 71f
  unenriched Pi, 71–72
isotopic analysis, GC–MS, 70
k_{-2}/k_3 ratio determination, intermediate exchange
  assay, reaction progress, 68
  description, 67–68
  enrichment, 67
  isotopic analysis, 68
  pH, 68
  pyruvate kinase and PEP, 67
medium Pi = HOH medium exchange
  description, 72
  net hydrolysis, 68–70
^{18}O-enriched water, unenriched ATP, 66–67
species distribution, 66–67
volatile triethyl phosphate, Pi derivatization, 68–70
water-derived oxygen, 66–67

P

Pateamine A (Pat A), 443
PKC. *See* Protein kinase C (PKC)
Polysome association, RNA helicase
  efficient isolation of protein, 416
  isolation from gradient fractions, 416–417
  ribosomal profile analysis, 415–416
p68/p72 sumoylation, cell lines
  description, 358
  myc-tagged p68/p72 and His-tagged SUMO, 359
  Ni^{2+} isolation
    steps and solutions, 359–360, 361
    Western blotting, 359, 360f
Premature translation termination codon (PTC), 256, 273
Pre-steady-state approaches, unwinding reactions
  DEAD-box helicases, 15
  differences, unwinding modes, 15–16
  implementation, 16
  local strand separation, 15, 16f
  multiple cycle regime
    amplitudes, measurement, 20
    annealing rate constant, 20
    description, 17f, 19
    rate constant, DEAD-box helicases, 19–20
  protocol
    multiple cycle, 20, 21f
    single cycle, 18–19
  single cycle regime
    absolute reaction amplitude, equation, 18
    description, 16–17, 17f
    rate constant, DEAD-box helicases, 17–18

scavenger, use, 16–17
strand separation, translocating helicases, 18
translocation-based unwinding, 14–15, 15f
Processing-bodies (P-bodies). *See* RNA helicases analysis, P-bodies and SGs
Protein complex analysis
  ab initio modeling, 208–209, 209f
  complex, Prp43–Pfa1, 209–210, 209f
  linear combination, scattering data, 206, 207b, 207f
  solution structure, eIF4A and eIF4G-MC, 206, 208f
  super-positioning, 207, 208f
Protein kinase C (PKC), 447
Protocols, RNA helicases
  annealing, 233
  ATPase activity
    advantage, 228
    considerations, 229
    nucleotide affinities, 230–231
    single-turnover ATPase assays, 229–230
  binding
    considerations, 224–225
    gel-shift results, 225–227
    gel-shift RNA binding assay, 225
    specificity, 227
  nucleotide affinities, 230–231
  preparation, RNA duplex, 231–232
  release rate constant, 227–228
  unwinding, 232–233
PTC. *See* Premature translation termination codon (PTC)

Q

Qiagen MinElute reaction cleanup kit, 285
Quantitation and normalization, NAIM signals
  conservative confidence, 162–163
  ImageQuantTL software, 161
  iodine-independent RNA cleavage, 162
  NαS modifications, 161–162
  normalized band intensities, 161
  phosphorothioate tags, 162
  standard deviation (SD), 162
Quantitative reverse transcription PCR (qRT-PCR), mRNA decay measurement
  description, 372–374, 373f
  reference RNA and primers, 374

R

Real-time enzyme-coupled assay
  analysis
    Briggs-Haldane equation, 44
    observed ATPase rates, 44
    RNA:DBP stoichiometry, 44
    RNA-stimulated ATPase activity, 44
  description, 41–42
  method, 43t

Real-time enzyme-coupled assay (cont.)
  NADH absorbance, 42–45
  stock solutions, 42t
  working solutions, 43t
RecA-like domains
  autoinhibitory/regulatory roles, 133
  auxiliary cofactors (Gle1), 133
  ssRNA/dsRNA, 133, 134f
  substrate recognition and helicase function, 133
RHA-deficient HIV-1 virions, 430–431
Rho hexamer
  "anchoring" RNA arm, 156–158
  initiation mixture, 159–161
  transcription termination, 156, 157f
Rho–RNA–nucleotide complex, 177
Ribosomal profile analysis, 415–416
Ribosome biogenesis (RB). *See also* Eukaryotic ribosome assembly, RNA helicases
  eukaryotes, 290
  evaluation, rRNA production
    gel electrophoresis and transfer, 300
    Northern analysis, 301–304
    RNA extraction, yeast cells, 299
    selection and culture, yeast strain, 297–298
    steps, 297
  helicase activity
    chromosomal insertion, GAL1 promoter, 304–306, 305f
    principle, 304
    protein depletion, 306–307
    protein mutagenesis and expression, 307–308
Ribozyme
  HIV NC protein, 114
  Saccharomyces cerevisiae, D135, 120–121
  Tetrahymena thermophila, 113–114
RNA-affinity chromatography
  biotinylation of RNA, 417–418
  cofactors isolation, 418–419
RNA annealing, 233
RNA-binding activity
  using EMSA
    expression and purification, recombinant proteins, 420–421
    gel electrophoresis, 421
    in vitro transcription, 420
  using FA
    measurements, 422–424
    template, 422t
RNA chaperones
  and ATP, 112–113
  catalytic activity, 113
  folding
    group I and group II introns, 114, 115f
    HIV NC protein, 114
    kinetic barriers, 114
    secondary and tertiary structures, 114

misfolded nonnative structures, 112–113
RecA domain, 112–113
RNA-dependent hexameric helicase
  ATPase sites, 172–173
  biochemical foundation
    ATP analogs, 174–175
    defined, primary and secondary site, 173–174
    motor domains, 173–174
    stoichiometric concentrations, 174
  crystallization (*see* Crystallization, RNA and adenosine nucleotides)
  description, 172
  ligand preparation and storage, 176
  mechanism, Rho's catalytic, 173
  nucleic acid substrate and nucleotide cofactors, 172
  protein crystallization experiment, 185–186
  protein expression, purification, and storage
    BL21 pLysS cells, 175
    cation-exchange column, 176
    defined, signal-to-noise ratio, 175
    freezing and two gel-filtration columns, 176
  pseudo-symmetric protein–ligand complex, 186
  Rho–RNA–nucleotide complexes, 177
  structural analysis, 187
  structure solution and refinement, 186–187
  substrate-centric crystal-screening strategy
    ADP and AMP-PNP mixtures, 177–178
    description, 177–178
    matrix, 177–178, 178t
RNA folding
  chaperone-assisted RNA (*see* RNA chaperones)
  native state, catalytic activity
    group I intron, 113–114
    oligonucleotide substrate, 113–114
    Tetrahymena thermophila, 113–114
RNA helicase assay
  generation, duplexes, 455
  helicase assay, 455
  hypericin effects, 453f, 454–455
  in vitro transcription, RNA-1, 454
  ^{32}P-labeling, RNA-11, 454
RNA helicases
  Anabaena sp. strain PCC 7120 transformation, 390–391
  ATP hydrolysis (*see* Oxygen isotopic exchange probes, ATP hydrolysis)
  biochemical and biophysical methods, 417–424
  cellular ultrastructure, 397–398
  confirmation, genetic alteration, 393–394
  cyanobacteria growth
    cautionary note, 388
    freshwater, 387–388
    preservation, 388

detection in virion preparations, 425–427
duplex unwinding analysis (see Stopped-flow
  fluorescence spectroscopy)
enzymatic and nonenzymatic functions,
  408–409
experimental design, 409
gene inactivation, 392
host and viral factors, 427–428
in situ RNA helicase localization, 398–400
inducible, 392–393
infectivity measurement, 429–431
molecule inhibitors
  hippuristanol, 442–443
  NTPase, 444
  Pat A, 443
  rocaglates, 443–444
Northern analysis, transcript levels
  electrophoresis and blotting, 394–395
  RNA loading control, 395
  transcript detection, 395
polysome association, 415–417
protein isolation, 389–390
RNA isolation, 388–389
role, virus replication, 407t
siRNA downregulation, 409–417
Synechocystis sp. strain PCC 6803
  transformation, 391
with target RNA/protein cofactors, 411–415
virus–host interface, 406–407
Western analysis, protein levels
  complexes detection, 396–397
  membranes blocking, 396–397
  protein loading controls, 397
RNA helicases analysis, P-bodies and SGs
  assembly and disassembly, 329, 330t
  ATPase-deficient mutants, overexpression,
    341
  cellular stress, 324
  cytoplasmic mRNP granules (see Cytoplasmic
    mRNP granules, RNA helicases)
  Ded1/DDX3 and Dhh1/RCK, 329–330
  glucose deprivation, budding yeast, 330–331
  mRNAs regulation, 324
  overexpression, DED1, 330
  proteins shuttle in and out, 324–325
  stress/genetic condition, 340–341
  Upf1, 331
RNA helicase UPF1, biochemical
  characterization
  ATPase assay
    charcoal-based, 267–268
    hydrolysis, 268
    reaction mix preparation, 268
  complex assembly
    helicases, 260
    pre-blocking affinity beads, 261–262
    protein coprecipitation, 260f
    protein-protein interaction, 262–263

RNA-protein coprecipitation, 263–264
RNA pull-down assays, 260–261
preparation
  CBP, 257
  cloning, 258
  expression, 258
  histidine purification, 257f
  purification, 258–259
RNAse protection assay
  assay procedure, 266–267
  substrate preparation, 264–266
unwinding assay, 268–273
RNA pull-down assay, 449–453, 453f
RNAs labeling
  body-labeling, ^{32}P-α-ATP, 222
  5'-end-labeling, ^{32}P-γ-ATP, 222–223
  in vitro transcription, 221
RNA substrate selection/design
  analytical ultracentrifugation, 36
  ATPase activity, 35
  ATP-dependent RNA unwinding, 35
  DBP-RNA binding
    affinity, 35
    interactions, 35–36
    specificity, 34–35
  structure, 35, 36f
RNA translocases
  autoinhibitory/regulatory roles, 133
  chemical and structural determinants, 132
  HCV NS3, 132–133
  NS3/NPH-II family, SF2 proteins, 134–137
  retinoic acid inducible gene I (RIG-I),
    132–133
  RIG-I/dicer family, 140–144
  SF2 proteins, 134–137
  structural and functional behavior, 132–133,
    132f
  UPf1 family, SF1 proteins, 138–140
RNA unwinding, 232–233
Rocaglates, 443–444
rRNA production
  gel electrophoresis and transfer, 300
  Northern analysis
    oligonucleotide probes, 301–304, 302t
    procedure, 301–304
  RNA extraction, yeast cells, 299
  selection and culture, yeast strain
    guidelines, 297–298
    YPH499, 297–298
  steps, 297

S

Saccharomyces cerevisiae, cytoplasmic mRNP
  granules
  induction
    glucose deprivation, 332–333
    high heat stress, 332–333
    sodium azide treatment, 332–333

Saccharomyces cerevisiae, cytoplasmic mRNP
granules (*cont.*)
  stresses, 332
  "wild-type" laboratory strains, 333–334
  markers
    budding yeast and plasmids, 331
    GFP fusions, 331–332
    P-body factors, 332
SAXS. *See* Small-angle X-ray scattering (SAXS)
SBHA. *See* Split beacon helicase assay (SBHA)
Self-splicing
  catalytic steps, 118
  chaperone-promoted changes
    intron folding, 117
    rate constant, 117
  folding and the catalytic steps, 115–116
  group I and group II introns, 118
  products, intron, 116–117, 116*f*
  substantial effects, 118
SF1 proteins
  autoinhibition mechanism, 138–139
  CH domain, 140
  conformational change, 140
  dynamic motions, Upf1, 138–139, 139*f*
  HEL1 and HEL2, 138
  NMD pathway, 140
  nonsense-mediated decay (NMD), 138
  RecA- like domains, 138
  SARS coronavirus, 138
SF2 proteins
  ATP binding and hydrolysis, 137
  ATP hydrolysis, 135
  β-hairpin, 137
  dynamic motions, HCV NS3,
    135–137, 136*f*
  HCV NS3, 135
  HEL1 and HEL2, 135
  single-stranded regions, RNA, 135
  tailed duplex, 135–137
  Thr269 and Thr411, 135–137
  Trp501, 137
  vaccinia virus, 134
  viral lifecycles, 134
Single molecule fluorescence resonance energy
  transfer (smFRET)
  C- and N-terminal domains, DEAD-box
    helicase core
    Dbp5p, 79
    description, 78–79
    Mss116p and Cyt-19, 79
    small-angle X-ray scattering (SAXS), 79
  conformational cycle, DEAD-box proteins
    catalytic cycle, YxiN helicase core, 96–98
    full-length YxiN structural model, 98–99
    guidance mechanism, eIF4A activation,
      99–102
    helicase modules, 102–103
    YxiN helicase core, 94–95

description, 80–82
distance histograms
  calculation, 94
  donor fluorescence, 94
  donor quantum yield, 92
  Förster distance, 91–92, 93
  half-cone angles, YxiN, 92
  lifetime determination, 94
  mean distances, 93
  model-free approaches, 93
  order parameter, 92
  orientation factor, 92
  overlap integral, 93
donor-acceptor pairs, 82
efficiency, 80–82
Förster mechanism, 80–82
helicase core, DEAD-box proteins
  description, 76, 77*f*
  motifs types, 76
  RNA duplexes, 76–78
  structures, 77*f*, 78
helicase modules
  eIF4A, 80
  nucleic acid processing, 80
  SecA, 80
histograms calculation, corrected smFRET
  β, correction parameter, 89
  α, correction parameter, 89
  data analysis, 88
  δ, correction parameter, 88–89
  effects, correction parameters, 89–90, 90*f*
  efficiency (E_{FRET}), 89–90
  efficiency distributions, 91
  excitation light, 87–88
  g, correction parameters, 89
  shot-noise limited accuracies, 90–91
  signal-to-noise ratio, 88
intensities, measurement, 83
preparation, DEAD-box proteins (*see* DEAD-
  box proteins preparation)
proximity ratio, 83
RNA conformational changes, 103–104
structure, 80–82, 81*f*
transfer efficiency, 82–83
Single-turnover ATPase assays, 229–230
Small-angle X-ray scattering (SAXS)
  dsRNA, 143
  measurements, 142*f*, 143
smFRET. *See* Single molecule fluorescence
  resonance energy transfer (smFRET)
Split beacon helicase assay (SBHA)
  HCV helicase, 468
  RNA-based
    HCV helicase assay, HTS, 478–479, 478*f*
    MBHA substrate, 478
    protocol, 478–479
Steady-state ATPase assay
  calculation, 246

Subject Index

mixtures #1 and #2, preparing reaction, 244–245
NADH oxidation, 244
performance, 245–246
Steady-state ATPase measurements
  DBP ATPase activity
    description, 39–41, 40f
    "Michaelis constant" (K_m), 41
    regulatory proteins, 41
    RNA-stimulated DBP ATPase cycle, 41
  real-time enzyme-coupled assay (see Real-time enzyme-coupled assay)
Stopped-flow fluorescence spectroscopy
  approach, 3
  ATP-driven RNA unwinding, 2
  background fluorescence and inner filter effects
    absorption, 7
    buffer components, use, 7
  data fitting
    description, 20, 21
    homogeneous first-order reaction, 21
    photobleaching, 22
    reaction amplitudes, 21–22
    reaction rate law, 22
    residuals, calculation, 22
    treatment, 22
  duplex concentration, 4–6
  duplex substrate design and preparation, 7–13
  electrophoretic mobility shift assays (EMSA), 3
  fluorescence changes, duplex unwinding
    photobleaching, 6–7
    strand separation, 6
  functional equilibrium constants
    affinity, enzyme substrate complex, 23–24
    description, 23
    Hill binding isotherm, 23
    multiple cycle experiments, 24
    unwinding rate constants and reaction amplitudes, 23
  measurements, 4
  optimal wavelengths identification, 13–14
  pre-steady-state approaches, 14–20
  protocol, 13, 14f
  RNA helicase in vitro, 2
  setup, 3, 4f
  unwinding rate constants, 22–23
Stress granules (SGs). See RNA helicases analysis, P-bodies and SGs
Structural analysis, RNA helicases
  analysis, protein complex (see Protein complex analysis)
  atomic resolution models (see Atomic resolution models)
  defined, SAXS, 192
  model-independent analysis, 194–198
  model-independent and dependent methods, 192
  NMR analysis, 192
  RNA helicases, 192
  shape determination
    ab initio modeling, 198–199, 198b
    compactness and connectivity, 198–199
    DAMAVER and DAMFILT, 200
    DAMMIF, 199
    GASBOR, 199–200
    solution structure, eIF4G-MC, 200, 201f
  solution scattering and initial data treatment
    interparticle interactions, 193–194
    monodisperse particles, 193
    practical points, 193–194, 194b
    principles, 193, 193b
    sample data, 194
Substrate cleavage
  catalysis stage
    Azoarcus evansii, 123
    burst amplitudes, 122
    discontinuous assay, 122
    dissociation rate constants, 122–123
    D135 ribozyme, 123
    rate measurements, 122
  discontinuous assay, chaperone-assisted RNA folding
    D135 ribozyme, 124, 125f
    folding curve, 124
    misfold, 123
    observed rate constant, 124
  discontinuous assays, 118–119, 119f
  folding stages
    D135 and *Azoarcus* ribozymes, 120, 121f
    native folding, 120
    normalization, 120
    proteinase K, 120
    rate constant, 120
    ribozyme formation, 121–122
  fraction, native ribozyme, 118–119
  intron domains, 118–119
Synechocystis sp. strain PCC 6803 transformation, 391

T

Tandem affinity purification (TAP)
  epitope tagging target proteins
    C-/N- terminal tagging, 310
    sequence incorporation, 310
  protocols
    antibody-conjugated magnetic beads, 312–313
    genome-wide studies, 311–312
    IgG binding domains, 312
    mammalian cells, 312–313
    procedure, 311f

Tandem affinity purification (TAP) (*cont.*)
RNA helicases, 312
TAP. *See* Tandem affinity purification (TAP)
Target sites identification, RNA helicase
adapter ligation and radioactive labeling
3' adapter ligations, 283
5' adapter ligations, 283
nickel beads, 283
CLIP, 276–277
coprecipitation techniques, 276
CRAC, 277
data analysis, 285–286
materials, 277–279
partial RNase digestion and denaturing nickel purification, 282–283
PCR and size selection
LA Taq polymerase (Lonza), 284
purification, 285
separation, 285
protocol adaptation and trouble-shooting, 286–287
reverse transcription, 284
SDS-PAGE and RNA isolation
cross-linked proteins, 283–284
proteinase K buffer, 284
radioactive cross-linked protein, 284
sequencing, 285
UV cross-linking (*see* UV cross-linking)
yeast strains, cross-linking and affinity purification
growth and cross-linking conditions, 280–281
tags, protein fusions, 279–280
Thin layer chromatography (TLC)
development, 447
inorganic phosphate and g-^{32}P-ATP, 446
TLC. *See* Thin layer chromatography (TLC)
Transcription termination
ring-shaped RNA helicase, 150–151
T7 RNAP, 166–167
Transient kinetic analysis, DBP ATPase cycle
fluorescent-labeled ATP and ADP, 46–51
measurements, biomolecules, 46
methylanthraniloyl (mant) nucleotide binding interactions, 46, 47*f*
Pi release data fitting and analysis
ATP dependence, 53–54
concentration, 53
description, 52, 53*f*
pre-steady-state lag phases, 52–53
quench-flow method, ATP hydrolysis, 51–52
RNA saturation, 46
stopped-flow, Pi release, 52
thermodynamic coupling, 46
Tripartite, preparation and purification
autoradiography, 158
NαS analogs, 158
PAGE, 158

U

Unwinding assay
contaminants, 272–273
helicase substrate preparation, 270
minimal enzyme concentration, 272–273
oligonucleotide radiolabeling, 270
RNA-DNA hybrid preparation, 271
RNA production, 270–271
unwound product fraction, 272
UPf1 family. *See* SF1 proteins
UV cross-linking
growing cells, culture, 282
in vitro, 281–282
resuspended cells, 282

V

Virion-associated RNA helicase, cultured mammalian cells
detection
HIV-1 particles, 424*f*
method, 425–427
RHA, 425
host and viral factors
protein–RNA interactions, 427
screen candidate virion-associated protein (s), 427–428
infectivity measurement
RHA-deficient HIV-1 virions, 430–431
RHA downregulation, 429–430, 429*f*

W

Western analysis, protein levels
complexes detection, 396–397
membranes blocking, 396–397
protein loading controls, 397
West Nile virus, 444, 464–465

Y

Yeast two-hybrid analysis (Y2H)
construction framework, 315–316
description, 314
GAL4-responsive promoters, 315
histidine selection, 316
RNA helicases, 315
selectable markers, 316
yeast mating system, 315
Yellow fever virus, 464–465
Y2H. *See* Yeast two-hybrid analysis (Y2H)
YxiN helicase core
catalytic cycle
activation step, 97–98
alanine residue, 96
analogues, ATP, 96

ATP hydrolysis and phosphate release, 97–98
description, 96, 97f
functional mutants, 96
interdye distances, comparison, 94–95
nucleotide-driven, 95
unimodal distributions, 94–95
variants, 95

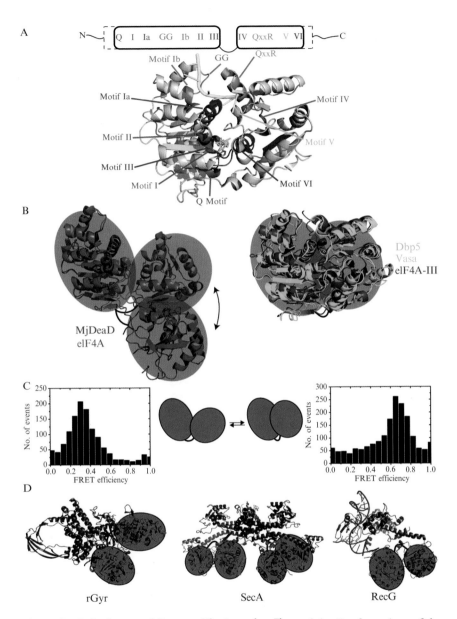

Alexandra Z. Andreou and Dagmar Klostermeier, Figure 4.1 Conformations of the DEAD-box protein helicase core. (A) Top: Scheme of DEAD-box proteins, composed of a helicase core formed by two RecA domains that carry the conserved sequence motifs. Bottom: Structure of eIF4A-III (PDB-ID 2j0s) in complex with single-stranded RNA (yellow) and ADPNP (green). Motifs involved in ATP binding and hydrolysis are depicted in red tones, motifs involved in RNA binding in blue tones, and motifs involved in coupling of ATP hydrolysis to duplex separation in purple. (B) Structures of DEAD-box proteins, superimposed on the N-terminal RecA domain. Left: unliganded, open conformations of eIF4A (red, PDB-ID 1fuu) and mjDeaD (blue, PDB-ID 1hv8); right: closed conformation of eIF4A-III (violet, PDB-ID 2j0s), Vasa (blue, PDB-ID 2db3), and Dbp5p (green, PDB-ID 3pew) in complex with nucleotide and RNA (not shown). (C) FRET histograms for donor/acceptor-labeled YxiN_C61/267A_A115/S229C in the absence (left) and in the presence of 153mer RNA and ADPNP (right) (Karow and Klostermeier, 2009). The FRET efficiency in the open conformation is much higher than expected from the wide-open conformation observed in the crystal structure of eIF4A (B, left), suggesting that the two RecA domains are closer in solution. The FRET efficiency for the closed conformation is in good agreement with the structures (B, right). (D) Helicase modules (red) are part of larger enzymes, such as reverse gyrase (rGyr, left, PDB-ID 2gku), SecA (middle, PDB-ID 3jv2), and RecG (right, PDB-ID 1gm5).

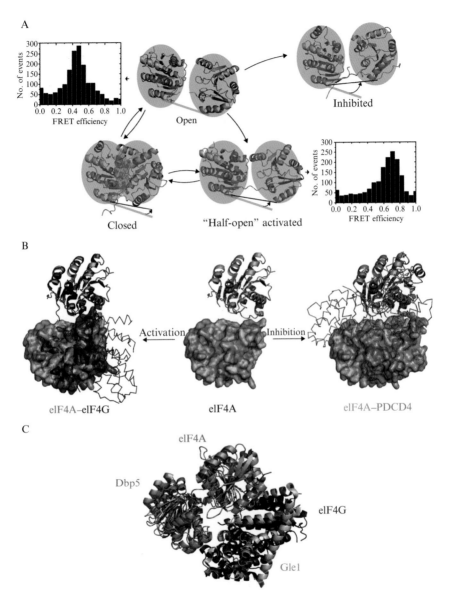

Alexandra Z. Andreou and Dagmar Klostermeier, Figure 4.7 Regulation of the helicase core through effects on conformation. Binding partners of eIF4A regulate its activity by affecting the conformation of the helicase core. (A) Regulation of eIF4A activity by eIF4G and PDCD4 via conformational changes. eIF4A is in an open conformation in the absence of ligands (Hilbert et al., 2011) with the two domains closer than in the crystal structure (Schutz et al., 2008). In the presence of eIF4G, eIF4A is stabilized in a more compact, "half-open" conformation. In the presence of RNA and ATP, the helicase core

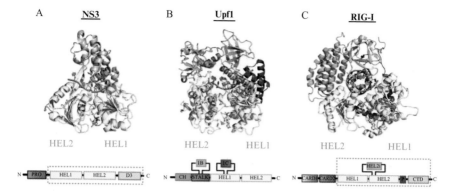

Steve C. Ding and Anna Marie Pyle, Figure 6.2 Structural attributes of translocating helicases. Structural snapshots of helicases on their respective RNA substrates are shown. These helicases are (A) HCV NS3, a $3' \to 5'$ ssRNA translocase, (B) Upf1, a $5' \to 3'$ ssRNA translocase, and (C) RIG-I, a dsRNA translocase. Boxed regions indicate the illustrated portion of the helicase. Domains outside the boxed regions were omitted for clarity. The two RecA-like domains are colored in yellow, and additional domains are colored according to the diagrams beneath each structure. ATP binds in the cleft between the two RecA-like domains. Regardless of the translocation direction, the ssRNAs (orange) lie in the same orientation atop the helicase motor: the 5′ end rests above HEL2 and the 3′ end rests above HEL1. This strand is called the tracking strand. For RIG-I, the tracking strand of the duplex is also colored in orange, and the activating strand is colored in purple.

---

closes (model based on the structure of eIF4A-III in the presence of RNA and nucleotide, PDB-ID 2hyi). Binding of the tumor suppressor programmed cell death protein 4 (PDCD4) stabilizes an eIF4A conformation with the two RecA domains closer than in the open conformation (PDB-ID 2zu6). PDCD4 binding has also been suggested to prevent closure (Chang *et al.*, 2009). (B) Molecular models of free (green), activated (red), and inhibited eIF4A (blue), superimposed on the C-terminal RecA domain, highlighting the conformational changes. The N-terminal domain RecA domain is shown in surface representation, the C-terminal domain in cartoon representation. eIF4G (red) and PDCD4 (green) are depicted in ribbon representation. (C) Superposition of the eIF4A–eIF4G (green/red, PDB-ID 2vso) and the Dbp5p-Gle1$_{InsP6}$ (blue/orange, PDB-ID 3rrm) complexes.

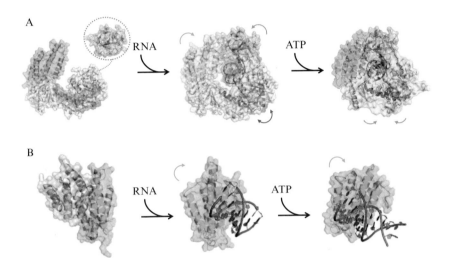

Steve C. Ding and Anna Marie Pyle, Figure 6.5 Dynamic motions of RIG-I, a dsRNA translocase. (A) As determined by SAXS measurements, RIG-I adopts an open conformation in the absence of dsRNA substrate. For clarity, we omitted the CARDs from the first structure. We model in a proposed position of the CTD in this figure (dotted circle) to account for the open conformation needed to allow dsRNA binding. HEL2i and the CTD surround the dsRNA substrate upon binding, which is likely facilitated by the V-shaped pincer domain. Additionally, the RecA-like domains close upon binding ATP, and HEL2 establishes additional contacts with the dsRNA backbone. Domain motions were interpreted based upon aligning all structures to HEL1 and dsRNA. (B) Major domain motions by HEL2i in response to dsRNA and ATP. In the *apo* conformation, the CARDs form a large hydrophobic interface with HEL2i and prevent HEL2i from interacting with the dsRNA. Likewise, the presence of the CARDs is inhibitory for processive translocative behavior. Upon binding dsRNA, HEL2i establishes a single, weak interaction with the backbone of the substrate using the face of a specific α-helix (green). Upon binding ATP, HEL2i moves further toward the dsRNA and establishes more contacts using this α-helix. This alternation between strong and weak interactions in response to ATP is likely the structural basis for dsRNA translocation.

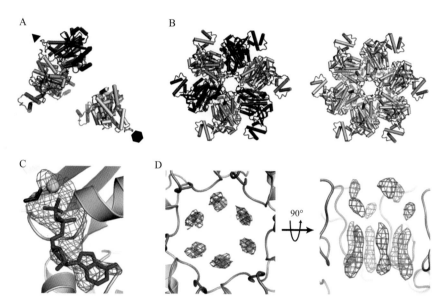

Nathan D. Thomsen and James M. Berger, Figure 8.2 Structure solution and ligand binding in crystal-form II. (A) Molecular replacement reveals one Rho dimer (blue/dark gray and green/gray) and one monomer (pink/light gray) per asymmetric unit. Crystallographic six- and threefold symmetry axes are indicated by a hexagon and triangle, respectively. (B) Crystallographic symmetry generates one threefold symmetric trimer-of-dimers (blue/dark gray and green/gray) and one sixfold symmetric hexamer (pink/light gray). These structures represented our first crystallographic views of a full-length, closed-ring Rho hexamer. (C) Initial refinement of molecular replacement solutions produced $F_o - F_c$ difference electron density maps revealing the presence of nucleotide (modeled as ADP·BeF$_3$) at all three ATP binding sites per asymmetric unit. ADP is colored magenta/gray, BeF$_3$ is colored black, and the Mg^{2+} ion is colored yellow-green/light gray. Electron density (green mesh) is contoured at 3σ. (D) Initial refinement of molecular replacement solutions produced $F_o - F_c$ difference electron density maps revealing the presence of RNA in the central channel of each Rho hexamer. The RNA density appeared to be averaged around the crystallographic symmetry axis, producing uninterpretable maps. Electron density (green mesh) is contoured at 2.5σ.

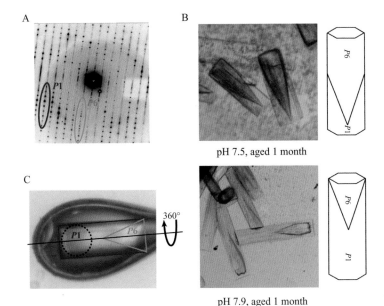

pH 7.5, aged 1 month

pH 7.9, aged 1 month

Nathan D. Thomsen and James M. Berger, Figure 8.3 Overcoming a vexing form of coincident nonmerohedral twinning. (A) Initial diffraction patterns obtained from crystal-form III (Fig. 8.1C), revealed two distinct lattices. While the $P1$ diffraction dominated the pattern (blue/dark gray), a $P6$ diffraction pattern corresponding to crystal-form II was present at low resolution (pink/light gray). Since crystal-forms I and II shared a coincident ∼69 Å unit cell edge, the diffraction patterns almost perfectly overlapped and prevented accurate measurement of low resolution $P1$ data. (B) Aging caused the $P1$ portion of each crystal to degrade at a faster rate than the $P6$ portion. When viewed under bright-field illumination, the two twin domains are clearly distinguishable. The $P6$ twin domain forms at the nucleating end of the crystals and has a "cone-shaped" protrusion that interleaved with the $P1$ twin domain (top). An X-ray beam aimed at the $P1$ tip of the crystal thus passes through a small fraction of the $P6$ twin domain, producing $P6$ diffraction only at low resolution as seen in panel A. Increasing the pH of crystal growth increased the relative size of the $P1$ twin domain (bottom). (C) Crystals grown at a pH of 7.9 were mounted in bendable cryoloop and oriented such that the axis of crystal rotation (solid line) was parallel with the long axis of the hexagonal rods. A 100 µM collimated beam (dashed circle) was directed into the $P1$ tip (blue/dark gray) of the crystal. Data collection in this orientation avoided contributions from the $P6$ twin domain (pink/light gray).

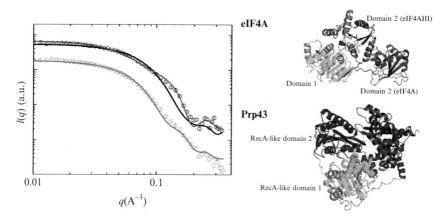

Manja A. Behrens et al., Figure 9.4 *Left*: Scattering data obtained for Prp43 (gray) and eIF4A (black) (scaled by a factor of 10) with their respective CRYSOL fits. For eIF4A, two structures are available with atomic resolution, an open conformation (PDB entry 1FUU) (magenta line) and a closed conformation (PDB entry 2HXY) (black line). For Prp43, the model obtained from the X-ray crystallography (PDB entry 3KX2) was used (solid gray line). *Right*: Crystal structure of eIF4A in the open and closed conformation. Here, domain 1 (the N-terminal RecA-like domain) is colored cyan, while domain 2 (the C-terminal RecA-like domain) in the open structure is magenta and orange in the closed structure. In the crystal structure for Prp43, the N-terminal RecA-like domain is colored green, the C-terminal RecA-like domain is colored blue, and the remaining part of the molecule is colored red.

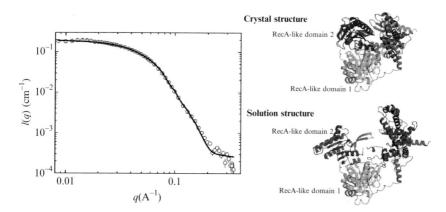

Manja A. Behrens et al., Figure 9.5 *Left*: Scattering data obtained for Prp43 (open circle) and best fit from rigid body modeling (black line). *Right*: Most representative model obtained from rigid body modeling (bottom) compared to the crystal structure model (top), where domain 1 in both structures is green and with the same orientation. Domain 2 is blue and domain 3 is red.

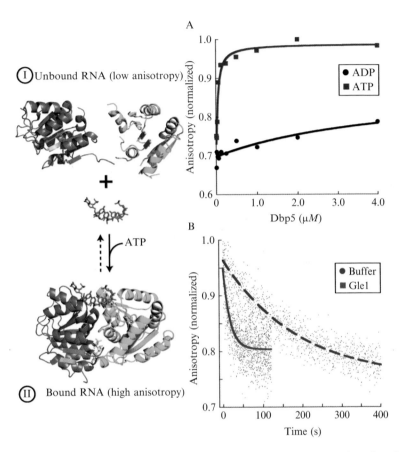

Ben Montpetit et al., Figure 11.2 Measurement of RNA binding and release by Dbp5 E240Q, a catalytically inactive version of Dbp5 that binds both ATP and RNA but is unable to hydrolyze ATP. (A) In the presence of ATP (squares), but not ADP (circles), Dbp5 is able to bind fluorescein-labeled RNA causing a measured change in anisotropy (complex I → II). (B) Stopped flow experiment measuring the release rate of RNA (complex II → I), which is accelerated by the presence of Gle1 (solid line) in comparison to reactions containing Dbp5 alone (dashed line). Crystal structure model in state I is PDB ID: 3FHO (Fan et al., 2009) and state II is PDB: 3PEY (Montpetit et al., 2011).